Research, Invention, and Innovation Congress (RI2C 2021)

Held online

E3S Web of Conferences Volume 302

Bangkok, Thailand
1 - 2 September 2021

Editors:

Malinee Sriariyanun
Yushen Cheng
Wawat Rodiahwati
Marttin P. Gundupalli

ISBN: 978-1-7138-3648-3

Printed from e-media with permission by:

Curran Associates, Inc.
57 Morehouse Lane
Red Hook, NY 12571

Some format issues inherent in the e-media version may also appear in this print version.

For additional information, please contact EDP Sciences – Web of Conferences
at the address below.

EDP Sciences – Web of Conferences
17, Avenue du Hoggar
Parc d'Activité de Courtabœuf
BP 112
F-91944 Les Ulis Cedex A
France

Phone: +33 (0) 1 69 18 75 75
Fax: +33 (0) 1 69 28 84 91

contact-edps@webofconferences.org

Additional copies of this publication are available from:

Curran Associates, Inc.
57 Morehouse Lane
Red Hook, NY 12571 USA
Phone: 845-758-0400
Fax: 845-758-2633
Email: curran@proceedings.com
Web: www.proceedings.com

TABLE OF CONTENTS

Preface

2021 Research, Invention, and Innovation Congress (Ri2C2021) is the second time for Ri2C and it is hosted by King Mongkut's University of Technology Bangkok (KMUTNB, https://www.kmutnb.ac.th/). Our president, Prof. Dr. Suchart Siengchin, and conference committees would like to invite academic researchers, engineers and industries worldwide to attend this event to share the novel finding and development of research in topic related to i) Materials and Mechanics (iMM), ii) Electricals and Electronics (iEE), and iii) Sustainable Sciences (iSS). With successful event and high quality of research presentations, all of papers presented in iSS-Ri2C2019 were published in E3S Web of conference and indexed in SCOPUS.

This year, it is a challenging time amidst pandemic to set up and host Ri2C2021, therefore we conduct all events with online system to welcome all participants worldwide. With enthusiasm of our participants, we receive 263 submissions, and after peer-reviewing process we accept 139 papers in Ri2C2021. In iSS section, the focused research topics include Environmental Science and Engineering and Energy. We received submission of 70 submissions, and 37 full papers from 9 countries are accepted based on peer-reviewing of 29 international reviewers to publish in E3S Web of conference. Interestingly, with online conference platform in Ri2C 2021, our accepted papers gained submissions from international authors to 36% during pandemic situation. We are enthusiastic to continue managing this event every year to contribute to the development of science and engineering.

Malinee Sriariyanun, Ph.D

Managing Editor

iSS-Ri2C2021

https://doi.org/10.1051/e3sconf/202130200002

Opening Address

Welcome Message from Advisory Chair

Dear friends and colleagues,

On behalf of the President of King Mongkut's University of Technology North Bangkok (KMUTNB), I would like to welcome you to the 2021 Research, Invention, and Innovation Congress (RI2C 2021) in Virtual Platform during 1 – 2 September 2021.

From the previous successful RI2C 2019 Conferences, The RI2C 2021 in each category will provide a wonderful forum for you to refresh your knowledge basis and explore the Invention and Innovation. As the theme of the conference is Invention and Innovation and we have over 150 research works accepted to present in this event, I would like to encourage the participants to find time to meet other delegates from three main categories, including i) Materials Engineering (ME), ii) Electricals and Electronics (EE), and iii) Sustainable Sciences (SS) who come for the RI2C 2021 comprising of professors, researchers, students, and government staff to discuss about their experiences and re-vow our relationship in academic and research affairs.

I would like to thank all committee members involving with the network and organizing committee of the conference. Especially I would like to thanks the staff members of the Science and Technology Research Institute, King Mongkut's University of Technology North Bangkok that is the main organizer of the event. I wish this academic conference to be a success and achieve all the objectives.

Faithfully yours,

Professor Dr.-Ing.habil. Suchart Siengchin

President, King Mongkut's University of Technology North Bangkok

Advisory Chair, Ri2C2021

E3S Web of Conferences **302**, 00002 (2021) https://doi.org/10.1051/e3sconf/202130200002
RI²C 2021

List of Conference Chairs

Conference Chair: 2021 Research, Invention, and Innovation Congress (RI2C2021)

1. Advisory Chair

   Suchart Siengchin, King Mongkut's University of Technology North Bangkok (KMUTNB), Thailand

2. Conference Chair

   Somrerk Chandra-ambhorn, KMUTNB, Thailand

3. General Secretary

   Wanlop Harnnarongchai, KMUTNB, Thailand

4. International Advisory Board

   Babak Nahid-Mobarakeh, University of Lorraine, France

   San Hoa Thang, Monash University, Australia

   Athanasios Tsolakis, The University of Birmingham, UK

   Pau Loke Show, University of Nottingham Malaysia, Malaysia

   Yushen Cheng, National Yunlin University of Science and Technology, Taiwan

   Wen-Chien Lee, National Chung Cheng University, Taiwan

   Rizalman Mamat, Universiti Malaysia Pahang, Malaysia

   Edmund Haberstroh, RWTH Aachen University, Germany

5. RI2C2019 Academic Chair

   Somrerk Chandra-ambhorn, KMUTNB, Thailand

   Kampanart Theinnoi, KMUTNB, Thailand

   Malinee Sriariyanun, KMUTNB, Thailand

   Titipong Lertwiriyaprapa, KMUTNB, Thailand

   Rungsima Yeetsorn, KMUTNB, Thailand

   Wanlop Harnnarongchai, KMUTNB, Thailand

   Sunanta Sodsee, KMUTNB, Thailand

   Peerapong Pornwongthong, KMUTNB, Thailand

   Nonchanutt Chudpooti, KMUTNB, Thailand

   Voranuch Somsongkul, KMUTNB, Thailand

   Sunanta Chuayprakong, KMUTNB, Thailand

   Watchareewan Jitsakul, KMUTNB, Thailand

   Chaniga Chuensangjun, KMUTNB, Thailand

Statement of Peer review

In submitting conference proceedings to *Web of Conferences*, the editors of the proceedings certify to the Publisher that

1. They adhere to its **Policy on Publishing Integrity** in order to safeguard good scientific practice in publishing.
2. All articles have been subjected to peer review administered by the proceedings editors.
3. Reviews have been conducted by expert referees, who have been requested to provide unbiased and constructive comments aimed, whenever possible, at improving the work.
4. Proceedings editors have taken all reasonable steps to ensure the quality of the materials they publish and their decision to accept or reject a paper for publication has been based only on the merits of the work and the relevance to the journal.

Title, date and place of the conference

Title: 2021 Research, Invention, and Innovation Congress (RI2C2021)
Date: 1-2 September 2021
Place: Online conference at King Mongkut's University of Technology North Bangkok

Proceedings editor(s):

Managing Editor: Malinee Sriariyanun King Mongkut's University of Technology North Bangkok, Thailand

Editorial Board Members:
1. Yushen Cheng, National Yunlin University of Science and Technology, Taiwan
2. Wawat Rodiahwati, University of New England, Australia
3. Marttin P. Gundupalli, Indian Institute of Technology Hyderabad, India

Date and editor's signature 15 Aug 2021

17, avenue du Hoggar - PA de Courtabœuf – BP 112 - 91944 Les Ulis Cedex A (France)
Tél. : 33 (0)1 69 18 75 75 - Fax : 33(0)1 69 07 45 17 – www.edpsciences.org

Performance modelling of the Darrieus wind turbine

Thuzar Mon[1], and *Supakit* Worasinchai[2,*]

[1]Department of Automotive Engineering, Thailand Advanced Institute of Science and Technology (TAIST), King Mongkut Institute of Technology Lad Krabang, Bangkok, 10520, Thailand
[2]Machinery Development Research Team, Material Processing and Manufacturing Automation Research Group, National Metal and Materials Technology Center, 12120, Thailand

Abstract. Three-dimensional numerical investigation of the Darrieus wind turbines equipped with different aerofoils is presented in this paper. In the modelling, the computational domain was divided into three different domains and they are blade, rotor, and tunnel domains. A cylindrical domain was created to cover the blade area so that a fine mesh can be applied. The Computational Fluid Dynamics (CFD) was employed to solve and analyze the flow field around the turbine. The Menter Shear Stress turbulence model was chosen in this investigation. Boundary conditions applied were velocity at the inlet, pressure opening at the outlet, and symmetry on other sides. Comparison of simulation results and experiments showed good agreement. The investigation of the effects of the rotor solidity and the aerofoil shape was performed. The simulation results reveal that the aerofoil shape has a significant impact on the turbine performance. For the rotor solidity of 0.7, the change from the NACA section to the S1046 leads to a reduction of power at low tip speed ratios but the performance improvement is observed when the tip speed ratio is greater than 1.5. With the lower solidity of 0.375, the effects of the aerofoil change is less pronounced at low tip speed ratios. Nevertheless, the maximum power coefficient increases for both cases. Further analysis shows that the S1046 is less sensitive to the wind speed change and is promising in the urban application where the wind speed is relatively low.

Keyword. Darrieus Wind Turbine, Performance modelling, Aerofoil shape, Rotor solidity

1 Introduction

As the world has become struggle with the fossil-fuel shortage and the environmental pollution, the renewable sources of energy have become increasing popular and the use of renewable sources to produce electricity is significantly increased. The wind energy is one of the sources and its capacity of installation has increased significantly in the last decade [1].

Wind turbines have been employed for decades to generate electricity. Its technology development leads to large-scale clean electricity production. Apart from the large-scale production, the idea of decentralizing of the energy sources is beneficial as it can provide power to the specific areas and increase the amount of the overall installed capacity. This leads to a wide interest in the vertical axis wind turbines for the small-scale power generation [2].

The vertical axis wind turbines have a number of advantages over the horizontal axis counterpart as it is insensitive to wind direction and needs no yaw mechanism [3]. Nevertheless, one of its drawback is that it has a relatively poor starting performance and its blade design needs further investigation to improve this performance [4].

Beri and Yao [5] had studied the self-starting performance of the Darrieus wind turbines. The study showed that the employment of asymmetrical aerofoils leads to an improved self-starting behavior with a reduction in peak efficiency. Mohamed et al. [6] further numerically investigated the effect of the aerofoil shape and it had been found that the LS(1)-0413 aerofoil gives maximum power coefficient of 0.415 with an increase around 10% compared to the conventional NACA 0018 section.

Many studies had investigated effects of the aerofoil thickness on the turbine performance [7-8]. Castelli and Benini [7] numerically evaluated the performance of the Darrieus turbine equipped with NACA0012 and NACA0021. Takahashi et al [8] carried out numerical simulations and wind tunnel experiments to study the effects of the aerofoil shape. The results showed that the NACA0024 that has a high value of thickness is the best aerofoil. Both studies indicated that, for the conventional NACA sections, the increase of the aerofoil thickness could improve the turbine performance.

A series of newly designed, low-speed aerofoils had been investigated for the Darrieus wind turbine [9-10]. Fang et al [9] investigated the employment of S-series aerofoils on the straight-bladed vertical axis wind turbine in comparison with the conventional aerofoil sections such as the S809 and the NACA0018. The results showed that the S1046 aerofoil section performs better than the conventional sections, implying that a good selection of the aerofoil shape can significantly improve the performance of the turbine. Hshem et al [10] investigated the aerodynamic performance of H-rotor wind turbine by using 24 different symmetric and non-

* Corresponding author: supakit.wor@mtec.or.th

symmetric aerofoils. It was found that S1046 aerofoil show the maximum power coefficient of 0.3463 which performs the best of the tip speed ratio of 2 to 7. This work presents a numerical investigation of the effects of the aerofoil shapes and the rotor configuration on the performance of the Darrieus wind turbine.

2 The turbine model

The turbine models investigated in this paper is three-bladed turbines with different rotor configurations. The first model is square and has the same width and height of 0.8 metre. The second model is rectangle and has the width and the height of 1.6 and 0.8 metre, respectively. Both models are investigated when they are equipped with two different aerofoils, which are the NACA0021 and the S1046. Both rotors have the same blade chord of 0.2 metre, leading to a solidity of 0.75 and 0.375, respectively. The geometry of two turbine models is summarised in Table 1.

Table 1. Turbine configurations

Geometry	Turbine 1	Turbine 2
Aerofoil section	NACA0021/S1046	NACA0021/S1046
Number of blades	3	3
Chord length	0.2 m	0.2 m
Rotor diameter, D	0.8 m	1.6 m
Blade span, H	0.8 m	0.8 m
Solidity	0.75	0.375

The NACA0021 aerofoil is a thick section with 21% thickness at the 30% of the chord. It possesses soft stall behaviour with a moderate maximum lift coefficient. It is commonly employed on low speed wind turbines and hydropower applications [11]. It is reported that the use of the NACA0021 leads to a relatively high performance, compared to other aerofoils with comparable thickness such as the DU91-W2-250 aerofoil [12]. The S1046 aerofoil is a symmetrical section with 17% thickness at 30.7% chord. It was designed for low wind speed application and might be promising for urban wind turbine applications. Mohamed et al [6] mentioned that the S1046 is a very promising aerofoil shape for wind energy generation compared to the other symmetric aerofoils.

3 Numerical methodology

3.1 Numerical setup

The three dimensional numerical analysis of the straight bladed Darrieus wind turbine is conducted by using the computational fluid dynamic (CFD).

The computational domain was divided into three parts and they are wind tunnel region, rotor region, and blade region as shown in Figure 1. Boolean operations were performed to subtract the three cylindrical blade domains from the cylindrical rotor domain and to subtract the cylindrical rotor domain form the outer wind tunnel domain. An interface boundary condition was assigned at the contact surfaces of the rotating domains and the outer fixed domain, as depicted in Figure 1a. The size of each regions were configured based upon the rotor diameter [13] as presented in Figure 1b and they are summarised in Table 2.

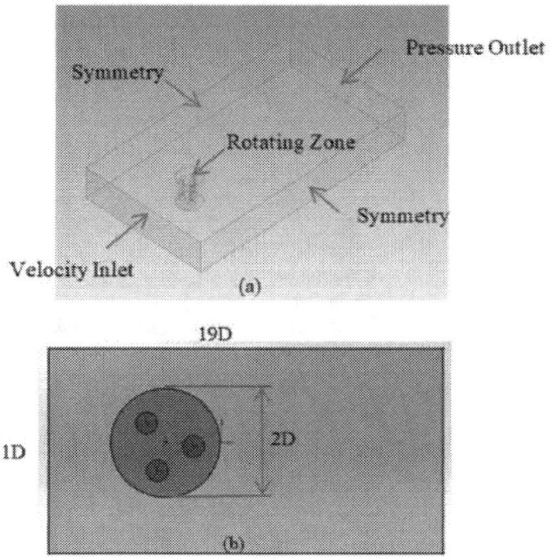

Fig. 1. (a) Boundary conditions (b) Plane view of three domains with specified dimensions

Table 2. Specified dimensions of CFD domain

Domain	Denomination	Value
Wind tunnel	Length	19D
	Width	11D
	Height	2D
Rotor	Diameter	2D
	Height	1.85D
Blade	Diameter	0.5D
	Height	1.5D

3.2 Numerical validation

The present simulation results were validated against the experimental measurement investigated by Elkhoury et al. [13].

3.2.1 Mesh and time step study

Mesh quality has an important influence in obtaining accurate CFD results. The most important factor to identify a good grid resolution is the ability to predict the time and place of flow separation around the turbine blades [14]. This prediction is the most important factor in the overall performance of turbine at low tip speed ratios. The characteristic cell size on both sides of the interface (rotor sub-grid and fix sub-grid areas and rotor

https://doi.org/10.1051/e3sconf/202130201001

sub-grid and blades sub-domains) has the same characteristic in order to gain faster convergence [13].

In order to ensure that the solution is independent of the mesh size, numerical simulations were conducted using three different mesh element number: 2717718, 3623942 and 4044312, which were defined as coarse, medium, and fine. The boundary layers of 15 and the surfaces growth rate of 1.2 were applied to the mesh inflation to ensure a sufficient boundary layer. Mesh quality was checked in terms of its skewness around the blades and the overall domain and the angle of the skewness of the cells 0.9. Mesh independency was made for the NACA 0021 at a wind speed of 8 m/s and at TSR 0.25. Table 3 shows the number of nodes, mesh elements, and the value of the average torque generated by the rotor blades for three cases. Figure 2a presents the overall picture of the final mesh where a coarse mesh was applied outside the rotor domain. The rotor domain in the Figure 2a has three cylindrical domains inside that cover each blade as shown in Figure 2b.

Table 3. Effect of Mesh on the torque at the rotor blades

	Nodes	Elements	Average torque (Nm)
Coarse	941749	2717718	1.9668
Medium	1188335	3623942	2.2035
Fine	1288912	4044312	2.1764

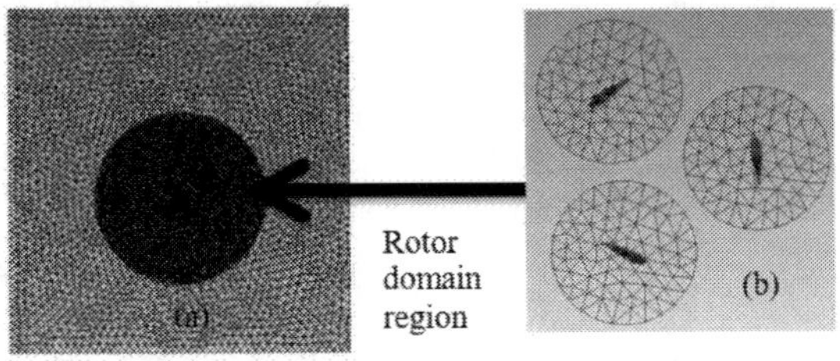

Fig. 2. Mesh near surface clustering of the CFD domain

Simulations results indicated that for mesh element numbers 2717718 and 4044312 shows a difference less than 0.5%. Therefore, in order to obtain the accurate results and to save the computational time, a coarse mesh was set up for the fixed domain. The finest unstructured mesh was used in the rotating domains with specific sizings and inflation layers to be accurate for solving near-wall flow field around the rotor blades.

A suitable time step is another important factor for good CFD simulations. Two different time steps were carried out for the turbine with the NACA0021 at the TSR of 0.25 and at the wind speed of 8m/s with the time steps of 0.00139626 and 0.125664 for one revolution respectively. The power coefficient of 0.020281, which is much more close to the predicted result, was obtained when the time step is 0.00139626.

3.2.2 Turbulence model validation

Two turbulence models were evaluated to predict the turbine performance: the Large Eddy Simulation (LES) model and the Shear Stress Transport (SST) model.

The Large Eddy Simulation (LES) is a turbulence model focusing on resolving eddies within the flow with an appropriate mesh size. It is based upon Kolmogorov's theory that the large eddies of the flow are dependent on the geometry while the smaller scales are more universal. This implies that large eddies can be solved implicitly and the smaller scales can be solved through subgrid-scale model [15].

The Shear Stress Transport (SST) k-ω model is a two-equation turbulence model developed by Menter [16]. It is actually a hybrid model that combines the k-ω formulation in the inner region of the boundary layer and the k-ε model in the bulk flow. The validation results are presented in Figure 3.

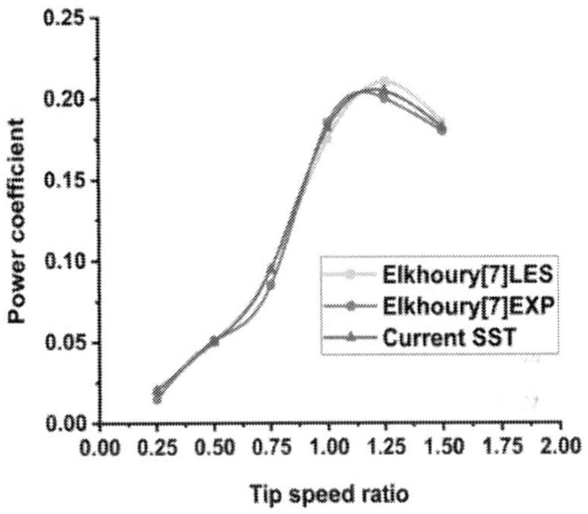

Fig. 3. Power coefficient validation

As can be seen from the Figure 3, the LES and the SST models can predict the experiments well for the whole range of the tip speed ratio. It is worth noting that the discrepancies between numerical and experimental data were calculated in terms of the normalized mean square error (NMSE). The maximum error occurred at the tip speed ratio of 0.25 with values of 8.33% and 9.16% for the LES and the SST models, respectively. Nevertheless, this deviation at the peak power output were 0.23% and 0.048% for the LES and the SST models, respectively. The SST model was then selected to evaluate the turbine performance in this course of study.

4 Results and discussion

The aerodynamic characteristics of H-rotor Darrieus are investigated in terms of power coefficient at different tip speed ratio.

The turbine power coefficient is defined as:

$$C_P = T\omega / \rho R h V^3 \quad (1)$$

and as a function of the tip speed ratio, defined as:

$$TSR = R\omega / V \quad (2)$$

 https://doi.org/10.1051/e3sconf/202130201001

Where C_P is the turbine power coefficient, T is the turbine torque, ω is the turbine rotational speed, ρ is the air density, H is the blade span length,, R is the turbine radius, and V is the inflow velocity.

Figure 4 shows the power coefficient of the rotor with high solidity of 0.75 for two aerofoil sections (the NACA0021 and the S1046) at the wind speed of 8 m/s. It can be observed that the maximum power coefficient and the operating tip speed ratio depends on the aerofoil shape. The maximum power coefficients are 0.2044 and 0.2103 for the NACA0021 and the S1046, respectively. In addition, while the turbine with the NACA0021 exhibit its maximum power at the tip speed ratio of 1.25, the turbine with the S1046 exhibits its peak power at a higher tip speed ratio of 1.50.

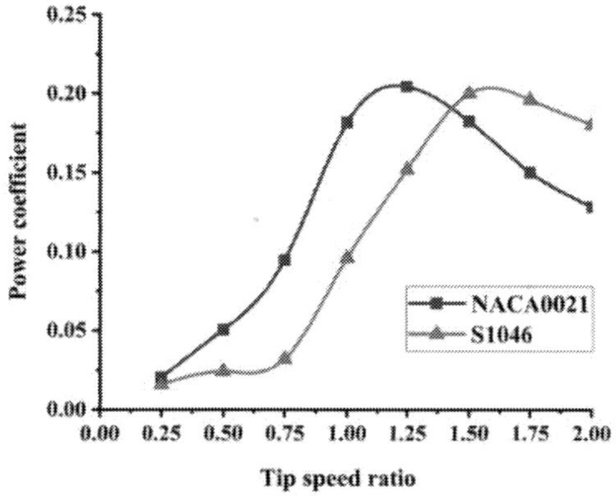

Fig. 4. Power coefficient at a wind speed of 8 m/s and the solidity of 0.75

It is worth noting that the power coefficient of the turbine equipped with the NACA0021 outperforms the turbine with the S1046 section at tip speed ratios lower than 1.5 and this is anticipated to be the effect of the thickness that promotes more suction peak at low TSRs [7].

The power coefficient of the rotor with a solidity of 0.375 is presented in Figure 5. It can be seen from Figure 5 that their maximum power coefficients increase with the reduction of the rotor solidity and they are 0.4 and 0.425 for the NACA0021 and the S1046, respectively. Their maximum operating tip speed ratios are increased to 2 and 2.25, respectively. After these TSRs, the power of the turbine with the NACA0021 case reduces sharply while the turbine with the S1046 section can continuously produce power and the reduction of the power occurs at a significantly slower rate. This small decrease after peak is the effect of the aerofoil thickness that contributes to a lower base drag at high, operating tip speed ratio. This characteristic is more pronounced when the solidity is lower [6].

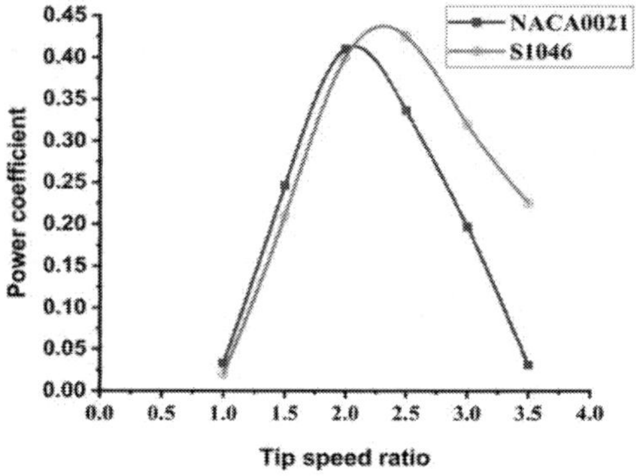

Fig. 5. Power coefficient at a wind speed 8 m/s and a solidity 0.375

It is worth mentioning that the turbine with the NACA0021 section slightly outperform the turbine with the S1046 section at tip speed ratios lower than 2. This is because, with the reduction of solidity, the induction effect that slows down the inflow velocity is reduced, causing both sections operate with a higher Reynolds number. With this high Reynolds number, the S1046 will be able to generate more lift and hence the total torque can be increased despite having a lower value of thickness.

In order to investigate their suitability for the urban application, the turbine performance at a wind speed of 6 m/s are investigated and the results are presented in Figure 6 and 7. It can be seen from Figures 6 and 7 that the reduction of the wind speed reduces the power coefficient slightly for both cases. Reduction of the peak power are 4.9% and 3.09% for the NACA0021 and the S1046 sections, respectively.

For the S1046 case, this small reduction of power illustrates that the employment of a specifically designed aerofoil can improve the overall performance of the Darrieus turbine and the turbine will be able to operate with a wider range of operation. This is considered benefical when the turbine has to operate in the urban areas where the average wind speed is relatively low.

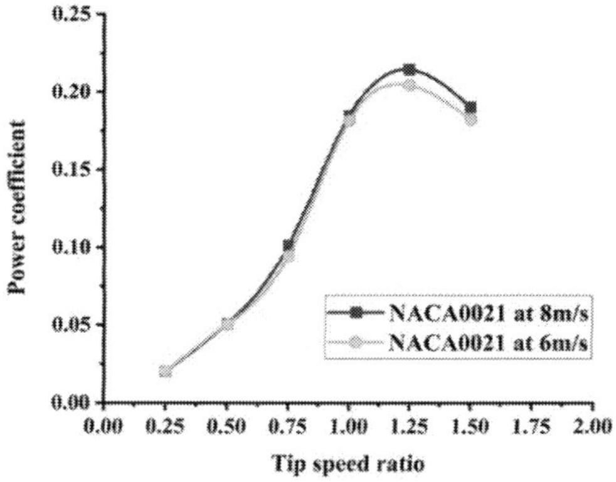

Fig. 6. Power coefficient of the NACA0021 at wind speeds of 6 m/s and 8 m/s

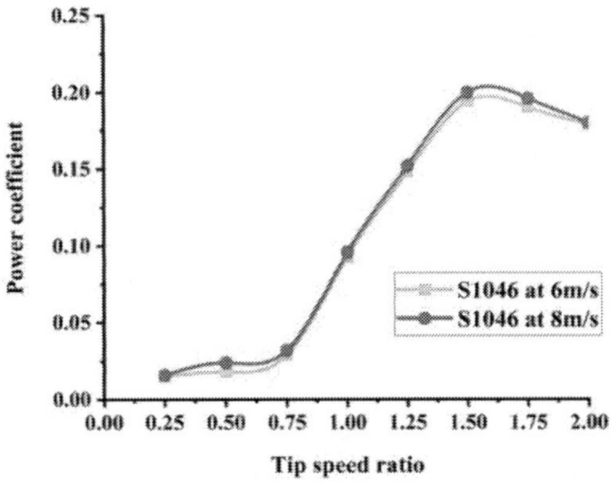

Fig. 7. Power coefficient of the S1046 at wind speeds of 6 m/s and 8 m/s

5 Conclusion

Performance modelling of the Darrieus wind turbine was conducted in this paper. The validation of the modelling showed good agreement with the experimental results. The validated model was then employed to study the effects of the aerofoil shape and the rotor solidities. The results showed that the low solidity wind turbine in general operate at higher tip speed ratios and exhibits higher peak power which is around 0.4 compared to about 0.2 in the high solidity case.

The power coefficient of the turbine also depends upon the aerofoil shape. The employment of the conventional NACA0021 section shows good performance in the low tip speed ratio range. Nevertheless, the employment of the S1026 aerofoil section can significantly improve the turbine performance at low tip speed ratios especially for the low solidity rotor, which generally possesses a high power peak property. It is then recommended that the S1046 aerofoil section should be utilized with the turbine that has low solidity especially if the turbine is expected to operate with a wide range of tip speed ratios.

References

1. M. Kaya, M. Aksoy, F. Kose, Renewable Energy in Turkey: Potential, Current Status and Future Aspects, The International Journal of Engineering, **15**, 1 (2017): 65-69
2. A. Bahaj, L. Byers, P. James, Urban Energy Generation: Influence of Micro Wind Turbine Output on Electricity Consumption in Building, Energy and Buildings, **39**, 2 (2007): 154-165
3. Y. Wang, X. Sun, X. Dong, B. Zhu, D. Huang, Z. Zheng, Numerical Investigation on Aerodynamic Performance of a Novel Vertical Axis Wind Turbine with Adaptive Blades, Energy Conversion and Management, **108**, (2016): 275-286
4. S. Worasinchai, Small Wind Turbine Starting Behaviour, PhD Thesis, Faculty of Engineering and Computing Sciences, The University of Durham, (2012)
5. H. Beri, Y. Yao, Effect of Camber Airfoil on Self-starting of Vertical Axis Wind Turbine, Journal of Environmental Science and Technology, **4**, 3 (2011): 302-312
6. M. Mohamed, Performance Investigation of H-rotor Darrieus Turbine with New Airfoil Shapes, Energy, **47**, 1 (2012): 522-530
7. M. Castelli, E. Benini, Effect of blade thickness on Darrieus Vertical-Axis Wind turbine performance, *The 2nd International Conference on Computer Modelling and Simulation*, (2011)
8. S. Takahashi, J. Hamada, Y. Takashi, Numerical and Experimental Studies of Airfoils Suitable for Vertical Axis Wind Turbines and An Application of Wind-energy Collecting Structure for Higher Performance, *Proceedings of the 19th National Symposium on Wind Engineering*, (2006) 327–330.
9. F. Feng, S. Zhao, C. Qu, Y. Bai, Y. Zhang, Y. Li, Research on Aerodynamic Characteristics of Straight-Bladed Vertical Axis Wind Turbine with S Series Airfoils, International Journal of Rotating Machinery, (2018)
10. I. Hashem, M. Mohamed, Aerodynamic Performance Enhancements of H-rotor Darrieus Wind Turbine, Energy, **142**, (2017), 531-545
11. M. Salam, V. Deshpande, S. Panday, N. Khan, M. Ali, Improvement in Aerodynamic Performance of NACA0021 Airfoil Using Moving Surface Boundary Layer: A computational study. AIP Conference Proceedings 2121, (2019).
12. M. Castelli, G. Grandi E. Benini, Numerical Analysis of the Performance of the DU91-W2-250 Airfoil for Straight-Bladed Vertical-Axis Wind Turbine Application, International Journal of Aerospace and Mechanical Engineering, **6**, 3 (2012)
13. M. Elkhoury, T. Kiwata, E. Aoun, Experimental and Numerical Investigation of a Three-dimensional Vertical-axis Wind Turbine with Variable-pitch," J. Wind Eng. Ind. Aerodyn. **139**, (2015), 111–123
14. P. Kozak, Effects of Unsteady Aerodynamics on Vertical-axis Wind Turbine Performance, M.S. Thesis, Illinois Institute of Technology, (2014)
15. H. Schlichting, H., *Boundary Layer Theory*, McGraw- Hill book company, New York, (1996)
16. F. Menter, Two-equation Eddy-viscosity Turbulence Models for Engineering Applications, American Institute of Aeronautics and Astronautics Journal, **32**, 8 (1994), 1598-1605

Coal/biomass co-combustion investigation by thermogravimetric analysis

Akarasingh Bampenrat[1,*], *Hussanai* Sukkathanyawat[1], and *Thidaporn* Seangwattana[1]

[1]Faculty of Science Energy and Environment, King Mongkut's University of Technology North Bangkok (Rayong Campus), Rayong, Thailand, 21120

Abstract. In this study, the thermal characteristics and kinetic parameters of coal/biomass blended fuels (75:25, 50:50 and 25:75 wt.%/wt.%) were investigated by using the thermogravimetric technique under atmospheric air. Three types of agricultural waste biomass including cassava root, palm kernel shell and rice husk were used as raw materials. The experiments were performed under different temperatures, ranging from 313-973 K with the heating rate of 5, 10, 20 and 40 K/min. The results show that the thermal decomposition of biomass exhibit three-four stages including moisture and some light volatile removal stage (up to 463 K), volatile oxidation stage (423-663 K), char combustion stage (663-823 K) and inorganic oxidation stage (803-953 K). Lignite on the other hand exhibits only two main peaks during the entire combustion process, corresponding to the moisture removal (up to 433 K) and the decomposition/oxidation (433-833 K), respectively. In addition, it was also found that the blending of biomass residues improved the ignition temperature of the blended fuels, indicating an improvement of devolatilization of coal. Kinetic studies show that the average apparent activation energies of the co-combustion of coal/cassava root, coal/palm kernel shell and coal/rice husk calculated from the Kissinger-Akahira-Sunose method are reported at ca. 105.25, 179.66 and 121.84 kJ/mol, respectively.

Keyword. Biomass, Coal, Co-combustion, Thermogravimetric analysis

1 Introduction

In Thailand, the domestic resource potentials of major biomass residues, namely cassava root, palm kernel shell and rice husk, are about 6.1, 0.5 and 8.2 million tons per year, respectively, according to the Department of Alternative Energy Development and Efficiency, Ministry of Energy. Most of these agricultural and agro-industrial residues are utilized for power generation. However, the electricity produced in the country is still based on fossil fuel power plants. In order to reduce the consumption of fossil fuels, the co-combustion or co-firing technology is one of the most promising options for applications with renewable fuels. Co-combustion of solid fossil fuel with biomass benefits in terms of cost and emission reductions. In addition, co-combustion of biomass with coal in conventional coal-fired boilers can provide a reasonability attractive option for utilization of biomass for power generation in which using the infrastructure that is associated with the existing fossil fuel-based power system and requires some capital investment [1].

Thermo-analytical methods, such as thermogravimetric analysis (TGA or TG), differential thermogravimetry (DTG), differential thermal analysis (DTA) and differential scanning calorimetry (DSC), have been used to evaluate the measurements of mass and energy variations which cover a wide range of applications in research and development. For example, the DTG thermograms were employed to determine the decomposition temperatures of biopolymer templates for photocatalyst preparation [2], the TGA/DTA thermograms were used to obtain the thermal stability information of antibacterial filler prepared from leather industrial waste [3], the TGA and DSC data were applied to analyze the reaction areas, characteristic temperatures and heat flow rates of pellets derived from blends of animal manure and lignocellulosic biomass [4], and so on. Among these techniques, TGA, along with DTG, is one of the most approved techniques to provide a rapid assessment of the thermal characteristics during the combustion of solid fuels. Besides decomposition behavior, quantitative methods can be applied to TGA profiles in order to obtain kinetic parameters as reported in the cases of co-combustion of coal with pine sawdust [5], coconut fibers and eucalyptus leaves [6], lignocellulosic and animal wastes [7] and wood waste and rice straw [8], etc. Thermal decomposition characteristics and kinetic parameters of biomass, coal, as well as their blends provide pivotal information to evaluate energy potential of renewable feedstocks. Up to date, numerous TGA experiments in the field of biomass/coal co-combustion have been studied continually. For instance, F. Guo *et al.* [9] investigated the co-combustion of biomass pellets and the two kinds of coal using a thermogravimetric analyzer. They found

* Corresponding author: akarasingh.b@sciee.kmutnb.ac.th

that by increasing biomass pellet ratio, the maximum combustion rate and combustion index increased, while the burnout temperature decreased, indicating that the combustion performance of coal can be improved. S. Qi *et al.* [10] evaluated the ignition and combustion characteristics of single particles of two biomass residues (corn straw and wheat straw) and three coals (lignite, bituminous and anthacite coals) in N_2/O_2 and CO_2/O_2 environments. They reported that the biomass and lignite particles ignited homogeneously, while the bituminous coal and anthracite particles ignited heterogeneously. Moreover, the biomass particles ignited earlier than the coal particles. For the three coals, the ignition delay time decreased almost linearly with the increase in their volatiles contents.

Due to the combustion characteristics and reaction of biomass, composed of hemicellulose, cellulose and lignin, are quite different. Therefore, the purpose of this research is to investigate the combustion characteristics of the three abundant biomass wastes in Thailand, i.e., cassava root, palm kernel shell and rice husk, as well as the influence of these residues when co-combusted with lignite coal. The experimental results of the co-combustion between lignite coal and such three biomass wastes discussed in this article would be helpful to promote the utilization of coal/biomass co-combustion for energy production and also to reduce the coal consumption (partially substitute for coal) in any coal-based thermal power plants in Thailand.

2 Materials and methods

2.1 Sample preparation and characterisation

Lignite and palm kernel shell samples were provided by an industry, while cassava root and rice husk are obtained from the local agricultural areas of Mabkha City, Rayong Province, Thailand. The biomass samples were washed with water, cut into small pieces and naturally dried for 3-5 days. Using a grinding machine, the sample was ground and was sieved to collect the fraction with particle size smaller than 250 μm for further study. The individual raw materials were named C100 (lignite coal), CR100 (cassava root), PKS100 (palm kernel shell) and RH100 (rice husk). Different mixtures of lignite coal and biomass residues were prepared by physical mixing and were denoted by CXCRY, CXPKSY and CXRHY, where X = 75, 50 and 25 ;Y = 25, 50 and 75 wt.%/wt.% for coal/cassava root, coal/palm kernel shell and coal/rick husk blends, respectively. The proximate analyses including moisture content M), volatile matter (VM), and ash content (A) were determined according to ASTM E870–82, while the fixed carbon (FC) was calculated by subtracting the percentages of moisture content, volatile matter and ash content from 100%. In addition, the calorific value or higher heating value (HHV) of samples was calculated by the data obtained from the oxygen bomb calorimeter (Parr 1341).

2.2 Thermogravimetric experiments

The combustion behavior of solid fuel samples was carried out with TA Instrument Q50 analyser under an air atmosphere. Prior to the TGA experiments, the fine sample powder, ca. 5-10 mg, was placed in a small platinum crucible and thermally purged under a nitrogen stream at 313 K for 10 min. Thermogravimetry (TG) and derivative thermogravimetric (DTG) profiles were attained by heating the sample from 313 to 973 K using an air flow of 60 ml/min, with the heating rate of 20 K/min. Analysis of each sample was repeated at least twice and the repeatability of experiments was more than 99%. In an attempt to obtained the kinetic parameters of coal/biomass co-combustion, the heating rates of 5, 10, 20 and 40 K/min were used for kinetic analysis.

2.3 Kinetic procedure

The reaction rate (r) of co-combustion of coal-biomass blends may be expressed by the following equation:

$$r = \frac{d\alpha}{dt} = k(T) f(\alpha) = A e^{(-E_a/RT)} f(\alpha) \quad (1)$$

where $k(T)$ is the reaction rate constant, T is the absolute temperature, $f(\alpha)$ is the function of reaction mechanism, A is the frequency factor, E_a is the activation energy, R is the universal gas constant and α is the fractional change in solid mass due to thermal decomposition at any time t or conversion, given by the equation:

$$\alpha = \frac{m_0 - m_t}{m_0 - m_f} \quad (2)$$

where m_0, m_t and m_f are the initial mass, the instantaneous mass and the final mass of the solid sample, respectively. Under constant heating rate, $\beta = dT/dt$, and integrating both sides of equation (1) give:

$$g(\alpha) = \int_0^{\alpha} \frac{d\alpha}{f(\alpha)} = \frac{A}{\beta} \int_{T_0}^{T} e^{(-E_a/RT)} dT = \frac{AE_a}{\beta R} p\left(\frac{E_a}{RT}\right) = \frac{AE_a}{\beta R} p(u) \quad (3)$$

where $g(\alpha)$ is the integral form of the conversion dependence function and $p(u)$ is the temperature integral function which is no analytical solution.

The most common soluion used to determine the kinetic parameters is the isoconversional method or model-free method. Isoconversional method has been proven effective particularly for kinetic analysis and has recently been strongly recommended by the Kinetics Committee of the International Confederation for Thermal Analysis and Calorimetry (ICTAC) [11]. By applying the Kissinger-Akahira-Sunose (KAS) method, one of the isoconversional method, in Equation (3), the KAS equation can be expressed as:

$$\ln\left(\frac{\beta}{T^2}\right) = \ln\left[\frac{AR}{E_a g(\alpha)}\right] - \frac{E_a}{RT} \quad (4)$$

Ploting $\ln(\beta/T^2)$ against $(1/T)$, the activation energy was determined from the slope of the resulting straight line. In order to find the frequency factor, the general thermal decomposition reaction represented by $f(\alpha) = (1-\alpha)$ was used in this study.

3 Results and discussion

3.1 Physical properties of sample

The basic characteristics, including M, VM, FC, A and calorific value, of lignite and biomass wastes are presented in Table 1. It was observed that C100 is dominated by fixed carbon content (32.04 wt.%), whereas the biomass wastes, CR100, PKS100 and RH100, are prevailed by volatile matter (54.96-75.25 wt.%) and relatively low ash content except rice husk. High amount of ash in the case of RH100 can be explained by high silica content [12]. Considering the key elements involved in the initiation of ignition, some research studies [13, 14] reveal that biomass contain higher volatile matter and volatile fuel ratio [VM/(VM+FC)] than coal. In this work, the values of volatile fuel ratio for biomass residues are in the range of 0.74-0.89 higher than that of lignite coal (0.63). This indicates that biomass residues have a higher reactivity and are easier to achieve ignition. Therefore, it can be expected that the blending of biomass waste and coal would enhance the ignition property of fuel.

Table 1. Characteristics of coal and biomass wastes.

Sample	Proximate analysis (wt.%)				HHV
	M	VM	FC	A	(MJ/kg)
C100	8.97	53.54	32.04	5.45	23.77
CR100	9.05	75.07	8.94	6.94	16.94
PKS100	7.62	72.25	13.52	6.61	21.25
RH100	7.98	54.96	19.29	17.78	14.29

Based on data obtained from the oxygen bomb calorimeter, the higher heating value of C100 is 23.77 MJ/kg, while the calorific values of biomass wastes are 16.94, 21.25, and 14.29 MJ/kg for CR100, PKS100, and RH100, respectively, which are located at the normal HHV range (i.e., 14-20 MJ/kg) of biomass [15].

3.2 Combustion behavior of individual fuels

Figure 1 and Table 2 show the typical TG curve, DTG curve and decomposition temperature intervals of lignite coal and biomass wastes decomposed under atmospheric air. According to the derivative weight loss, agricultural and agro-industrial residues exhibit 3-4 degradation stages reflecting the existence of structures of different reactivity. Whereas lignite coal shows only two characteristic peaks indicating more homogeneous material.

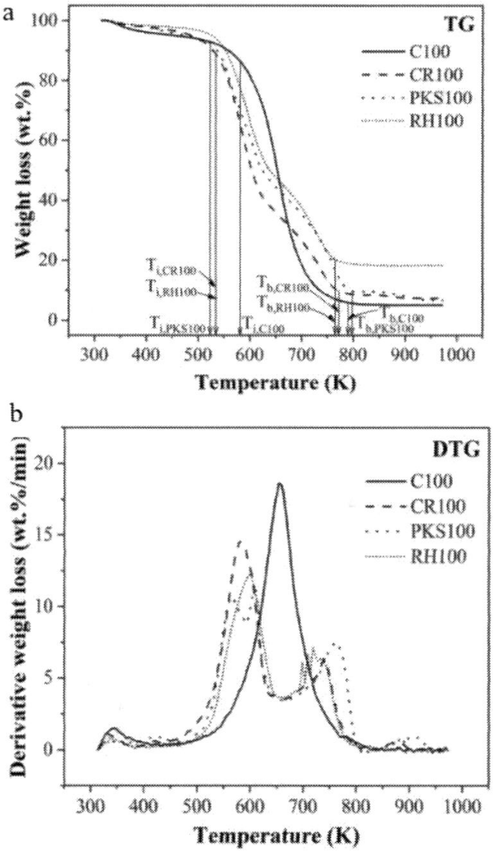

Fig. 1. a) TG and b) DTG profiles of coal and biomass wastes combustions with the heating rate of 20 K/min.

Table 2. Temperature intervals for different mass loss regions.

Sample	Temperature range (K)			
	Stage 1	Stage 2	Stage 3	Stage 4
C100	300-433	433-833	-	-
CR100	300-433	433-663	663-803	803-933
PKS100	300-463	463-663	663-823	823-953
RH100	300-423	423-663	663-823	-

 https://doi.org/10.1051/e3sconf/202130201002

The mass loss during the first stage of both the lignite coal and biomass wastes are normally due to the evaporation of free moisture, inherent moisture as well as low-boiling volatile release occurred at the temperature up to 463 K. For C100 sample, the main peak occurs in the second stage between 433 and 833 K with the maximum weight loss rate (R_{max}) of 18.67 wt.%/min, which was caused by the simultaneous combustion of volatile and solid carbon. In contrary to coal, biomass residues, CR100, PKS100 and RH100, exhibit 2-3 more decomposition stages after evaporation stage with the maximum weight loss rates between 11.29 and 14.57 wt.%/min. Stage 2 occurs between 423 and 663 K and stage 3 occurs in the temperature range of 663-823 K, which are ascribed as the volatilization of three main constituents (hemicellulose, cellulose and lignin) of biomass and the combustion of char residue, respectively. Nevertheless, some researchers [9, 16] claimed that the mass loss during the second stage is attributed to the total decomposition of hemicellulose and cellulose and the partial decomposition of lignin and during the third stage by the decomposition of remaining lignin and combustion of char residue. Another thermal decomposition stage (stage 4) in the temperature range of 803-953 K can be observed in the cases of CR100 and PKS100, which is attributed to the inorganic material decomposition [14]. It is noteworthy that the maximum weight loss rate of C100 is highest. Moreover, all biomass wastes lost their mass rapidly in the temperature below 650 K, while lignite coal sample decomposed rapidly at the temperature above 650 K. These circumstances imply that the incorporation of biomass wastes into lignite coal might not be totally benefit in terms of combustion performance.

Ignition temperature (T_i) and burnout temperature (T_b) are important properties of solid fuels for their industrial application. A number of methods have been developed to figure out the ignition and burnout temperatures of fuels, i.e., intersection method, derivation method, and conversion method. In this study, the ignition and burnout temperatures were determined following the intersection method as described by Lu and Chen [17]. Figure 2 illustrates the schematic of the ignition and burnout temperature determinations for CR100 sample by using the intersection method. Two points, A and B, on a TG curve are first identified. Point A is the point at which a vertical line from the highest value in the DTG curve crosses the TG curve. Point B is the point at which devolatilization begins (about 378-383 K). A tangent at point A on the TG curve and the horizontal line through point B are drawn. The corresponding temperature at the intersection of two lines is identified as the ignition temperature. In order to determine the burnout temperature of fuel sample, points C and D are located. Point C is the position on the TG curve at which a vertical line from the second peak of the DTG curve crosses the TG curve, while point D is the location at which the TG curve becomes steady. A temperature corresponding to the intersection of the tangent on the TG curve at C and the horizontal line through D is defined as the burnout temperature.

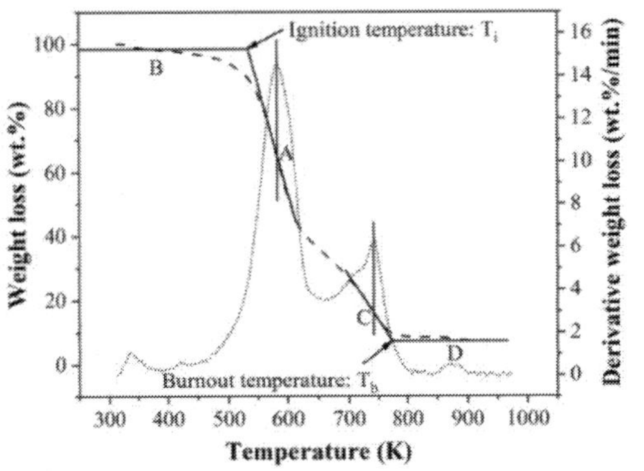

Fig 2. Schematic of the ignition and burnout temperature determinations.

The result of combustion characteristic parameters of individual sample is given in Table 3. It is noted that in the cases of cassava root and palm kernel shell the weight loss occurred above 800 K, due to the inorganic substance decompositions, were not considered for burnout temperature determination. It was found that the ignition temperatures of biomass residues are lower by 45-57 K than that of lignite coal. This could be enhanced the ignition performance of blended fuels caused by the blending of biomass wastes into coal. In terms of the burnout temperature, however, palm kernel shell shows higher value than lignite (796 K > 790 K). This implies that some volatiles released during the early combustion of PSK100 inhibit the char or heavy volatile combustion and thereby caused a longer complete combustion. The total burnout of C100, CR100, PKS100 and RH100 are about 94.85, 92.64, 93.32, and 81.70 wt.%, respectively.

Table 3. Characteristic parameters of coal and biomass wastes.

Sample	T_i (K)	T_b (K)	R_{max} (wt.%/min)	Total burn out (wt.%)
C100	580	790	18.67	94.85
CR100	535	772	14.57	92.64
PKS100	523	796	11.29	93.32
RH100	535	764	12.16	81.70

Note: T_i, ignition temperature; T_b, burnout temperature; R_{max}, maximum mass loss rate.

3.3 Co-combustion of fuels

The TG and DTG profiles of co-combustion of coal/biomass blended fuels with different percentages are shown in Figure 3-5. A similar trend can be seen for all biomass blended where the combustion of samples takes place with combined weight loss lied between that of the individual fuel. Similar phenomena were also found in the case of other blending samples [13, 14]. With considering the example of coal/biomass co-combustion in present study, C50CR50 (equally

https://doi.org/10.1051/e3sconf/202130201002

C25CR75) decomposes rapidly and more reactive than C100 at the temperature below 660 K. However, C75CR25 exhibits the most reactivity at the temperature above 660 K but it still lower than that of C100. In addition, the maximum weight loss rate of the combustion of coal component shifts to higher temperature (from 654 K to 678 K) with increasing of biomass portions. For coal-palm kernel shell blended fuels, C50PKS50 losses its weight abruptly at the temperature below 650 K, but at the temperature above 650 K C75PSK25 shows the most reactive for combustion. The maximum weight loss rate of the combustion of coal component, in addition, shifts to higher temperature from 655 K to 670 K with increasing of palm kernel shell portions. In the case of coal-rice husk blended fuels, C50RH50 exhibits the most reactivity at the temperature below 644 K, while C75RH25 shows the most reactivity at the temperature above 644 K. Moreover, the maximum combustion weight loss rate of coal component shifts to higher temperature from 656 K to 686 K with increasing of rice husk portions. These phenomena indicate that the combustion processes of the blended fuels are complex, which could be attributed to the interaction combustion behavior between biomass residues and lignite coal.

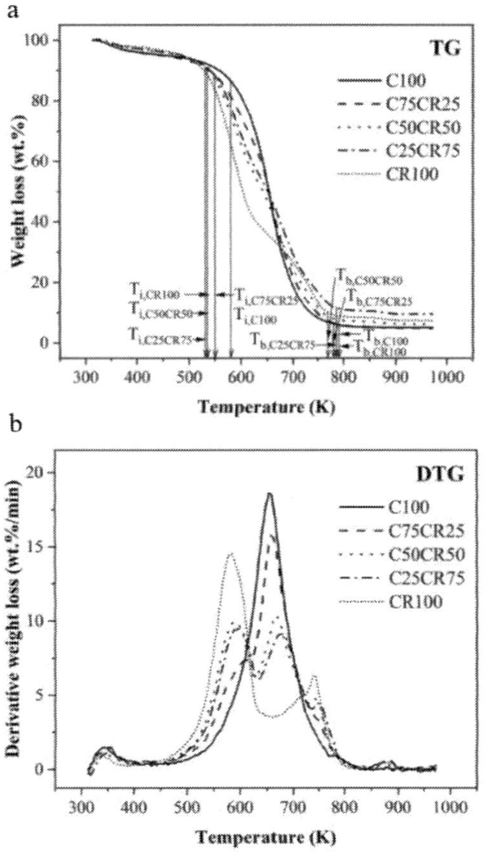

Fig. 3. a) TG and b) DTG profiles of coal/cassava root co-combustion.

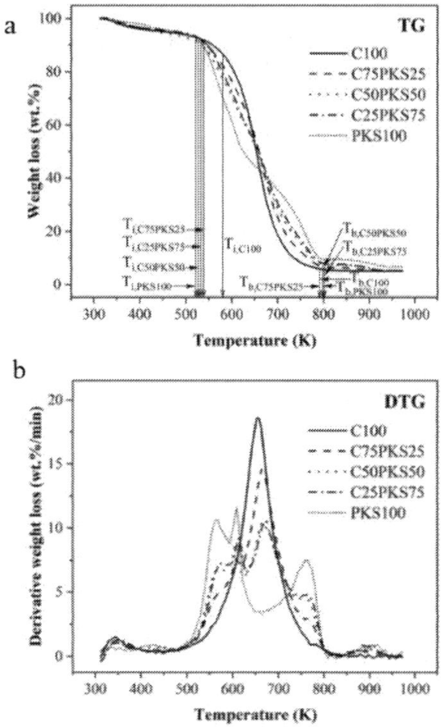

Fig. 4. a) TG and b) DTG profiles of coal/palm kernel shell co-combustion.

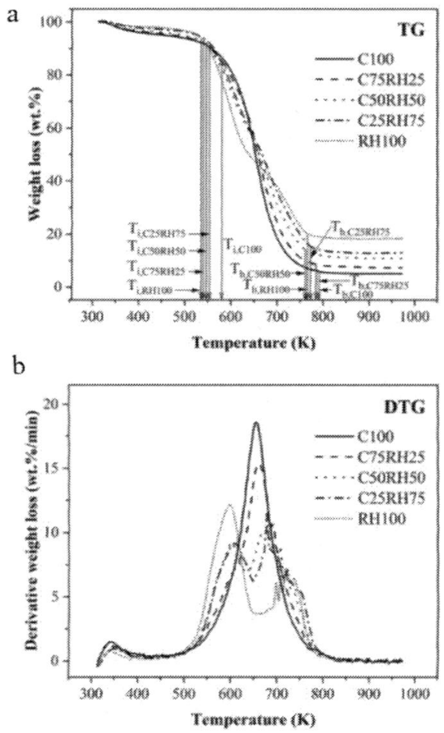

Fig. 5. a) TG and b) DTG profiles of coal/rice husk co-combustion.

It can be briefly suggested as: (i) biomass residues have high content of volatile matters which easily decompose and ignite. Therefore, they could combust rapidly and release the energy preheating coal at the low temperature region, e.g., < 660 K for C25CR75. In addition, the volatile compounds released abundantly consume a great amount of oxygen and thus hindered the oxygen molecules to reach the lignite coal surface, resulting in a lower reactivity of coal decomposition. So, this stage is dominated by the volatilization and volatile matter combustion (ii) when the temperature is higher (> 660 K), most of volatile compounds released in the early stage are combust completely, accompanied with coal has enough energy to ignite and combust, the rapidly mass loss due to coal combustion could be observed beyond 660 K. At this stage the fixed carbon combustion of coal is dominated.

In order to evaluate the combustion performance of coal/biomass co-combustion, two parameters, i.e., ignition index (D_i) and combustion index (S_i), have been used for various fuel systems [16, 18-20]. The ignition and combustion index can be determined by the following equations:

$$D_i = \frac{R_{max}}{t_i t_m} \tag{5}$$

$$S_i = \frac{R_{max} R_{mean}}{T_i^2 T_b} \tag{6}$$

where R_{max} and R_{mean} are maximum and average mass loss rates, respectively. T_i and T_b stand for the ignition and burnout temperatures, respectively. t_i and t_m are the ignition time and the time which corresponds to maximum mass loss rate, respectively.

The result of characteristic combustion parameters calculated from the thermal decomposition profiles is shown in Table 4. It was unexpectedly found that although biomass residues improve the ignition temperature of the blended fuels (528-553 K) compared to coal (580 K), the ignition index of all blended (7.08×10^{-4}-12.28×10^{-4}) are lower than that of C100 (12.81×10^{-4}). Furthermore, the combustion index of the blended fuels are also lower than pure lignite coal, i.e., 27.42×10^{-8}-47.48×10^{-8} are lower than that of 53.95×10^{-8}. This might be because the maximum weight loss rates of the blended fuels are lower than that of coal. In fact, it can be remarked that most of research works [5, 9, 13-14, 21-23] studied on the low-grade coals, which have the lower maximum combustion rate than the studied biomass and less thermal decomposition reactivity.

Although, both of ignition and combustion index of coal/biomass-blended fuels in present study are lower than lignite coal, the utilization of biomass wastes approximately 25 wt.% instead of coal could be recommended in which the combustion performances are lower than coal in the range of 11.45-19.35%. The application in equal parts, however, the combustion index would be diminished by almost 50%.

3.4 Kinetic analysis

The TG and DTG profiles of the C75CR25, C75PKS25 and C75RH25 samples at different heating rates are shown in Fig. 6-8. It can be seen that the TG curves shift to high temperatures without a significant change in their shapes while the heating rate increases, suggested that the entire combustion process slows down with the increase of heating rate. Additionally, the increasing in heating rate made the maximum mass loss rate shift toward higher values, suggesting an improvement of combustion intensity. Similar phenomena were also found in the cases of other biomass residues [8, 13, 16]. Figures 9-11 show the KAS plots for kinetic parameter calculations and Table 5 summarises the kinetic parameters for coal/biomass co-combustion. It was found that the activation energies of coal/biomass co-combustion are in the range of 91.92-113.20, 141.05-230.60 and 98.42-129.63 kJ/mol, which the average values are about 105.25, 179.66 and 121.84 kJ/mol for C75CR25, C75PKS25 and C75RH25, respectively. It should be noticed that the activation energy of C75PKS25 is larger than those of C75CR25 and C75RH25, suggesting more energy requirement for coal/palm kernel sell co-combustion system. The variation of activation energy with respect to conversion, furthermore, can be observed, implying the complex reactions took place during the coal/biomass co-combustion process. The values of calculated frequency factors also show in the wide range: 6.62×10^5 – 6.26×10^7, 1.62×10^8 – 6.49×10^{16}, 4.13×10^6 – 9.60×10^8 s^{-1}, for C75CR25, C75PKS25 and C75RH25, respectively.

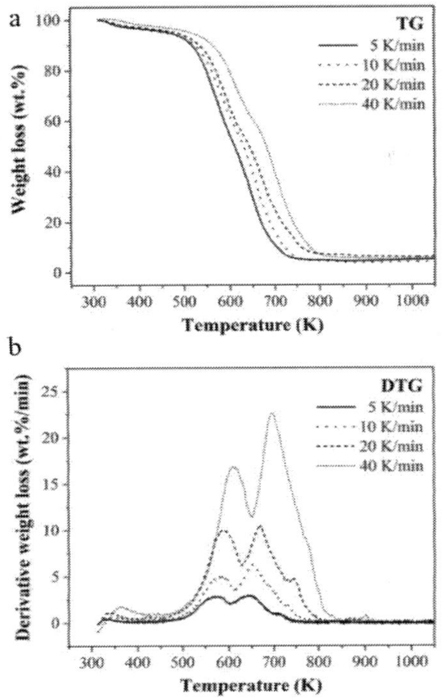

Fig. 6. a) TG and b) DTG profiles of C75CR25 co-combustion with different of heating rates.

https://doi.org/10.1051/e3sconf/202130201002

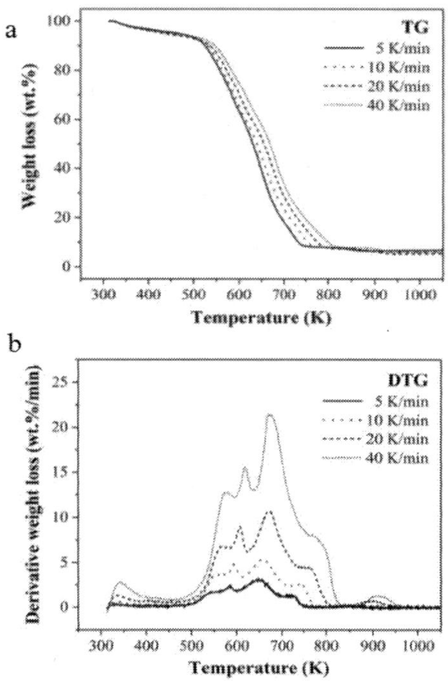

Fig. 7. a) TG and b) DTG profiles of C75PKS25 co-combustion with different of heating rates.

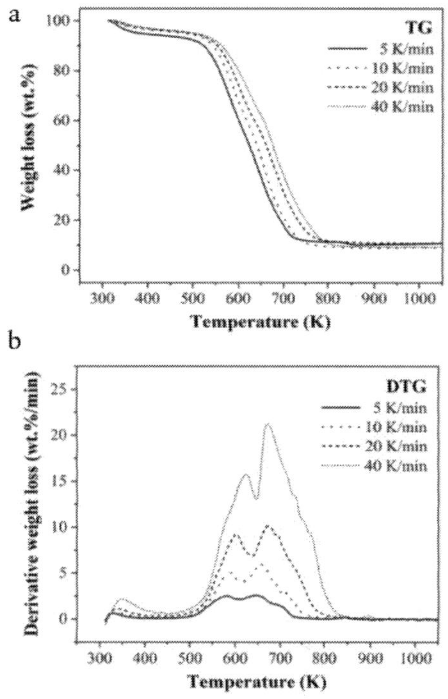

Fig. 8. a) TG and b) DTG profiles of C75RH25 co-combustion with different of heating rates.

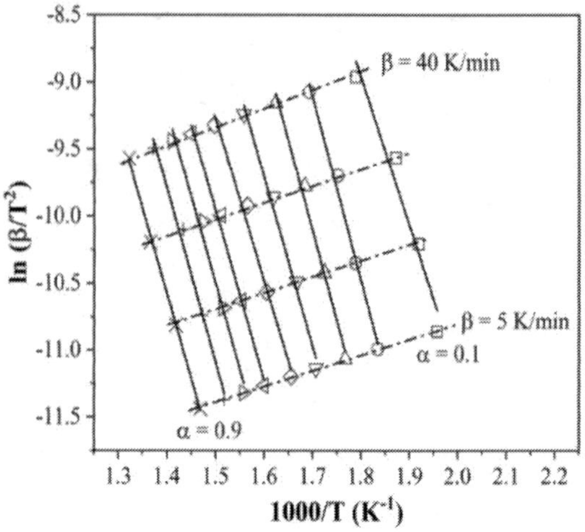

Fig. 9. KAS plot for C75CR25 co-combustion.

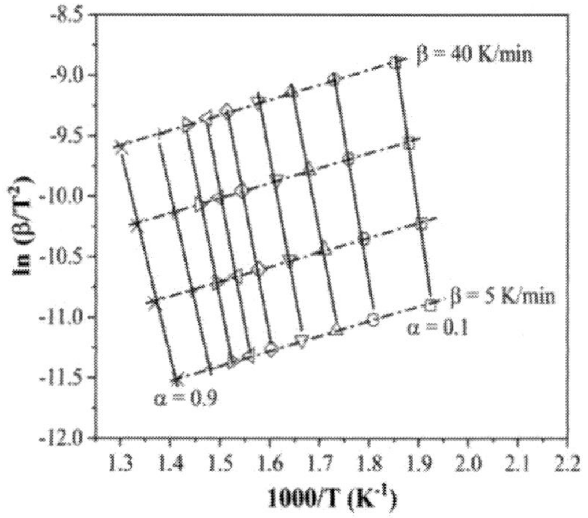

Fig. 10. KAS plot for C75PKS25 co-combustion.

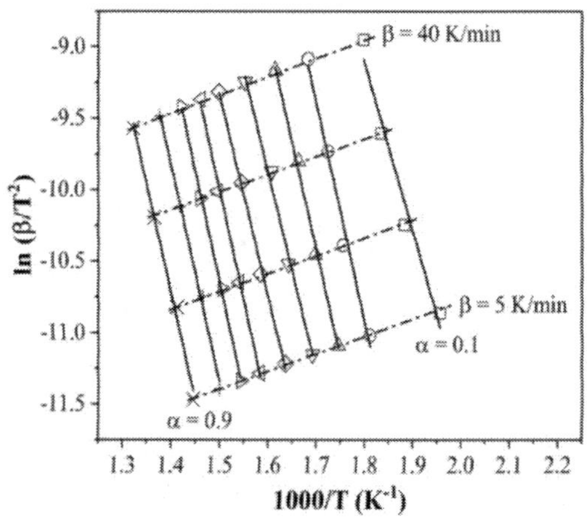

Fig. 11. KAS plot for C75RH25 co-combustion.

Table 4. Characteristic parameters of co-combustion.

Sample	T_i (K)	T_b (K)	R_{max} (wt.%/min)	t_i (min)	t_b (min)	t_m (min)	Total Burnout (wt.%)	D_i ($\times10^{-4}$)	S_i ($\times10^{-8}$)
C100	580	790	18.67	13.84	24.35	17.55	94.85	12.81	53.95
C75CR25	550	780	15.81	12.31	23.83	17.75	95.04	12.06	47.78
C50CR50	536	769	10.58	11.55	24.55	18.05	93.75	8.46	30.41
C25CR75	532	781	9.484	11.43	23.88	14.35	90.16	9.64	27.42
CR100	535	790	14.57	11.55	23.42	13.95	92.64	15.07	43.91
C75PKS25	536	789	14.81	12.03	24.67	18.45	94.66	11.12	43.51
C50PKS50	528	798	10.75	11.23	24.78	18.35	94.41	8.69	29.81
C25PKS75	530	801	10.42	11.35	24.91	18.15	94.74	8.43	28.90
PKS100	523	796	11.29	10.87	24.54	15.15	93.32	11.43	31.42
C75RH25	540	788	11.83	11.83	24.21	17.80	92.39	12.28	45.46
C50RH50	549	762	12.29	12.29	22.93	18.45	88.96	7.47	31.98
C25RH75	553	773	12.52	12.52	23.49	18.75	86.80	7.08	29.49
RH100	535	764	11.65	11.65	23.10	14.75	81.70	11.79	35.76

Note: T_i, ignition temperature; T_b, burnout temperature; R_{max}, maximum mass loss rate; t_i, time at T_i; t_b, time at T_b; t_m, time at R_{max}; D_i, ignition index; S_i, combustion index.

Table 5. Kinetic parameters for coal/bio mass co-combustion by KAS method.

α	C75CR25		C75PKS25		C75RH25	
	E_a (kJ/mol)	A (s^{-1})	E_a (kJ/mol)	A (s^{-1})	E_a (kJ/mol)	A (s^{-1})
0.1	91.92	1.09×10^{6}	230.60	6.49×10^{16}	98.42	4.13×10^{6}
0.2	113.20	6.26×10^{7}	198.01	8.04×10^{15}	127.30	9.60×10^{8}
0.3	111.48	2.56×10^{7}	177.74	2.55×10^{13}	125.75	3.93×10^{8}
0.4	105.19	4.04×10^{6}	183.66	2.59×10^{13}	116.33	3.26×10^{7}
0.5	99.87	8.63×10^{5}	178.80	3.11×10^{12}	116.71	2.01×10^{7}
0.6	104.90	1.50×10^{6}	180.81	2.24×10^{12}	128.49	1.25×10^{8}
0.7	106.79	1.65×10^{6}	173.90	3.37×10^{11}	129.63	1.07×10^{8}
0.8	108.20	1.59×10^{6}	152.40	3.18×10^{9}	127.20	4.47×10^{7}
0.9	105.67	6.62×10^{5}	141.05	1.62×10^{8}	126.71	2.39×10^{7}
Average	105.25		179.66		121.84	

Note: R^2 are between 0.96 and 0.99.

4 Conclusions

It can be concluded that lignite coal mainly decomposes only two stages, i.e., evaporation and the simultaneous volatile and char combustions. Biomass residues, namely cassava root, palm kernel shell and rice husk, decompose three to four stages which are corresponding to the evaporation, volatilization, volatile and char combustions and inorganic decomposition, respectively. Moreover, the coal/biomass blended fuels show the combined weight loss decomposition behaviors. Even though the incorporation of biomass wastes into lignite coal improved the ignition temperature of the blended fuels, the ignition and combustion index are worsened. This might be because biomass wastes, used in this present work, possess a lower values of maximum combustion rate and the less combustion reactivity in high temperature region, compared to lignite coal. From the kinetic studies, the average apparent activation energies of C75CR25, C75PKS25 and C75RH25 co-combustion calculated from KAS method are reported at 105.25, 179.66 and 121.84 kJ/mol, respectively. In order to maintain the combustion performance with regarding to the concerned environmental issues, it would be recommended that the co-combustion of coal/biomass, 75/25 wt.%/wt.% in particular, can be utilized for energy production.

From the present experimental data, the outcome of this research would be useful for biomass waste utilization in terms of energy production and environmental concerns. In practical, the next step should be to evaluate all costs involved in the entire logistics chain, e.g., collecting, milling, storage, and transportation, in order to obtain the information for an economic point of view. It is also recommended that, moreover, the investigation of other biomass wastes as well as the solid products obtained from the biomass thermochemical conversion processes, e.g., torrefied biomass and biochar, should be accounted for coal/biomass waste co-combustion system.

Acknowledgements

The authors would like to thank Faculty of Science Energy and Environment, King Mongkut's University of Technology North Bangkok. This research was funded by King Mongkut's University of Technology North Bangkok. Contract no. KMUTNB-64-DRIVE-22.

References

1. S.G. Sahu, N. Chakraborty, P. Sarkar, Coal-biomass co-combustion: An overview, Renewable and Sustainable Energy Reviews, **39**, (2014): 575-586
2. A. Hamisu, U.I. Gaya, Bi-template assisted sol-gel synthesis of photocatalytically-active mesoporous anatase TiO_2 nanoparticles, Applied Science and Engineering Progress, **Online First**, (2021): 1-15
3. B. Ashok, M. Umamahesh, N. Hariram, S. Siengchin, A.V. Rajulu, Modification of waste leather trimming with *in situ* generated silver nanoparticles by one step method, Applied Science and Engineering Progress, **14**, (2021): 236-246
4. M. Wzorek, R. Junga, E. Yilmaz, P. Niemiec, Combustion behavior and mechanical properties of pellets derived from blends of animal manure and lignocellulosic biomass, Journal of Environmental Management, **290**, (2021): 112487
5. M.V. Gil, D. Casal, C. Pevida, J.J. Pis, F. Rubiera, Thermal behaviour and kinetics of coal/biomass blends during co-combustion, Bioresource Technology, **101**, (2010): 5601-5608
6. Z. Liu, A. Quek, S. Kent Hoekman, M.P. Srinivasan, R. Balasubramanian, Thermogravimetric investigation of hydrochar-lignite co-combustion, Bioresource Technology, **123**, (2012): 646-652
7. A. Toptas, Y. Yildirim, G. Duman, J. Yanik, Combustion behavior of different kinds of torrefied biomass and their blends with lignite, Bioresource Technology, **177**, (2015): 328-336
8. F. Gyo, Z. Zhong, Optimization of the co-combustion of coal and composite biomass pellets, Journal of Cleaner Production, **185**, (2018): 399-407
9. F. Guo, Y. He, A. Hassanpour, J. Gardy, Z. Zhong, Thermogravimetric analysis on the co-combustion of biomass pellets with lignite and bituminous coal, Energy, **197**, (2020): 117147
10. S. Qi, Z. Wang, M. Costa, Y. He, K. Cen, Ignition and combustion of single pulverized biomass and coal particles in N_2/O_2 and CO_2/O_2 environments, Fuel, **283**, (2021): 118956
11. S. Vyazovkin, A.K. Burnham, J.M. Criado, L.A. Perez-Maqueda, C. Popescu, N. Sbirrazzuoli, ICTAC kinetics committee recommendations for performing kinetic computations on thermal analysis data, Thermochimica Acta, **520**, (2011): 1-19
12. P. Jongpradist, W. Homtragoon, R. Sukkarak, W. Kongkitkul, P. Jamsawang, Efficiency of rick husk ash as cementitious materials in high-strength cement-admixed clay, Advances in Civil Engineering, **2018**, (2018): Article ID 8346319, 11 pages
13. G. Wang, J. Zhang, J. Shao, Z. Liu, G. Zhang, T. Xu, J. Guo, H. Wang, R. Xu, H. Lin, Thermal behavior and kinetic analysis of co-combustion of waste biomass/low rank coal blends, Energy Conversion and Management, **124**, (2016): 414-426
14. X.G. Li, Y. Lv, B.G. Ma, S.W. Jian, H.B. Tan, Thermogravimetric investigation on co-combustion characteristics of tobacco residue and high-ash anthracite coal, Bioresource Technology, **102**, (2011): 9783-9787
15. A. Dashti, A.S. Noushabadi, M. Raji, A. Razmi, S. Ceylan, A.H. Mohammadi, Estimation of biomass higher heating value (HHV) based on the proximate analysis: Smart modelling and correlation, Fuel, **257**, (2019): 115931

16. D. Vamvuka, S. Sfakiotakis, Combustion behaviour of biomass fuels and their blends with lignite, Thermochimica Acta, **526**, (2011): 192-199
17. J.-J. Lu, W.-H. Chen, Investigation on the ignition and burnout temperatures of bamboo and sugarcane bagasse by thermogravimetric analysis, Applied Energy, **160**, (2015): 49-57
18. S.Y. Luo, B. Xiao, Z.Q. Hu, S.M. Liu, Y.W. Guan, Experimental study on oxygen-enriched combustion of biomass micro fuel, Energy, **34**, (2009): 1880-1884
19. Y. Lin, Y. Liao, Z. Yu, S. Fang, X. Ma, The investigation of co-combustion of sewage sludge and oil shale using thermogravimetric analysis, Thermochimica Acta, **653**, (2017): 71-78
20. K. Jayaraman, M.V. Kok, I. Gokalp, Combustion mechanism and model free kinetics of different origin coal samples: Thermal analysis approach, Energy, **204**, (2020): 117905
21. C. Chen, S. Qin, F. Chen, Z. Lu, Z. Cheng, Co-combustion characteristics study of bagasse, coal and their blends by thermogravimetric analysis, Journal of the Energy Institute, **92**, (2019): 364-369
22. K. Cong, F. Han, Y. Zhang, Q. Li, The investigation of co-combustion characteristics of tobacco stalk and low rank coal using a macro-TGA, Fuel, **237**, (2019): 126-132
23. B. Mahanta, A. Saikia, U.N. Gupta, P. Saikia, B.K. Saikia, J. Jayaramudu, P.S. Sellamuthu, E.R. Sadiku, Study of low-rank high sulphur coal fine with biomass, Current Research in Green and Sustainable Chemistry, **3**, (2020): 100023

Relative Efficiency Analysis of Biomass Agricultural Plants using Data Envelopment Analysis

Kasin Ransikarbum[1,*] and *Rapeepan* Pitakaso[1]

[1]Industrial Engineering, Ubonratchathani University, Ubonratchathani, 34190, Thailand

Abstract. Renewable energy has recently been a promising interest as a substitute for fossil fuels due to an increasing energy demand as well as a rising concern over the environmental impact of fossil fuel consumption around the globe. Biofuel, in particular, is a type of renewable energy, which can be derived from various biomass types. In this research, we analyze relative efficiencies using Data Envelopment Analysis (DEA) technique from three types of energy-related plants in the Northeastern region of Thailand, which are cassava, sugarcane, and palm. The relative efficiency of each province is further analyzed during 2017 to 2019 for a comparative study. Next, the input criteria are collected including allowable planting area, labor cost, and rainfall amount; whereas the included output criterion is the quantity of harvested product. Our initial analysis using CCR, BBC, and Scale Efficiency (SE) models of DEA provides the baseline of efficient provinces to be benchmarked and directions for improving inefficient provinces, given desired input and output criteria in this study.

Keyword. Relative Efficiency, Data Envelopment Analysis, Biomass, Renewable Energy

1 Introduction and Motivation

Renewable energy, such as biomass, solar, and wind has recently been a promising interest as a substitute for fossil fuels, such as oil and coal, due to an increasing energy demand as well as a rising concern over the environmental impact of fossil fuel consumption around the globe. Many countries have taken a variety of actions through strategic policies aiming at meeting energy needs more securely and sustainably. For example, the United States mandates to have more than 20 billion gallons of biofuel under the Energy Security Act by 2022. The European Union (EU) also aims to achieve 20% of energy from renewable sources by 2020. Also, China issues a long-term development plan of renewable energy aiming to increase the capacity of biomass power generation for 30 million Kilowatt (kW) by 2020 [1-2]. Thailand has also promoted a new economic model towards Industry 4.0 development plan by focusing on 10 targeted, S-curve industries – three of them are agricultural, logistics, and biofuel sectors [3].

Biomass, in particular, can be obtained from several sources including edible crops, non-edible crops, crop residues, forests, and waste. In comparison to fossil fuels, biomass is easy to grow and replace quickly without depleting natural resources. The advantages of using biomass are noted for its ability to be stored and used on demand, clean energy, renewable, and no carbon dioxide side effect. In addition, biomass also has the potential to reduce the dependency on fossil fuels, which are the main source of carbon dioxide release in the atmosphere [4-7].

Biofuel supply chain, in particular, involves a number of stakeholders, including farms providing feedstocks from biomass, pre-processing facilities, transshipment depots, bio-refinery plants, fuel-blending facilities, and demanding points of gas stations. Thus, biomass can be viewed as the upstream of the biofuel supply chain, in which the efficiency evaluation needs to be properly addressed. Fig. 1 illustrates the differences and similarities between traditional industrial and bioenergy supply chain.

In this research, we collect and analyze biomass data of three major feedstock for biofuel in the Northeastern region of Thailand. In particular, energy plants are collected for cassava, sugarcane, and palm during 2017 to 2019. Then, the relative efficiency of each province is further analyzed using Data Envelopment Analysis (DEA) for a comparative study. The input criteria are collected including allowable planting area, labor cost, and rainfall amount; whereas the included output criterion is the quantity of harvested product. Our initial analysis using CCR, BBC, and Scale Efficiency (SE) models of DEA provides the baseline of efficient provinces to be benchmarked and directions for improving inefficient provinces, given desired input and output criteria in this study.

2 Related Studies and Method

2.1. Biomass and Bioenergy Supply Chain in Thailand

* Corresponding author: kasin.r@ubu.ac.th; kasinphd@gmail.com

https://doi.org/10.1051/e3sconf/202130201003

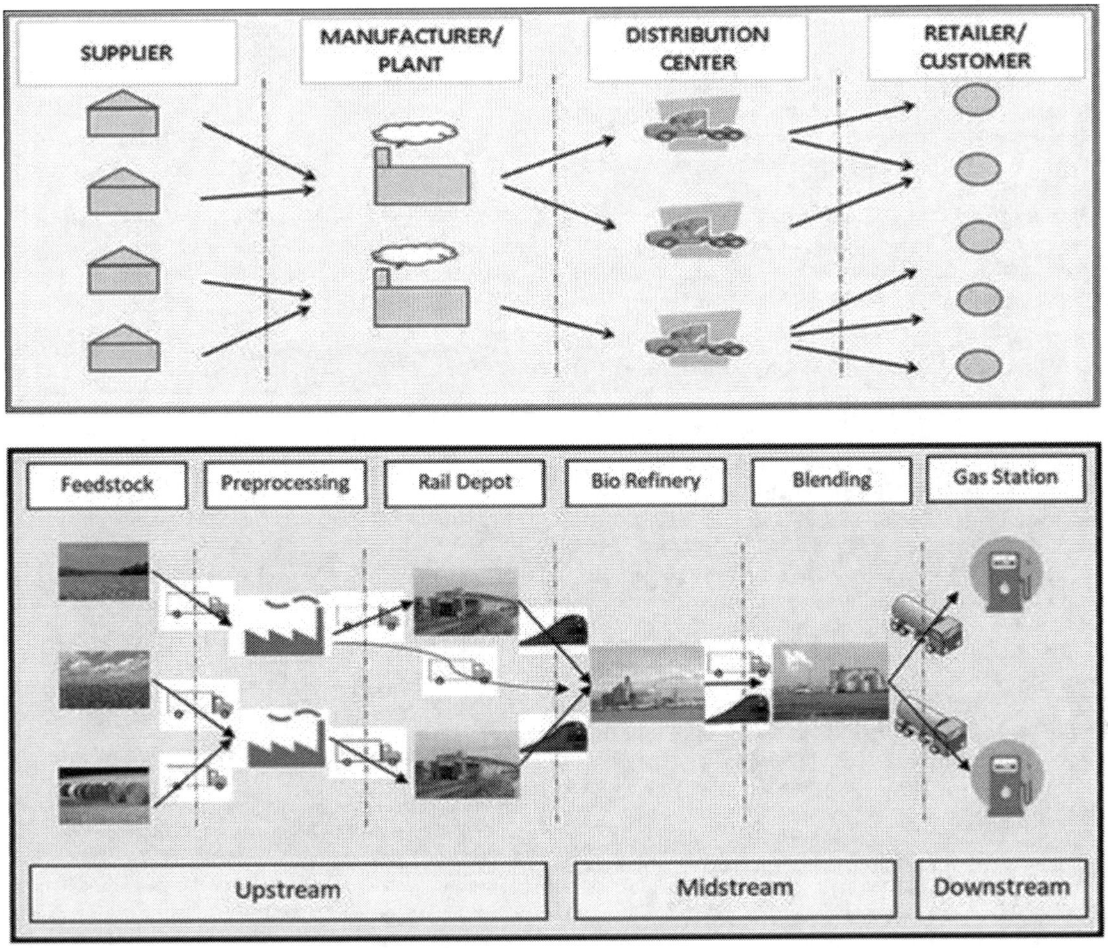

Fig. 1. (Top) Traditional industrial logistics; (Bottom) Bioenergy logistics

Renewable energy has attracted the attention of researchers around the globe for ensuring future energy security and sustainability. Biofuel energy, in particular, is one of the renewable energy that has gained ground in this regard. According to REN21 [5], biofuels employment attracted around 2 million jobs in 2018, in which most of these jobs are in the agricultural supply chain in developing countries, especially in the case of Southeast Asia, including Thailand. REN21 (2019) also estimates annual capacity and production of ethanol production in 2018 and finds that the top five countries are United States, Brazil, China, Canada, and Thailand, respectively. Besides, the top five countries for biodiesel production are United States, Brazil, Indonesia, Germany, and Argentina, respectively. Thus, Thailand also has a high potential to enhance its economics through bioethanol process.

In Thailand, the Department of Alternative Energy Development and Efficiency (DEDE [8]) plays a key role, in which a mission is to promote and support sustainable and worthy energy consumption and production for exporting and domestic use and to build collaborative network for bringing the country into the knowledge based society with sustainable economic stability and social well beings. Two key performance-related projects noted are 1) developing community-based biomass power plants and 2) developing the biomass potential database in Thailand.

In addition, the report by DEDE [9] for Research and development (R&D) studies of renewable energy in Thailand suggests that there are four groups of research studies going on in Thailand – 1) the research on the potential of materials focusing on assessing the overall potential of biomass as raw materials; 2) research on biomass preparation process focusing on finding a way to improve the quality of biomasses, such as chipping, grinding, pelletizing, and humidity reduction; 3) research on electricity and heat production technologies for improving production process and quality of technologies in producing power and heat from biomass; and 4) research on economics and environmental impacts of biomass. The authors also note that most of the research in Thailand falls under group 3 and there is a need to pursue studies in other research areas.

With regard to biomass and biofuel status in Thailand, according to DEDE [8], Thailand has the target for ethanol production in 2036 to be 11.3 million liters per day. The actual ethanol production, however, is below the target (i.e., 3.51 million liters in 2015, 3.67 in 2016, 3.94 in 2017, and 4.20 in 2018, respectively). Obviously, the trend of ethanol production in Thailand will continuously grow in the future and thus a proper evaluation of biomass efficiency for each agricultural regional area is required. Regarding biomass types in Thailand, studies from DEDE [8] also show the potential of the biomass for varied types of feedstocks with more or less capacity, in which cassava, sugarcane, and palm are among the top potential biomass types.

2.2. Efficiency Study with DEA applications

Multi-Criteria Decision Analysis (MCDM) is a sub-discipline of operations research and management science (OR/MS) that explicitly considers multiple criteria in a decision-making environment and has been used to support decision-makers facing decision and planning problems that a unique optimal solution does not exist and/or decision-maker's preferences are involved. Common methods, specifically, include various tools, such as Analytic Hierarchy Process (AHP), Data Envelopment Analysis (DEA), Technique for Order of Preference by Similarity to Ideal Solution (TOPSIS), Multi-Attribute Utility Theory (MAUT), Multi-Objective Mathematical Programming (MOMP), and Goal Programming (GP). These tools have been applied and extended in a number of applications (e.g., [10-19]).

DEA, particularly, is a Linear Programming (LP) methodology to measure relative efficiency of multiple Decision-Making Units (DMUs) or so-called alternatives when the problem is presented with multiple input and output criteria. After the DEA linear programming model is solved, a particular DMU will be considered efficient if it obtains a score of one, whereas scores that are lesser than one imply relative inefficiency. It is also possible that more than one alternatives are found to be efficient. According to survey study from Liu et al. [20], the DEA literature's size is expected to continue to grow at least double the size of the existing literature. In addition, the DEA method has been applied in various applications [21-22]. We next discuss the three prevalent DEA models commonly used in the literature and the DEAP computer program.

2.2.1 CCR Model

The CCR model was early developed and named after the three researchers (Charnes, Cooper and Rhodes [23] to measure the overall technical efficiency ($TE_{overall}$), in which a Constant Return to Scale (CRS) assumption holds. That is, the CRS assumption holds true when the DMUs are operated under the condition of the optimal size and perfect competition. In particular, equations (1)-(5) present the CCR model of DEA in a linear programming form.

Sets

I: Set of inputs, indexed by i
J: Set of outputs, indexed by j
K: Set of DMUs, indexed by k

Parameters

x_{i,k_0} *: Amount of input data for input i of DMU k*

y_{j,k_0} *: Amount of output data for output j of DMU k*

Decision variables

U_i *: The weight assigned to input i*

V_j *: The weight assigned to output j*

Mathematical model

$$\text{Maximize} \quad TE_{overall} \quad \sum_{j \in J} y_{j,k_0} V_j \tag{1}$$

$$\text{Subject to} \quad \sum_{i \in I} x_{i,k_0} U_i = 1 \tag{2}$$

$$\sum_{j \in J} y_{j,k} V_j - \sum_{i \in I} x_{i,k} U_i \leq 0 \quad ;\forall k \in K \tag{3}$$

$$U_i \geq 0 \quad ;\forall i \in I \tag{4}$$

$$V_j \geq 0 \quad ;\forall j \in J \tag{5}$$

2.2.2 BCC Model

The BCC model was later developed by and named after Banker, Charnes, Cooper [24] to measure the pure technical efficiency (TE_{pure}) of DMUs. The BCC model is formulated by extending from the dual model of the primal CCR model, which transforms the primal maximization to dual minimization problem. In contrast to CCR model, the BCC allows DMUs to be operated under imperfect condition and not necessarily at optimal size, which is more practical in real situations. That is, the Variable Return to Scale (VRS) assumption holds for the BCC model of DEA (Equations (6)-(10)), where θ is the relative efficiency and λ_k is the dual decision variable for each DMU.

Mathematical model

$$\text{Minimize} \quad TE_{pure} \quad \theta \tag{6}$$

$$\text{Subject to:} \quad \sum_{k \in K} \lambda_k x_{i,k} \leq \theta x_{i,k_0} \quad ;\forall i \in I \tag{7}$$

$$\sum_{k \in K} \lambda_k y_{j,k} \geq y_{j,k_0} \quad ;\forall j \in J \tag{8}$$

$$\sum_{k} \lambda_k = 1 \tag{9}$$

$$\lambda_k \geq 0 \tag{10}$$

https://doi.org/10.1051/e3sconf/202130201003

2.2.3 SE Model

The SE can be computed to express whether a particular DMU is operating at optimal size (i.e., similar to the CRS assumption) or whether at imperfect condition (i.e., similar to the VRS assumption). That is, if the latter holds true, the value of SE can be used to indicate whether the DMU operates under Increasing Return to Scale (IRS) (i.e., the size is too large) or Decreasing Return to Scale (i.e., the size is too small). In particular, the SE can be computed as a ratio between the relative efficiency obtained from the CCR model and the BCC model as shown in Equation (11).

$$\frac{TE_{overall}}{TE_{pure}} \tag{11}$$

2.2.4 DEAP Computer Program

We next discuss the Data Envelopment Analysis Program (DEAP). The program consists of the instruction file, the data file, and the output file; in which the CCR model, the BCC model, and the SE model can be simultaneously computed to obtain relative efficiencies of DMUs of interest. The program is also capable of computing how much the input criteria should be decreased for inefficient DMUs to be efficient (i.e., input-oriented) and how much the output criteria should be increased for inefficient DMUs to be efficient (i.e., output-oriented) for benchmarking purpose. In this research, the computer program DEAP Version 2.1 is used for analyzing related efficiency data.

3 Case Study of Biomass Feedstock and Analysis

3.1. Data Collection

We next discuss the case study of biomass data obtained from the Office of Agricultural Economics (OAE) of Thailand, in which the mission is to provide suggestions for policy development plans related to agricultural trade and international agricultural economic cooperation [25]. In particular, data are chosen from the Northeastern region of Thailand inclusive of 20 provincial areas as follows: A1) Loei, A2) Nong Bua Lamphu, A3) Udon Thani, A4) Nong Khai, A5) Bueng Kan, A6) Sakon Nakhon, A7) Nakhon Phanom, A8) Mukdahan, A9) Yasothon, A10) Amnat Charoen, A11) Ubon Ratchathani, A12) Sisaket, A13) Surin, A14) Buriram, A15) Maha Sarakham, A16) Roi Et, A17) Kalasin, A18) Khon Kaen, A19) Chaiyaphum, and A20) Nakhon Ratchasima (Fig. 1). Next, information is gathered for cassava, sugarcane, and palm during 2017 to 2019 as shown in Tables 1-3, respectively. The input criteria are inclusive of I1) allowable planting area, I2) labor cost, and I3) rainfall amount; whereas the output criterion is the quantity of harvested product for energy crop of O1) cassava, O2) sugarcane, and O3) palm, respectively.

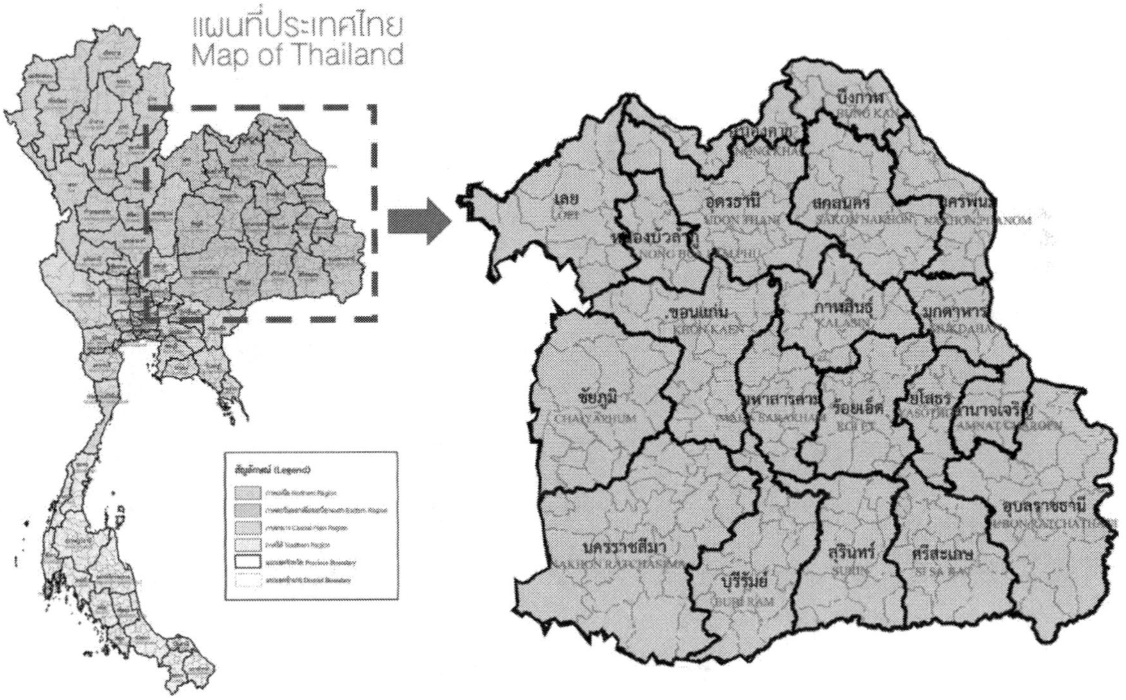

Fig. 2. Case study of the Northeastern region of Thailand (Adapted from [25])

 https://doi.org/10.1051/e3sconf/202130201003

Table 1. Collected data for biomass during 2017 [25]

DMUs	I1 Planting Area Unit (Rai)	I2 Labor Cost Unit (Baht)	I3 Rainfall Unit (mm.)	O1 Cassava Unit (Ton)	O2 Sugarcane Unit (Ton)	O3 Palm Unit (Ton)
A1	617,721	305	1,545	25,574	3,583,762	1,044,323
A2	377,254	305	1,608	3,486	3,725,326	223,764
A3	969,116	305	1,860	26,297	8,343,466	925,753
A4	100,822	305	2,087	23,187	810,936	51,926
A5	29,958	305	1,963	38,080	26,072	24,372
A6	207,176	305	2,369	16,022	845,440	370,380
A7	41,328	305	2,479	5,323	77,698	96,670
A8	361,492	305	2,057	2,702	2,636,511	507,843
A9	175,474	305	1,685	3,390	966,951	334,562
A10	163,373	305	1,720	5,657	833,413	317,478
A11	475,635	305	1,760	21,857	181,485	1,575,033
A12	176,030	305	1,663	7,544	386,678	496,034
A13	323,631	305	1,427	2,881	2,651,375	386,948
A14	445,348	305	1,563	7,343	2,472,410	952,503
A15	289,063	305	1,891	72	1,994,215	383,242
A16	213,411	305	1,691	1,635	1,876,815	176,199
A17	672,759	305	1,714	5,037	5,021,133	875,616
A18	857,249	308	1,561	975	7,587,787	737,716
A19	1,151,083	305	1,136	5,797	7,183,845	1,840,241
A20	2,155,739	308	1,386	10,021	7,893,730	5,514,475

Table 2. Collected data for biomass during 2018 [25]

DMUs	I1 Planting Area Unit (Rai)	I2 Labor Cost Unit (Baht)	I3 Rainfall Unit (mm.)	O1 Cassava Unit (Ton)	O2 Sugarcane Unit (Ton)	O3 Palm Unit (Ton)
A1	372,127	315	1,126	31,827	3,524,287	1,115,774
A2	378,728	310	1,309	5,236	3,663,764	222,125
A3	959,020	315	1,424	30,700	8,204,646	871,482
A4	97,624	320	1,707	25,896	797,524	41,243
A5	31,540	315	1,667	42,774	25,638	20,044
A6	193,094	318	1,735	17,495	831,326	333,302
A7	31,131	315	2,757	6,301	76,397	59,590
A8	356,668	318	1,929	3,508	3,592,793	481,759
A9	172,657	315	1,796	4,364	950,766	319,787
A10	171,582	310	1,804	7,239	819,523	343,245
A11	474,030	320	1,831	27,197	178,420	1,561,688
A12	178,295	310	1,436	9,017	380,201	503,287
A13	324,530	315	1,078	4,523	2,607,745	390,032
A14	446,794	310	792	9,923	2,431,963	969,635
A15	289,968	310	1,225	121	1,960,807	383,371
A16	210,337	315	1,210	2,583	1,845,455	196,194
A17	668,980	318	1,109	6,830	4,936,871	856,325
A18	215,376	320	1,168	2,437	7,460,999	735,135
A19	1,133,900	310	912	7,506	7,063,600	1,787,325
A20	2,073,375	320	1,095	12,489	7,762,504	5,298,895

Table 3. Collected data for biomass during 2019 [25]

DMUs	I1 Planting Area Unit (Rai)	I2 Labor Cost Unit (Baht)	I3 Rainfall Unit (mm.)	O1 Cassava Unit (Ton)	O2 Sugarcane Unit (Ton)	O3 Palm Unit (Ton)
A1	82,966	320	681	34,657	3,519,084	1,124,470
A2	423,113	315	1,024	5,565	3,832,243	252,398
A3	1,029,612	320	1,286	30,790	8,309,719	950,032
A4	84,622	325	1,545	26,936	798,363	42,509
A5	35,510	320	1,525	46,146	53,982	20,069
A6	208,786	323	1,252	21,398	941,496	332,932
A7	38,375	320	2,111	7,082	115,670	65,623
A8	371,272	323	1,587	4,027	2,638,052	492,214
A9	162,174	320	1,587	4,225	1,052,388	347,932
A10	197,762	315	1,598	6,800	1,021,304	355,759
A11	491,856	325	1,619	27,869	169,503	1,700,045
A12	185,095	315	1,194	9,855	349,185	524,574
A13	328,609	320	1,271	5,923	2,273,529	481,394
A14	488,575	320	1,042	11,813	2,494,540	1,022,085
A15	312,340	315	1,228	120	1,929,941	438,884
A16	227,862	320	1,628	2,695	1,912,407	194,487
A17	721,754	323	1,300	6,813	5,296,986	904,922
A18	894,427	325	1,018	2,406	7,257,231	786,745
A19	1,229,592	315	775	7,078	6,808,992	2,169,264
A20	2,115,752	325	727	12,979	7,277,088	5,325,614

3.2. DEA Analysis and Results

We next analyzed results from the DEAP computer program as shown in Tables 4-6 for data from 2017-2019, respectively. The overall technical efficiency from the CCR model, the pure technical efficiency from the BCC model, and the analysis from the SE model are presented. Provincial DMUs with either IRS (too-large size) or DRS (too-small size) are also analyzed. Regardless, other techniques (i.e., heuristics, simulation) can also be used and integrated to solve the linear programming problem of DEA model as well [26-34].

Table 4. DEA analysis for 2017 data

DMUs	CCR Model ($TE_{overall}$)	BCC Model (TE_{pure})	SE	Type
A1	1.000	1.000	1.000	-
A2	1.000	1.000	1.000	-
A3	1.000	1.000	1.000	-
A4	0.974	1.000	0.974	irs
A5	1.000	1.000	1.000	-
A6	0.833	1.000	0.833	irs
A7	0.870	1.000	0.870	irs
A8	0.973	1.000	0.973	irs
A9	0.943	1.000	0.943	irs
A10	0.925	1.000	0.925	irs
A11	1.000	1.000	1.000	-
A12	0.981	1.000	0.981	irs
A13	1.000	1.000	1.000	-
A14	1.000	1.000	1.000	-
A15	0.919	1.000	0.919	irs
A16	0.963	1.000	0.963	irs
A17	0.976	1.000	0.976	irs
A18	1.000	1.000	1.000	-
A19	1.000	1.000	1.000	-
A20	1.000	1.000	1.000	-

Table 5. DEA analysis for 2018 data

DMUs	CCR Model ($TE_{overall}$)	BCC Model (TE_{pure})	SE	Type
A1	1.000	1.000	1.000	-
A2	0.509	1.000	0.509	irs
A3	1.000	1.000	1.000	-
A4	0.669	0.980	0.683	irs
A5	1.000	1.000	1.000	-
A6	0.561	0.979	0.573	irs
A7	0.679	1.000	0.679	irs
A8	0.499	0.979	0.510	irs
A9	0.554	0.985	0.562	irs
A10	0.608	1.000	0.608	irs
A11	1.000	1.000	1.000	-
A12	0.852	1.000	0.852	irs

A13	0.431	0.989	0.436	irs
A14	0.778	1.000	0.778	irs
A15	0.404	1.000	0.404	irs
A16	0.292	1.000	0.292	irs
A17	0.686	0.977	0.703	irs
A18	1.000	1.000	1.000	-
A19	1.000	1.000	1.000	-
A20	1.000	1.000	1.000	-

Table 6. DEA analysis for 2019 data

DMUs	CCR Model ($TE_{overall}$)	BCC Model (TE_{pure})	SE	Type
A1	1.000	1.000	1.000	-
A2	0.738	1.000	0.738	irs
A3	1.000	1.000	1.000	-
A4	0.623	0.981	0.635	irs
A5	1.000	1.000	1.000	-
A6	0.523	0.981	0.533	irs
A7	0.225	1.000	0.225	irs
A8	0.527	0.976	0.540	irs
A9	0.282	0.988	0.285	irs
A10	0.273	1.000	0.273	irs
A11	0.887	0.983	0.902	irs
A12	0.397	1.000	0.397	irs
A13	0.477	0.985	0.485	irs
A14	0.560	0.986	0.568	irs
A15	0.417	1.000	0.417	irs
A16	0.450	0.987	0.456	irs
A17	0.800	0.979	0.818	irs
A18	1.000	1.000	1.000	-
A19	1.000	1.000	1.000	-
A20	1.000	1.000	1.000	-

As illustrated in Table 4, provincial DMUs that operate with an efficient condition (i.e., the score for relative efficiency of 1) and with the optimal size (i.e., the score for SE of 1) during 2017 and should be considered the benchmark units are A1, A2, A3, A5, A11, A13, A14, A18, A19, and A20. In addition, data analyzed for 2018 (Table 5) show that efficient DMUs with optimal sizes are A1, A3, A5, A11, A18, A19, and A20. Next, based on the 2019 data obtained in Table 6, efficient DMUs with optimal sizes are found to be A1, A3, A5, A18, A19, and A20, respectively.

Clearly, an operation for some provincial DMUs fluctuates during 2017-2019, whereas certain provincial DMUs can operate with all efficient conditions for three years. Additionally, the IRS condition for certain provincial DMUs suggest that scale inefficiency exists, in which the size is considered too large when comparing to other DMUs. These analyzed results are also categorized for CCR model, BCC model, and SE model across all progressive years to illustrate the trend with respect to time as shown in Figs 2-4, respectively.

3.3. Discussion

Analyzed results from the CCR model, the BCC model, and the SE model obtained earlier suggest that A1) Loei, A3) Udon Thani, A5) Bueng Kan, A18) Khon Kaen, A19) Chaiyaphum, and A20) Nakhon Ratchasima are efficient across three years from 2017 to 2019. This is due to that the analyzed relative efficiency scores are shown to be 1.00 for three consecutive years in Figs 3-5. Overall, these efficient provinces are found to utilize lesser inputs (e.g., planting area, labor cost, rainfall) and/or obtain higher outputs (e.g., tons of products) when comparing to other peers. Thus, these provinces have been operated with efficient condition, in which they should be further used as a benchmark DMUs for other provinces.

In addition, other provinces operated at inefficient condition can consider whether a particular input criterion should be decreased with a fixed output requirement or a particular output criterion can be increased under a fixed input.

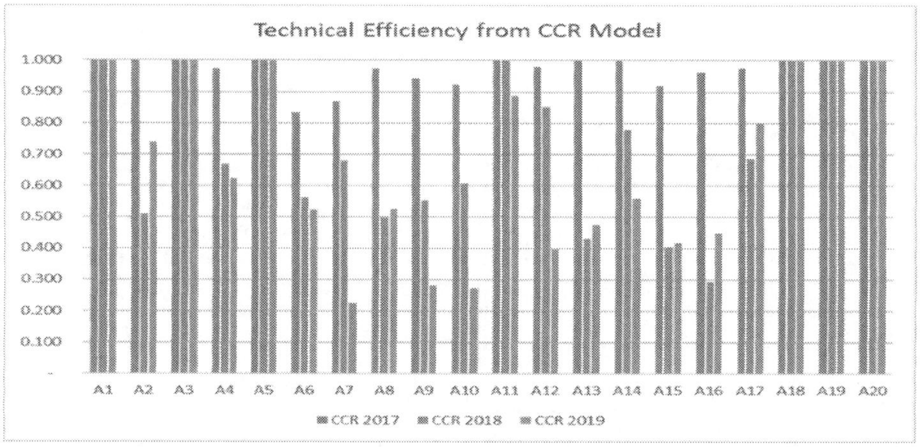

Fig. 3. Trend of CCR model's technical efficiency from 2017-2019

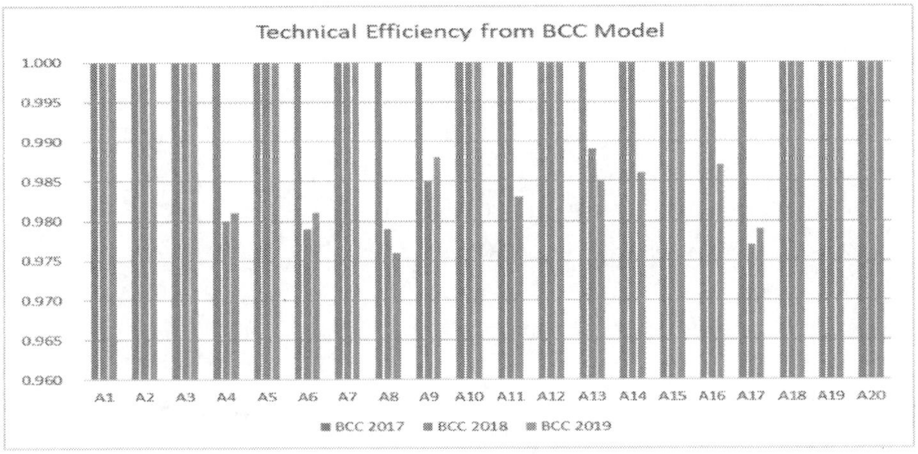

Fig. 4. Trend of BCC model's technical efficiency from 2017-2019

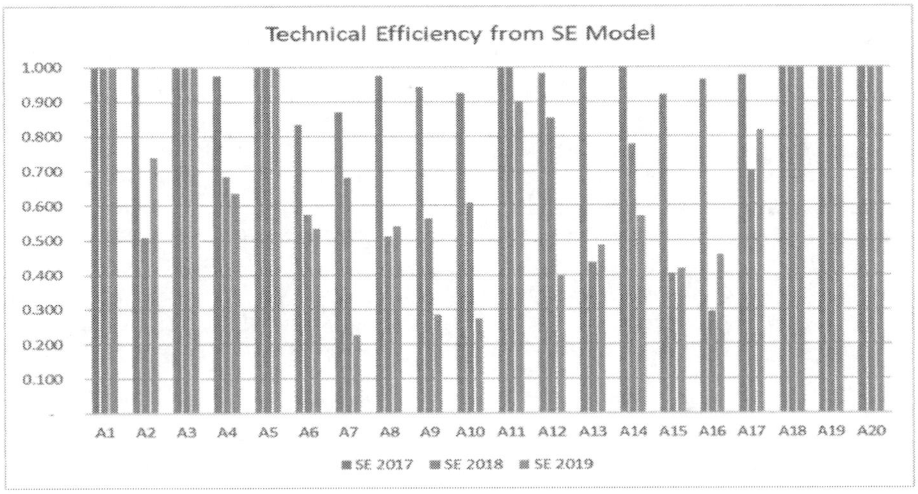

Fig. 5. Trend of SE model from 2017-2019

4 Conclusion and Future Research

Biomass represents a significant source of biofuel, which is a type of renewable energy getting attention from many countries nowadays. In this research, biomass data of three major feedstocks for biofuel in the Northeastern region of Thailand were collected and analyzed using DEA to analyze each provincial efficiency. The input criteria of allowable planting area, labor cost, and rainfall amount as well as the output criterion of the quantity of harvested product were, in particular, collected for the top energy crops of cassava, sugarcane, and palm during 2017 to 2019. Accordingly, the relative efficiency of each provincial alternative was analyzed using DEA analysis of CCR model, BCC model, and SE model, respectively.

Analyzed results showed that, among 20 provinces of the Northeastern region of Thailand, there were six provinces that operated efficiently under the selective criteria. These provinces were found to be Loei, Udon Thani, Bueng Kan, Khon Kaen, Chaiyaphum, and Nakhon Ratchasima, respectively. Thus, these efficient provinces could be further used as benchmark DMUs for other provinces. Regardless, it is important to note that the analyzed results are dependent on selected criteria for inputs and outputs, in which the caution should be noted.

Directions for future research of this study include 1) expanding the case study for other regional areas in Thailand for further comparative study, 2) exploring other types of crops related to energy feedstock, 3) investigating other time spans for different years or with other time units, such as monthly basis, and 4) assessing other criteria types inclusive of both inputs and outputs. That is, other economic aspects can be further included for the input criteria. In addition, outputs concerning the sustainability index can also be enhanced. Additionally, we note that this study is the first phase of our research framework to investigate the upstream process of the bioenergy supply chain. That is, the results obtained from this study will be used as input for further supply chain modelling study.

Acknowledgement

This research was supported by the Ministry of Higher Education, Science, Research and Innovation (MHESI) under research grant RGNS63-245 '*Development of Decision Support System for Biofuel Logistics under Uncertainty Consideration*'. We note that the opinions expressed are those of the authors and do not necessarily reflect the views of the funding agencies.

References

1. A. Mecke, I. Lee, J.R. Baker jr., M.M. Banaszak Holl, W. Wen, Q. Zhang, A design of straw acquisition mode for China's straw power plant based on supply chain coordination, Renewable energy, **76** (2015): 369-374
2. F. Rosillo-Calle, L. Pelkmans, A. Walter, A global overview of vegetable oils, with reference to biodiesel. A report for the IEA Bioenergy Task, **40** (2009)
3. Thailand Board of Investment, Thailand 4.0 – a new value-based economy. (2019) URL: https://www.boi.go.th/ [Accessed 15 May 2020]
4. A. Raychaudhuri, S. K. Ghosh, Biomass Supply Chain in Asian and European Countries. Procedia Environmental Sciences, **35** (2016): 914-924
5. REN21 - Renewables global status report. (2019) URL: https://www.ren21.net/ [Accessed 15 March 2020]
6. Y.S. Cheng, P. Mutrakulcharoen, S. Chuetor, K. Cheenkachorn, P. Tantayotai, E.J. Panakkal, M. Sriariyanun, Recent Situation and Progress in Biorefining Process of Lignocellulosic Biomass: Toward Green Economy. Applied Science and Engineering Progress, **13**, 4 (2020): 299-311
7. A. Martsri, N. Yodpijit, M. Jongprasithporn, S. Junsupasen, Energy, Economic and Environmental (3E) Analysis for Sustainable Development: A Case Study of a 9.9 MW Biomass Power Plant in Thailand. Applied Science and Engineering Progress, (2021): DOI: 10.14416/j.asep.2020.07.002
8. DEDE - Department of Alternative Energy and Energy Conservation. (2019) URL: http://www.dede.go.th/ [Accessed 15 July 2020]
9. DEDE - Research and Development in the field of Energy conservation and renewable energy in Thailand (2012) URL: http://weben.dede.go.th/ [Accessed 15 July 2020]
10. K. Ransikarbum, S.J. Mason, Multiple-objective analysis of integrated relief supply and network restoration in humanitarian logistics operations, International Journal of Production Research, **54**, 1 (2016): 49-68
11. K. Ransikarbum, S.J. Mason, Goal programming-based post-disaster decision making for integrated relief distribution and early-stage network restoration, International Journal of Production Economics, **182** (2016): 324-341
12. K. Ransikarbum, S. Ha, J. Ma, N. Kim, Multi-objective optimization analysis for part-to-Printer assignment in a network of 3D fused deposition modelling, Journal of Manufacturing Systems, **43** (2017): 35-46
13. K. Ransikarbum, R. Leksomboon, Criteria Analysis for Additive Manufacturing-based Healthcare Educational Media using Analytic Hierarchy Process. KKU Research Journal (Graduate Studies), **21**, 1 (2021): 133-144
14. C. Puchongkawarin, K. Ransikarbum, An Integrative Decision Support System for Improving Tourism Logistics and Public Transportation in Thailand, Tourism Planning & Development, (2020): 1-16
15. K. Ransikarbum, R. Pitakaso, N. Kim, A Decision-Support Model for Additive Manufacturing Scheduling Using an Integrative Analytic Hierarchy Process and Multi-Objective Optimization, Applied Sciences, **10**, 15 (2020): 5159
16. C. Chaiyaphan, K. Ransikarbum, Criteria analysis of food safety using the Analytic Hierarchy Process (AHP)-a case study of Thailand's fresh markets, In E3S Web of Conferences (EDP Sciences), **141**, (2020): 02001
17. P. Khamhong, C. Yingviwatanapong, K. Ransikarbum, Fuzzy Analytic Hierarchy Process (AHP)-based Criteria Analysis for 3D Printer Selection in Additive Manufacturing, In 2019 Research Invention, and Innovation Congress (IEEE), (2019): 1-5
18. N. Wattanasaeng, K. Ransikarbum, Cost Optimization Model for Plant Assignment in Industrial Estate Planning, In 2019 Research Invention and Innovation Congress (IEEE), (2019): 1-6
19. K. Ransikarbum, N. Kim, Multi-criteria selection problem of part orientation in 3D fused deposition modeling based on analytic hierarchy process model: A case study, In 2017 IEEE International Conference on Industrial Engineering and Engineering Management (IEEE), (2017): 1455-1459
20. J.S. Liu, L.Y. Lu, W.M. Lu, B.J. Lin, Data envelopment analysis 1978–2010: A citation-based literature survey, Omega, **41**(1), (2013): 3-15
21. K. Ransikarbum, N. Kim, Data envelopment analysis-based multi-criteria decision making for part orientation selection in fused deposition modelling, In 2017 4th International Conference on Industrial Engineering and Applications, (2017): 81-85
22. K. Ransikarbum, R. Pitakaso, N. Kim, Evaluation of Assembly Part Build Orientation in Additive

Manufacturing Environment using Data Envelopment Analysis, In MATEC Web of Conferences (EDP Sciences), **293** (2019): 02002

23. A. Charnes, W.W. Cooper, and E. Rhodes, Measuring the efficiency of decision making units. European journal of operational research, **2**, 6 (1978): 429-444
24. R.D. Banker, A. Charnes, W.W. Cooper, Some models for estimating technical and scale inefficiencies in data envelopment analysis, Management science, **30**, 9 (1984): 1078-1092
25. OAE – Office of Agricultural Economics. (2019) URL: http:// www.oae.go.th/ [Accessed 15 July 2020]
26. M. Mirmozaffari, A. Boskabadi, G. Azeem, R. Massah, E. Boskabadi, H.A. Dolatsara, A. Liravian, Machine learning clustering algorithms based on the DEA optimization approach for banking system in developing countries. European Journal of Engineering and Technology Research, **5**, 6 (2020): 651-658
27. K. Ransikarbum, N. Kim, S. Ha, R.A. Wysk, L. Rothrock, A highway-driving system design viewpoint using an agent-based modeling of an affordance-based finite state automata. IEEE Access, **6** (2017): 2193-2205
28. J. Kim, K. Ransikarbum, N. Kim, E. Paik, Agent-based simulation modeling of low fertility trap hypothesis. In Proceedings of the 2016 ACM SIGSIM Conference on Principles of Advanced Discrete Simulation, (2016): 83-86
29. A. Apornak, S. Raissi, M.R. Pourhassan, Solving flexible flow-shop problem using a hybrid multi criteria Taguchi based computer simulation model and DEA approach. Journal of Industrial and Systems Engineering, **13**, 2 (2021): 264-276
30. A. Tayal, A. Solanki, S.P. Singh, Integrated frame work for identifying sustainable manufacturing layouts based on big data, machine learning, meta-heuristic and data envelopment analysis. Sustainable Cities and Society, **62** (2020): 102383
31. N. Wattanasaeng, K. Ransikarbum, Model and Analysis of Economic-and Risk-Based Objective Optimization Problem for Plant Location within Industrial Estates Using Epsilon-Constraint Algorithms. Computation, **9**, 4 (2021): 46
32. K. Ransikarbum, P. Khamhong, Integrated Fuzzy Analytic Hierarchy Process and Technique for Order of Preference by Similarity to Ideal Solution for Additive Manufacturing Printer Selection. Journal of Materials Engineering and Performance, (2021): 1-12
33. W. Chanthakhot, K. Ransikarbum, Integrated IEW-TOPSIS and Fire Dynamics Simulation for Agent-Based Evacuation Modeling in Industrial Safety, Safety, **7**, 2 (2021): 47
34. K. Ransikarbum, R. Pitakaso, N. Kim, J. Ma, Multicriteria decision analysis framework for part orientation analysis in additive manufacturing, Journal of Computational Design and Engineering, **8**, 4 (2021): 1141-1157

Effect of injection timing on particle size distribution of a diesel engine fueled with biodiesel-diesel blends

Theeranan Phatai[1], *Raschanon* Kumpiranon[1], *Kritsana* Ounnarut[1], *Nattapol* Pongrasri[1,2], *Pubet* Meenaroch[1] and *Pomprab* Sriumpunpuk[1,2*]

[1]Department of Mechanical and Automotive Engineering Technology (MAet), Faculty of Engineering and Technology, King Mongkut's University of Technology North Bangkok (Rayong Campus), Nonglalok, Bankhai, Rayong 21120, Thailand.
[2]Research Centre for Combustion Technology and Alternative Energy (CTAE), Science and Technology Research Institute, King Mongkut's University of Technology North Bangkok, Bangkok 10800, Thailand.

Abstract. The purpose of this study is to investigate injection timing affected particle size distribution of a diesel engine fueled with biodiesel/diesel blends. The biodiesel blending ratios include 7, 10 and 20%vol. The experiments were operated at engine speed 1,500 rpm under 25, 50 and 75% engine loads and two different injection timing (25.5 CAD BTDC and 27.5 CAD BTDC). The effects of biodiesel addition and injection timing on the changes in black smoke and particle size distribution were discussed. Results showed that the standard injection timing for the addition of biodiesel, especially for 20% biodiesel blended with diesel fuel (B20) had a significant effect of increasing particle concentration and thus black smoke emission. The advanced injection timing for B20 could improve combustion efficiency and reduce small particles emission in nucleation-mode (Dp 5 – 50 nm) and particles in accumulation-mode (Dp 50 – 1,000 nm) became the smaller particle size, compared to standard and retarded injection timing. Moreover, the oxygen content in biodiesel could improve combustion efficiency and reduce the emission but it is necessary to modify the engine especially fuel injection timing.

Keyword. Compression ignition engine; Particle size distribution; Biodiesel

1 Introduction

Nowadays, the popularity of compression ignition engines is growing in both the transportation and power generation sectors because of its high efficiency, fuel economy and durability compared to spark ignition engines [1-2]. Due to the declining trend of fossil energy sources in the future, it is necessary to find alternative energy that can be used instead of diesel fuel [3-4]. In addition, air pollution problems arise from combustion process of compression ignition engines, especially particulate matter (PM). Most particulate matter consists of solid carbon particles absorbing organic matter from the fuel, lubricants and sulfates in the exhaust particles, all of which are produced by the combustion of diesel fuel [5]. Diesel combustion with a C-C bond in the thick mixture region forms very small particles which can be agglomerated with other organic matter to become the larger particle size. Over the past few decades, researchers have studied the dust particle distribution. The diesel engine particulates in the exhaust gas are classified according to the particle diameter, which contains nucleation-mode particles (diameter of the particle, 5 nm < D_p < 50 nm), accumulation-mode particles (50 nm < D_p < 1,000 nm) and coarse-mode particles (D_p > 1,000 nm) [6]. These particles can impact on the human respiratory system, so it is important to find solutions to reduce the amount of particulate matter produced by the combustion of diesel fuel in compression ignition engines.

Previous research examines that oxygen-rich diesel can promote sufficient oxygen for combustion and increase combustion efficiency, resulting in reduced soot [2,7]. Biodiesel, one of the most widespread renewable fuels, can be substituted for diesel without minor engine modifications. The process for biodiesel production is the transesterification of triglycerides of vegetable oils, animal fats and waste cooking oils [4]. The study of biodiesel properties found that biodiesel has similar properties to diesel. The advantages of using biodiesel over diesel are its higher cetane numbers and high oxygen content, resulting in a reduction in the amount of combustion dust. Chiatti et. al., [8] compared the results of the particles size distributions on an unmodified diesel engine using ultra-low sulfur diesel (ULSD) blended

* Corresponding author: pomprab.s@eat.kmutnb.ac.th

with biodiesel produced from waste cooking oil at ratios of 20 and 40% to standard ultra-low sulfur diesel fuel.

The total particle number concentration results showed that when using both types of biodiesels, the overall particle concentration decreased compared to the ultra-low sulfur diesel in all test conditions. This is due to the high cetane number of the biodiesel, which shortens the ignition delay, as a result, a longer combustion of the mixed control range. The oxygen content of the biodiesels also promotes efficient combustion and contributes to soot oxidation, particularly during high engine loads. The results of the particle number size distributions at engine operating conditions with high load and speed explain that as the ratio of biodiesel increases, the number of very fine particles (nucleation-mode particles, DP < 50 nm) increases. On the other hand, the number of accumulation and coarse-mode particles decreases compared to the standard diesel. This means that the biodiesels make the size of particulate matter to become smaller. Agarwal et al. [9] studied the effect of increasing biodiesel ratio blended with diesel. They found that when increasing the biodiesel content at 20% and 100%, the benzene soluble organic fraction (BSOF) in the exhaust composition is increased compared to the exhaust from the diesel. The BSOF is the primary component of particulate matter and directly affects particle size. In contrast, oxygen content in biodiesel can reduce soot generation, especially solid carbon particles, and results in an increasing number of 30 - 40 nm particles. Likewise, Puzun et al. [10] reported that when using biodiesel from rapeseed oil, the particles in the range of nucleation mode particles increase while particles larger than 50 nm (accumulation-mode particles) reduce.

Compared to the particulate matter in the diesel exhaust, it had the single peak distribution dominating accumulation-mode particles. From the previous research, most of the experiments were conducted without any engine modifications. Because the fuel properties have been changed, it resulted in certain engine operating conditions causing an increase in emissions especially the number of particles in the exhaust. Therefore, it is necessary to study the minor engine modifications in order to accommodate the use of biodiesel and diesel fuel to be more efficient and reduce emissions. Kim et al. [11] investigated the effect of biodiesel and injection timing on ignition delay and engine emission. The results shown that advance injection timing can be improved the maximum cylinder pressure and good for premixed combustion period. Advance injection timing also reduced a number of particles matter due to evaporation characteristic of fuel better than retard injection timing. It is resulted in high temperature and pressure of combustion. Reddy et al. [12] are investigated the effect of waste mango seed biodiesel at different injection timing. The results shown that the brake thermal efficiency in high proportion biodiesel is reduce when compare with diesel fuel at STD injection timing. It is led to increase the black smoke. However, the advance injection timing for 2 CAD is applied to the fuel injection system. The effect of advance injection timing shown the high efficiency of combustion and enhancement in fuel and air mixing effected to good combustion and low smoke emission.

In current study, the PM emission of conventional diesel fuel and bio diesel/diesel blends were tested on single cylinder diesel engine. The effect of biodiesel blending ratios and injection timing on total particles concentration and particle number size distribution were discussed by using a particle size spectrometer. Smoke emission was analyzed using smoke meter.

2 Experimental apparatus

2.1 Engine test rig

The experimental engine in this study was a single cylinder, four strokes, water cooled, spherical type combustion system diesel engine and the engine specifications are shown in Table 1. The engine speed and engine load were controlled by asynchronous motor engine dynamometer (Gunt Hamburg CT 110). Exhaust, air inlet and fuel temperature were measured by type K thermocouples. The opacity of black smoke emission was measured by Horiba Mexa-600s opacimeter. The total number and particle size distribution were measured by engine exhaust particle size spectrometer (TSI EEPS 3090), which can measure diameter range is from 5.6 – 560 nm. Dekati TI-1000 double stage diluter which total dilution ratio was 73.27:1 (8.56 each diluter) used for dilute exhaust concentration before the particles pass through the EEPS. The First stage diluter and air dilution were heated to 190°C for avoid condense of water and hydrocarbon in exhaust. Time duration for collecting data of particle size distribution was 60 s. Schematic diagram shows in Fig. 1.

Table 1. Specifications of engine

Item	Specification
Model	Kubota EA300 E2 NB1
Type	4 Stroke Diesel engine
Number of cylinder	1
Bore/Stroke	75 mm/70 mm
Displacement volume	0.309 L
Combustion system	Spherical type
Max. Output (kW)	5.2 kW @ 3000 RPM
Injection timing	25.5 CAD BTDC
Fuel injection system	Mechanical

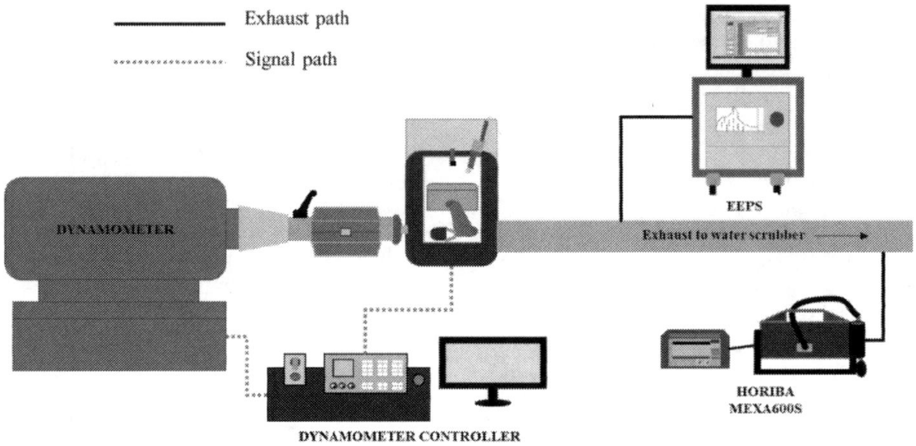

Fig. 1. Schematic diagram of the experimental setup.

2.2 Fuels

Three different fuels were a commercial diesel fuel (93% diesel and 7% Vol. bio-diesel blends, B7), blend of 90% diesel with 10% bio-diesel and 80% diesel with 20% bio-diesel, labelled as B7, B10 and B20, respectively. The heating value of all fuels were evaluated by bomb calorimeter (IKA C6000). The heating value of the test fuels are given in Table 2.

Table 2. Heating value of the test fuels

Fuel	Heating value (J/g)
Commercial diesel (B7)	45,352
Biodiesel (B100)	41,240
B 10	44,386
B 20	44,145

2.3 Test conditions and procedure

The effect of biodiesel blends (B7, B10 and B20) on smoke emission, total number of particle concentration and particulate number size distribution were analyzed under engine speed of 1,500 RPM and engine loads were set at 25%, 50% and 75% of engine full load without exhaust gas recirculation (EGR) system. The injection timing was set at 25.5 crank angle degree before top dead center (CAD BTDC) as standard injection timing and advance injection timing for 2 CAD (27.5 CAD BTDC) was considered.

The temperatures of engine coolant and engine oil were controlled in ranges of 85 ± 3° C and 87 ± 2° C respectively [10, 13]. The engine was run for 15 min when the fuel was changed to new fuel to avoid the previous fuel remaining in the system. The data of smoke emission and heating value of fuels were recorded three times. The results of particle size distribution were recorded 60 times in a minute and their average values were taken for the discussions.

3 Results and discussion

3.1 Effect of biodiesel blends fuel on smoke emission and particulate size distribution

In this section, the results of smoke emission and particle size distribution were considered under three different engine loads of 1,500 RPM engine speed. Fig. 2 shows that the addition of biodiesel into diesel fuel had a greater effect on smoke emissions and when the engine load increased, the effect of smoke emissions also increased significantly. Compared with conventional diesel (B7), the opacity of smoke emission of B10 and B20 increased by 80.3% and 204.6% respectively at 75% engine load condition. Because of the high viscosity of biodiesel, when biodiesel was injected to combustion chamber, it was difficult to atomize and evaporate. This resulted in a longer ignition delay [9], which promoted rich region of fuel spray and soot precursor.

Fig. 3 presents effect of biodiesel blending ratio on particulate number size distribution at 25% (3a), 50% (3b) and 75% (3c) engine load of 1,500 RMP engine speed and standard injection timing. From Fig 3a. at light engine load (25%), the results demonstrate that particle concentration and size distribution present in two curves. The first peak was at particle diameter 10 – 20 nm as well as the second peak was 50 nm represented in nucleation-mode particle. Fig. 3b, at medium engine load, shows a double peak which occurred in nucleation-mode particles. An increase in the biodiesel ratio, especially in the high ratio of biodiesel (B20), resulted in an increase in concentration of particles due to high viscosity of fuel deteriorated the fuel atomization, lager droplet size and longer ignition delay compares with diesel fuel [8].

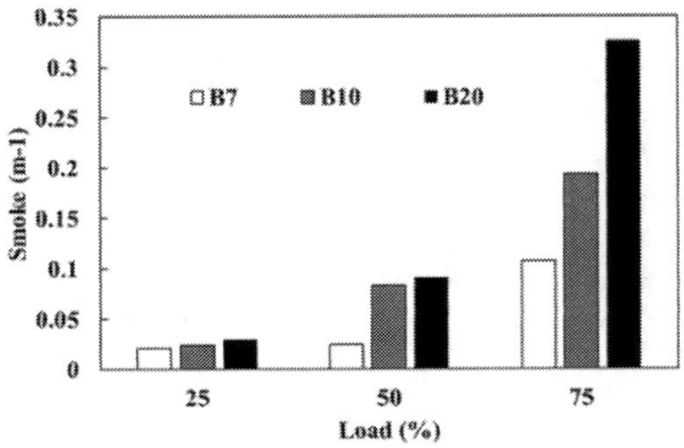

Fig. 2. Effect of biodiesel blending ratio on smoke emission for various engine loads.

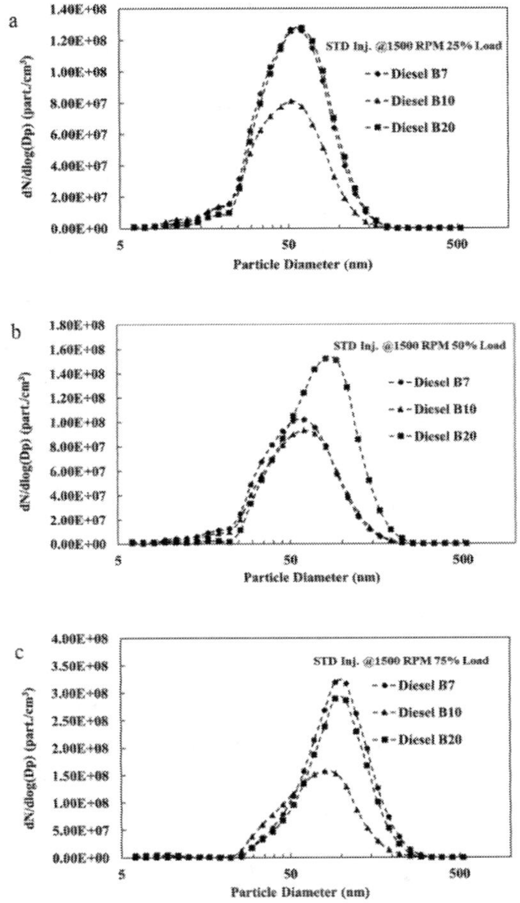

Fig. 3. effect of biodiesel blending ratio on particulate number size distribution.

Fig. 3c presents the variation of particles concentration at high engine load (75%). The particles were shown as a single peak in accumulation-mode at 100 – 150 nm due to the reduction of air-fuel mixing and incomplete combustion form rich combustion resulting in higher soot formation and concentration [8]. On the other hand, the diameter of particles was gradually decreased with addition of biodiesel into diesel fuel because the higher oxygen content of biodiesel promoted the soot oxidation and high percentage of benzene soluble organic fraction (BSOF) in biodiesel fuel released with engine exhaust. These particles, consisting of organic products, contribute to the increase of nucleation mode particles [9].

Fig. 4 shows the total number concentration of particles from conventional diesel and biodiesel blends. The results showed that B10 had a reduction in particle concentration compared to B7 fuel because its higher oxygen ratio in molecules encouraged the combustion efficiency and can reduce particle emission. It also improved diffusion combustion and promoted the post oxidation process of solid carbon, which formed during combustion. On the other hand, when the proportion of the biodiesel was 20% (B20), there was an increase in the total particle number concentration, especially particles in nucleation mode [9]. This means that the use of this biodiesel proportion had a greater effect on the total concentration than accumulation-mode particles because of the highly change of physical properties in B20 fuel such as high viscosity and density.

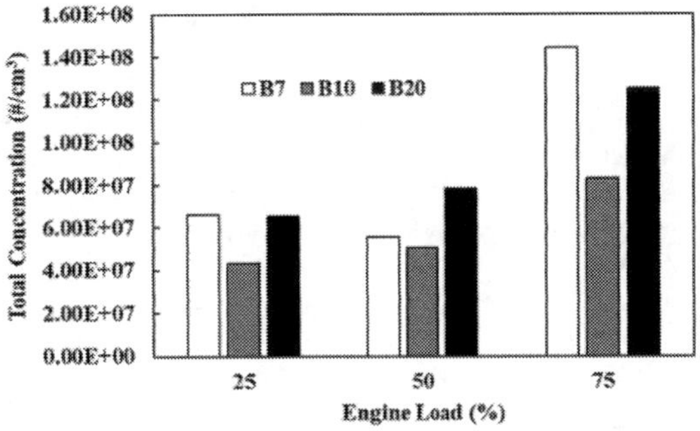

Fig. 4. Effect of biodiesel blending ratio on total particulate number concentration.

The increment of biodiesel blending ratio (B20) resulted in high smoke emission and high particles concentration. To improve the combustion performance and reduce smoke and PM emission, it is necessary to modify the engine such as injection timing. An influence of advance injection timing for 2 CAD BTDC will discuss in next chapter.

3.2 Effect of injection timing and biodiesel blends fuel on smoke emission and particulate size distribution

This section shows the results of smoke emission, total concentration, particulate size distribution of B20 fuel with different engine load and advance injection timing (27.5 CAD BTDC) versus standard injection timing (25.5 CAD BTDC). Fig. 5 shows that advance injection timing led to decrease the opacity of black smoke by

53.3%, 28.5% and 44.4% with 25%, 50% and 75% engine load, respectively. Generally, soot formation in combustion process is form due to uneven mixture of air and fuel. Also, the high viscosity of biodiesel is difficult to fuel atomization and evaporation, that effect to longer ignition delay. The advance injection timing can increase air-fuel mixing efficiency, which decrease rich mixture in diffusion combustion stage [1, 12].

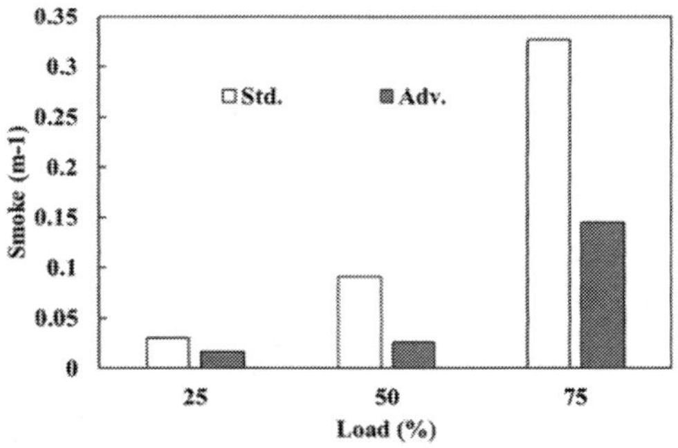

Fig. 5. Effect of advance injection timing with B20 on smoke emission for various engine loads.

The particulate number size distribution of advance and standard injection with B20 fuel in various engine load are shown in Fig.6. It is obviously shown that at low and middle engine load, particle size distributions occurred the bimodal structure in nucleation-mode and accumulation-mode and particles in high engine transformed to accumulation-mode only. Fig. 6a showed advance injection timing had no significant effect on smallest particles (nucleation-mode), while accumulation-mode particles decreased and also reduced peak of particle concentration from 1.3E+08 to 8.0E+07 part. /cm³. Fig. 6b showed particulate number size distribution with 50% engine load. The trend of peak and distribution of particles was not significant different compared to 25% engine load. In the meantime, advance injection timing affected to accumulation mode particles reduced, and the peak area shifted to the smaller particle size. This is due to the sufficient atomization of fuel when it is injected into combustion chamber and the longer air-fuel mixing time before combustion compared to standard injection timing [11-12]. The high engine load (75%) in Fig. 6c presents single peak particles size distribution occurred in accumulation mode only. When comparing this particle size distribution result to low and middle load, it was found that the particle sizes generated by high engine loads were larger.

Fig. 7 shows the total number concentration of particles from B20 exhaust in standard and advance injection timing. With the advance injection timing, the proportion of particles concentration was significantly decreased, which is obtained by particle size distribution in Fig. 6. Advance injection timing can reduce total number concentration by 35.5%, 35.1% and 44.5% with 25%, 50% and 75% engine load, respectively. The reason for the decrease in particles concentration, especially accumulation-mode particles, is that the B20 fuel with advance injection timing has longer time to have good evaporation and atomization characteristics. On the other hand, the delayed fuel injection increases the combustion duration and affects more the collision and coagulation time to increase accumulation particles [11, 13-14].

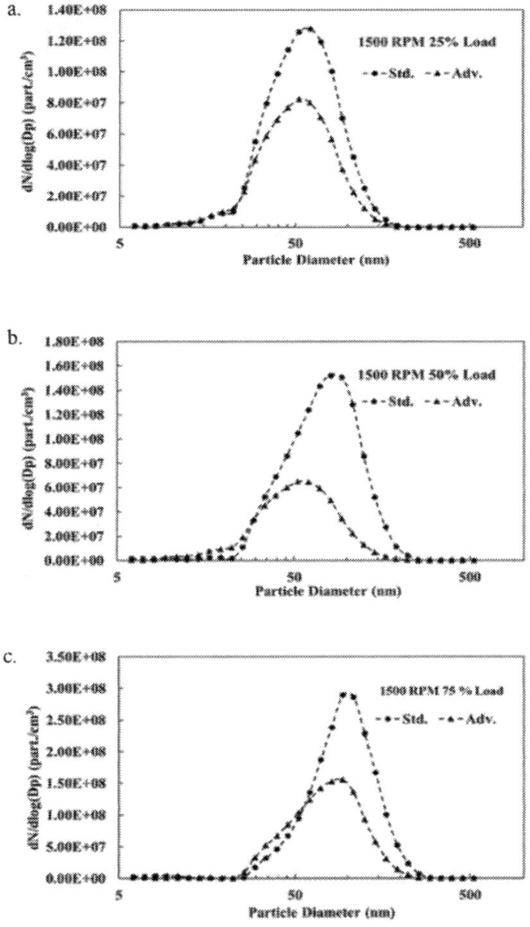

Fig. 6. Effect of injection timing on particulate number size distribution.

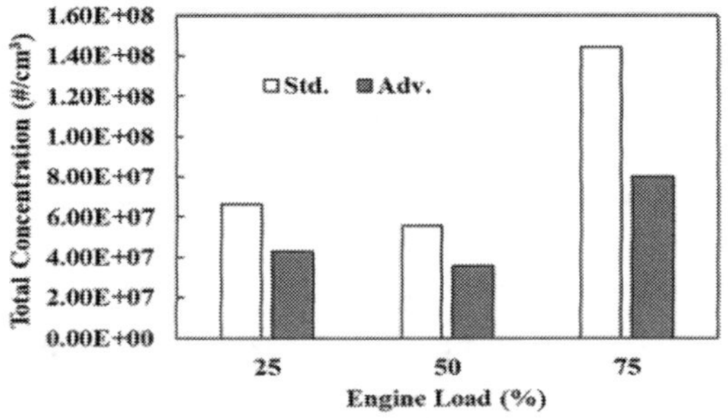

Fig. 7. Effect of injection timing on total particulate number concentration.

4 Conclusions

In current study, the smoke emission and the particle size distribution of exhaust were carried out on a single cylinder mechanical diesel engine with biodiesel blends at various engine loads and injection timing. The main results are concluded as follows:

Biodiesel/diesel blends can effectively reduce the smoke emission and particle size distribution. On the other hand, the increasing proportion of biodiesel (B20) results in changes in the properties of fuel such as viscosity and cetane number, which trends to increase the opacity of black and high concentration of particles in exhaust. B10 is appropriate fuel for this experiment engine due to its higher oxygen content than conventional diesel fuel, which contributes to self-oxygen supply during combustion, especially in diffusion combustion phase. Due to the high biodiesel blends (B20) with advance injection for 2° CA, the particle distribution moves towards the direction of small particle size, and the ratio of accumulation-mode particles are decreased. The different engine loads show different particle size distributions. For low and middle engine, the peak of particles concentration appears a bimodal curve (nucleation and accumulation mode particle) and converts to single curve in accumulation particle. Finally, the trend shifts to larger particle sizes with high engine loads.

Future investigations are necessary to validate with the effect of high proportion of biodiesel or other fuel additive blends with diesel fuel in commercial diesel engine such as electric control common rail diesel engine with after treatment system on performance and emission characteristic.

Acknowledgement

The equipment of this work was supported by the Automotive Performance Testing Laboratory (Enhancement of Safety Practice of Research Laboratory in Thailand, ESPRel No. 2-0870-000-3), Faculty of Engineering and Technology, King Mongkut's University of Technology North Bangkok (Rayong Campus) and was funded by the Special Project Fund, King Mongkut's University of Technology North Bangkok. We would like to express our great appreciation for their valuable sponsorship.

References

1. A.K. Agarwal, T. Gupta, P.C. Shukla, A. Dhar, Particulate emissions from biodiesel fuelled CI engines, Energy conversion and management, **94**, (2015): 311-330
2. J. Liu, Z. Liu, L. Wang, P. Sun, H. Ma, P. Wu, Effects of PODE/diesel blends on particulate matter emission and particle oxidation characteristics of a common-rail diesel engine, Fuel processing technology, **212** (2021): 106634
3. L. Geng, Y. Xiao, S. Li, H. Chen, X. Chen, Effects of injection timing and rail pressure on particulate size-number distribution of a common rail DI engine fueled with fischer-tropsch diesel synthesized from coal, Journal of the energy institute, **95** (2021): 219-230
4. M. Hajbabaei, G. Karavalakis, K.C. Johnson, J. Guthrie, A. Mitchell, T.D. Durbin, Impacts of biodiesel feedstock and additives on criteria emissions from a heavy-duty engine, Fuel processing technology, **126** (2014): 402-411
5. D.B. Kittelson, Engine and nanoparticles: A review, Journal of aerosol science, **29** (1998): 575-588
6. J. Schneider, N. Hock, S. Weimer, S. Borrmann, U. Kirchner, R. Vogt, V. Scheer, Nucleation particles in diesel exhaust: Composition inferred from in situ mass spectrometric analysis, Environmental science & technology, **39** (2005): 6153-6161
7. J. Liu, J. Yang, P. Sun, W. Gao, C. Yang, J. Fang, Compound combustion and pollutant emissions characteristics of a common-rail engine with ethanol homogeneous charge and polyoxymethylene dimethyl ethers injection, Applied energy, **239** (2019): 1154-1162
8. G. Chiatti, O. Chiavola, F. Palmieri, Impact of waste cooking oil in biodiesel blends on particle size distributions from a city-car engine, Journal of the Energy Institute, **91** (2018): 262-269
9. A.K. Agarwal, T. Gupta, A. Kothari, Particulate emissions from biodiesel vs diesel fuelled compression ignition engine, Renewable and Sustainable Energy Reviews, **15** (2011): 3278-3300
10. A. Puzun, S. Wanchen, L. Guoliang, T. Manzhi, L. Chunjie, C. Shibao, *Characteristics of Particle Size Distributions About Emissions in A Common-rail Diesel Engine with Biodiesel Blends* (2011 2nd International Conference on Challenges in Environmental Science and Computer Engineering, Haikou, 2011)
11. H.J. Kim, S. Jo, J.T. Lee, S. Park, Biodiesel fueled combustion performance and emission characteristics under various intake air temperature and injection timing conditions, Energy, **206** (2020): 118154
12. S.R. Reddy, G. Murali, A.A. Shaik, V.D. Raju, M. B. S. S. Reddy, Experimental evaluation of diesel engine powered with waste mango seed biodiesel at different injection timings and EGR rates, Fuel, **285** (2021): 119047
13. L. Xinling, X. Zhen, G. Chun, H. Zhen, Effect of injection timing on particle size distribution from a diesel engine, Fuel, **134** (2014): 189-195
14. T. Wang, J. Liu, P. Sun, Q. Ji, W. Gao, C. Yang, Influence of injection parameters on combustion, gaseous emissions and particle size distribution of a CRDI diesel engine operating with PODE/diesel blends, Fuel, **281** (2020): 118733

The Comparative Study on Compressed Natural Gas (CNG) and Compressed Biomethane Gas (CBG) Fueled in a Spark Ignition Engine

Tanat Limpachoti[1], and *Kampanart* Theinnoi[1,2,*]

[1]College of Industrial Technology, King Mongkut's University of Technology North Bangkok, 1518 Pracharat 1 Road, Wongsawang, Bangsue, Bangkok 10800, Thailand.
[2]Research Centre for Combustion Technology and Alternative Energy (CTAE), Science and Technology Research Institute, King Mongkut's University of Technology North Bangkok, Bangkok 10800, Thailand.

Abstract. Compressed biomethane gas (CBG) is an interesting alternative fuel that can be used to solve fossil fuel crisis problems. Nevertheless, it is concerned about engine performance and exhaust gas emission. In additions, CBG can be produced from agricultural harvesting residues and the food industry that is a good opportunity for utilizing biogas as a vehicle fuel. The aims of this research are comparative analysis of the engine performance and exhaust emission on a CBG and compressed natural gas (CNG) fuelled in a spark ignition engine on engine performance and emissions. The engine has been modified to operate with both fuel gases under constant engine load (50% maximum load) and engine speed (1500 - 3500 rpm). The results show that the engine operated with CBG has higher thermal efficiency with lower the nitrogen oxide (NO_x) and hydrocarbon (HC) emissions compared with CNG. Thus, the CBG fuel can be used as an alternative fuel to substitute CNG fuel in the spark ignition engine. In additions, the results give the useful information for developed and optimised the engine operated with compression biogas to impact the higher demand of automotive fuels in the future.

Keyword. Compressed biogas (CBG), Spark ignition engine, Engine performance, Emission

1. Introduction

The scarcity of fossil fuel and the extremely high demand and become an important issue of a national industrial and living. Especially, the fuels that will be used in the automotive sector, which are important role in driving the economy in both production and transportation. Thus, the alternative fuels that will be used to solve this problem should be renewable fuel. In addition, the properties of alternative fuel should be similar efficiency to the fossil fuels. That could be provided the same or better engine performance without modification. Furthermore, fuel has emitted less engine out pollution than the fossil fuel. The automotive pollutions issue is another important factor such as greenhouse gas emissions which damages to the Earth's atmosphere and climate from internal combustion engine. There is exigency discreet to use for conserve fossil fuel and replace it by other alternatives wherever possible.

Nowadays, the CNG generally used as an alternative fuel due to the advantage of CNG are higher thermal efficiency with lower exhaust gas emissions as compared SI engines fuelling with gasoline fuel. In additions, CNG can obtain significantly lower fuel prices than gasoline fuel. However, the non conservable energy development which affected by natural gas demand are extremely increase while natural gas which existed are decrease that effect to consumed in different applications such as automotive, industrial, power generation, domestic fuel, fertilizer industry, petrochemicals. [1,2]

The biogas commonly known as alternative fuel or renewable fuel, which has been proposed to substitute the fossil fuel to solve that problem due to it have many advantages such as natural gas which commonly used as vehicle fuel. CBG can be produced from both agriculture and industrial wastes [3]. That can be produced both urban and countryside area [4]. Furthermore, the biogas is suitable fuel for heat and power production [5]. Biogas has high potential to replace natural gas due to CBG has properties approximately to CNG [1]. The CBG compositions mainly consists of methane and carbon dioxide compositions in gas phase. It was compressed and stored in container under high pressure for the vehicles purpose. In fact, biogas has not been widely applied as a renewable substitute for internal combustion engines due to the limitation of biogas properties such as the large number of CO_2 concentration, calorific value, flame velocity and flammability are lower when compare with natural gas [6], the metal parts in engines can corrosion which cause by a small percentage of H_2S in biogas

* Corresponding author: kampanart.t@cit.kmutnb.ac.th

[4,7]. Therefore, the retrofit of biogas for better quality that suitable for used in vehicle engine by eliminated carbon dioxide, moisture and other impurities which will adversely affect to engine parts or engine performance [8-10]. Moreover, biomethane could compressed into a tank force as compressed natural gas (CNG) to compressed biomethane gas (CBG) for used as a transport fuel in a CNG vehicle furthermore that is conveniently to storage and reduce costs to transportation.[4] As mentioned by a series of published theoretical and experimental works , the combustion mechanisms and applications in internal combustion engines which used biogas are concerned thus the multiple operating conditions of engine while used biogas as fuel ought to be immensely studied. [11] The previous research which refer to intensity of emission impact such as Carbon monoxide (CO), Hydrocarbon (HC) and Nitrogen oxides (NO_x) that motivate to studies effect of SI engine to emission while used alternatives fuel so as CNG and CBG. [12,13]

In this experimental study will comparative analysis of the engine performance and exhaust gas emission on a compressed biomethane gas (CBG) and compressed natural gas (CNG) fuelled in spark ignition engine. The research will enrich the understanding and provide a practical guide to suitable CBG application.

2. Experimental

The TOYOTA model 4A-FE gasoline engine with an eddy-current dynamometer which multi-cylinder engine fixed with test-rig was used in this study with various engine speed (1500, 2000, 2500, 3000, 3500 rpm) and fixed an engine load (50% of maximum load). The engine specification is list in Table 1.

Table 1. Specifications of test engine.

	Specification
Engine series	TOYOTA 4A-FE Spark ignition
Engine type	4 cylinders, in line
Displacement	1,587 cc
Bore X Stroke	81 x 77 mm
Compression ratio	9.5 : 1
Maximum output	100 HP @ 5,600 rpm
Maximum torque	101 ft.lbs @ 4,400 rpm
Valves mechanism	4 Valves, DOHC, Belt & Gear Drive

The engine is modified to operate with compressed natural gas (CNG) and compressed biomethane gas (CBG). The gas fuel entrains to the engine via multipoint electronically controlled port injection system. The in-cylinder pressure was recorded using a national instruments data acquisition board that connected through with a Kistler 5018A charge amplifier and a Kistler 6056A pressure transducer (1% measurement accuracy) mounted flush was installed in the cylinder head. A digital shaft encoder is used to measure the crankshaft position. Thermocouples which measure air, oil, inlet manifold and exhaust temperatures and pressure gauges which fixed at relevant spot are comprise in test-rig. In addition to that, Normal engine test bed safety features are comprised. During tested Atmospheric conditions such as humidity, temperature and pressure are monitor. The lay out of the experimental as illustrated in Figure 1.

In this work, the exhaust gas in each operating condition is measured via Horiba model MEXA-584L was used to measure exhaust gas include carbon dioxide, unburned hydrocarbons (NIDR-non-dispersive infrared), and NO_x (CLD chemiluminescence detection). CBG and CNG gas in this study are commercial fuel in market. Properties of CNG and CBG are showed in Table 2.

Table 2. Properties of CNG and CBG [14].

Fuel properties	CNG	CBG
Methane (%Vol.)	77.87	73.78
Nitrogen (%Vol.)	14.88	15.99
Carbon dioxide (% Vol.)	7.08	7.18
Oxigen (% Vol.)	-	2.95
Hydrogen (% Vol.)	-	-
Hydrogen sulfide(mmg/m^3)	-	-
Heating value (KJ/kg)	31.59	28.98
Density (kg/cm^3) at 1 atm	0.8014	0.7593
Wobbe number (MJ/m^3)	41.9-44	40-41

3. Results and discussion

3.1 The effect of CBG and CNG on engine performance

3.1.1 The effect of engine speed on in-cylinder peak pressure

The in-cylinder pressure during combustion from engine fuelling with CNG and CBG as main fuel are shown in Fig. 2 and Fig. 3. The higher engine speed led to the in-cylinder pressure are increased for both fuels. Due to the gas fuels are injected with high concentration to produce the power for engine. The pressure in combustion chamber was dropped under 3000 rpm for CNG combustion however, it has no significant effect on engine using with CBG. That could be the engine is not optimise setup to operate with gas fuel. Thus, the optimisation is required for the engine to provide better performance when operated with CNG and CBG fuel.

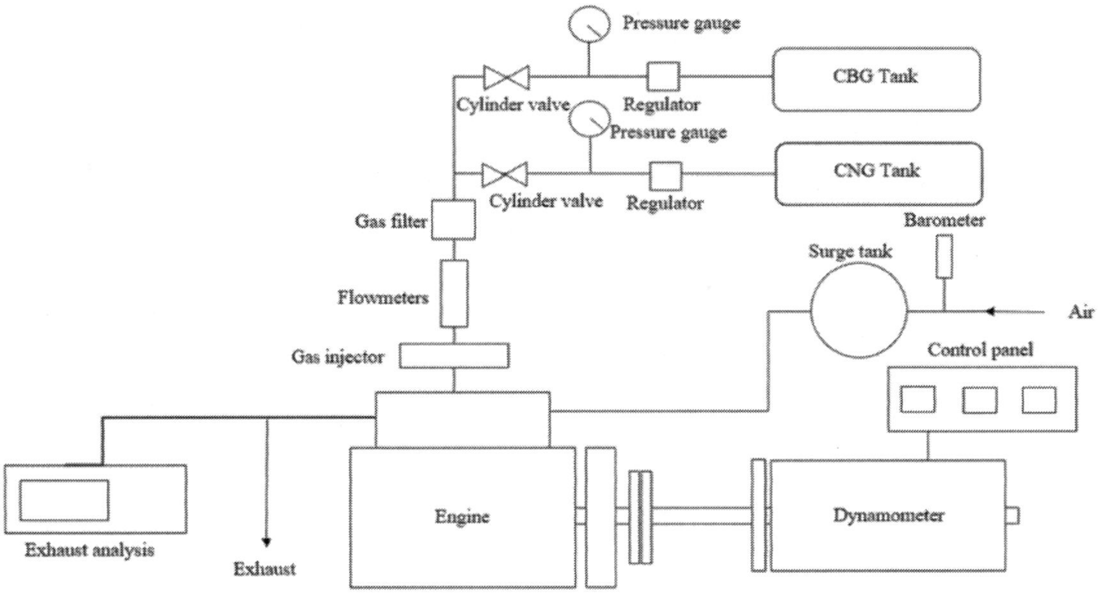

Fig. 1. Schematic diagram of experimental test setup.

The comparison of in cylinder peak pressure by using CNG and CBG as shown in Fig.4. The in-cylinder pressure are increased with the higher engine speed, however the in-cylinder peak pressure of CBG fuel is lower than CNG fuel at various engine speeds that mainly due to the CNG density is higher than the CBG as shown in Table 2. The fuel density effected to the fuel mass fraction inside the combustion chamber due to the higher density can be produce the high burning rated of fuel, including shown high temperature during the ignition phase that resulted to higher in-cylinder pressure [15].

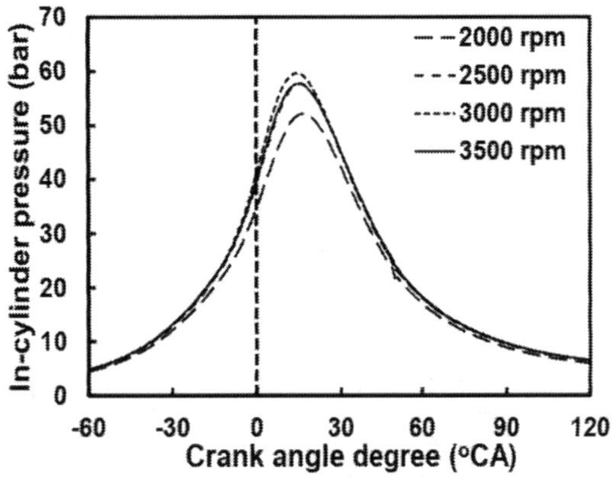

Fig. 2. The effect of engine speed on in-cylinder pressure from combustion of CNG fuelling.

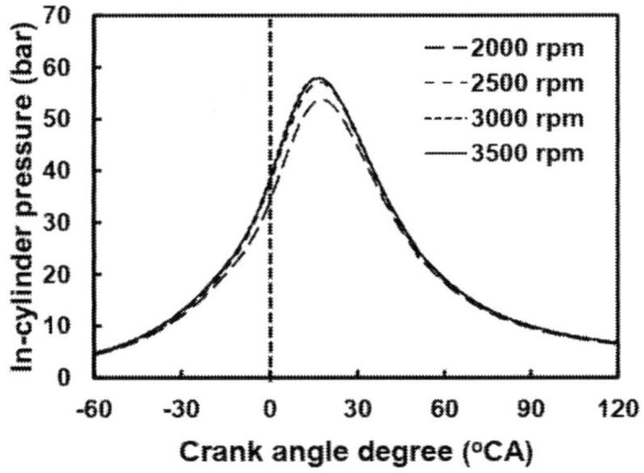

Fig. 3. The effect of engine speed on in-cylinder pressure from combustion of CBG fuelling.

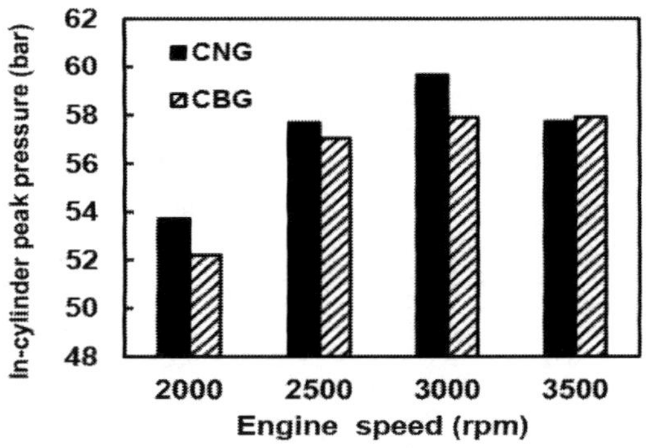

Fig. 4. The effect of engine speed on In-cylinder peak pressure from combustion of CBG and CNG fuelling.

3.1.2 The effect of engine speed on exhaust gas temperature

The effect of engine speed on exhaust gas temperature from combustion of CNG and CBG as show in Fig.5. The exhaust gas temperature that increases respectively as increases engine speed both used CNG and CBG. Moreover the engine was running using the CBG, exhaust gas temperature was lower than using CNG throughout the engine speed That can be explained by that result to flame propagation speed of the CNG which slower than CBG provided the combustion to conduct until the end of the expansion stroke that raised the exhaust gas temperature for the CNG operation.[16]

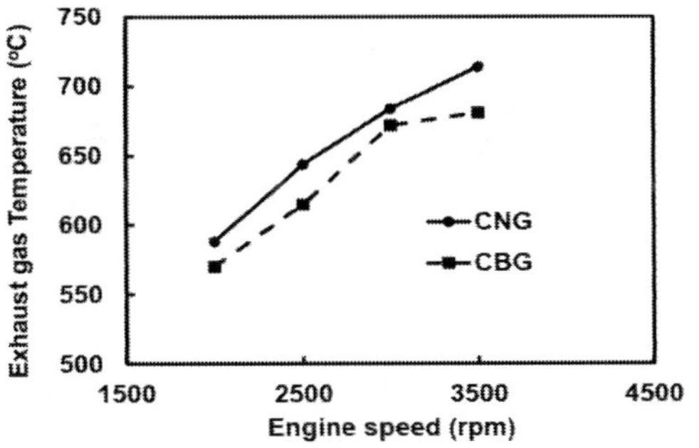

Fig 5. Exhaust gas temperature in various engine speed while used CNG and CBG as fuel.

3.1.3 The effect of engine speed on engine brake power

The effect of engine speed to engine brake power both using the CBG and the CNG as fuel show in Fig. 6. The brake power are increased with higher engine speed for CNG and CBG fuelling in the engine. Moreover, the engine was fuelled with CBG gas shown lower brake power compared with the engine fuelling with CNG. In order to the higher methane (CH_4) contents in CNG fuel compositions [15] as shown in table 1. The high heating value also dominated for higher brake power compared with CBG. However, that can observe the brake power in each engine speed has shown no significant effect for both fuels.

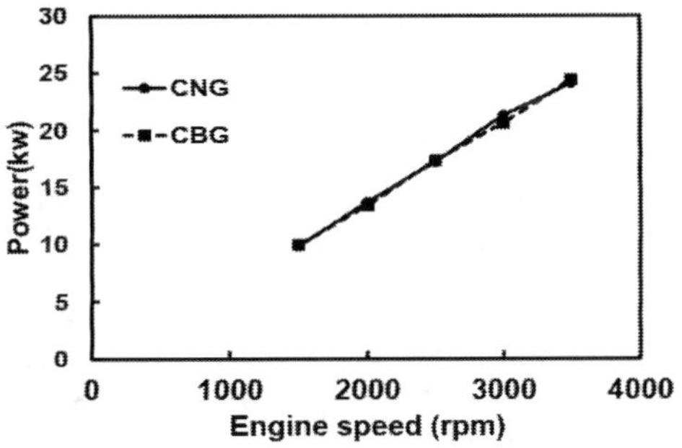

Fig. 6. Effect of engine speed when using CNG and CBG as fuel on the brake power output.

3.1.4 The effect of engine speed on brake specific fuel consumption.

The variation of brake specific fuel consumption (BSFC) in the engine speed between 1500 to 3500 rpm under constant engine load as shown in Fig.7. BCG has low heating value lead to show the higher BSFC for all engine speed compared to CNG. In addition, CBG shows high burning rate compare to CNG [16]. The overall of BSFC are respectively decreases with higher engine speed both CNG and CBG fuel that is mainly due to the higher heat loss from combustion chamber to chamber walls.

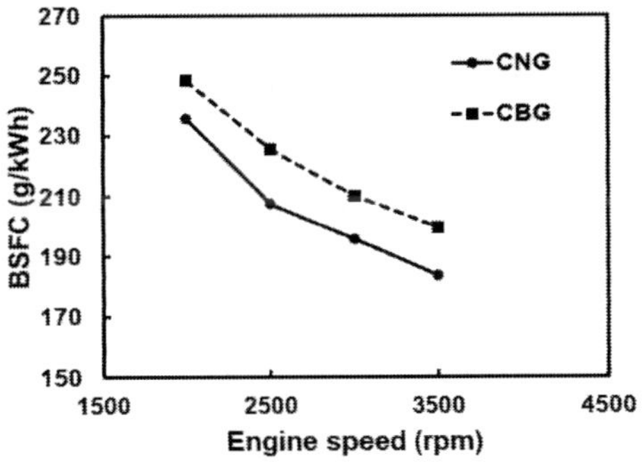

Fig. 7. Effect of engine speed when using CNG and CBG as fuel to Brake specific fuel consumption.

3.1.5 The effect of engine speed on brake thermal efficiency

The brake thermal efficiency (BTE) of SI engine was used CNG and CBG as fuel in various engine speeds that show in Fig.8. The engine fueled with CNG was shown the lower BTE than using CBG which mainly due to CBG have oxygen content which effect to provides better combustion performance. In additions, CBG has shown the faster laminar flame propagation and higher heat release rate in properties [17], However the higher engine speed (2500 rpm) has no significant effect on BTE for both CNG and CBG fuel.

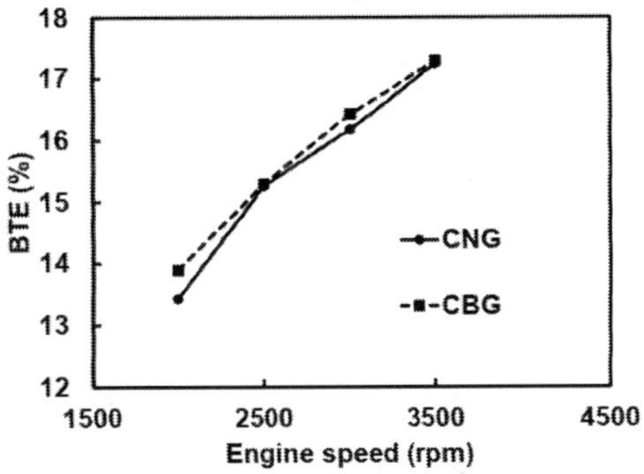

Fig. 8. Effect of engine speed when using CNG and CBG as fuel on brake thermal efficiency.

 https://doi.org/10.1051/e3sconf/202130201005

3.2 The effect of CNG and CBG on exhaust gas emissions

3.2.1 Carbon monoxide (CO) emission

CO emission while used CNG and CBG as fuel in various engine speeds as shown in Fig.9. The CO concentration is decreases respectively with higher engine speed. Furthermore, CNG has less CO emission than CBG mainly due to the higher combustion temperature was reason of the low CO emission cause by the CO will converts to CO_2. The result are matched with Fig.5. CNG can be combustible than the CBG, which effected CNG has lower CO emission than CBG [16].

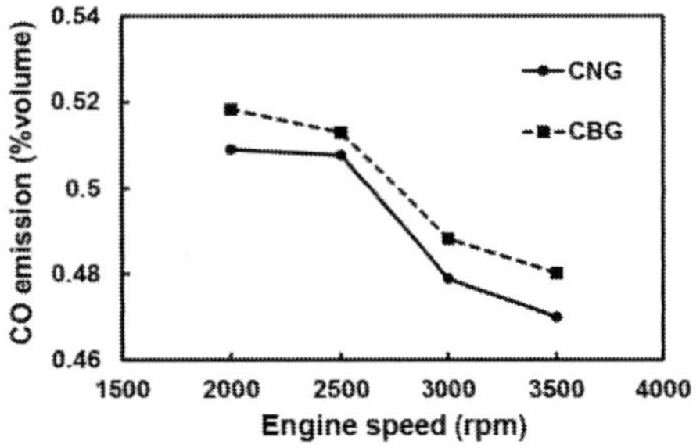

Fig. 9. The effect of engine speed on carbon monoxide concentration from combustion of CNG and CBG fuelling.

3.2.2 Hydrocarbon (HC) emission

The hydrocarbon emission of CNG and CBG are reduced with increases engine speed as seen in Fig.10. The engine was operated with CBG show the lower amount of HC emission emitted in exhaust gas than fuelling with CNG due to the formation of unburnt hydrocarbon (UBHC) during incomplete combustion under mixing between air and fuel (rich or lean mixture). CNG show higher HC emission than CBG especially in low engine speed due to poor oxidation of unburnt hydrocarbon (UBHC) during the warm-up period [1]. However, the high engine speed (>3000 rpm) found that the HC emission level of CBG is usually matched tend of CNG that has no significant effect on higher engine speed.

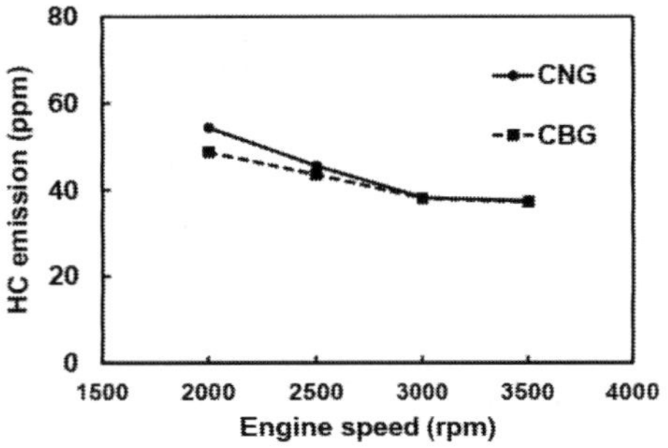

Fig. 10. The effect of engine speed on hydrocarbon concentration emissions from combustion of CNG and CBG fuelling.

3.2.3 Nitrogen oxides (NO_x) emission

NOx concentration from combustion of CNG and CBG fuelling under various engine speeds shown in Fig.11. The overall NO_x emission from both fuels increases with higher engine speed. The combustion of CBG fuel resulted in substantial reduction of NO_x compared to CNG fuel due to CNG has better combustion, as seen in peak cylinder pressure of CNG are higher compared with CBG [16,18]. In additions, the thermal NO_x is generally generated from high temperature (high in-cylinder pressure) in engine were significantly reduced in the exhaust from the combustion of CNG. In additions, CNG was reduced the formation rate of NO_x after engine speed 3,000 rpm, however the NO_x concentration from the combustion of CBG was not significantly reduced among measured, with high engine speed.

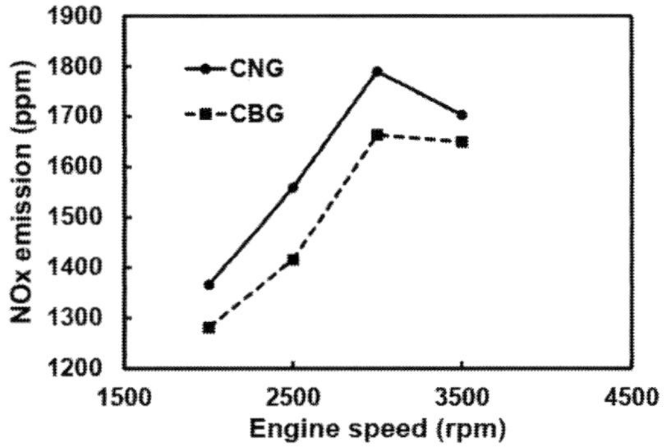

Fig. 11. The effect of engine speed on nitrogen oxides concentration emissions from combustion of CNG and CBG fuelling.

4. Conclusions

The comparison study between CBG and CNG have been shown to alternative fuel for spark ignition engine to unlock potential performance synergies in the fuel properties, improving the engine performance with low NO_x emission. The engine fueling with CBG produces higher BSFC and BTE compared to those of CNG. In additions, the use of CBG show lower brake power than the CNG but there are not significantly different range. In terms of exhaust gas emission found that CBG dramatically released lower HC and NOx emission than CNG fuel under higher engine speed. In additions, NOx concentration from combustion of CBG fuel are emitted evidently less than CNG whereas CNG released obviously lower CO emission than CBG fuel.

The CBG is one of alternative fuel that impact to spark ignition engines and can be used for substituted CNG. In additions, CBG can be reduce HC and NO_x emission levels. Therefore, the design of CBG in spark ignition engine presented in this work could be considered as giving an engine performance with environmental benefits. However, the optimising engine operation such as fuel injection strategy, EGR additions and turbocharger) to improve engine performance and efficiency with CBG fuelling SI engine.

Nomenclature

Abbreviation

BSFC	Brake specific fuel consumption
BTE	Brake thermal efficiency
CBG	Compressed biomethane gas
CH_4	Methane
CLD	Chemiluminescence detection
CNG	Compressed natural gas
CO	Carbon monoxide
CO_2	Carbon dioxide
HC	Hydrocarbon
NIDR	Non-dispersive infrared
NO_x	Nitrogen oxides
SI	Spark ignition
UBHC	unburnt hydrocarbon

References

1. K.A. Subramanian, V.C. Mathad, V.K. Vijay, P.M.V. Subbarao, Comparative evaluation of emission and fuel economy of an automotive spark ignition vehicle fuelled with methane enriched biogas and CNG using chassis dynamometer. Applied Energy, **105** (2013): 17–29
2. C. Jury, E. Benetto, D. Koster, B Schmitt, J. Welfring, Life cycle assessment of biogas production by monofermentation of energy crops and injection into the natural gas grid. Biomass Bioenergy, **34**, 1 (2010):54–66.
3. J.B. Holm-Nielsen, T. Seadi A.L., P. Oleskowicz-Popiel, The future of anaerobic digestion and biogas utilization. Bioresource Technology, **100** (2009), 5478–5784.
4. D. Deublein, A. Steinhauser, Biogas from Waste and Renewable Resources: An Introduction. Wiley-VCH Verlag GmbH & Co KGaA, Weinheim, Germany (2008)
5. M. Larsson, S. Grönkvist and P. Alvfors, Upgraded biogas for transport in Sweden : effects of policy instruments on production, infrastructure deployment and vehicle sales, Journal of Cleaner Production, **112** (2016): 3774-3784.
6. E. Porpatham, A. Ramesh, B. Nagalingam, Effect of compression ratio on the performance and combustion of a biogas fuelled spark ignition engine, Fuel, **95** (2012): 247–256.
7. E. Porpatham, A. Ramesh, B. Nagalingam, Investigation on the effect of concentration of methane in biogas when used as a fuel for a spark ignition engine, Fuel, **87** (2008): 1651–1659.
8. N. Dussadee, K. Reansuwan, R. Ramaraj, Potential development of compressed bio-methane gas production from pig farms and elephant grass silage for transportation in Thailand, Bioresource Technology, **155** (2014): 438–441.
9. A. Demirbas, Political economic and environmental impacts of biofuels: a review, Applied Energy, **86** (2009): 108–117.
10. N. Johansson, Production of liquid biogas, LBG, with cryogenic and conventional upgrading technology, Lund Institute of Technology (2008)
11. Q. Yong, S. Shuzhou , J. Dehao , S. Xinxing , L. Xingcai , Review of the state-of-the art of biogas combustion mechanisms and applications in internal combustion engines, Renewable and Sustainable Energy Reviews, **69** (2017): 50–58.
12. R. Janta, K. Sekiguchu, R. Yamaguchi, K. Sopajaree, B. Plubin, T. Chetiyanukornkul, Spatial and Temporal Variations of Atmospheric PM10 and Air Pollutants Concentration in Upper Northern Thailand During 2006–2016, Applied Science and Engineering Progress **13**, 3 (2020): 256-267
13. S.N.K. Reddy, M.M. Wani, A Comprehensive Review on Effects of Nanoparticles-Antioxidant Additives-Biodiesel Blends on Performance and Emissions of Diesel Engine, Applied Science and Engineering Progress, **13**, 4 (2020): 285-298
14. A. Santiarpa, T. Wongchang, B. Sawatmongkhon, K. Theinnoi, C. Chaitongrat, Impact of Compressed Biogas on Combustion and Exhaust Emission Characteristic from Spark Ignition Engine, 2018 Third International Conference on Engineering Science and Innovative Technology (ESIT), (2018): 1-4, doi: 10.1109/ESIT.2018.8665329.
15. M.M. Tahir, M. S. Ali, M.A. Salim, R.A. Bakar, A.M. Fudhail, M.Z. Hassan, M.S. Abdul Muhaimin, Performance analysis of a spark ignition engine using compressed natural gas (CNG) as fuel.
16. M.I. Jahirul, H.H. Masjuki, R. Saidur, M.A. Kalam, M.H. Jayed, M.A. Wazed, Comparative engine performance and emission analysis of CNG and gasoline in a retrofitted car engine, Applied Thermal Engineering, **30** (2010): 2219-2226
17. C. Jung, J. Park , S. Song, Performance and NOx emissions of a biogas-fueled turbocharged internal combustion engine, Energy, **86** (2015): 186-195
18. J.B. Heywood, Internal combustion engine Fundamental, Automotive Technology Series, (1998).

The Electrical Energy Modeling of the Modular Off-Grid PV System for Aquaculture Application in the EEC Region

Sukkhi Buakaew[1], *Prayut* Jiamrittiwong[1], and *Nutdechatorn* Puangngernmak[1,*]

[1]Faculty of Science, Energy and Environment, King Mongkut's University of Technology North Bangkok Rayong Campus 19 Moo 11 Nong Lalok, Ban Khai, Rayong 21120

Abstract. Renewable energy as a solar photovoltaic system represents an essential role in improving and promoting energy sustainability in agricultural/aquaculture-related activities. The "Crab Bank" is the program on practicing the caught gravid females of Blue Swimming Crab (BSC) in supporting conditions, allowing them to spawn and then release the zoea and young crabs back to the sea where are in the Eastern Economic Corridor (EEC). Onward with the project, many scientific studies assess the BSC stocks condition, evaluate the fishing habitat, and other relevant issues as the energy used in a micro-indoor system as Recirculating Aquaculture Systems (RAS). The off-grid solar photovoltaic system is used in the system cause the installed location was near the seaside area. The system's large size was operating, including solar panels, inverter, charger, and batteries to maintain the AC motor pump, affecting much energy loss in the system. The off-grid solar system design for modular systems is represented in this paper to describe the new structure of the micro modular RAS that reduces cost and increases the solar system's efficiency in modular concepts comparing the traditional system. The new structure model represents the calculation result in the term of electrical energy and experiment to scale for replacement in the aquaculture area in Rayong or Jantaburi in the EEC region. This research is a cooperation between KMUTNB Rayong and the Provincial Energy Office Of Rayong.

Keyword. Photovoltaic system, Off-grid solar system, Recirculating Aquaculture System, Eastern Economic Corridor (EEC), Modular solar system.

1 Introduction

The Crab Bank is a sustainable crab conversation project first implemented in Pakklong, Pathew district, Chumphon province, Thailand, for a scheming conservative crab source [1-3]. This concept was a decent case project on practicing the caught gravid females of Blue Swimming Crab (BSC) in supporting conditions, allowing them to spawn and then releasing the zoea and young crabs back to the sea area the Eastern Economic Corridor (EEC).

Onward with the project, many scientific studies assess the BSC stocks condition, evaluate the fishing habitat, and other relevant issues as the energy using in a micro-indoor system as Recirculating Aquaculture Systems (RAS). The system was operating in a condition of feed seawater flow, oxygen feed rate that controls with electrical pump and air compressor. The source of energy supply is the off-grid solar photovoltaic system and installed location near the seaside area. However, renewable energy use promoting energy sustainability term. The efficiency should be a concern for the actual operation.

The traditional system was a simple off-grid design for an extensive system located on Chantaburi, helping for the research in area [4-5], as shown in Figure 1.

It was operating, including solar panels, inverter, charger, and batteries to maintain the AC electrical machine, affecting much energy loss in the system.

A high-efficiency inverter and an increasing number of solar panels and batteries required operating 24 hours a day with an electrical power of 565 Watt for all electrical machines in the extensive system.

Fig. 1. The photo of the traditional system.

The traditional system includes an extensive system that is a high production scale for nursing the spawn and zoea. The support for a middle and large village uses high energy consumption and system installation and maintenance costs. The purpose system is the modular off-grid system that is compact, low energy consumption, and costs. The energy model using to design the module system with the relation of the

* Corresponding author: nutdechatorn.p@sciee.kmutnb.ac.th

https://doi.org/10.1051/e3sconf/202130201006

traditional system and operation parameters. The data of electrical measurement between an operation of the extensive and the module system using for developing the energy model.

1.1 The PV off-grid energy model of the system

1.1.1 The traditional system

The traditional system is an extensive system that includes two AC motors connecting to an air compressor (Supercharger) with redundancy mode operation and one AC motor pump for oxygen and water circulation of the micro-indoor system as RAS. The type of motor of the air compress and water pump is a single-phase 220 Vac motor.

One unit of the extensive system can divide into three subsystems contains the total productivity for a maximum of 10.5 kg of spawn and used 3,780 lits of seawater. The diagram and photo of the extensive system and represented in Figures 2 and 3.

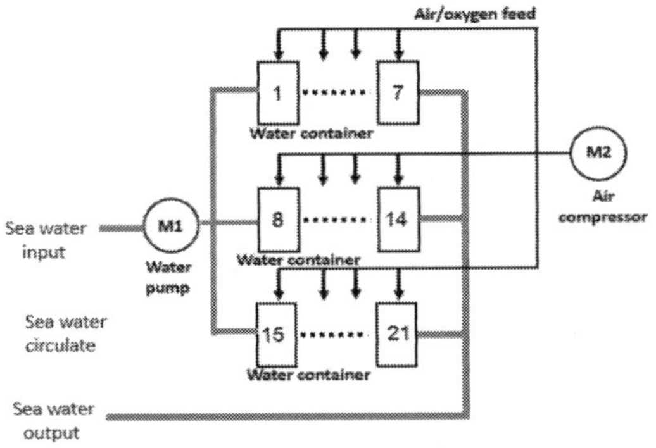

Fig. 2. The diagram of the extensive system.

Fig. 3. The photo an one of three-part in the extensive system.

1.1.2 The traditional off-grid PV system

The traditional off-grid PV system consists of solar panels, a battery charger, and an inverter [6-8]. The solar panels and batteries design support the inverter and energy-souring to the system. The diagram of the PV system represents in Figure 4.

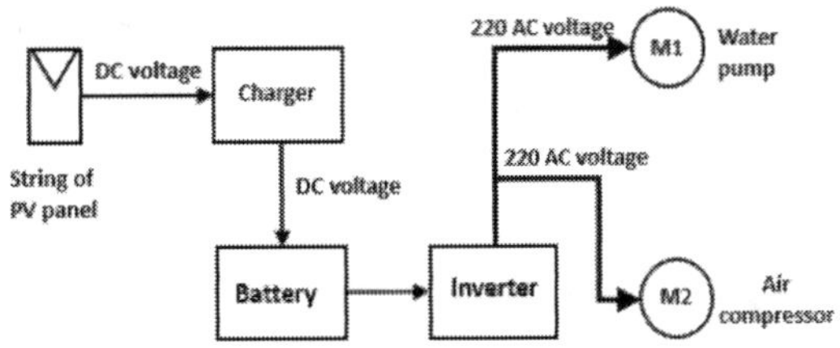

Fig. 4. The diagram of the PV system.

The electrical generate-time of solar panels depends on daylight time, around 5 hours/day. Therefore, the number and size of the panel and battery should be enough for energy-using in the application. The maintaining issue is another point of interest for design and cost-optimized.

1.2 The PV off-grid modular concept

1.2.1 Concept of PV off-grid modular unit

The concept of PV off-grid modular is proposed by the Provincial Energy Office Of Rayong for studies and optimize the efficiency and cost. Because of PV energy source for the traditional system was inefficient and needed more maintenance cost. For the small size of Crab bank, the extensive system is not appropriate and expensive in terms of production scale. Therefore, the modular concept using to the scale-down system into small production.

The modular unit consists of one dc motor pump and three motor air compressors. The diagram and photo of the modular concept represent in Figures 5 and 6.

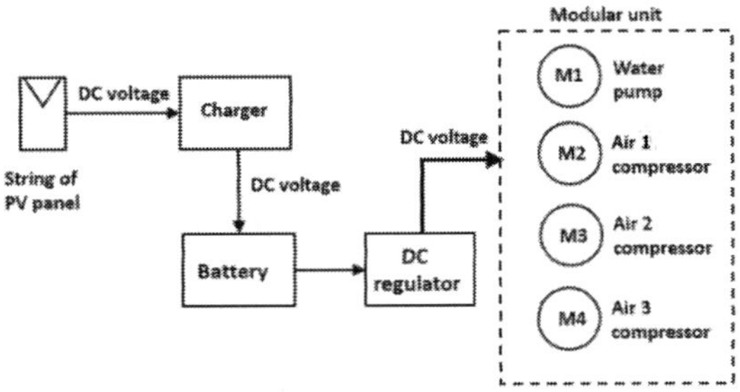

Fig. 5. The diagram of the modular system.

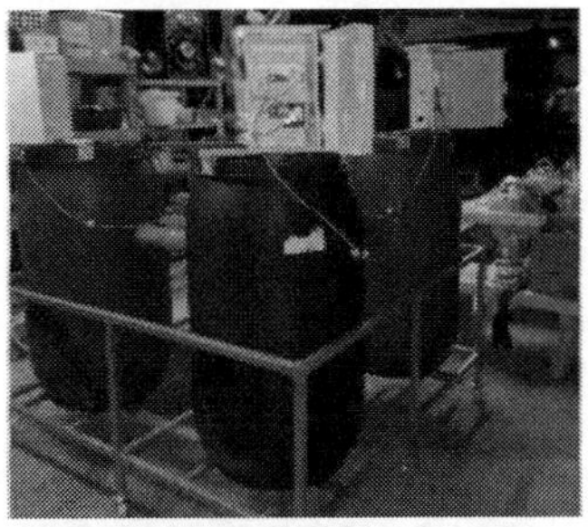

Fig. 6. The photo of the four modular system

1.2.2 The replacement in the traditional system

The segment of the traditional system can be replaced by the proposed modular system, as shown in Figure 7, operated with three containers.

Each module has one water pump, and three air compressors operate independently with a single dc voltage source.

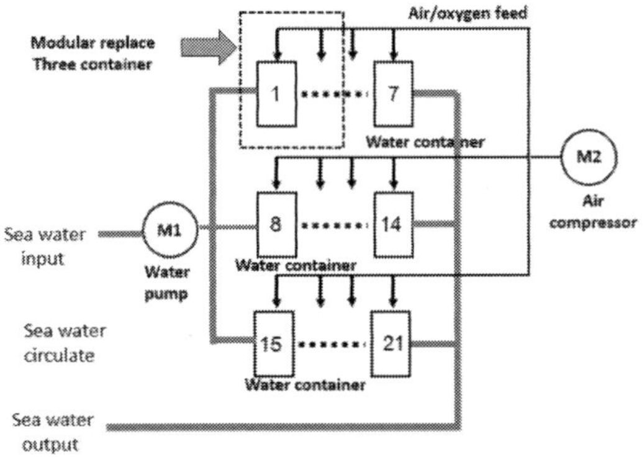

Fig. 7. The modular replacement in the traditional system

2 Modeling the modular unit

The energy model of the modular can be applying from the data of the extensive system. The description of electrical machines as datasheet or nameplate use to understand the relation of electrical energy used.

The working load of operation is used to create the reference load for optimizing and calculating the modular compared with the extensive. The electrical measurement processes are used for collecting the energy and re-calculate the model into the extensive.

2.1 Methodology

2.1.1 Electrical energy of the extensive system

First of all, the electrical instrument, power meter, and flow meter are used to collect the energy used in the extensive system.

The workload condition is setting with the onsite operation, and the instrument is collecting the data for one week of operation time. The condition of workload is measure with a flow meter as define in Table 1. The electrical power use of the machine can represent in Table 2. The total power is 565 Watt.

The measurement result is the data for comparing and modeling the modular unit in the term of power.

Table 1. The condition workload of the extensive system.

Parameter	value	unit
Water flow (circuilation)	4.5	lit/minute
Air flow (Oxygen feed)	183.75	lit/minute

Table 2. The measurement of electrical power of the AC machine.

AC machine	value	unit
The motor of the water pump	135	Watt
The motor of the air compressor	430	Watt

2.1.2 Electrical energy of the modular unit

Similar to the extensive measurement process, the workload condition set up and fractions and scaling to the small and electrical instruments are used to collect the energy data. The DC machine (motor and air compressor) is used to apply the water flow and oxygen feed system. The working load condition defines in Table 3, and the electrical power represents in Table 4.

Table 3. The condition workload of the modular unit.

Parameter	value	unit
Water flow (circuilation)	1.5	lit/minute
Airflow (Oxygen feed)	26.25	lit/minute

Table 4. The measurement of electrical power of DC machine.

DC machine	value	unit
A dc motor of water pump	4	Watt
A dc motors of air compressor	8	Watt

2.2 Electrical power and energy modeling

Electrical power and energy relation between the extensive and the modular represented with equations (1) and (2).

The factor of energy is the operation time and power of the machine. The difference between both systems is the number and type of electrical machines that affect the energy consumption.

$$E_{system} = P_{total} \times t_{operation} \quad (1)$$

$$P_{total} = \sum \text{Power of electrical machine} \quad (2)$$

where

E_{system} : energy used in the system (W-hour)
P_{total} : total power of all machine in system (W)
$t_{opearation}$: operation time of system (hour)

3 The electrical energy modeling of the modular unit

The modular unit's electrical power and energy models are written by equations (3) and (4). One of the modular systems consists of components are a dc pump and three dc compressors units.

$$Power_{modular} = Power_{DC\ Pump} + \sum Power_{DC\ Comp} \quad (3)$$

$$E_{modular} = Power_{modular} \times t_{operation} \quad (4)$$

where

$Power_{modular}$: power of the modular used (W)
$Power_{DC\ Pump}$: power of dc pump unit (W)
$Power_{DC\ Comp}$: power of dc compressor unit (W)
E_{system} : energy used in the system (W-hour)

The relation of scale production between the modular system and the extensive system can be represented by equation (5) with the production factor (k) is 14.3 % that calculate from the total productivity for a maximum of the modular (1.5 kg) and the extensive (10.5 kg) .

$$Y_{modular} = k \times Y_{extensive} \quad (5)$$

where

$Y_{modular}$: weight of product from the modular (kg)
$Y_{extensive}$: weight of product from the extensive (kg)
k : production factor

This model is used to optimize the system when replacing the extensive system with the modular system. Although the proposed model is only an electrical machine, it is to simplest to optimize the modular system.

4 Result and conclusion

The calculation power from proposing a modular model with one motor pump and three compressors is 28 Watt. The factor of production weight the modular to the extensive is 14.3 %, the recalculation of electrical power used in 100 % is 195.80 Watt. The differential power between the extensive and the modular at same weight production is 565 – 195.80 = 374.2 Watt. Therefore, the power and energy used will reduce to 65.35 %.

However, the proposed modular contains 1.5 kg per three containers, affect the power reduced to 374.2 Watt. The trend of power used and production factor is represented by table 5 and figure 8.

Table 5. The power uesd of modular varied with number of compressors and container.

No. of compressers and containners	Power used of modular (W)	Production factor (%)
3	28	14.3
4	36	19.0
5	44	23.8
6	52	28.6
7	60	33.33

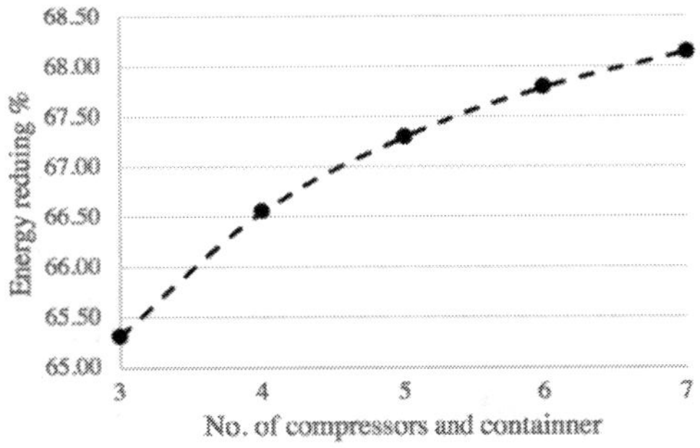

Fig. 8. The relationship between numbers of compresser/container and the percent of enegy reduction.

References

1. W. Arkronrat, V. Oniam, N. Hengcharoen, Crab bank implementation: Case study of the blue swimming crab bank in Prachuap Khiri Khan province, Thailand, Kasetsart University Fisheries Research Bulletin (Thailand), **37**, 1 (2013) : 30-39
2. P. Suanrattanachai, T. Suppanirun, S. Etoh, V. Sulit, The Role of Crab Bank System in Securing Fisheries Livelihood and Resources Conservation and Management, Southeast Asian Fisheries Development Center, **7** , 1 (2009): 24-30
3. D. Thiammueang, R. Chuenpagdee, K. Juntarashote, The "Crab Bank" Project: Lessons from the Voluntary Fishery Conservation Initiative in Phetchaburi Province, Thailand, Kasetsart J. (Natural Sciences.) **46** (2012): 427-439
4. C. Kunsook, N. Gajaseni, N. Paphavasit, The Feeding Ecology of the Blue Swimming Crab, Portunus pelagicus (Linnaeus, 1758), at Kung Krabaen Bay, Chanthaburi Province, Thailand, Tropical Life Sciences Research, **25** (2014): 13-27
5. C. Kunsook, N. Gajaseni, N. Paphavasit, A Stock Assessment of the Blue Swimming Crab Portunus pelagicus (Linnaeus, 1758) for Sustainable Management in Kung Krabaen Bay, Gulf of Thailand, Tropical Life Sciences Research, **25** (2014): 41-59
6. G.K. Singh, Solar power generation by PV (photovoltaic) technology: A review, Energy, **53** (2013): 1-13
7. A. Ajan, J. Nirmala, Performance Evaluation of On-Grid and Off-Grid Solar Photovoltaic Systems, International Journal Of Innovative Research In Electrical, Electronics, Instrumentation And Control Engineering, **3** (2015): 20-23
8. M.L. Edson, B.L. Carine, T. Raymond, Performance Parameters Of An Off-Grid Building Integrated Photovoltaic System In South Africa, 33rd European Photovoltaic Solar Energy Conference and Exhibition

Formation of multiple combustion fronts in an imbert downdraft gasification reactor

Kittipass Wasinarom[1], *Sarawut* Sungworagarn[2], *Prasan* Sathitruangsak[2] and *Kasemsil* Onthong[3,*]

[1]School of International and Interdisciplinary Engineering Programs, School of Engineering, King Mongkut's Institute of Technology Ladkrabang, 1 Soi Chalong Krung 1, Chalong Krung Road, Ladkrabang Bangkok, 10520, Thailand
[2]Department of Mechanical Engineering, Faculty of Engineering and Technology, Mahanakorn University of Technology 140 Cheum-Sampan Road, Nong-Chok, Bangkok 10530, Thailand
[3]Department of Teacher Training in Mechanical Engineering,King Mongkut's University of Technology North Bangkok, 1518 Phacharat 1 Road., Bang Sue, Bangkok 10800, Thailand

Abstract. The experimental study of downdraft gasification was performed in this paper. The operation which led to the formation of the second combustion front was pointed out. In this situation, both combustion fronts will lose their intensity and finally be extinguished. The operation was unintentionally stopped. It was revealed that the combustion front propagated upward in the reactor after starting the test. While it was about to reach the air inlet nozzle, the second combustion front was detected by an abrupt temperature rise of the thermocouple above the air supply nozzle. After the formation of the second combustion front, both fronts started to lose their intensity which indicated by the decrease in temperature corresponding with their locations. It was possible that the second combustion front would dilute the oxygen concentration supplied to the first combustion front. The decreasing temperature of the first combustion front reduced the heat transfer rate to the second combustion front. Finally, both combustion fronts were extinguished. The operation was unintentionally stopped.

Keyword. Gasification, Imbert type, Downdraft, biomass, Reactor.

1 Introduction

Solid fuel combustion is a highly complex process. Beginning with the pyrolysis process, the fresh fuel undergoes thermal decomposition which releases the combustible gas and condensable heavy hydrocarbon. Then, the released gas is reacted with air to produce thermal energy. The continuous combustion process can be sustained by utilizing a portion of the produced thermal energy back to supply the pyrolysis process of the remaining fresh fuel. After releasing all volatile gas, the surface heterogeneous combustion is continued on the left-over fixed carbon (char). Finally, the fixed-carbon is then consumed leaving the non-combustible ash as the residual.

Gasification is considered a more efficient means of solid fuel combustion compared with direct combustion. It has more capability to improve producer gas heating value before reacting with air. In gasification, the pyrolysis gas undergoes a shift reaction within the reduction layer. The combustible CO was increased by the consumption of non-combustible CO_2. Heavy hydrocarbons (tar) are also converted to lighter gas molecules in the reduction layer. These resulted in the increase of synthesis gas heating value.

The combustion front propagation in fixed bed gasification was studied by many researchers [1] [2]. In downdraft gasification, the propagation was explained by diffusion transport of heat from the reaction front and the generated pyrolysis gas in counter direction with the supply air. Therefore, the reaction front moves upward at a finite speed. The speed of the moving front depends on the rate of heat transfer from the combustion front and the rate of combustible pyrolysis gases accumulation above the combustion front. When the air-fuel mixture and the temperature is falling within the flammability range. The mixture will auto-ignite with the incoming oxygen resulting in the combustion front moving upward. The front propagation will never happen in a downward direction for downdraft configuration because there is no oxygen left below the front. The propagation speed depends on many factors, for example, the fuel type and size, air mass flow rate, the fuel moisture composition [3].

The homogeneous combustion front of fixed bed gasification is identical to the 1D gas-phase lamina flame except that the reaction front occurs inside the pore space of the stacked fuel bed. These reaction fronts are very thin. It has a length in the order of 1 mm. The heat transfer around the combustion front inside the fuel bed is governed by radiation and convection of hot combusted mixture to the fuel particle surface that makes up the pore space. The heat conduction among stacked fuel particles has little effect compared to other heat

* Corresponding author: kasemsilo@kmutnb.ac.th

transfer modes due to the high contact conduction resistance between stacked particles and the high thermal resistance of the wooden biomass [4]. It was found that the temperature profile along the height of the fuel bed has a milder gradient and spreads over a longer distance around the combustion front compared to the conventional gas-fueled combustion front. The temperature profile and heat transfer mechanism are identical to the combustion in the inert porous media which was illustrated by [5].

In this paper, the experimental study of downdraft gasification has been performed. The operation which led to the formation of the second combustion front was pointed out. This allows operators to understand the insight of the events.

2 Experimental procedure

A cylindrical shape Imbert type gasification reactor was used in this test [6]. It is made of steel with an inner diameter of 75 mm and 500 mm height as shown in figure 1. The door was attached along with the height of the reactor which allowed the operators to investigate the fuel evolution in different layers after finishing the test as shown in figure 2. The reactor was well insulated during the test. The three inlet supply air nozzles were installed at 230 mm from the bottom where the fuel grate was located. Type-K thermocouples were installed to measure temperature distribution along with the reactor height. The supply air mass flow rate was measured by an air rotameter. The reading temperature was recorded by the MW-100 data logger. Testo 350 flue gas analyzer was used to measure O_2, CO, and CO_2 gas composition.

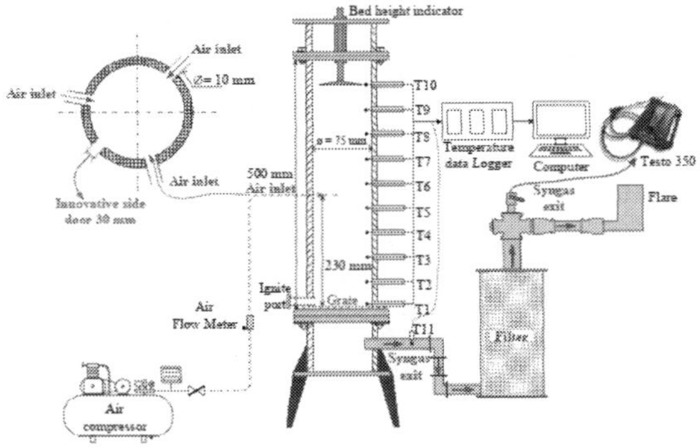

Fig. 1. Schematic diagram for downdraft gasification.

The 250 g of rice husk was filled in the reactor which corresponded to the 450 mm height of the fuel bed. The feedstock property is shown in table 1. Air was introduced to the reactor at the desired flow rate by three nozzles. The flow rate was regulated by a ball valve to satisfy each of the test conditions in table 2. Then, the portable butane burner was put in the ignite port. The reading T_1 was monitored until the reading T_1 raised to 100 C. After that, the ignition port was closed. The temperature and gas composition were started recorded for every 2 minutes interval until the test was finished. Finally, the recorded data were interpreted as a series of events.

Table 1. Proximate and Ultimate Analyses of Rice Husk.

Ultimate Analysis (%, as received)				
C	H	O	N	S
38.0	4.55	32.4	0.69	0.60
Proximate Analysis (%, as received)				
Volatile	Fixed carbon	Moisture	Ash	HHV (MJ/kg)
55.6	20.1	10.3	14.0	15.0

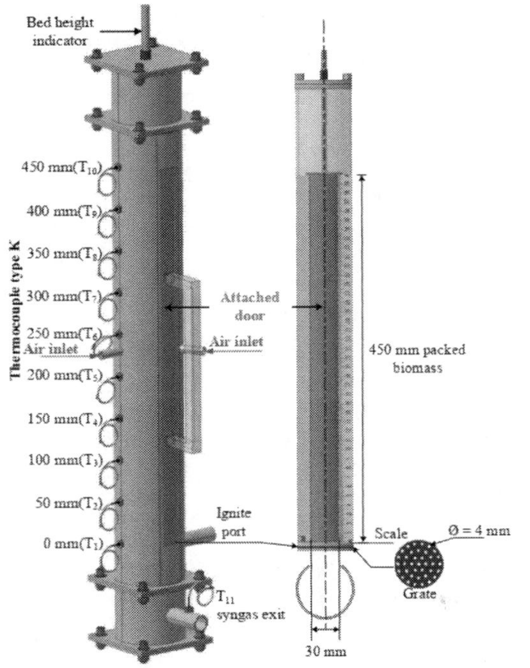

Fig. 2. Configuration of the packed bed reactor used for downdraft gasification.

Table 2. Experimental condition.

Air flow rate (l/min)	Air mass flux (kg/m²-h)
15	255
20	340
30	680

3 Result and discussion

For all cases, the measured temperature showed that the combustion front propagated upward to the top of the reactor as the peak temperature continually moved upward during the test. This was in agreement with other research papers on downdraft front propagation [1] [7] [8]. The increasing peak temperature along the testing period (Fig 4a, 4b, and 4c) reflected the increase in combustion intensity. For air mass flux 340 and 680 kg/m²-h, the CO was increased with the increasing time but the CO_2 was decreased along with the increasing time (Fig 3b and 3c). This could be explained by the intensification of the shift reaction that occurred just below the combustion front. The evidence of the shift reaction was pronounced in the case with the higher air mass flux because the higher combustion gas temperature boosted the shift reaction and tars cracking below the reaction front. There were researchers who mention the direct relation between peak combustion temperature at the reaction front and CO concentration in producer gas [8] [9].

For all cases of the input air mass flux, when the combustion front was about to reach the supply air nozzle location, the second combustion front was formed above the air supply nozzle (T_6 location). The forming of the second combustion front could be detected by the raising of the measured T_6 as shown in the figure 4a, 4b, and 4c. This could be described by many possible incidents that led to the presence of oxygen, combustible gas (pyrolysis gas) and temperature that allowed the ignition to be taking place at that location. The first incident that led to the presence of oxygen above the air supply nozzle is the high dispersion of the flow inside the cavity of the porous media [10]. The flow inside porous media is identical to the flow inside the pore space of the stacked fuel bed. Air was injected into the reactor by the supply nozzle. The high dispersion flow allows the inlet air to transport over a wide area of the inlet location. Some portion of the inlet air is transported to the area above the inlet nozzle resulting in a high concentration of oxygen. The second and third incidents that led to the presence of the combustible gas (pyrolysis gas) and high temperature above the air inlet nozzle is the diffusion transport of pyrolysis gas and heat energy that produced from the first combustion front. Which at that time, propagated to the location just below the inlet air nozzle. The second reaction front is the updraft configuration since it consumes oxygen that disperses upward from the air nozzle.

After the ignition of the second combustion front, both fronts started to lose their intensity which indicated by the decreasing in temperature corresponding with their locations. It was possible that the second combustion front released non-combustible gas and consumed some portion of the supplied oxygen. Which results in the dilution of the oxygen that flowed down to supply the first combustion front. The falling temperature of the first combustion front resulted in decreased heat transfer rate to the second combustion front. Therefore, the second combustion front was extinguished without enough heat supply. Finally, both combustion fronts were extinguished. The operation was unintentionally stopped.

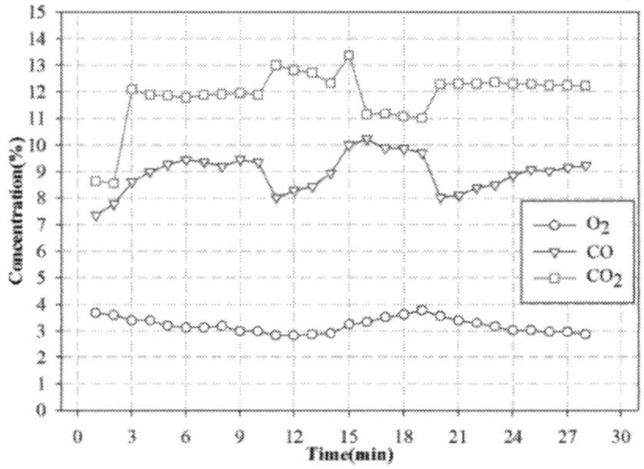

(a) 255 kg/m²-h

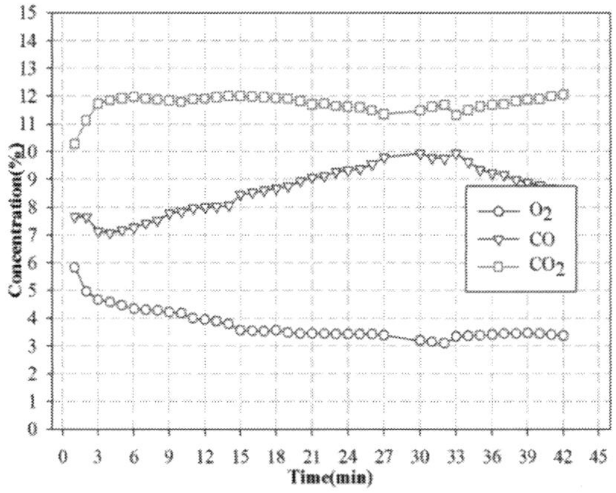

(b) 340 kg/m²-h

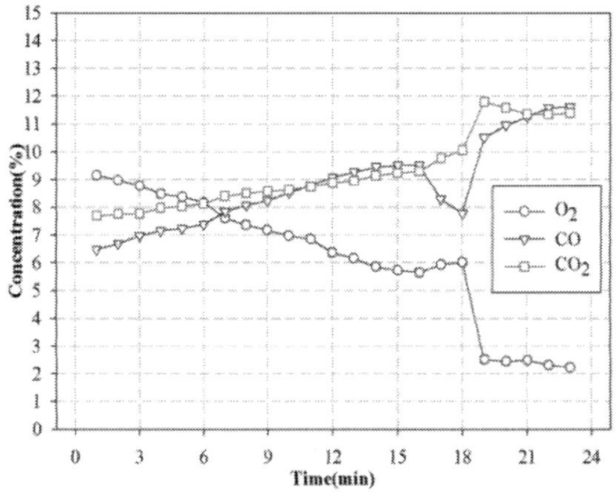

(c) 680 kg/m²-h

Fig. 3. Gas composition at the different time.

https://doi.org/10.1051/e3sconf/202130201007

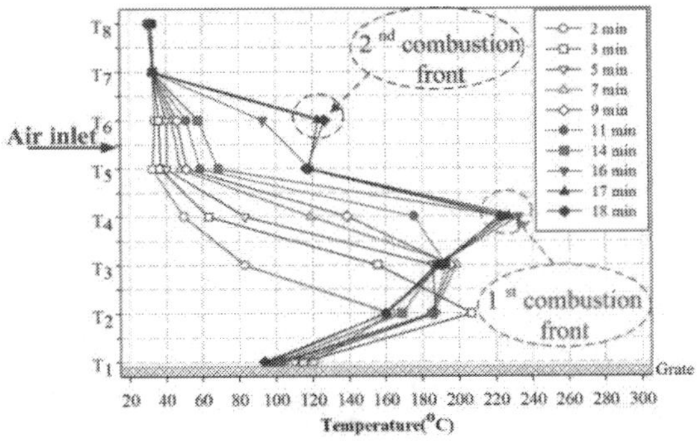

(a) 255 kg/m²-h

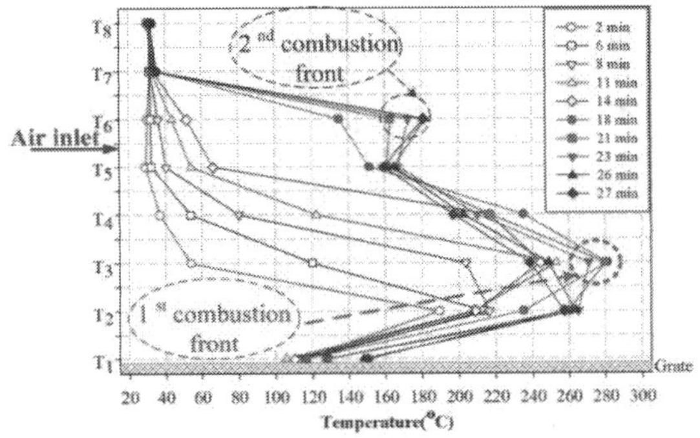

(b) 340 kg/m²-h

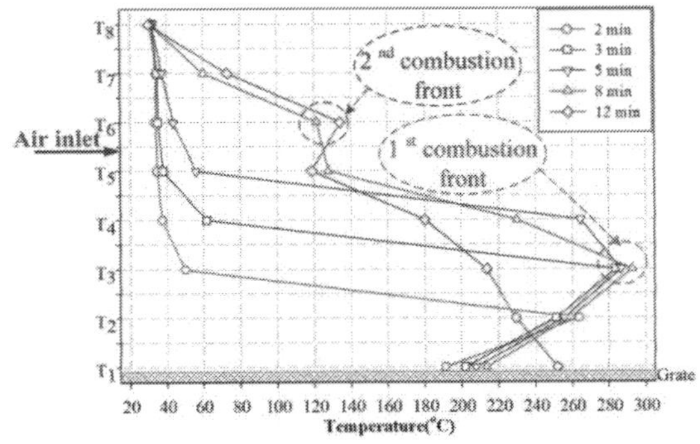

(c) 680 kg/m²-h

Fig. 4. Temperature profile inside the reactor at the different time.

After the flame extinguished, the reactor was left until it cooled down to a temperature below 50 °C. The attached door was opened to investigate the fuel evolution in different layers as shown in tablrure 5. The black char residual above the air inlet nozzle proved that the pyrolysis combustion has occurred by consuming the dispersed air above the air nozzle.

Fig. 5. Residual investigation by opening the attached door.

4 Conclusion

The experimental study of downdraft gasification was performed in this paper. The operation which led to the formation of the second combustion front was pointed out. In this situation, both combustion fronts will lose their intensity and finally be extinguished. Air was injected by nozzles at the throat area of the stacked fuel bed. In the beginning, the combustion front was ignited underneath the reactor. After that, the combustion front propagated upward in the reactor during the test. While it was about to reach the air inlet nozzle, the second combustion front was detected by an abrupt temperature rise of the thermocouple above the air supply nozzle. After the formation of the second combustion front, both fronts started to lose their intensity indicated by the decrease in temperature corresponding with their locations. It was possible that the second combustion front would dilute the oxygen concentration supplied to the first combustion front. The decreasing temperature of the first combustion front reduced the heat transfer rate to the second combustion front. Finally, both combustion fronts were extinguished. The operation was unintentionally stopped.

Acknowledgement

This work is supported by King Mongkut's Institute of Technology Ladkrabang [KREF186414].

References

1. C. Ryu, A.N. Phan, V.N. Sharifi and J. Swithenbank ,Co-combustion of textile residues with cardboard and waste wood in a packed bed, Experimental Thermal and Fluid Science, **32**, (2007): 450-458.
2. K. Onthong, J. Chareonsuk, A New Method for Zone Development Observation for Updraft Rice Husk Gasification, BioResources, **14**, 3(2019): 5080 – 5096
3. S. Mahapatra and S. Dasappa, Experiments and analysis of propagation front under gasification regimes in a packed bed, Fuel Processing Technology., **121**, (2014): 83-90.
4. M. Jingwen and B. Heng, Measurements and theoretical modeling of effective thermal conductivity of particle beds under compression in air and vacuum, Case Studies in Thermal Engineering., **10**, (2017): 423–433.
5. K. Wasinarom, J. Charoensuk and L. Visarn, Non-equilibrium numerical modeling for combustion of LPG within porous media, International Journal of Heat and Mass Transfer., **143**, (2019): 118551.
6. T. Madhiyanon, P. Sathitruangsak and S. Soponronnarit, Effect of swirl secondary air injection on rice husk combustion in a short-combustion-chamber fluidized bed combustor using nozzle-type air distributor, Journal of Research and Applications in Mechanical., **1**,(2011): 77-82.
7. K. Wasinarom, J. Charoensuk, Experiment and Numerical Modeling of Stratified Downdraft Gasification Using Rice Husk and Wood Pallet, BioResources, **14**, 3(2019): 5235 – 5253.
8. C.D. Blasi, Dynamic behavior of stratified downdraft gasifiers, Chemical Engineering Science., **55**, (2000): 2931-2944.
9. S. Dasappa and P.J. Paul, Gasification of Char particles in packed beds: analysis and results, International Journal of Energy Research., **25**, (2001): 1053-1072.
10. J. Amir, J. Gunnar I. Hellstrom and T. Staffan Lundstrom, Numerical Derivation of Dispersion Coefficients for Flow through Three-Dimensional Randomly Packed Beds of Monodisperse Spheres, AIChE Journal., **60**, 2(2014): 749-761.

Soot Oxidation in Diesel Exhaust on Silver Catalyst Supported by Alumina, Titanium and Zirconium

Punya Promhuad[1,2,*] and *Boonlue* Sawatmongkhon[1,2]

[1]College of Industrial Technology, King Mongkut's University of Technology North Bangkok, 1518 Pracharat 1 Road, Wongsawang, Bangsue, Bangkok 10800, Thailand.

[2]Research Centre for Combustion Technology and Alternative Energy (CTAE), Science and Technology Research Institute, King Mongkut's University of Technology North Bangkok, Thailand.

Abstract. Diesel Particulate Filter (DPF) is used to limit the emission of particulate matter (PM). The operation of DPF has two consecutive functions which are filtration of PM and regeneration. Performance of DPF is reduced by clogging of the filter. This problem is improved by soot oxidation in the regeneration process. The soot is completely oxidized by oxygen when temperature is higher than 600 °C. However, the exhaust gas temperature in normal operating of the diesel engine is lower than the temperature of soot complete oxidation. The problem of low temperature in soot oxidation is improved by oxidation catalyst because the oxidation catalyst is used to reduce light of temperature in soot oxidation. The study's purpose is to compare the oxidation activity of silver catalyst supported on alumina (Al_2O_3), Titanium oxide (TiO_2), and Zirconium oxide (ZrO_2). The compression of soot oxidation on silver catalyst loaded on several support which showed silver base on alumina was the best of soot oxidation compared with titanium oxide and zirconium oxide. The behaviour of soot oxidation in silver base on titanium oxide and zirconium oxide were similar activity.

Keyword. Soot oxidation, Silver catalyst, TiO_2, ZrO_2

1 Introduction

Air pollution from the emissions of diesel vehicles, such as particulate matter (PM) and nitrogen oxides (NOx), has been caused harm to human health [1]. There are many technologies to reduce the pollution such as using the biodiesel as fuel in diesel engine, and installation of exhaust gas after-treatment system. The diesel engine used biodiesel as fuel that it showed low emission such as hydrocarbon, carbon monoxide, smoke[2].Currently, diesel particulate filter (DPF) is the device used to control effectively PM emissions[1, 3-4]. The operating of DPF consists of two sequential processes, which are trapping of soot and regeneration. First, the trapping of soot is the process of soot to deposit in porous walls of DPF [3]. Next, the regeneration of DPF is required to reduce the problem of DPF blockages. In normal, soot is oxidized by oxygen at a temperature higher than 600 °C [5–7], which is over the exhaust gas temperature in diesel engine.

The oxidation of soot was promoted using an oxidation catalyst, which assisted to burn soot at lower then temperatures of exhaust gas in diesel engine. Noble-metal (such as Ag, Au, Pd, Ru, and Pt etc.) were used as the active metal in oxidation catalyst [8-9]. Silver was interested as the oxidation catalyst because it prompted active oxygen species [10–12] and showed high stability [3]. Moreover, silver is cheaper than other noble metals (e.g., Pt) [13]. The oxygen species or active oxygen may be classified in to peroxide (O^-) and superoxide (O^{2-}). The active oxygen transfer to soot which was considered to be an important for soot oxidation activity [4, 14] .

Soot oxidation activity was promoted by not only the active metal but also the support materials, such as, Al_2O_3 TiO_2, ZrO_2, and CeO_2. Alumina was widely used as the support material of oxidation catalysts, such as, Pt/Al_2O_3 [15] and Ag/Al_2O_3 [3, 12]. Silver catalyst 16 wt% coated on alumina showed best soot oxidation [3]. The Ag/ZrO_2 showed a promising candidate for oxidizing soot with O_2 when the mole fraction of adding Ag to ZrO_2 was 30 % in citric acid-assisted sol-gel (SG) methods [11]. Moreover, it showed very active catalysts in a fresh or aged state [12]. The titanium oxide, which used with silver catalyst, showed high oxidation efficiency of soot in the tight contact between catalyst and soot [16]. From, several works of literature caused interest in the oxide materials to use with silver metal such as Al_2O_3, TiO_2, and ZrO_2.

The thermogravimetric (TG) methods was used to determine kinetics parameter such as activation energy (Ea) and frequency factor (f(α)) etc. Normally, the activation energy is important kinetic parameter to explain the reaction process and identify the reaction mechanism [17]. The reaction rate of a heterogeneous reaction The Arrhenius equation, which depends on temperature, was used to describe the rate of soot oxidation [18]. The rate of reaction in the case of constant heating rate can be explained by the following equation [19] :

* Corresponding author: s6303026912014@kmutnb.ac.th

$$\frac{d\alpha}{dT} = \frac{A}{\beta}\exp(\frac{-Ea}{RT})f(\alpha) \quad (1)$$

when A and Ea are the frequency factor and activation energy, f(α) is the differential conversion function, β is heating rate and R is the universal gas constant. The mass faction of the soot oxidation decomposition (α) can be calculated by the following equation:

$$\alpha = \frac{m-m_0}{m-m_f} \quad (2)$$

when m is the instantaneous mass of soot, m_0 is the initial mass, and m_f is the final mass. In this work used the Kissinger–Akahira–Sunose to determine activation energy to the following equation[20] :

$$\ln(\frac{\beta}{T^{1.92}}) = Const - 1.0008(\frac{Ea}{RT}) \quad (3)$$

The activation energy can be obtained by slope from plot of $\ln(\beta/T^{1.92})$ against 1/T for different heating rates at a constant α [20]. From equation (3) can be calculated the activation energy by equation as the following:

$$Ea = \frac{(\text{slope at } \alpha)\times R}{-1.0008} \quad (4)$$

The aim of the following study was to compare the soot oxidation activity of silver catalyst loaded on three metal-oxide supports (e.g., Al_2O_3, TiO_2, and ZrO_2) and find the activation energy from soot oxidation. The activity of soot oxidation was used by thermogravimetric (TG) methods in the three heating rates which were 5, 7.5, and 10 °C/min. The activation energy was calculated by KAS method.

2 Experimental

2.1. Catalyst and soot preparation

The silver catalyst used on different supporters was prepared via incipient wetness impregnation that silver nitrate to obtain 16 wt% was dissolved in an appropriate quantity of distilled water [3]. Then, the precursor solution was added dropwise to the powdered supports: γ-alumina, titanium oxide, and zirconium oxide. Then, the solvent was removed by drying at 110 °C for 8 h in an oven. Then, the dry sample will be calcined in static air at 600 °C with a heating rate of 10 °C/min for 2 h.

Soot was directly taken from the exhaust gas of a traditional diesel engine. The engine was operated at a speed of 1000–2000 rpm and load of 25–75% of the maximum load. A stain less steel mesh was rolled at one round and placed inside the exhaust pipe (50 mm ID) then, the trapped soot was collected and dried in a furnace at 110 °C for 8 h [3]. SOF was removed in furnace at 250 °C for 2 h.

2.2 Catalyst activity tests

Catalytic oxidation activity of soot was examined using the TGA method (PerkinElmer® Pyris 1). Soot (10 mg) was physically mixed with the catalyst at the designed weight ratio for 5 min in a stainless steel mortar to obtain the tight contact mode or was shaken for 5 min in a glass tube for the loose contact conditions. Then, the mixture (≈10 mg) was sampled and placed in a ceramic crucible and heated from room temperature to 110 °C with a heating rate of 10 °C/min under N_2 and, then, the purge gas was switched to O_2 until temperature 700 with heating rates of 5 °C/min, 7.5 °C/min, and 10ȯ °C/min and the contact of soot on the catalyst is tight contact. Purified oxygen (99.999% purity) with a constant flow rate of 50 cm³/min was used as the oxidiser. The sample weight was recorded continuously with the change in the temperature. The catalytic activity was assessed by using the following: T_{20}, defined as the onset temperature at which 20% of soot was removed; T_{50}, defined as the temperature at which of soot 50% was completely consumed; T_{90}, defined as the temperature at which 90% of soot oxidation.

3 Result and Discussion

3.1 Diesel soot oxidation

The thermogram profile and its first derivative (DTG) for soot oxidation under the tight contact mode are presented in Fig. 1-3. The x-axis shows the temperature, the y-axis shows the percentage of soot oxidation in figure (a), and the y-axis shows the first derivative of soot oxidation in figure (b). An oxidation reaction was the only interaction of soot (solid carbon) because the SOF was removed in the soot preparation stage. The DTG showed the maximum reaction rate in the oxidation process [3]. In case of pure soot in all the heating rates showed the peak of DTG at 610 °C. It explained the best of oxidation reaction of pure soot at this temperature which confirmed the soot is oxidized by oxygen at a temperature higher than 600 °C[5–7].

In the case of soot used in conjunction with silver supported by alumina. The oxidation of soot was well promoted by Ag/Al_2O_3 as observed from the experimental results from all heating rates. The heating rate was 5 °C/min. The peak of DTG showed the maximum oxidation rate temperature was reduced from 610 °C, in the case of non-catalyst soot, to 353 °C (shows in Fig.1) In case of the heating rate was 7.5 °C/min and 10 °C/min. The result showed similarly the oxidation of soot with case of 5 °C/min. In the case of the heating rate was 7.5 °C/min and 10 °C/min. The result showed similarly the oxidation of soot with a case of 5 °C/min from the peak of DTG at 359 °C and 405 °C in cases 7.5 °C/min and 10°C/min, respectively (shows in Fig. 2-3).

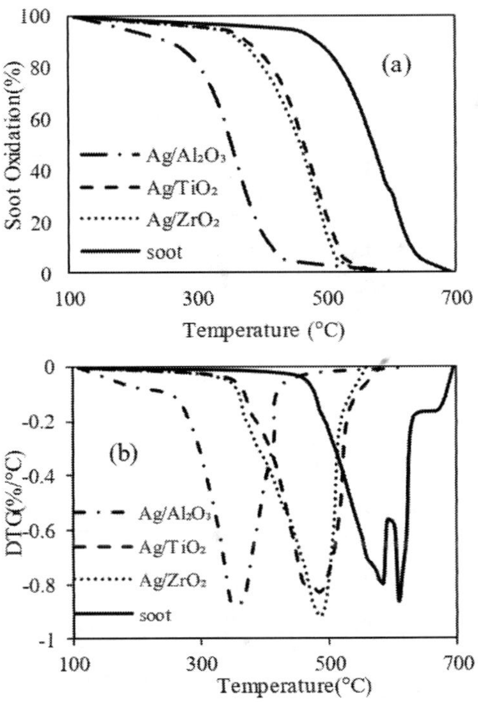

Fig. 1. (a) Thermogravimetric and (b) first derivative profiles of soot oxidation; soot/catalyst weight ratio of 1/5 and heating rate of 5°C/min.

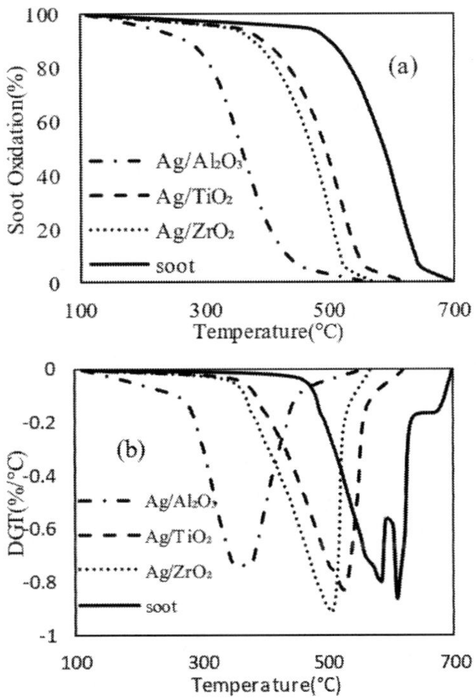

Fig. 2. (a) Thermogravimetric and (b) first derivative profiles of soot oxidation; soot/catalyst weight ratio of 1/5 and heating rate of 7.5°C/min.

The cases of Ag/TiO_2 and Ag/ZrO_2 showed a similar promotion in soot oxidation due to similar soot oxidation temperatures. Ag/ZrO_2 showed the peak of DTG at temperatures values were 485 °C, 508 °C, and 541 °C for heating rates of 5, 7.5, and 10, respectively (shows in Fig.1-3). Ag/TiO_2 showed the peak of DTG at temperature values were 485 °C, 528°C, and 528 °C for heating rates of 5 °C/min, 7.5°C/min, and 10°C/min, respectively (shows in Fig. 1-3). However, the comparison of Ag/TiO_2 and Ag/ZrO_2 showed that Ag/ZrO_2 had a better-promoting effect on oxidation because the temperature at peak of DTG in the case of Ag/ZrO_2 represented a lower temperature than the case Ag/TiO_2.

Fig.4-6 show T_{20}, T_{50} and T_{90} from the experiment had three heating rates which were 5 °C/min, 7.5 °C/min, and 10°C/min. According to Fig.4-6 explained the temperature of percenter of soot decomposition since it was oxidized by oxygen. The T_{20} explained the 20% of soot oxidation. T_{20} may be indicated as a light of temperature which was the start of soot oxidation. In case of Ag/Al_2O_3 presented excellent to reduce the light of temperature for soot oxidation from 523 °C in case of pure soot was reduced to 305 °C with heating rate 5 °C/min. The results at heating rate 7.5 °C/min and 10 °C/min showed to similar in case of 5 °C/min, they were T_{20}=310 °C and T_{20} = 342 °C with heating rates of 7.5 °C, and 10°C, respectively. In case of Ag/TiO_2 presented T_{20} were 417 °C, 438 °C, and 452 °C with heating rates of 5°C/min, 7.5°C/min, and 10 °C/min, respectively. The result from Ag/ZrO_2 showed T_{20} were 404 °C, 416 °C, and 437 °C with heating rates of 5 °C/min, 7.5°C/min, and 10 °C/min, respectively. The result in several cases in the experiment presented shows light of temperature for soot oxidation as follows: Ag /Al_2O_3< Ag /ZrO_2< Ag /TiO_2< pure soot. It showed the effect of metal oxides supporter to promote soot oxidation because light of temperature for soot oxidation was reduced from pure soot oxidation. In the case of soot oxidation from Ag/TiO_2 and Ag/ZrO_2. The T_{20} were similar values, it explained the light of soot oxidation which show similar activity in all the heating rates.

The soot oxidation performance was explained by T_{50} and T_{90}. They were temperature at soot decomposition at 50%(T_{50}) and 90%(T_{90}) from initial mass. The results of T_{50} in case of Ag/Al_2O_3 was shown in Fig.4-6. The T_{50} were 353 °C with a heating rate of 5 °C/min. In the case of heating rates were 7 °C/min and 10°C/min that they showed T_{50} = 359 °C and T_{90}= 398 °C, respectively. The T_{90} were 353 °C with a heating rate of 5 °C/min, 359 °C with 7 °C/min and 398 °C with 10 °C/min. The results of a silver catalyst supported by alumina showed a good assister to promote soot oxidation. As it reduced the T_{50} and T_{90} by half from the case of pure soot. The T_{50} and T_{90} in the case of a silver catalyst supported by titanium oxide is shown in Fig.4 - Fig.6.

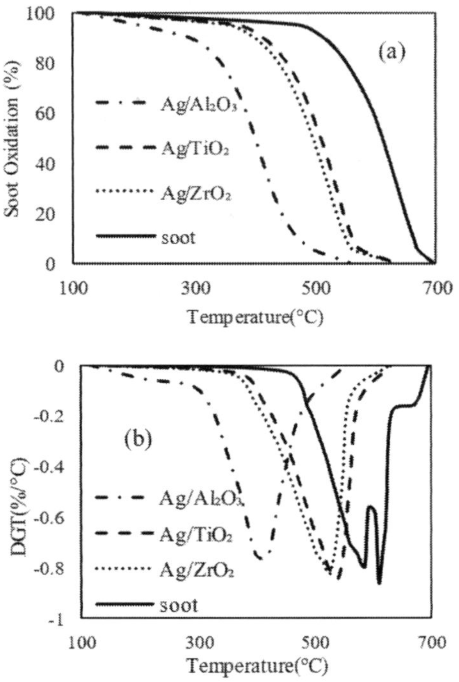

Fig. 3. (a) Thermogravimetric and (b) first derivative profiles of soot oxidation; soot/catalyst weight ratio of 1/5 and heating rate of 10°C/min.

In these cases, the T_{50} were 466 °C, 495 °C, and 509 °C with heating rates of 5 °C/min, 7.5°C/min, and 10 °C/min, respectively. T_{90} were 519 °C, 549 °C, and 562 °C with heating rates of 5 °C/min, 7.5°C/min, and 10 °C/min, respectively. In case of silver catalyst supported by zirconium oxide presented T50 = 461 °C with heating rate 5 °C/min, T_{50} = 473 °C with heating rate 7.5 °C/min, and T_{50} = 495 °C with heating rate 10 °C/min. The T90 from Ag/ZrO_2 were 519 °C, 549 °C, and 562 °C with heating rates of 5 °C/min, 7.5°C/min, and 10 °C/min, respectively. The results of T50 and T90 from Ag/TiO_2 and Ag/ZrO_2 showed a similar performance to promote soot oxidation. Because T50 and T_{90} in the case of Ag/ZrO_2 and Ag/TiO_2 were slightly different temperatures. It confirmed the similar activity in soot oxidation. The comparison of soot oxidation performance in several case showed as follows: Ag $/Al_2O_3$< Ag $/ZrO_2$< Ag $/TiO_2$< pure soot. From the results of silver catalyst supported by alumina showed high performance for soot oxidation from the result. Because of it presented to low T50 and T_{90} which compared with Ag/TiO_2, Ag/ZrO_2, and pure soot in all heating rates.

3.2 Activation energy

The results of the analysis for the determination of Ea used KAS method, it is necessary to find the slope of the plot of $\ln(\beta/T^{1.92})$ against 1/T for different heating rates at a constant α=0.1 to α=0.9 which showed soot decomposition.

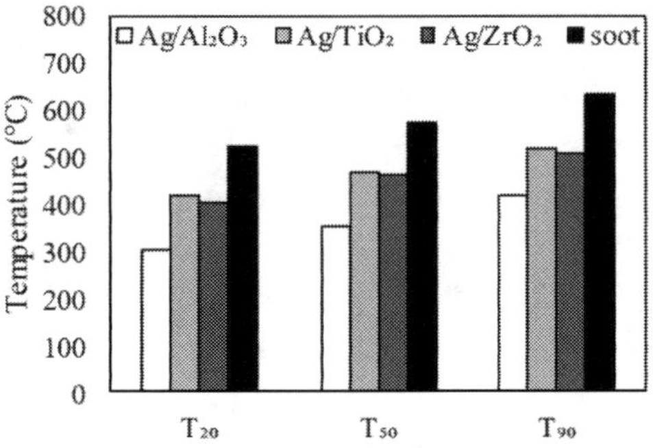

Fig. 4. T_{20}, T_{50}, and T_{90} of soot oxidation (heating rate5°C/min)

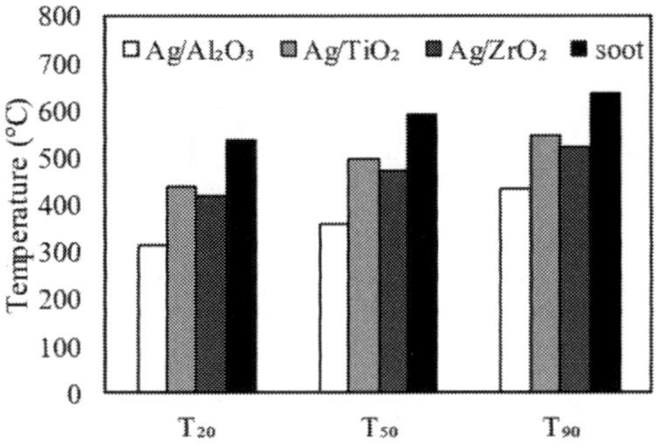

Fig. 5. T_{20}, T_{50}, and T_{90} of soot oxidation (heating rate7.5°C/min)

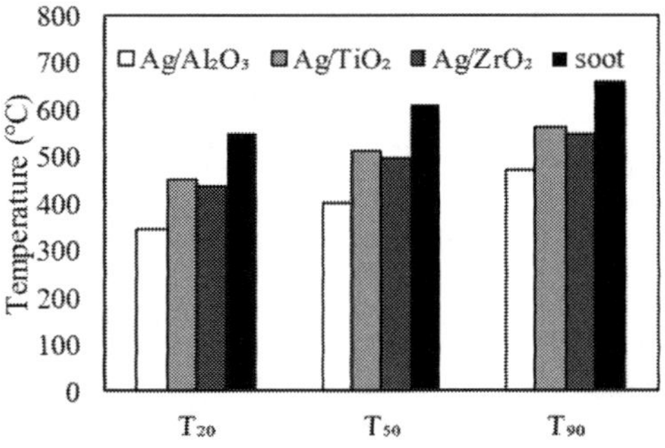

Fig. 6. T_{20}, T_{50}, and T_{90} of soot oxidation (heating rate 10°C/min)

The case of soot oxidation on the catalyst base on alumina supporter was presented in Fig. 7. it showed α for determine slope of the plot of ln(β/T1.92) against 1/T. From result of calculation in this condition made to activation energy were between 33-49 kJ/mol.Fig.8 and Fig.9 show the plot of $\ln(\beta/T^{1.92})$ against 1/T and in case of silver catalyst base on titanium oxide can be calculate the activation energy were between 64-79 kJ/mol and case of silver catalyst base on zirconium oxide can be calculate the activation energy were between 71-81 kJ/mol. In case pure soot, the plot of $\ln(\beta/T^{1.92})$ against 1/T is shown in Fig.10, can be fined the activation energy were between 103-128 kJ/mol. Fig.11 presents the comparison of the activation energy from soot oxidation in four condition. From results of activation energy were one to explain the performance of soot oxidation. The several cases of silver catalyst base on different metal oxides support showed different activation energy too. This result showed highest of activation energy in pure soot. However. soot loaded on silver catalyst supported by oxide metal. It showed to reduce the activation energy. In the best case of soot oxidation was a silver catalyst supported by alumina because the result of maximum rate oxidation reaction, T_{20}, T_{50}, T_{90}, and activation energy which the most reduced form pure soot oxidation. It attributed alumina well to promote soot oxidation activity. In case of silver catalyst base on titanium oxide and zirconium oxide showed similar of the result to reduce the activation energy, T20, T50, and T90 from the oxidation of pure soot.

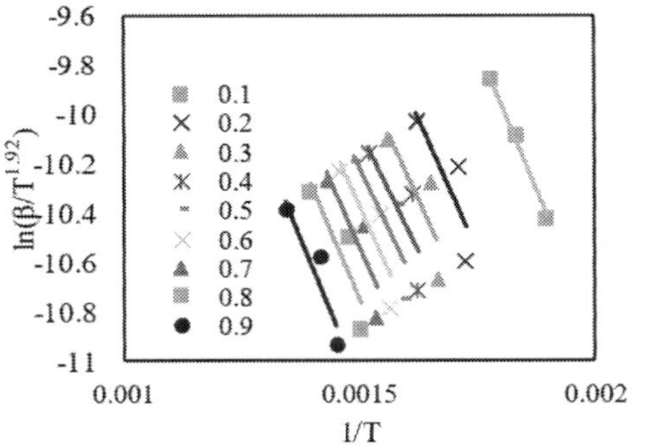

Fig. 7. The plot of $\ln(\beta/T^{1.92})$ against 1/T for different heating rates at a constant α=0.1 to α=0.9 in Ag/Al_2O_3.

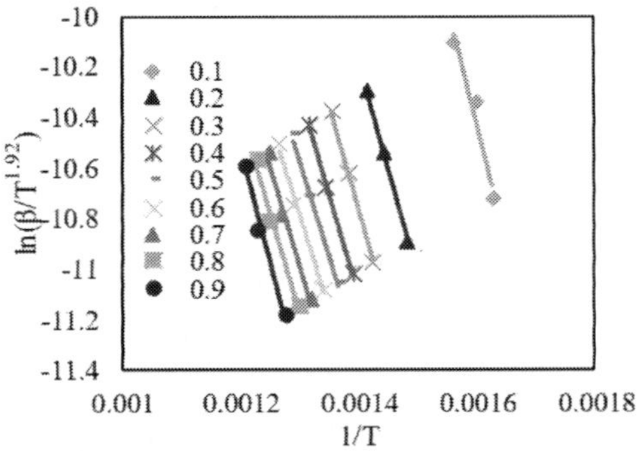

Fig. 8. The plot of $\ln(\beta/T^{1.92})$ against 1/T for different heating rates at a constant α=0.1 to α=0.9 in Ag/TiO_2.

4 Conclusion

The oxidation activity of silver catalyst supported on alumina (Al_2O_3), titanium oxide (TiO_2), and zirconium oxide (ZrO_2) were tested by the thermogravimetric analysis (TG) to compare soot oxidation performance. The result silver catalyst support by alumina, titanium, and zirconium. It was success to promoted the soot oxidation because the temperature of soot oxidation reaction, maximum of oxidation rate and activation

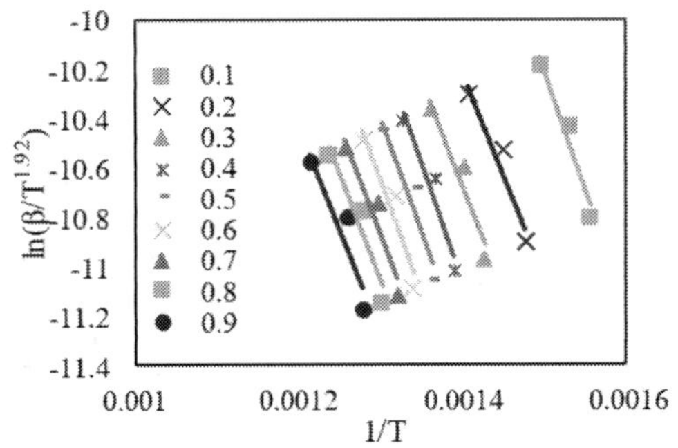

Fig. 9. The plot of $\ln(\beta/T^{1.92})$ against 1/T for different heating rates at a constant α=0.1 to α=0.9 in Ag/ZrO_2.

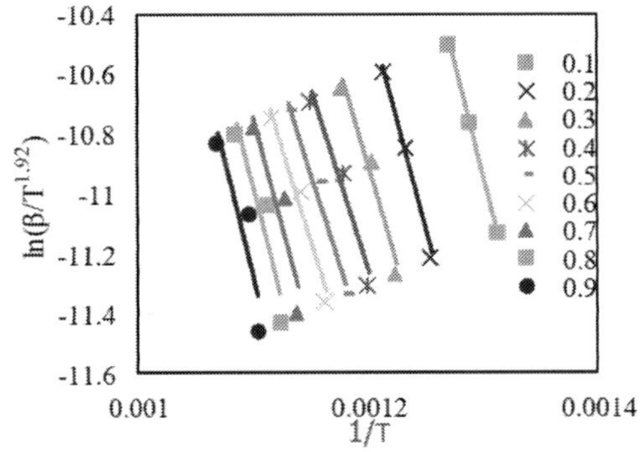

Fig. 10. The plot of $\ln(\beta/T^{1.92})$ against 1/T for different heating rates at a constant α=0.1 to α=0.9 in pure soot.

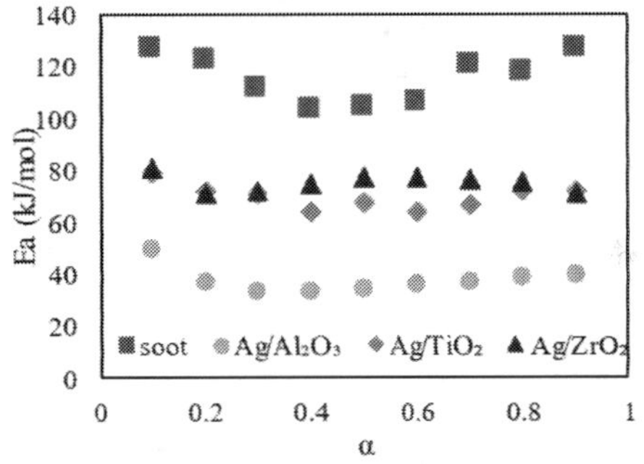

Fig. 11. the comparison of the activation energy from soot oxidation

energy were decrease from temperature in normal soot oxidation. However, the different support in metal oxide group showed the different soot oxidation activity because the activation energy was different which make to different soot oxidation. Ag/Al_2O_3 presented Ea = 33-49 kJ/mol, it showed good assister for soot oxidation. Ag/TiO2 presented Ea = 64-79 kJ/mol and Ag/ZrO_2 presented Ea = 71-81 kJ/mol that the rank of activation energy were similar values, it showed similar of the soot oxidation performance. In future work, the researchers will study the structural characterizations in several catalysts supported by the oxide metal such as specific surface area, crystal structure, and electric state of surface area, etc because some the characteristics of surface properties may assist to promote soot.

Acknowledgement

This research was supported the budget grant from KMUTNB-NSTDA High Performance Post Graduate Scholarship in Science and Technology (KMUTNB-NSTDA HiP Scholarship) and King Mongkut's University of Technology North Bangkok. The researcher would like to show my appreciation in being supported for this research project until it was accomplished very well.

References

1. S. Li *et al.*, Oxidative reactivity enhancement for soot combustion catalysts by co-doping silver and manganese in ceria, Applied Catalysis A: General., **570**, (2019):299–307,
2. S. N. K. Reddy and M. M. Wani, A comprehensive review on effects of nanoparticles-antioxidant additives-biodiesel blends on performance and emissions of diesel engine, Applied Catalysis A: General., **13**, 4(2020): 285–298
3. B. Sawatmongkhon *et al.*, Catalytic oxidation of diesel particulate matter by using silver and ceria supported on alumina as the oxidation catalyst, *Appl.* Applied Catalysis A: General., **574** , (2020): 33–40
4. K. Krishna, A. Bueno-López, M. Makkee, and J. A. Moulijn, Potential rare earth modified CeO2 catalysts for soot oxidation. I. Characterisation and catalytic activity with O2, Applied Catalysis B: Environmental., **75**, 3–4(2007):189–200, 2007,
5. J. Zokoe and P. J. McGinn, Catalytic diesel soot oxidation by hydrothermally stable glass catalysts, Chemical Engineering Journal, **262**, (2015): 68–77,
6. R. Matarrese, L. Castoldi, and L. Lietti, Oxidation of model soot by NO2 and O2 in the presence of water vapor, Chemical Engineering Science., **173**, 2(2017): 560–569
7. A. Bueno-López, Diesel soot combustion ceria catalysts, Applied Catalysis B: Environmental., **146**, (2014): 1–11
8. T. S. Nguyen, F. Morfin, M. Aouine, F. Bosselet, J. L. Rousset, and L. Piccolo, Trends in the CO oxidation and PROX performances of the platinum-group metals supported on ceria, Catalysis Today, **253**, (2015): 106–114
9. M. S. Chen, Y. Cai, Z. Yan, K. K. Gath, S. Axnanda, and D. W. Goodman, Highly active surfaces for CO oxidation on Rh, Pd, and Pt, Surface Science., **601**, 23(2007): 5326–5331
10. K. ichi Shimizu, H. Kawachi, and A. Satsuma, Study of active sites and mechanism for soot oxidation by silver-loaded ceria catalyst, Applied Catalysis B: Environmental., **96**, 1–2(2010): 169–175
11. L. Nossova, G. Caravaggio, M. Couillard, and S. Ntais, Effect of preparation method on the performance of silver-zirconia catalysts for soot oxidation in diesel engine exhaust, Applied Catalysis B: Environmental., **225**, (2018): 538–549
12. E. Aneggi, J. Llorca, C. de Leitenburg, G. Dolcetti, and A. Trovarelli,Soot combustion over silver-supported catalysts, Applied Catalysis B: Environmental., **91**, 1–2(2009): 489–498
13. L. Zeng, L. Cui, C. Wang, W. Guo, and C. Gong, Ag-assisted CeO2 catalyst for soot oxidation, Frontiers of Materials Science., **13**, 3(2019): 288–295
14. M. Machida, Y. Murata, K. Kishikawa, D. Zhang, and K. Ikeue, On the reasons for high activity of CeO2 catalyst for soot oxidation, Chemistry of Materials., **20**, 13(2008): 4489–4494
15. C. M. Álvarez-Docio, R. Portela, J. J. Reinosa, F. Rubio-Marcos, and J. F. Fernández,Pt mechanical dispersion on non-porous alumina for soot oxidation, *Catal. Commun.*, **140**, (2020) : 105999
16. M. J. Kim *et al.*, CeO2 promoted Ag/TiO2 catalyst for soot oxidation with improved active oxygen generation and delivery abilities, *J. Hazard. Mater.*, **384**, (2020): 121341
17. H. Shimokawa, Y. Kurihara, H. Kusaba, H. Einaga, and Y. Teraoka, Comparison of catalytic performance of Ag- and K-based catalysts for diesel soot combustion, Catalysis Today, **185**, 1(2012): 99–103
18. N. Zouaoui, M. Issa, D. Kehrli, and M. Jeguirim, CeO 2 catalytic activity for soot oxidation under NO / O 2 in loose and tight contact, Catalysis Today, **189**, 1(2012): 65–69
19. R. López-Fonseca, I. Landa, U. Elizundia, M. A. Gutiérrez-Ortiz, and J. R. González-Velasco,A kinetic study of the combustion of porous synthetic soot, Chemical Engineering Journal., **129**, 1–3(2007): 41–49
20. S. Vyazovkin, A. K. Burnham, J. M. Criado, L. A. Pérez-maqueda, C. Popescu, and N. Sbirrazzuoli, Thermochimica Acta ICTAC Kinetics Committee recommendations for performing kinetic computations on thermal analysis data, Thermochimica Acta, **520**, 1–2(2011): 1–19

Optimization of Microwave-Assisted Extraction of Residual Soybean Oil from Spent Bleaching Earth

Chanatip Dejkajorn[1], *Panawan* Suttiarporn[1], *Hussanai* Sukkathanyawat[1], *Kittisak* Wichianwat[1], *Saichon* Sriphan[1] *and Thanarak* Srisurat[1],*

[1]Faculty of Science, Energy and Environment, King Mongkut's University of Technology North Bangkok, Rayong Campus, Rayong 21120, Thailand

Abstract. Spent bleaching earth (SBE) which is generated from bleaching process is a valuable industrial waste of edible vegetable oil production because of residual edible vegetable oil absorbed. The residual oil in spent bleaching earth can be recovered and reused for application in the industries such as the production of biodiesel and lubricant. Currently, microwave-assisted extraction (MAE) technique is widely used because this method has a shorter extraction time and less solvent consumption when compared with traditional methods. In this study, MAE combined with solvent reflux was optimized using solvent screening experiments and response surface methodology (RSM) to obtain the highest yield of MAE extraction of residual soybean oil from spent bleaching earth. The extraction yield of residual soybean oil obtained from selected solvent were hexane-ethanol mixture (2:1 v/v, 10.19%) > hexane-ethanol mixture (1:1 v/v, 10.00%) >hexane-ethanol mixture (1:2 v/v, 9.98%) > hexane-ethanol mixture (1:3 v/v, 9.83%) > hexane-ethanol mixture (3:1 v/v, 8.59%) > hexane (8.17%) > acetone (7.73%). The regression coefficient (R-squared = 0.9852) expresses the accuracy of the regression and indicates the relationship between experimental data and predicted result, with high regression coefficient close to 1, which is remarkably desired. The experiment conditions for optimal MAE extraction of residual soybean oil from SBE were hexane-ethanol mixture (2:1 v/v), liquid to solid ratio 15.56:1 mL/g, extraction time 12.22 min, and microwave power 350 W. Under such conditions, the highest predicted value of the extraction yield of residual soybean oil was 10.43%.

Keyword. microwave-assisted extraction, residual soybean oil, spent bleaching earth, response surface methodology.

1 Introduction

To remove impurities, raw vegetable oils are purified. The bleaching process is one of the purification processes in which soap materials and other impurities in the oil are removed by absorbents. The main absorbents are completely activated earth and activated carbon. It is collectively referred to as bleaching soil. After the bleaching earth is inactivated by impurities and clogging of its pores, it becomes a spent bleaching earth (SBE). SBE is a flammable waste material that can cause a fire if not properly stored or treated [1]. The waste contains 20 to 40% oils, fats, and pigments by weight, depending on various factors such as the processing technology and the type of oil [2]. The presence of oil in the clay, the potential leakage of fats into the water path, and the possibility of spontaneous combustion contribute to problems, hazards, and safety concerns, when SBE is disposed of in landfills. Moreover, the oil in the SBE results in dangerous pollution, constitutes an economic loss, and damages the environment. Furthermore, due to the rapid growth of industries and the concomitant increase in the rate of waste material production, the disposal of SBE has become a potential problem in countries that produce this waste. The recovery of oil from SBE reduces the cost of SBE disposal significantly, and the recovered oil may be reused. Due to its aromatic qualities, it may be reused in the refining process.

Oil extraction from SBE is applied to a wide range of techniques, such as solvent extraction, enzyme-assisted aqueous extraction, and supercritical carbon dioxide extraction [3]. However, solvent extraction can extract a higher percentage of residual oil (21% W_{oil}/W_{SBE}) [4]. Soxhlet extraction with n-hexane solvent is commonly used for vegetable oils extraction. Because of its high yield and extraction efficiency, Soxhlet extraction is widely used to extract oil [3, 5]. Soxhlet extraction has longer extraction times, higher energy consumption, possible compound degradation, lower extraction rates due to lower mechanical pressure, and may have a negative environmental impact [6].

Microwave-assisted extraction (MAE) is a modern method for extracting essential oils, fats, and oils [7]. The application involves the direct action of waves on materials, which can convert a component of the absorbed electromagnetic energy to heat energy. Unlike traditional heating methods, which rely on conduction and convection, this causes a long extraction time. As a

*Corresponding author : thanarak.sciee.kmutnb.ac.th

result, thermal energy is lost to the environment. When compared to soxhlet extraction, this heating process can significantly minimize extraction time (typically less than 30 minutes) [8]. The principle of heating using a microwave is based upon its direct impact on polar samples and solvents [9]. Ethanol is produced by fermenting biological materials and is derived from renewable resources. Therefore, ethanol has been considered as a possible green solvent alternative to hexane because it has more polarity than hexane and has long been used for oil extraction in the production of sesame seed cake, rice bran, and radish seed. Moreover, ethanol has been offered as an environmentally friendly extraction solvent of plant oil because it produces higher extraction rates and uses less solvent [10].

In this context, it represents a challenge for the development of more efficient methods of extracting residual oil from spent bleached earth. Meanwhile, microwave technologies have proven to be effective at accelerating the extraction of solvent from matrices and using less solvent. Moreover, the residual oil that was removed from the spent bleaching earth can be used to produce biodiesel or consumed after a deodorized process [11]. On the other hand, after removing the residual oil from the spent bleached earth, SBE can be reused in the bleaching process again [12] or as heavy metal absorbents [13]. The objective of this research was to use microwave-assisted extraction (MAE), which is an efficient and environmentally friendly technique, and select an alternative solvent for residual soybean oil extraction, as well as to use the response surface methodology for variable optimization in the MAE process to achieve maximum extraction yield of residual oil. Then, the optimum residual soybean oil response obtained from MAE was compared with that of the soxhlet traditional technique in terms of oil yield profile.

2 Material and Methods

2.1 Materials

Spent bleaching earth (SBE) received from Sime Darby Oils Morakot Pub Co., Ltd., Thailand. Hexane, acetone, and ethanol, used as extraction solvents, and other chemicals are analytical-grade reagents. They were purchased from DUKSAN (ASIA Chemie Co.,Ltd., Thailand).

2.2 Selection of solvent in MAE

A 1,000 mL flat-bottom flask containing different solvents (150 mL) and spent bleaching earth (10 g) is fitted with a microwave-assisted extractor (EME2024MW, Electrolux Thailand Co., Ltd., Thailand). The heating was turned off after 10 minutes, and the remaining soybean oil was recovered from the solution with a rotary vacuum evaporator. The residual soybean oil was then dried in an oven at 105 °C until it reached a constant weight, removing any remaining solvent and moisture. Finally, the oil was weighed after cooling in a desiccator. The MAE yield of residual soybean oil was calculated as equation 1.

$$Y = (W_{oil}/W_{SBE}) * 100 \quad (1)$$

Where, Y is the MAE yield of residual soybean oil (%); W_{oil} is weight of oil extraction and W_{SBE} is the weight of SBE.

2.3 Oil extraction methods

2.3.1 Soxhlet extraction

The total residual soybean oil content of spent bleaching earth was determined by soxhlet extraction, by the Association of Official Analytical Chemists (AOAC). Briefly, 200 mL of hexane and 10 g of the spent bleaching earth were heated to reflux for 8 hours. The heating was then turned off, and the residual oil was recovered using a rotary vacuum evaporator. The residual soybean oil was then dried in an oven at 105°C until it reached a constant weight, removing any remaining solvent and moisture. Finally, the oil was weighed after cooling in a desiccator [14]. This method extracted total oil content from the SBE 1.13 ± 0.38 g (11.30% of the SBE by weight).

2.3.2 MAE-single factor experiments

Microwave-assisted extractor was performed at ambient in a 1,000 mL flat bottom flask. Various extraction time (5, 10, 15, 20, and 30 min), liquid to solid ratios (10:1, 15:1, 20:1, 25:1, and 30:1 mL/g), and microwave power (210, 280, 350, 420, and 490 W) were applied. To obtain consistent residual soybean oil yields, the mixtures were dried using a rotary evaporator to remove the solvent. The extracted oil is then dried in an oven to keep a constant residual soybean oil. All experiments were performed in triplicate. The extraction yield of residual soybean oil was assessed according to equation 1.

2.4 Optimization of MAE parameters by RSM

Operational conditions were optimized using response surface methodology (RSM) coupled with the Box-Behnken design (BBD) to further evaluate the main factors affecting the MAE process. Based on single-factor tests, extraction time (X_1: 5 – 15 min), liquid to solid ratios (X_2: 10:1 – 20:1 mL/g), and microwave power (X_3: 280 – 420 W) were used as the MAE independent variables (Table 1). Residual soybean oil yield (Y) was selected as a dependent variable. Extraction yields were obtained under different conditions according to the 15 runs of experimental design. The experiments were performed in triplicate. The second-order polynomial model was fitted to MAE experiment data. The quadratic equation was given as equation 2.

$$Y=b_0+b_1X_1+b_2X_2+b_3X_3+b_{11}X_1^2+b_{22}X_2^2+b_{33}X_3^2+b_{12}X_1X_2+b_{13}X_1X_3+b_{23}X_2X_3 \quad (2)$$

where, Y is the response variable; X_1, X_2, and X_3 represents the MAE independent variables; b_0 is the intercept; b_1, b_2, and b_3 are linear coefficients; b_{11}, b_{22}, and b_{33} are quadratic coefficients; b_{12}, b_{13}, and b_{23} are interaction coefficients. MAE conditions were optimized using the software Minitab 18 (trial version, Minitab Inc., State College, PA, USA).

2.5 Statistical analysis

The extraction of the residual soybean oil was performed three times. All experimental data is represented using mean and standard deviation. The ANOVA procedure was used to perform analysis of variance, as well as Duncan's multiple comparison test to determine significant differences ($p < 0.05$) among the means using SPSS (version 26, IBM Institute Inc., USA). Minitab 18 was used to operate RSM analysis. To determine the model's significance and the best conditions for extracting residual soybean oil from spent bleaching earth, an analysis of variance (ANOVA) was used. The significance level was set at $p < 0.05$.

Table 1. Levels of the independent variables in the Box-Behnken designs (BBD)

Independent variables	Code	Level		
		-1	0	1
Extraction time (min)	X_1	5	10	15
Liquid to solid ratio (mL/g)	X_2	10:1	15:1	20:1
Microwave power (W)	X_3	280	350	420

2.6 Free fatty acid analysis

Free fatty acid value in oil from SBE was estimated by the titration AOCS Official Method Ca 5a-40 [15]. A 250-mL Erlenmeyer flask was filled with extracted oil soluble in ethanol solvent at 60 °C, and phenolphthalein solution (3– 5 drops) was added to the sample solution and stirred. The solution was titrated with KOH 0.1 N drop by drop until the pink color appeared for 10 seconds, which was the endpoint. FFA content in fats and oils was calculated as a percentage. The free fatty acid was calculated as equation 3.

$$\%FFA = (V \times N \times 28.2) / W_{SBE} \quad (3)$$

Where, V is the alkali volume; N is the alkali normality and W_{SBE} is the weight of SBE.

3 Results and discussion

3.1 Selection of solvent in MAE

The effects of different solvents on the extraction yield of oil are show in Fig. 1. As alternative oil extraction options, hexane, acetone, ethanol, and hexane mixed ethanol at various volume ratios were considered, while all other factors were kept constant (10 g SBE with 150 mL solvent in a microwave for 10 min at power 350 W). Using mixture of hexane and ethanol (2:1 v/v) gave a higher extraction yield of oil shown in Fig. 1. This is due to the fact that the type of solvent employed in MAE is important. The microwave absorption capacity of a solvent is determined by type, and the solubility of the target molecule should be considered when selecting a solvent for MAE. Although ethanol has a high polarity which always allows microwave radiation to be absorbed, residual oil has limited solubility. Hexane has a weak polarity, but its increased residual oil solubility [16]. The effect of a mixture of hexane and ethanol on oil extraction yield was investigated in order to improve oil extraction yield. The extraction yield of oil obtained with a mixture of hexane and ethanol was higher than that achieved with a single solvent, with the maximum extraction yield of oil obtained by a 2:1 volume ratio of hexane and ethanol. According to the previous study, using a mixture of hexane and ethanol in the MAE of peony seed oil increased oil yield significantly when compared to using only one solvent [17]. As a result, a mixture of hexane and ethanol (2:1 v/v) was utilized for further testing to maximize the MAE.

3.2 Optimization of single-factor extraction conditions

3.2.1 Effect of extraction time on the oil yield

One of the important variables influencing the extraction yield obtained from the MAE process is extraction time. Over-exposure to microwave radiation causes the extraction system to overheat, as well as the lengthened process time might raise the production cost [18]. The impact of extraction time on the MAE yield of residual soybean oil from SBE is shown in Fig. 2a. It can be observed that the extraction time increases from 5 to 10 minutes, the oil yield increases rapidly. It could be due to the rapid dissolution of oil solutes emanating from the surface of SBE. The highest oil yield was 10.43 ± 0.06%, which was observed at 10 minutes. After that, the oil yield remained constant over time.

3.2.2 Effect of liquid to solid ratio on the oil yield

The liquid to solid ratio has a significant impact on the oil MAE process. The extraction of oil yield from SBE is influenced by solvent volume, which can affect residual soybean oil yield in two different ways. Fig. 2b. shows the yield of oil from SBE extracted using various solvent to SBE ratios from 10:1 to 30:1 mL/g. With a liquid to solid ratio of 15:1 mL/g (10.33 ± 0.05%), the extraction yield of residual soybean oil from SBE was found to be the highest and remained constant as the liquid to solid ratio was increased. This was significant because it meant that, while a certain amount of solvent was

https://doi.org/10.1051/e3sconf/202130201009

desirable for SBE oil extraction, excessive solvent might result in higher costs [19].

3.2.3 Effect of microwave power on the oil yield

The quantity of energy transmitted per unit of time is measured as microwave power. Microwave power was used to heat the bleaching earth. The oil and matrix were disrupted by its driving force, allowing the oil to diffuse out and dissolve in the solvent. When microwave radiation is higher, it causes the excessive temperature in the extraction process. In Fig. 2c., when the microwave power was raised from 280 W to 350 W, the oil extraction yield increased from 9.10 ± 0.31% to 10.02 ± 0.03%, followed by a decrease in the oil yield at 420 W and 490 W. An increase in microwave power usually improves and heats, resulting in higher oil yields. On the other hand, increased microwave radiation power causes the extraction system to overheat, causing the solvent to evaporate [20].

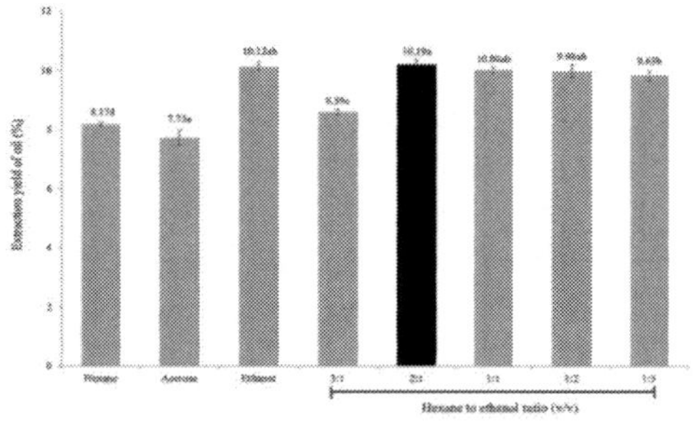

Fig. 1. Effect of solvent type on extraction yield of residual soybean oil.

3.3 Optimization of parameters by RSM

A Box-Behnken design was applied to develop a correlation between the extraction time (X_1), solvent to SBE ratio (X_2), and microwave power (X_3) to optimize the yield of residual soybean oil. Table 2 shows the variations in the corresponding coded values of three MAE parameters, as well as the MAE yield response based on experimental runs of oil extraction. The extraction yield of residual soybean oil was in the range of 5.52 to 10.40% in the experiments. The polynomial equation of MAE yield with the coefficients of the full regression model equation and their statistical significance were investigated using the software Minitab 18 for multiple regression analysis. The final model for extraction yield of oil in terms of coded values is presented in equation 4. The positive symbol in front of the terms expressed a synergistic effect, while the negative symbol expressed an antagonistic effect.

$$Y=10.060 + 1.724X_1 + 0.097X_2 + 0.413X_3 - 1.889X1_2 - 0.334X_2^2 - 0.963X_3^2 - 0.046X_1X_2 + 0.234X_1X_3 + 0.008X_2X_3 \quad (4)$$

Where Y is the extraction yield of oil, X_1 is the extraction time, X_2 is the liquid to solid ratio, and X_3 is the microwave power, respectively.

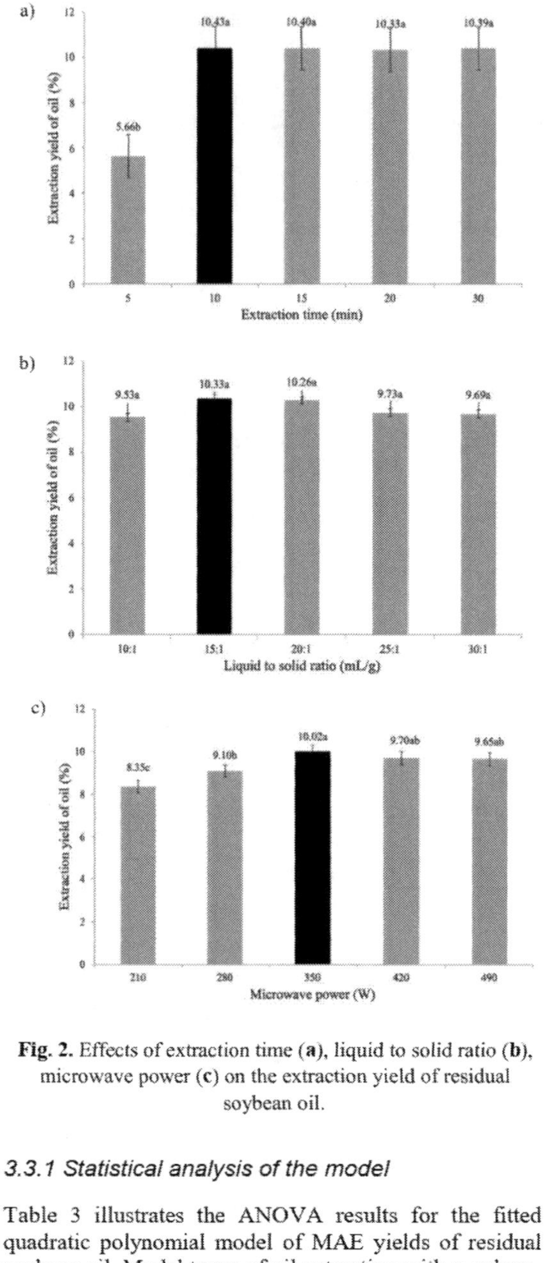

Fig. 2. Effects of extraction time (**a**), liquid to solid ratio (**b**), microwave power (**c**) on the extraction yield of residual soybean oil.

3.3.1 Statistical analysis of the model

Table 3 illustrates the ANOVA results for the fitted quadratic polynomial model of MAE yields of residual soybean oil. Model terms of oil extraction with p-values less than 0.05 are significant: in this model, X_1, X_3, X_{12} and X_{23} are significant terms. The lack-of-fit p-value was 0.09 (p-value > 0.05), indicating that the model equation for oil extraction is appropriate for predicting residual soybean oil yield from SBE. For residual soybean oil yield, the coefficient of determination (R^2) was 0.985, suggesting good model effectiveness: the predicted R^2 of 0.822 agrees with the study R^2 of 0.958 [21]. As a result, the regression equation can be used to decide which one

https://doi.org/10.1051/e3sconf/202130201009

the optimum condition is for the MAE process of residual oil from spent bleaching earth.

3.3.2 Analysis of the contour plot and response surface

The response surfaces of MAE oil yield can be plotted as three-dimensional plots and contour plots, as shown in Fig. 3-5., which shows the response surface for oil extraction yield as a function of two parameters with a fixed third variable. The interaction of extraction time (X_1) and solvent to SBE ratio (X_2) on oil extraction yield is shown in Fig. 3. The oil yield increased to its maximum when the extraction time was between 10 and 15 minutes and the liquid to solid ratio was between 10:1 and 20:1 mL/g, as shown by the quadratic surface. However, increased energy losses come at a cost, and a large solvent volume would make it difficult to remove the solvent and waste it, so it was critical to maintain a proper extraction time and liquid to solid ratio [22]. Fig. 4 illustrates the surface plot and contour plot for the interactive effect of solvent to SBE ratio (X_2) and microwave power (X_3) on oil extraction yield, liquid to solid ratio and microwave power provided obvious secondary effects, and oil yield increased to the maximum when the solvent to SBE ratio was between 13:1 and 18:1 mL/g and microwave power was between 350 and 390 W. The surface plot and contour plot for the interactive effect of extraction time (X_1) and microwave power (X_3) on the extraction yield of oil can be seen in Fig. 5.

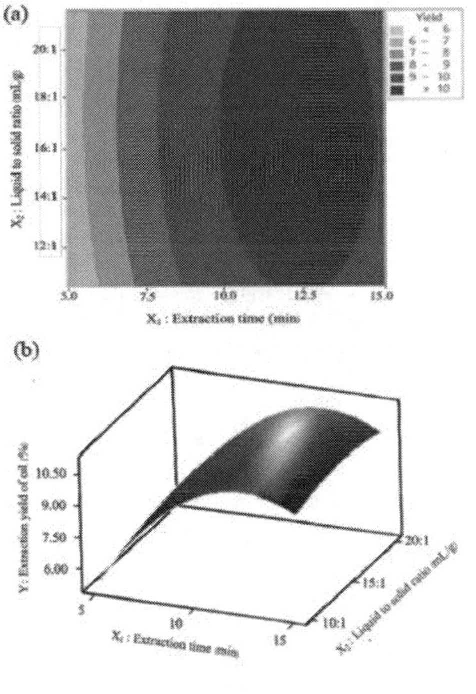

Fig. 3. Contour (a) and 3D surface (b) plots for the effects of extraction time and liquid to solid ratio on the yield of oil.

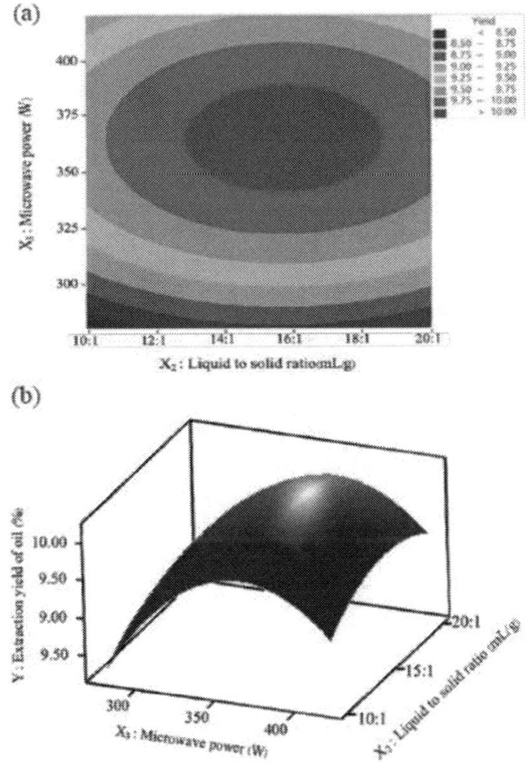

Fig. 4. Contour (a) and 3D surface (b) plots for the effects of liquid to solid ratio and microwave power on the yield of oil.

3.4 Optimal conditions and verification of the models

One of the main goals of this research was to optimize conditions for extracting residual soybean oil from SBE The extraction oil yield was found to be significantly affected by extraction time (X_1) and microwave power (X_3). In the range of extraction time (10 – 15 mins) and microwave power (325 – 400 W), the highest extraction oil yield was obtained when both variables were increased. generated during the purification of edible oil. Using Minitab 18, the fitted model was optimized to select the optimal conditions for extracting oil from SBE using mixed solvents (hexane: ethanol 2:1 v/v). As shown in Table 4, the maximum extraction oil yield was 10.43% (actual value) for mixture of hexane and ethanol (2:1 v/v) with an extraction time of 12.22 minutes, a solvent to SBE ratio of 15.56 mL/g, and a microwave power of 350 W, For the MAE of residual soybean oil, the second-order polynomial model suggested a maximum oil yield of 10.46%. The desirable value of 1.000 was used to measure the optimum conditions and predicted value. The verification of the oil yield model is shown in Table 4. with a 0.286% error, the actual oil yield was close to those predicted by the models, according to the validated results. This indicated that the defined response model was accurate at predicting oil yield.

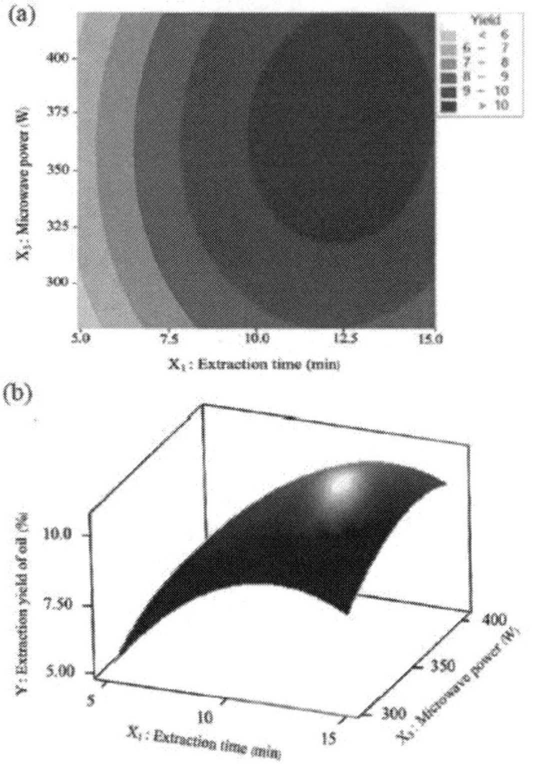

Fig. 5. Contour (a) and 3D surface (b) plots for the effects of extraction time and microwave power on the yield of oil.

3.5 Free fatty acid analysis

Table 5 shows the properties of residual oil obtained MAE process. The amount of free fatty acids present, which are products of oil and fat hydrolysis. Oils with a high content of free fatty acids are prone to oxidation and odor. The extraction of residual oils with ethanol has a darker color. It contains a lot of free fatty acids. While the residual oil extracted from non-polar solvent (hexane) are yellow, and contains low free fatty acids [14, 23]. The residual soybean oil, on the other hand, must be deodorized before it can be consumed. To reduce the amount of free fatty acids to less than 1% [24]. It must be done if it is to be used as a raw material for biodiesel production. Before entering the transesterification process, esterification is used to reduce the amount of free fatty acids to less than 2% [25].

3.6 Comparison of MAE with the Soxhlet method

The comparison of yield profiles of residual soybean oil extracted using the MAE and soxhlet processes is presented in Fig. 6. At a five-minute, MAE extraction yields greater than soxhlet extraction (5.62% and 1.91%, respectively). This is due to the fact that microwave use a more efficient heat current. Microwave, unlike conventional conductive heating technology, may heat all samples practically simultaneously and at a faster rate [26]. The residual soybean oil yield from SBE increases with increasing microwave irradiation time until the plateau or equilibrium is reached. The residual soybean oil from 9.37 ± 0.10% extracted SBE was recovered in 15 minutes by MAE, whereas the soxhlet requires 45 minutes (8.83 ± 0.11%) or more to reach equilibrium. These findings show significant time savings.

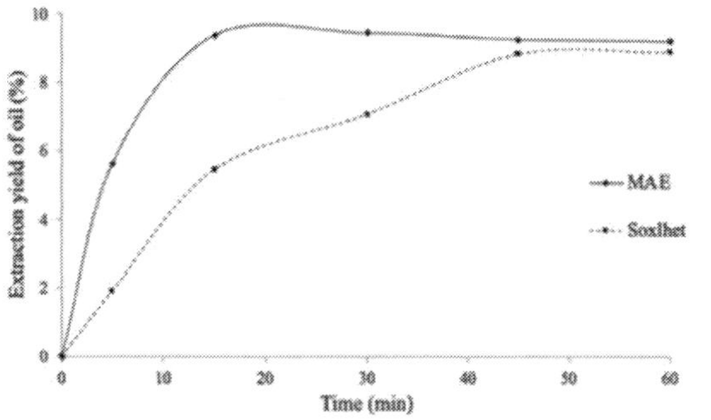

Fig. 6. Comparison of the yield profiles of microwave-assisted method (MAE) and soxhlet method.

4 Conclusion

From the selection of solvents, the mixture of hexane and ethanol (2:1 v/v) is a high extraction yield of oil. The optimal condition for oil extraction was extraction time 12.22 min, liquid to solid ratio 15.56:1 mL/g, and microwave 350 W, which has the amount of extraction yield of oil is 10.43% (92.30%). When comparing microwave-assisted extraction to the soxhlet extraction method, it was discovered that microwave-assisted extraction takes less time. The residual soybean oil extracted from spent bleaching earth can be recovered and reused in applications such as biodiesel and lubricants, as well as the reuse of spent bleaching earth in the bleaching process. It can also be utilized as a heavy metal absorbent.

Acknowledgements

This research was funded by Faculty of Science, Energy and Environment, King Mongkut's University of Technology North Bangkok. Contract no. SCIEE-KMUTNB-64-005. The authors gratefully thank to Sime Darby Oils Morakot Pub Co., Ltd., Thailand for supporting spent bleaching earth.

Table 2. MAE yield responses of BBD.

Runs	Extraction Time (X_1, min)	Liquid to solid ratio (X_2, mL/g)	Microwave power (X_3, W)	Extraction yield of oil (Y, %)
1	1(15)	1(20)	0(350)	9.64±0.31
2	-1(5)	0(15)	1(420)	5.79±0.11
3	0(10)	1(20)	1(420)	9.47±0.37
4	1(15)	0(15)	-1(280)	8.15±0.20
5	1(15)	0(15)	1(420)	9.35±0.74
6	0(10)	-1(10)	1(420)	8.97±0.05
7	-1(5)	-1(10)	0(350)	5.94±0.29
8	0(10)	-1(10)	-1(280)	8.06±0.04
9	1(15)	-1(10)	0(350)	9.83±0.23
10	0(10)	0(15)	0(350)	10.40±0.11
11	0(10)	0(15)	0(350)	9.90±0.91
12	0(10)	0(15)	0(350)	9.87±0.13
13	0(10)	1(20)	-1(280)	8.53±0.01
14	-1(5)	0(15)	-1(280)	5.52±0.03
15	-1(5)	1(20)	0(350)	5.93±0.04

Table 3. ANOVA statistics for response surface quadratic model of MAE yields of residual soybean oil.

Source	Sum of squares	df	Mean square	*F*-value	*p*-value	Remarks
Model	41.1776	9	4.5753	37.10	0.000	Significant
Linear	25.2266	3	8.4089	68.19	0.000	Significant
X_1	23.7843	1	23.7843	192.86	0.000	Significant
X_2	0.0753	1	0.0753	0.61	0.470	
X_3	1.3670	1	1.3670	11.09	0.021	Significant
Square	15.7228	3	5.2409	42.50	0.001	Significant
X_1^2	13.1731	1	13.1731	106.82	0.000	Significant
X_2^2	0.4130	1	0.4130	3.35	0.127	
X_3^2	3.4269	1	3.4269	27.79	0.003	Significant
2-Way interaction	0.2282	3	0.0761	0.62	0.633	
X_1X_2	0.0086	1	0.0086	0.07	0.802	
X_1X_3	0.2193	1	0.2193	1.78	0.240	
X_2X_3	0.0003	1	0.0003	0.00	0.963	
Error	0.6166	5	01233			
Lack-of-fit	0.4385	3	0.1462	1.64	0.400	Not significant
Pure error	0.1781	2	0.0891			
Total	41.7942	14				
$R^2 = 0.985$, adj $R^2 = 0.958$, pred $R^2 = 0.822$						

Table 4. The optimal condition of MAE process and verification of the model.

Solvent	Predicted)%(	Actual)%(	Optimal conditions			Error)%(
			Extraction time)min(	Liquid to solid ratio)mL/g(	Microwave power)W(	
Hexane to ethanol ratio)2:1 v/v(	10.46	10.43	12.22	15.56:1	350	0.286

Table 5. Properties of soybean and extracted residual oil.

Property	Soybean oil	Oil extracted from SBE (soybean oil)	Oil extracted from SBE (soybean oil)	
			hexane	H: E (2:1 v/v)
Color	Yellow	Dark brown	Yellow-brown	Dark brown
Free fatty acid (%)	< 1	24.1	24.19	42.18
Reference	[20]	[21]	This Work	

References

1. Hoy, Y.H. Edible Fats and Oils, (2000): 211-257
2. Z. Werner, Spent bleaching earth- practical solutions, INFORMS Journal on Applied Analytics, **5**, 1375 (1994): 1375-1377
3. J. Zhong, Y. Wang, R. Yang, X. Liu, Q. Yang, X. Qin, The application of ultrasound and microwave to increase oil extraction from Moringa oleifera seeds, Industrial Crops and Products, 120, 1 (2018): 1-10
4. L.S. Kheang, C.S. Foon, C.Y. May, M.A. Ngan, A study of residual oils recovered from spent bleaching earth: their characteristics and applications, American Journal of Applied Sciences, 3, 2063 (2006): 2063-2067
5. F. Anwar, U. Rashid, Physico- chemical characteristics of Moringa oleifera seeds and seed oil from a wild provenance of Pakistan, Pak. Journal of Botany, 39, 1443 (2007): 1443-1453
6. D. B. Nde, D. Boldor, C. Astete, P. Muley, Z. Xu, Oil extraction from sheanut (Vitellaria paradoxa Gaertn CF) kernels assisted by microwaves, Journal of food science and technology, 53, 1424 (2016) : 1424-1434
7. P. Cintas, E. Calcio- Gaudino, G. Cravotto, Pharmaceutical and nutraceutical compounds from natural matrices, Microwave-assisted extraction for bioactive compounds, 1, 181 (2012): 181-206
8. C.W. Huie, A review of modern sample-preparation techniques for the extraction and analysis of medicinal plants, Analytical and bioanalytical chemistry, 373, 23 (2002): 23-30
9. M. Letellier, H. Budzinski, L. Charrier, S. Capes, A. Dorthe, Optimization by factorial design of focused microwave assisted extraction of polycyclic aromatic hydrocarbons from marine sediment, Fresenius' journal of analytical chemistry, 364, 228 (1999): 228-237
10. M.C. Capellini, L. Chiavoloni, V. Giacomini, C.E. Rodrigues, Alcoholic extraction of sesame seed cake oil: Influence of the process conditions on the physicochemical characteristics of the oil and defatted meal proteins, Journal of Food Engineering, 240, 145 (2019): 145-152
11. B. P. Lim, G. P. Maniam, S. A. Hamid, Biodieselfrom adsorbed waste oil on spent bleaching clay using CaO as a heterogeneous catalyst. European Journal of Scientific Research. 33, 2 (2009) : 347-357.
12. M. Shahi, M. R. Sabour, A. Amiri, Cleaning spent bleaching clay through using solvent extraction method and RSM statistical approach. Cumhuriyet Üniversitesi Fen-Edebiyat Fakültesi Fen Bilimleri Dergisi, 36, 7, (2015): 23-40.
13. P. E. Dim, M. Termtanun, Treated Clay Mineral as Adsorbent for the Removal of Heavy Metals from Aqueous Solution, Applied Science and Engineering Progress, Online First (2021): 1-14.
14. A. Al. Zahrani, Y. Alhamed, Oil removal from spent bleaching clay by solvent extraction, Journal of Environmental Science & Health Part A, 35, 1577 (2000): 1577-1590
15. Official Methods and Recommended Practices of the American Oil Chemists' Society, 4th edn., edited by D. Firestone, American Oil Chemists' Society, Champaign (1989): Ca 5a-40
16. H. Y. Zhou, C. Z. Liu, Microwave- assisted extraction of solanesol from tobacco leaves, Journal of Chromatography A, 1129, 135 (2006): 135-139
17. X. Sun, W. Li, J. Li, Y. Zu, C. Y. Hse, J. Xie, X. Zhao, Process optimisation of microwave- assisted extraction of peony (Paeonia suffruticosa Andr.) seed oil using hexane– ethanol mixture and its characterisation, International Journal of Food Science & Technology, 51, 2663 (2016) 2663-2673
18. O. R. Alara, N. H. Abdurahman, C. I. Ukaegbu, N. H. Azhari, Vernonia cinerea leaves as the source of phenolic compounds, antioxidants, and anti-diabetic activity using microwave- assisted extraction technique, Industrial Crops and Products, 122, 533 (2018): 533-544
19. B. Yang, M. Zhang, H. Weng, Y. Xu, L. Zeng,

Optimization of ultrasound assisted extraction (UAE) of kinsenoside compound from Anoectochilus roxburghii (Wall.) lindl by response surface methodology (RSM), 25, 193 (2020): 193

20. Q. Chen, W. Dong, C. Wei, R. Hu, Y. Long, Combining integrated ultrasonic- microwave technique with ethanol to maximise extraction of green coffee oil from Arabica coffee beans, Industrial Crops and Products, 151, 112405 (2020) : 112405
21. K. Afshari, V. Samavati, S. A. Shahidi, Ultrasonic-assisted extraction and in- vitro antioxidant activity of polysaccharide from Hibiscus leaf, International journal of biological macromolecules, 74, 558 (2015): 558-567
22. T. J. Mason, F. Chemat, M. Vinatoru. The extraction of natural produxts using ultrasound or microwaves, Current Organic Chemistry, 15, 2 (2011): 237-247
23. C. Lee, C. Seng, K. Liew, Solvent efficiency for oil extraction from spent bleaching clay, Journal of the American Oil Chemists' Society, 77, 1219 (2000) : 1219-1223
24. I. H. Rukunudin, P. J. White, C. J. Bern, and T. B. Bailey, A Modified Method for Determining Free Fatty Acids from Small Soybean Oil Sample Sizes, Journal of the American Oil Chemists Society, 75, 5 (1998): 563-568
25. Y. P. Huang, J. I. Chang, Biodiesel production from residual oils recovered from spent bleaching earth, Renewable Energy, 35, 1 (2010):269-274
26. B. Kaufmanm, P. Christen, Recent extraction techniques for natural products: microwave-assisted extraction and pressurised solvent extraction, Phytochemical Analysis: An International Journal of Plant Chemical and Biochemical Techniques, 13, 2 (2002): 105-113his research was funded by Faculty of Science, Energy and Environment.
27. A. Zamora, Fatty acid composition of some common edible fats and oils, Bethesda, MD: Scientific Psychic, (2005).
28. Y.P. Huang, J.I. Chang, Biodiesel production from residual oils recovered from spent bleaching earth, Renewable Energy, 35, 269 (2010): 269-274

The Prototype of Non-thermal Plasma After treatment System for Simultaneous Reduction of Nitrogen Oxide Emission in Flue Gas

Dararat Laohalertdecha[1], *Kampanart* Theinnoi[1,2,*] and *Sak* Sittichompoo[3]

[1]College of Industrial Technology, King Mongkut's University of Technology North Bangkok, 1518 Pracharat 1 Road, Wongsawang, Bangsue, Bangkok 10800, Thailand.
[2]Research Centre for Combustion Technology and Alternative Energy (CTAE), Science and Technology Research Institute, King Mongkut's University of Technology North Bangkok, Bangkok 10800, Thailand.
[3] Mechanical Engineering, School of Engineering, University of Birmingham, Edgbaston, Birmingham B15 2TT, United Kingdom.

Abstract. Nowadays, global warming is the main environmental problems all over the world. The air pollutants mainly from the burning of fossil fuels and coal in power plants, transportation, and automobiles. There are release major point emission of the atmosphere. The nitrogen oxides are the most relevant for air pollution that contribute to the formation of photochemical smog and acid rain. Numerous methods have been studied to eliminate the nitrogen oxides such as the use low-nitrogen fuels technology, the selective catalytic reduction (SCR), wet scrubbing. The aim of this research is investigated non-thermal plasma (NTP) techniques offer an innovation to eliminate both nitrogen oxide (NOx) and soot emissions from combustion. This study is used to selectively transfer input electrical energy to electrons without expending this in heating the entire gas flow which creates free radicals in the flue gases. The simulated flue gas from combustion process is applied to the system. The results showed that the prototype of nonthermal plasma system is shown the highly efficient of NOx removal was achieved. However, the optimised of NTP operating conditions are required to enhance the NOx reduction activities.

Keyword. Non-Thermal plasma (NTP), Nitrogen Oxide, Flue Gas, Emissions

1 Introduction

Nitrogen oxides are toxic emissions produced from diesel combustion and consist of nitric oxide (NO) and nitrogen dioxide (NO_2). The nitrogen oxides from combustions are mainly air pollution generated by fuel combustion from stationary and mobile sources in the atmosphere. NO_x emission can cause a series of health and problems such as eye and throat irritation and headache and environment issues such as toxic chemical smog and acid rain [1]. The post combustion technologies, selective catalytic reduction (SCR) has become dominant in controlling NOx emission from industries i.e., utility boilers and industrial furnace stacks [2-3]. However, there are many improvements aimed to reducing the SCR systems cost have been achieved, due to the space requirements and operating cost are still high cost [4]. The generation of ammonia emissions and the fouling of equipment with ammonium sulphate.

Given the health of diesel emission, power plant and combustion, there is an explicit to regard novel technologies to remove the tailpipe emissions. The highly traffic jam in urban areas is a big main problem of Nitrogen Oxide [5-6]. Thus, Non-thermal plasma (NTP) is an alternative chance for after treatment application to control emission [7]. Plasma is term that used to identify to fourth state of matter apart from the three well known states [8]. The three states are the state of matter that temperature increases which molecules are enormous energy and break down form solid, liquid, gas into plasma. Stage of plasma is more likely gaseous which its atoms and other free moving charge disassociated and collision between charges is observed. Commonly, plasma is involved electric and very conductivity greater than metal. There are two type of plasma that are thermal plasma and non-thermal plasma. Thermal plasma is normally affected very high power and reaction temperature from 10,000K to 100,000K. NTP offers lower implementation cost for system as it has better species selective characteristic with higher energy conversion efficiency. A novel power supply is used, which enables high voltage and high frequency operation for optimized electrical power input into the plasma. NTP can be operated at room temperature and atmospheric pressure [9].

Among the emerging technologies for post-combustion treatment, non-thermal plasma (NTP) is an advance oxidation method. During the NTP process, many active radicals, such as O, OH and O_3 are generated. These oxidize gaseous pollutants such as volatile organic compound (VOC), NOx and particulate matter (PM) at room temperature and atmosphere pressure [10]. NTP

* Corresponding author: kampanart.t@cit.kmutnb.ac.th

holds the advantages high oxidation efficiency, low floor area, no chemical addition, low initial cost, low temperature and at atmospheric pressure, and energy consumption [11]

Non-thermal plasma (NTP) as a useful method for NOx removal has been investigated in many years. It is almost primary pollution and has a good application prospect. This method is used to selectively transfer input electrical energy to the electrons without heating temperature the whole gas flow which generates free radicals through collisions and supports the preferred chemical changes in the exhaust gases. The created active species react with the pollutant molecules and followed by decomposition of pollutants.

This work is focused on the design of prototype of nonthermal plasma reactor for nitrogen oxide reduction in flue gas. The aim of the study is to obtain a high influence of applied voltage and electrode configuration parameters on the optical and electrical performance on the plasma reactor. In addition, the electrode configuration of plasma reactors is also investigated.

2 Experimental Apparatus and Procedure

The prototype NTP reactor was designed and developed in laboratory scale. The experiment schematic diagrams as shown in Fig.1 is composed of NTP reactor, a high voltage power supply, a high voltage probe, a digital oscilloscope, and a flue gas analyser. The simulated flue gas was feed under atmospheric pressure to nonthermal plasma (NTP) reactor which gas flow rate was controlled by a digital gas mass flow controller. Then the supernatant was analysed through flue gas analyser.

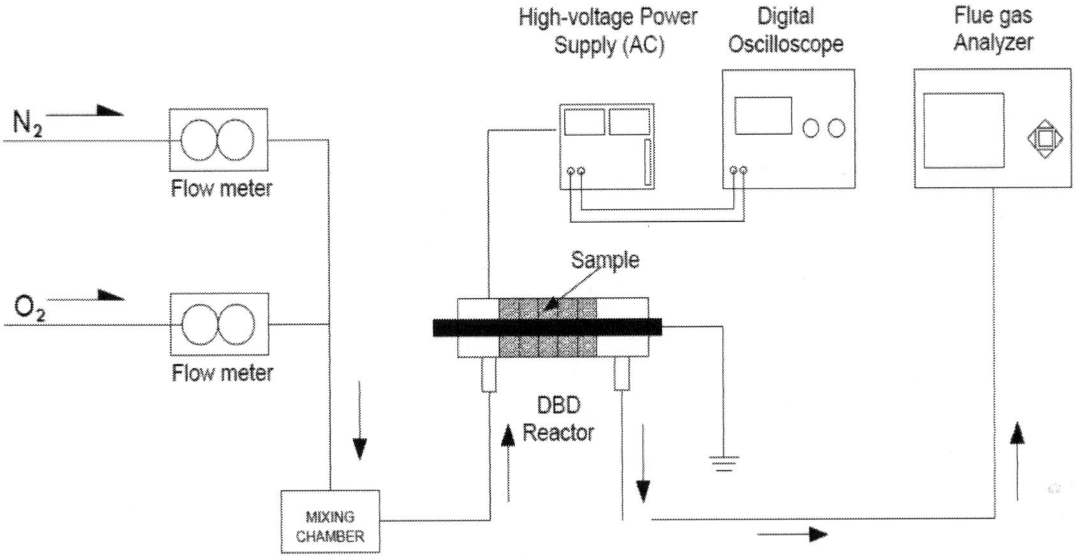

Fig. 1. Schematic diagram of the experimental setup for the non-thermal plasma Non-thermal plasma (NTP).

2.1. Experiment system and Conditions

A commercial grade nitrogen gas and oxygen-based nitrogen with a purity of 99.5 % and nitrogen oxide gas in nitrogen with 500 ppm concentration were use as simulated flue gas from combustion. The total simulated gas flow rate was controlled about 10 L/min via a mass flow controller to keep resident time constant. The simulated flue gas was passed in to NTP reactor. The flue gas was discharged by a high voltage (HV) power supply and plasma was generated at the same time. It flowed out of the reactor and according to the flue gas analyser's measurement. Then the plasma power source is turned on to generate NTP. The output power of the power source was adjusted by input voltage and input current. And the concentrations of the reactor outlet gas at different powers were monitored. The input voltage of the plasma power source was controlled by the frequency adjustment knob of the plasma power source.

2.2 Non-thermal plasma (NTP) reactor

The non-thermal plasma reactor was designed in length 200 mm and width 100 mm as shown in Fig.2 (a) and (b). It consists of 40, 20, 10, 5 copper sheets of electrodes that was connected to HV electrode. The copper metal plate dimensions are the length (200 mm) and the width (100 mm) and the thickness (1 mm). The downstream region for measurement of the voltage variation and isolate form the ground. The dynamic variation of the metal-plate voltage was measured via oscilloscope and HV probe. The metal-plate voltage was monitored under ionization wave reaching to the metal plate. The high voltage input of the metal plate was kept constant at the range of 10 kV, which implies the change of the space potential in the plasma jet under z-axis positions and discharge voltage conditions [12]. The effective discharge length is approximately 200 mm. The discharge gap in this study is 20, 40 and 60 mm.

(a)

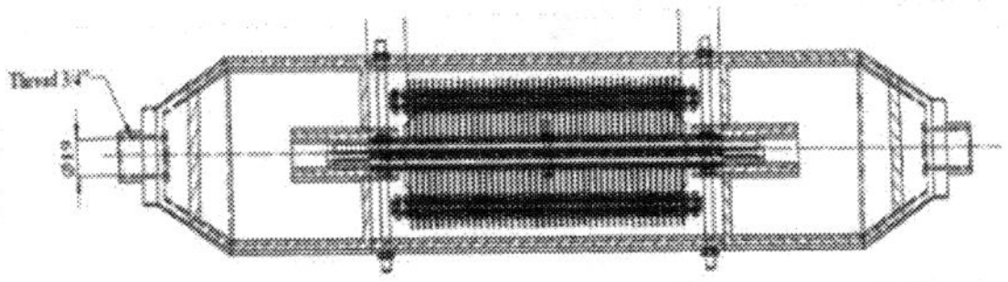

(b)

Fig 2. (a) The non-thermal plasma reactor, (b) the non-thermal plasma drawing.

2.3 Electrodes

Electrode is an important part of the conductivity of a reaction. An electrode in an electrochemical cell is referred to as either an anode or a cathode [13-14]. It has both an anode and a cathode side. The copper material has been selected as electrodes due to the properties of copper has the low electrical resistance (1.93 x 10^{-9} Ωcm) that has the high current flow [15-16].

2.4 Electric measurements

The important key parameters to evaluate a pollution control system is the energy consumption. It was determined form simultaneous measurements of the discharge current and applied voltage. The DC Voltage applied to the electrode is measured by using an HV probe (Testec TT-HVP15 HF 500MHz 10 kV). The electrical wave form was monitored via a fast digital oscilloscope (Keysight Technologies DSOX1204).

2.5 Electric measurements

All experiments were conducted under ambient temperature and pressure. Space velocity is the ratio between gas flow rate through reactor and effective volume of the reactor as shown in Eq.1

$$SV = \frac{Qgas}{Veff} \quad (h^{-1}) \quad (1)$$

Where Q_{gas} (m^3/h) is the exhaust gas volumetric flow rate and V_{eff} (m^3) is the reactor Volume. Increasing the space velocity results in high gas flow rate pass through the reactor lead to less time for reaction. The experiment conditions with NTP were designed to obtain results which affected by position electrode, reactor discharge gap and reactor high voltage input power as shown in Table1.

Table 1. Electrode configuration parameters investigated in this work.

Item Value No.	Number of copper	Discharge gap (mm)
1	40	20
2	40	40
3	40	60
4	20	20
5	20	40
6	20	60
7	10	20
8	10	40
9	10	60
10	5	20
11	5	40
12	5	60

2.6 Flue gas concentration measurements

Gas analysis model MEXA-584L from Horiba includes measurement of carbon dioxide, carbon monoxide and unburned hydrocarbons (NIDR-non-dispersive infrared), oxygen (magneto pneumatic sensor), and NO_x (CLD-chemi-luminescence detection). In this work, the flue gas in each operating condition is measured and reported.

3 Result and discussion

The aim of this study to optimize the electrode position, which are the number of copper electrodes and discharge gap between electrode, to finding the best NO_x reduction efficiency in NTP reactor. The power loss in system is also investigated. The reactor sizing to estimate the length and diameter of the reactor according to the empirical formula.

3.1 The effect of electrode configurations on the power input

This experiment was optimized the effective range of NTP reactor. The NTP power input are dependent on many parameters inclusive of the number of electrode copper sheets (5 sheets to 40 sheets), electrodes gap (40 mm to 5 mm) and fixed discharge gap at 20 mm. All electrode configurations are shown in Fig.3

The input voltage is generally related to the ionization degree to reaction with feed gas in NTP reactor. In additions, the electrical energy, discharge gap and number of coppers of the plasma reactor are all parameters that affect the number of energetic electrons [12]. The increasing input plasma power may effectively

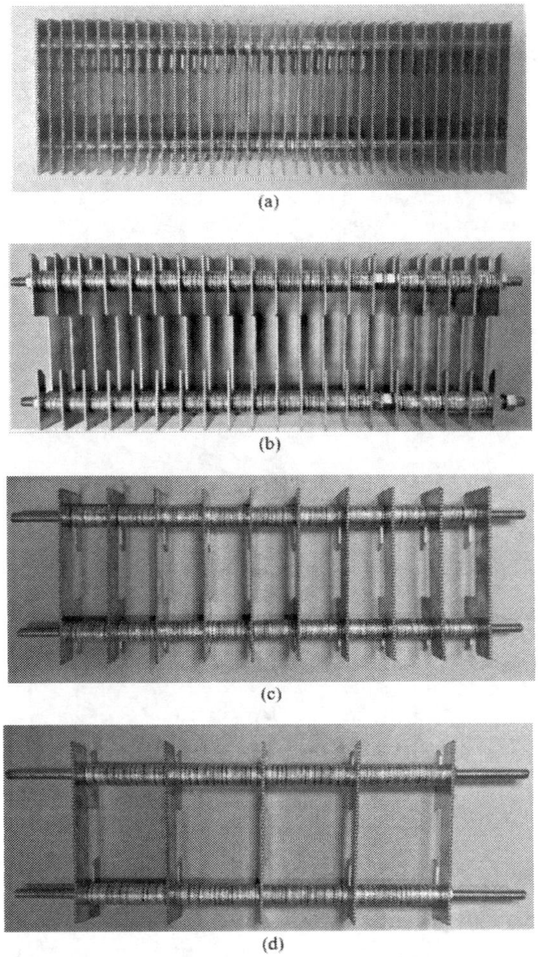

Fig. 3. Schematic diagram of the amount of copper (a) 40 sheets (electrodes gap 5 mm) (b) 20 sheets (electrodes gap 10 mm) (c) 10 sheets (electrodes gap 20 mm) and (d) 5 sheets (electrodes gap 40 mm)

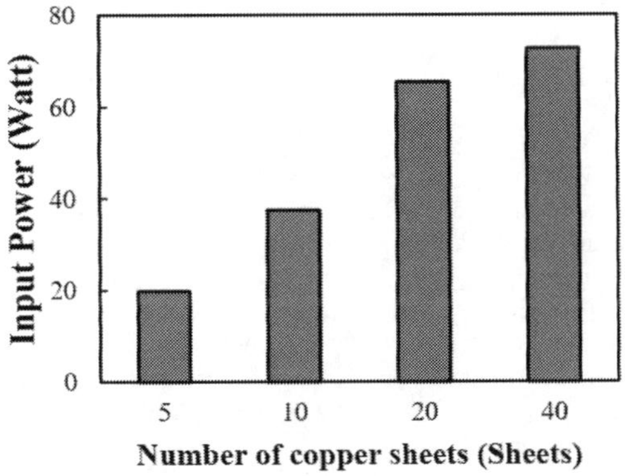

Fig. 4. Effect of electrode to apply input power

enhance the electric field, electron density under a constant excitation frequency. This also provides high energy to electrons consequently which can be enhanced the pollution removal efficiency [17]. The high amount of cooper sheets is increasing power input (high power consumption) under the same reaction zone (200 cm^2) as seen in Fig.4. To increasing the resistance with higher copper sheets. In addition, the higher number of copper sheets are requiring high level of energy source to generate plasma beam. The applied voltage supply is required to overcome the electrical impedance with high resistant (high amount of copper sheets). The plasma per reaction area has not work full efficiency, thus optimization of number of sheets for operating conditions was found to be crucial in enhancing plasma beam efficiency and applications. The 20 copper sheets with 10 cm electrode gap were selected in this experiment due to the power output is suitable to for NO_x reduction application. The input power for 40 copper sheets with 5 cm electrode gap has slightly higher than 20 copper sheets due to the small electrode heat losses.

3.2 The effect of discharge gap on input power

The influence of discharge gap on plasma beam performance has been studied to the optimum energy utilization in normal thermal plasma beam generation. The electrode configurations are following; discharge gap is 20, 40 and 60 mm with fixed copper 40 sheets.

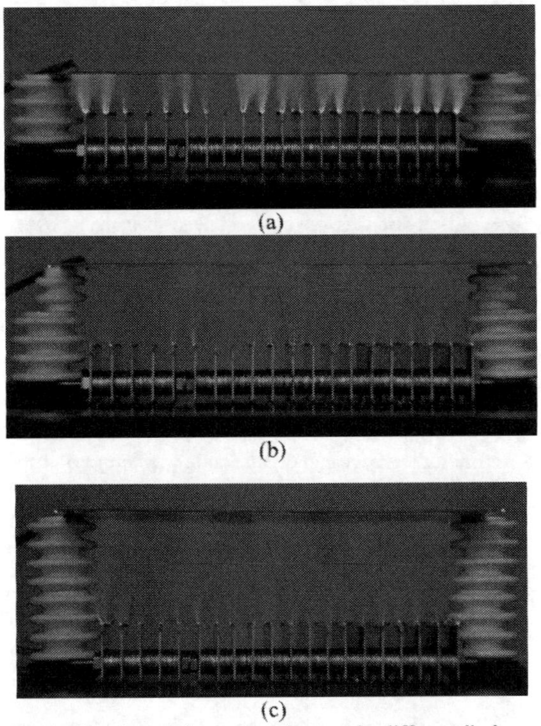

Fig. 5. Side view of plasma beam layer under different discharge gap (a) discharge gap 20 mm, (b) discharge gap 40 mm and (c) discharge gap 60 mm

Fig.5 shows the various discharge gap on plasma beam and the input power under atmospheric conditions (without simulated gas feed). The plasma beam density was dependent on the discharge gap due to its different power input and followed the general trend: discharge gap

20 mm > discharge gap 40 mm > discharge gap 60 mm as shown in Figure 5. The plasma beam in atmospheric condition show phenomena similar with uniform electric field under atmospheric condition as seen in Fig.5. The high electric field density occurring under the voltage lower than break down voltage. In additions, the electric field density was reduced under the increasing discharge gap because the ionization reaction between two electrodes is decreased. During begin, the discharge inception of semiconductor to produce external plasma beam is controlled by the specific conductivity of electrode. Thus, the capacity per electrode area unit, also called effective capacity, are very important for plasma density. Therefore, the voltage drop across the high gas gap leads to high electrical breakdown [18].

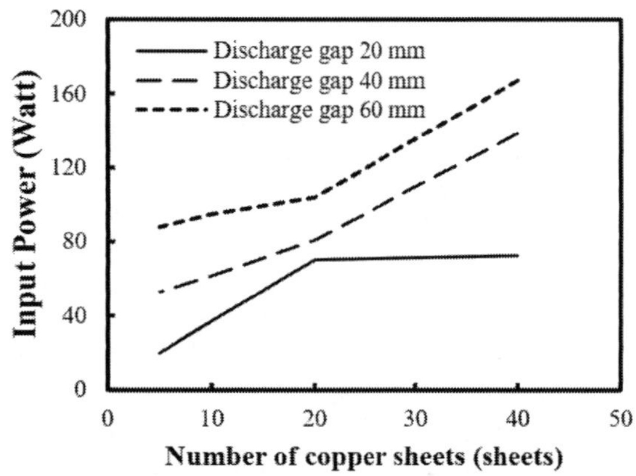

Fig. 6. Effect discharge gap of electrode to apply input power

The discharge gap is an important factor affecting electrical power input. The electron field behavior generally moves from a higher potential pole to a lower potential pole. The free electron can be move easily to lower potential pole under small discharge gap thus, the maximum electric field with small power input as shown in Fig.6. The highest power is 167.2 watts at 40 electrodes, discharge gap 60 mm and the minimum power is 19.8 watts at 5 electrodes, discharge gap 20 mm. The results can analyze that number of electrodes and the discharge affect to power consumption. In additions, the power generated is less than the other discharge gap in all cases. The discharge gap can have a significant influence on the number of micro-discharges. Therefore, the discharge gap can generate significant influence on the optical and electrical characteristics of the plasma reactor [19-20].

Furthermore, the number of electrodes and discharge is a very important factor to power consumption. The high number of electrodes will affect to higher power input. The input power is related to the electrode material resistance. Therefore, the high number of electrodes with high electric resistance are required the higher power to overcome electrodes resistant.

3.3 The influence of NO_x, NO, NO_2 removal activity over NTP reactor

This section presents the prototype of NTP reactor activities on NO_x, NO, NO_2 removal efficiency. The prototype configuration was selected as following; discharge gap 60 mm, 40 sheet of copper electrode sheets and approximate 168 watt of input-power that is the maximum condition for NTP reactor in this study. The simulated flue gas was feeding rate at 10 L/min. The result found that NTP reactor can be NO_x removal of approximately 78% and consist of nitric oxide (NO) removal 99% and nitrogen dioxide (NO_2) production 93% as shown in Fig.7.

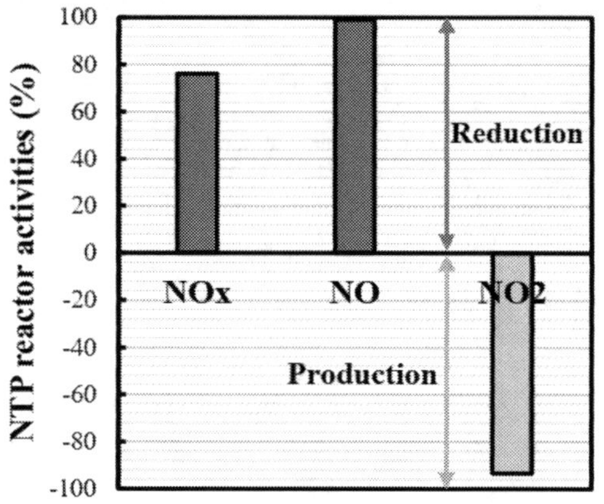

Fig. 7. NO_x NO, NO_2 reduction activities over NTP reactor

Reactor with NTP has increase the electric filed strength and electron density and provides more energy to electrons consequently leads to high NOx removal efficiency, supports high removal efficiency in this reactor [21]. In addition, the plasma reaction can reduce NO as the main component of NOx pollution, which then transforms to NO_2 > 99% that are active in a discharge zone.

4 Conclusion

The preliminary study of prototype non-thermal plasma reactor design for NOx reduction presents a solution to the current challenge of air pollution over the combustion. The influencing working parameters such as the applied voltage, discharge length, interval between adjacent outer electrodes and electrode width were all considered to obtain a further and deeper understanding of the discharge characteristics of the multi-electrode plasma reactor. The efficient length and fringe effect were found to be the two significant reasons for the change of the optical and electrical performance. In additions, the prototype of NTP reactor could significantly enhanced 78% of NOx reduction activity. This benefit could be utilized with the nonthermal plasma technology to purpose continuous NOx reduction. Moreover, contribute to an improved post

https://doi.org/10.1051/e3sconf/202130201010

treatment (e.g., SCR, Lean-NOx trap) aiming to achieve future emissions regulations that include control of NOx.

References

1. G. Xiao, W. Xu, Non-Thermal plasmas for VOCs abatement, Plasma Chemistry and Plasma Processing, **34**, (2014): 1033–1065
2. F. Gholami, M. Tomas, Technologies for the nitrogen oxides reduction from flue gas: A review, Science of The Total Environment, **714**, (2020): 136-712
3. Meunier, F., Breen, J., Zuzaniuk, V., Olsson, M., Ross, J., Mechanistic Aspects of the Selective Reduction of NO by Propene over Alumina and silver, Journal of Catalysis, **187**, (1999): 493-505
4. F. Gholami, Technologies for the nitrogen oxides reduction from flue gas: A review, Science of the Total Environment, Science of The Total Environment, **714**, (2020): 136
5. K.H. KIM.S.-B. Lee, S.H. Woo, NOx Profile Around a Signalized Intersection of Busy Roadway, Atmospheric Environment, **97**, (2014): 141-154
6. T. Rungratanaubon, Vertical Variation of Nitrogen Oxide (NOx) Concentration Using a Backward Air Mass Trajectories Model in an Urban Area of Bangkok, Thailand, KMUTNB International Journal of Applied Science and Technology, **11**, (2018): 73-81
7. L. Zhang, X.Zhu, Effects of water pressure on plasma sparker's acoustic characteristics, International Journal of Plasma Environmental Science and Technology, **11**, (2017): 60-63
8. M. Gallagher, Patial Oxidation and Autothermal Reforming of Heavy Hydrocabon Fuels with Non-Equilibrium Gliding Arc Plasma for Fuel Cell Applications, iDEA: Drexel Libraries E-Repository and Archives, (2010): 1-5
9. N. Takamura, D. Wang, D. Seki, T. Namihira, K Yano, H Saitoh, H Akiyama, Protein Transduction into Eukaryotic Cells using Non-thermal Plasma, International Journal of Plasma Environmental Science and Technology, **6**, (2012): 59-62
10. S. Li, Y. Huang, F. Wang, J. Liu, F. Feng, X. Shen, Fundamentals and environmental applications of non-thermal plasma: multi-pollutants emission control from coal-fired flue gas, Plasma Chem Plasma Process, **34**, (2014): 579-603
11. S. Ma, Y. Zhao, J. Yang, S. Zhang, C. Zhang, Research progress of pollutants removal from coal-fired flue gas using non- thermal plasma, Renewable and Sustainable Energy Reviews, **67**, (2017): 791-810
12. Y. Setsuhara, G. Uchida, K. Kawabata, A. Nakajima, K. Takenaka, Analysis of dynamic discharge characteristics of plasma jet based on voltage and current measurements using a metal plate, IEEE Transactions on Plasma Science, **43**, (2015): 3821-3826
13. W. William, History of the Inductive Sciences: From the Earliest to the Present Times, Cambridge University Press, (2010)
14. M. Moscosa-Santillan, Design of a DBD wire-cylinder reactor for NOx emission control: experimental and modelling approach, Journal of Cleaner Production, **16**, (2008): 198-207
15. F. Pawlek, D. Rogalla, The electrical resistivity of silver, copper, aluminium, and zinc as a function of purity in the range 4-298 °K, Cryogenics, **6**, (2016): 14-20
16. L. Haiyun, L. Zhuo, W. Xinxin, Homogeneous dielectric barrier discharge in nitrogen at atmospheric pressure, Journal of Physics D Applied Physics, **43**, (2010): 15
17. M. Laroussi, I. Alexeff, J. Richardson, Effects of nonequilibrium atmospheric pressure plasmas on the heterotrophic pathways of bacteria and on their cell morphology, Applied Physics Letters, **81**, (2002): 772-774
18. H. Nguyen, K. Kyo-Seon, Combination of plasmas and catalytic reactions for CO_2 reforming of CH_4 by dielectric barrier discharge process, Catalysis Today, **256**, (2015): 88–95
19. L.M. Portsel, Y.A. Astrov, I. Reimann, E. Ammelt, H.G. Purwins, High speed conversion of infrared images with a planar gas discharge system, Journal of Applied Physics, **85**, 3960 (1999)
20. L.S. Wei, D.K. Yuan, G.P. Dong, H.U. Zhao-Ji, Z.H. Tan, Experimental study on ozone generation using high-frequency coaxial-cylinders DBD in oxygen, CNKI: Nanchang University, **37**, (2013): 2209–2211
21. R. Dawson, J. Little, Characterization of nanosecond pulse driven dielectric barrier discharge plasma actuators for aerodynamic flow control, Journal of Applied Physics, **113**, (2013): 605

Study of Parameters Affecting Temperature of Journal Bearing and Shaft in Sugar Mill

Kumpanat Sirivedin[1,*], and *Keittipong* Choojit[1]

[1]TGGS, The Sirindhorn International Thai-German Graduate School of Engineering, King Mongkut's University of Technology North Bangkok, Bangkok, Thailand

Abstract. In the sugar mill machine, equipment is operated under high temperature. This high temperature will affect to the roller shaft and journal bearing. Additionally, the equipment damages could be caused by the various factors that affected directly to roller shaft and journal bearing. In this research, the operations of equipments were modelled using the finite element analysis program (FEA), to determine the maximum temperature of the roller shaft and journal bearing after a specified period of time. An experiment with the physical roller shaft and journal bearing was carried out to obtain the actual temperature readings. Verification of the simulated results was performed by comparing the simulated values with the actual values obtained in the experiment. According to this research, the monitored parameters affected temperature and damages of the shaft and bearings during pressing oil from the hydraulic top cap. Finally, this research applied the FEA in conjunction with engineering and environmental management in sugar factory to reduces machine damange and to save energy during sugar mill operation.

Keyword. Sugar mill machine, Finite element analysis, Journal bearing.

1 Introduction

A sugarcane juice extraction is the first process in a raw sugar production to produce white or brown sugars and other by-products. It extracts sugarcane juice by crushing and breaking up hard nodes and flatten the stems by roller crushers. The process of extracting sugarcane juice by squeezing will create a pressure load on the surface area of the left and right journal bearings (Fig.1). It is pressed with a hydraulic cylinder through the top bearing with high pressure to the sugarcane in order to release as much as possible of the sweetness. Therefore, roller shafts and journal bearings are the important mechanical parts in the roller crusher (Fig. 1). When hydraulic pressure exceeds the critical value, work load or stress on journal bearing and shaft also exceed yield point and then plastic deformation occurs. Temperature on both journal bearing and shaft are accumulated and increases with hydraulic pressure. This will directly cause the damage on shaft and bearings and leads to failure of the roller crusher. Fig. 2 shows possible damage of journal bearing and shaft.

In a production line, if the roller shafts and journal bearings are damaged, the sugar production process will be interrupted immediately [1]. This affects on the production costs and waste of energy in the following evaporating process and drying process when roller crusher is under maintenance [2]. Therefore, this paper focused on parameters affecting temperature of shaft and journal bearing in the sugarcane mill machine by focusing on the critical point of shafts and bearings.

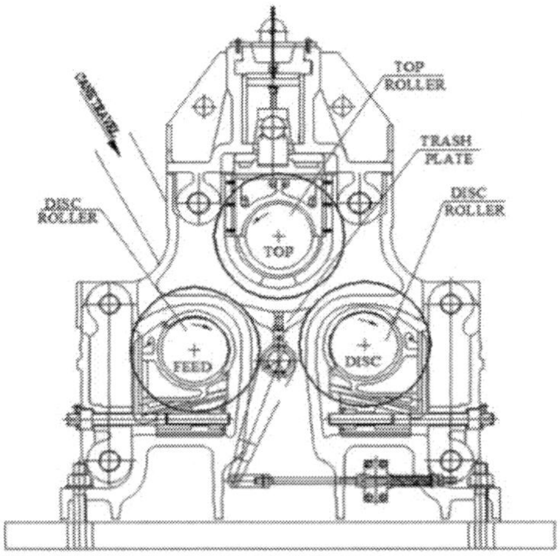

Fig. 1. Assembly of a conventional three-roller sugar mill.

2 Materials and Methods

In this research, the roller shaft and journal bearing were modelled using computer-aided design (CAD) software, SOLIDWORKS, and the finite element analysis program, ANSYS Workbench, to determine the maximum temperature of the roller shaft and journal bearing after a specified period of time [3]. Suggested geometry of shaft which showed minimum stress concentration was studied by Rathod *et.al.*[4]. Analytical

* Corresponding author: kumpanat.s@tggs.kmutnb.ac.th

method and FEM were used to analyze the force on the support points according to the specified hydraulic pressure, in order to obtain the appropriate hydraulic pressure. The apply load come from the transferring pressure on the hydraulic top cap to the bearing [5]. The testing process can be divided into two steps.

(a) Journal bearing failure

(b) Roller shaft failure

Fig. 2. The damages occur on (a) a journal bearing and (b) a roller shaft.

2.1. Captions/numbering

For thermal analysis of journal bearing, it is compulsory to collect the data for the bearing of the top roller (Fig.3).

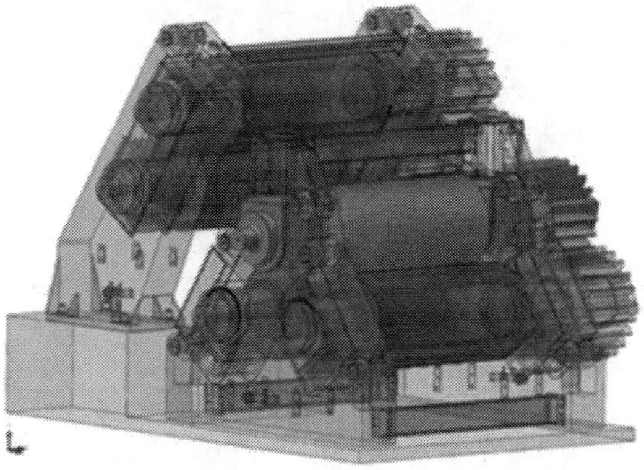

Fig. 3. The installation and position of top roller (3D-Model).

The specified pressure force in this study were 2600, 2800, 3000, and 3200 psi. The temperature distribution data was collected under specified pressure from the hydraulic top cap set by using an infrared thermometer during operating conditions at the shaft and journal bearing of sugarcane crushing machine (Fig.4). The collected values were plotted between temperature and time under the specified pressure.

Fig. 4. Setting of the pressure load and the temperature measurement.

2.2 The model analysis with the finite element method

The finite element analysis (FEA) of the shaft and journal bearing were carried out. The simulation setup was as follows. The CAD model from SOLIDWORKS was exported into Ansys engineering simulation software, ANSYS Workbench, where the meshing and thermal simulation were carried out.

2.2.1 Calculation for input parameters

Input parameters for FEA were calculated by analytical method according to the technical data collected from the operating manual of the company. The technical data is given in Table 1.

Table 1. Technical specification of a conventional three-roller sugar mill.

Details	Value
Power [p] (kw)	900
Roller speed [n] (rpm)	5
Roller diameter (mm)	1250
Shaft diameter at roller (mm)	700
Shaft diameter at bearing support (mm)	620
Shaft diameter at pinion (mm)	625
Pitch circle diameter of crown pinion (mm)	1136
Journal bearing size D x L (mm)	620 x 740

The top roller shaft (Fig. 5) is a critical component because the drive torque, hydraulic load, crushing load came from it. The forces acting on the mill roller increased shearing, bending, torsion and compressive stresses. Therefore, simple model of the forces acting on it must be considered [6].

 https://doi.org/10.1051/e3sconf/202130201011

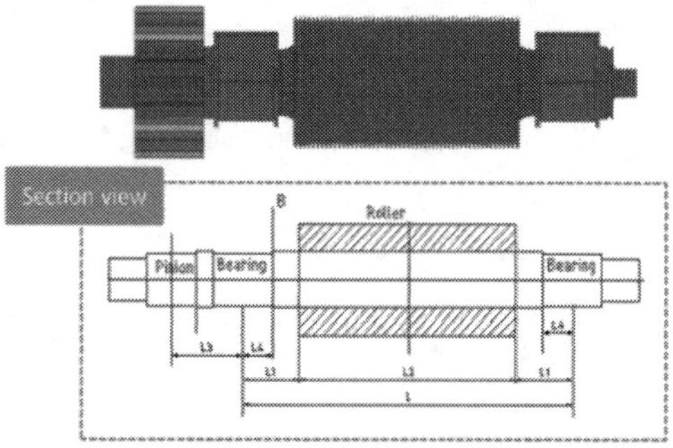

Fig. 5. Section view of the top roller shaft.

From Table 1, power = 900 kw and speed = 5 rpm, a drive torque yields

$$\text{Drive torque} = \frac{\text{Power} \times 9.55}{\text{speed (rpm)}} \tag{1}$$

$= 1719.000$ kN-m.

The 50% of torque is taken by top roller.

$$\text{Torque of the top roller} = \frac{\text{Drive torque}}{2} \tag{2}$$

$= 859.500$ N-m.

$= 859.500 \times 10^3$ kN-mm.

Tangential component (Horizontal) of force due to torque, F_t, where P.C.D is Pitch circle diameter of crown pinion = 940 mm

$$F_t\ (F_{ph}) = \frac{Torque}{(P.C.D)/2} \tag{3}$$

$= 1513.204$ kN

Radial component (Vertical) of force due to torque, F_r, where pressure angle (θ) is 25 deg.

$$F_r\ (F_{pv}) = F_t \tan(\theta) \tag{4}$$

$= 705.619$ kN

Load of crushing from discharge roller acting at an angle of 53 deg.

From right side = 505.714 tons.
= 5057.143 kN

The vertical component of load

= Vertical load x sin (θ)
(Discharge Roller) = 5057.143 sin (53)
= 4038.814 kN

The horizontal component of load

= Horizontal load x cos (θ)
(Discharge Roller) = 5057.143 cos (53)
= 3043.465 kN

The load of crushing from feed roller acting at an angle of 55 deg.

From left side = 235.714 tons.
= 2357.143 kN

The vertical component of load

= Vertical load x sin (θ)
(Feed Roller) = 2357.143 sin (55)
= 1930.858 kN

The horizontal component of load

= Horizontal load x cos (θ)
(Feed Roller) = 2357.143 cos (55)
= 1352.002 kN

The weight of top roller is 32 tons., which acts downwards.

The total vertical load acting on the roller

= (Load from discharge roller + Load from feed roller) - Weight of top roller
= 5649.672 kN

The total horizontal load acting on the roller

= (Load from discharge roller - Load from feed roller)
= 1691.463 kN

The load of crushing acts on the surface of roller. So, it will be shown as uniformly distributed load (U.D.L) on the loading diagram.

Intensity of U.D.L. in vertical loading diagram

$$(W_v) = \frac{\text{Total vertical load due to crushing}}{\text{Length of shell of roller}} \tag{5}$$

$= 2.456$ kN/mm.

Intensity of U.D.L. in horizontal loading diagram

$$(W_h) = \frac{\text{Total horizontal load due to crushing}}{\text{Length of shell of roller}} \tag{6}$$

$= 0.735$ kN/mm.

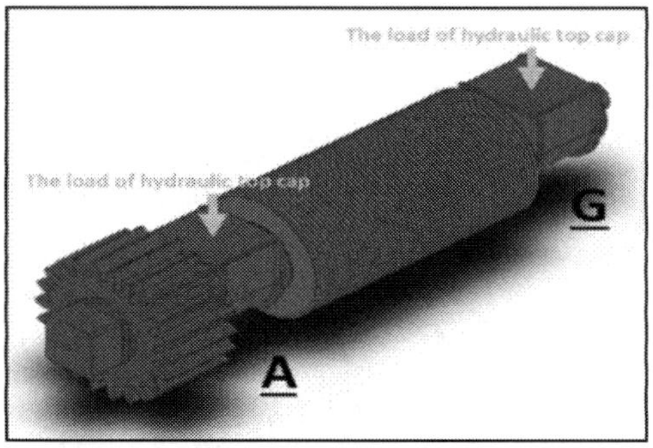

Fig. 6. Action force on the top roller.

The top roller was pressed with the hydraulic top cap at the top of the bearings on both sides (Fig.6). It worked by adjusting the hydraulic pressure from the power unit according to the desired value.

Diameter of hydraulic top cap = 460 mm.
Oil pressure = 3000 psi.
= 2.109 kg/mm²

Total hydraulic load acting on the top roller

= 2 x Area of piston x Oil pressure (7)

= 7010.599 kN

Hydraulic load on each bearing (F_h)

$$= \frac{\text{Total hydraulic load acting on the top roller}}{2} \quad (8)$$

= 3505.299 kN

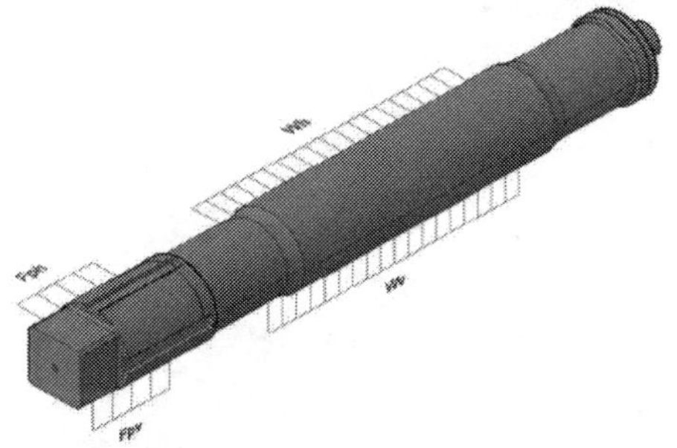

Fig. 7. Diagram of Load acting on the shaft.

Simple model analysis with equilibrium principle by determining the load of the force acting on the shaft was conducted. Two supports at points A and G in Fig. 6 were simplified into a simple model analysis with equilibrium principle (Fig.7). The simple model was later used to calculate the vertical and horizontal pressure load of journal bearing (Fig 8-9).

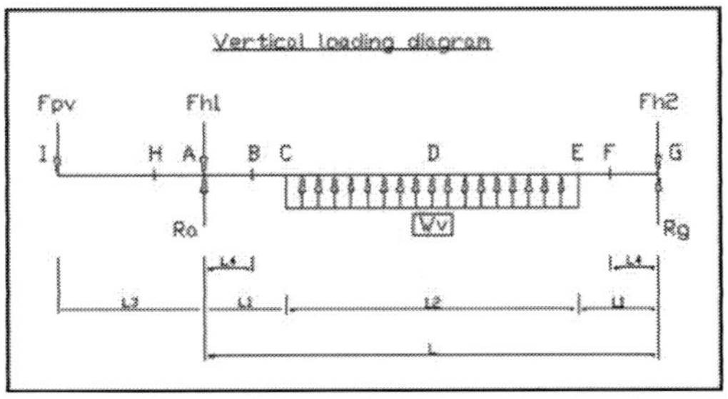

Fig. 8. Vertical loading diagram.

From Fig.5,

L1 = 650 mm.

L2 = 2300 mm.

L3 = 950 mm.

L4 = 375 mm.

Fh1 and Fh2 are hydraulic load on each bearing

= 3505.299 kN

Vertical component of force due to torque

F_r (Fpv) = 705.619 kN

Intensity of U.D.L (Wv) = 2.456 kN/mm.

Ra and Rg are reaction forces from bearing.

$\Sigma M_A = 0$;

Rg (vertical) = 494.258 kN

$\Sigma F_y = 0$;

Ra (vertical) = 1572.287 kN

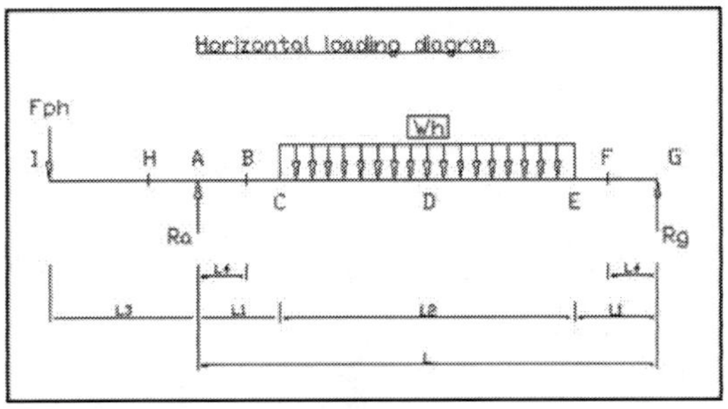

Fig. 9. Horizontal loading diagram.

Horizontal component of force due to torque

F_t (Fph) = 1513.204 kN

Intensity of U.D.L (Wh) = 0.735 kN/mm.

Ra and Rg are reaction forces from bearing.

$\Sigma M_A = 0$;

Rg (horizontal) = 446.414 kN

$\Sigma F_y = 0$;

Ra (horizontal) = 2758.253 kN

Resultant of horizontal and vertical loading:

∴ Resultant reaction at 'A'

= Ra (vertical) + Ra (horizontal)

= 3174.909 kN

∴ Resultant reaction at 'G'

= Rg (vertical) + Rg (horizontal)

= 666.015 kN

The determination result of the force on the shaft and the journal bearing under pressure load obtained from the hydraulic top cap pressure adjustment was found that the reaction force acting at the support points on both sides was different, which are shown in Table 2. It was found that the reaction force at point A was significantly higher than that of the point G. Therefore, the reaction force under the specified pressure load at point A was used as the input in FEA in order to calculate the maximum stress and deformation.

Table 2. The reaction force between point A and point G.

The load of hydraulic top cap/Oil pressure (psi)	Resultant reaction at 'A'(kN)	Resultant reaction at 'G'(kN)
2600	3042.581	540.214
2800	3108.744	602.761
3000	3174.909	666.015
3200	3241.076	729.793

2.2.2 The model analysis

Design the model (CAD model) was done by using computer-aided design with a mathematical model before entering the analysis process. Subsequently, a static analysis of top roller was done for observing maximum stresses and deformation. Static structural analysis was done using ANSYS Workbench [7]. Fig.10 shows a boundary condition in FEA model.

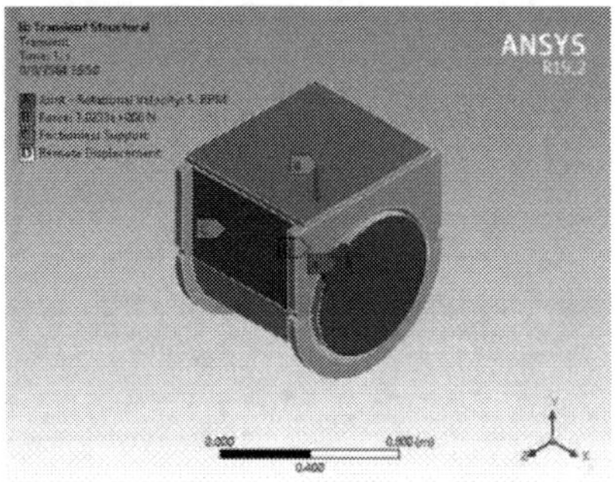

Fig. 10. Setting up boundary condition in details.

2.2. 3 Analysis of simulation

The objective of performing the thermal simulation was to determine the maximum temperature and define its location (Figs.11-14).

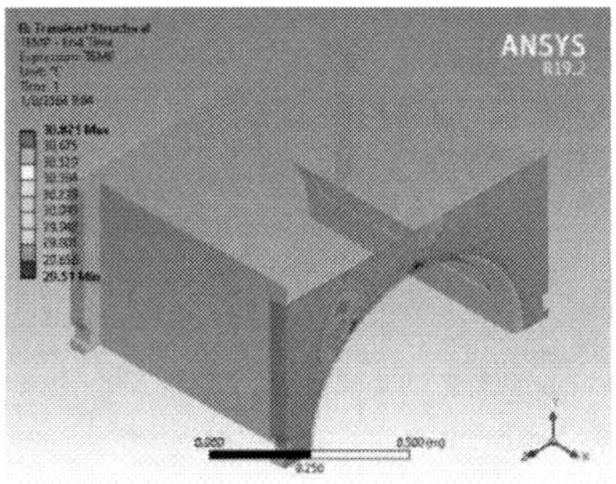

Fig. 11. Maximum temperature of journal bearing under pressure load 2600 psi.

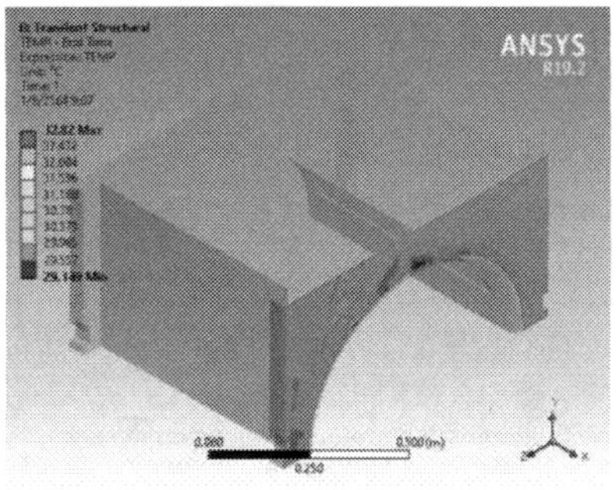

Fig. 12. Maximum temperature of journal bearing under pressure load 2800 psi.

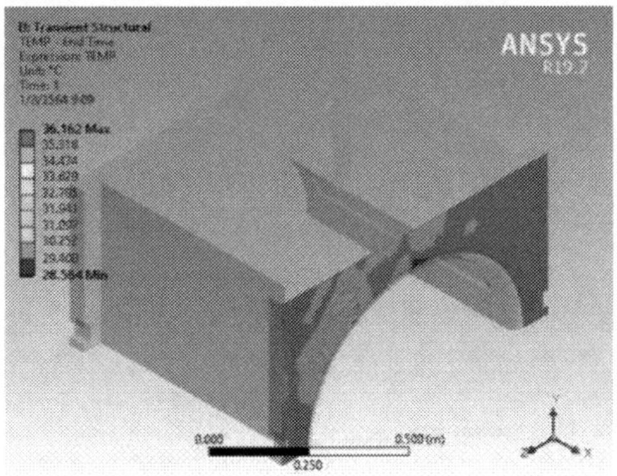

Fig. 13. Maximum temperature of journal bearing under pressure load 3000 psi.

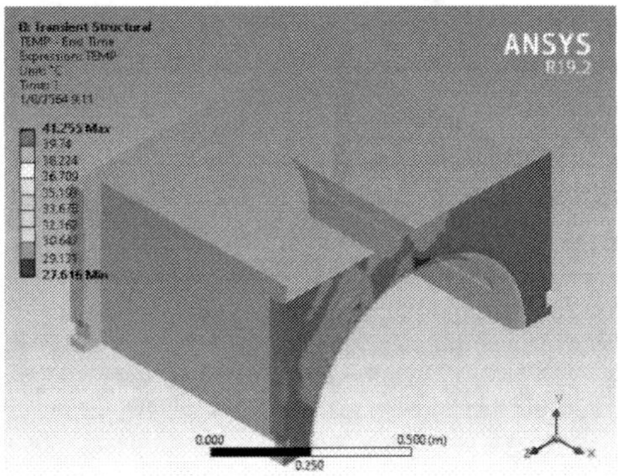

Fig. 14. Maximum temperature in journal bearing under pressure load 3200 psi.

3 Results and Discussion

The results from finite element method (FEM) with Ansys engineering simulation and 3D design software were compared with the experimental data. The results of surface temperature from FEM and experiments at pressure load of 2600, 2800, 3000, 3200 psi are shown in Figs.15-18, respectively.

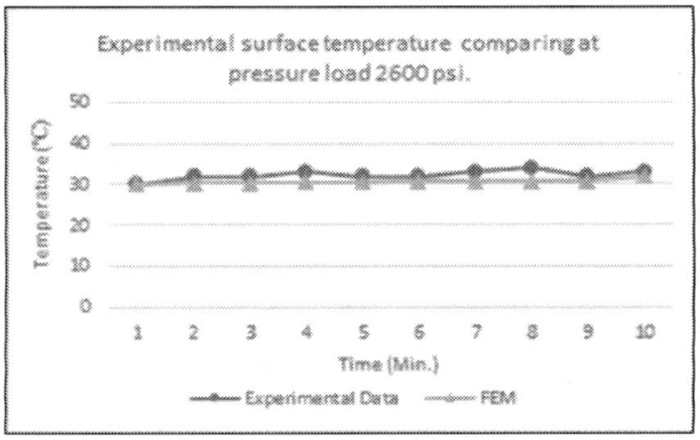

Fig.15. Comparison of temperature obtained from experimental data and FEM analysis at pressure load 2600 psi.

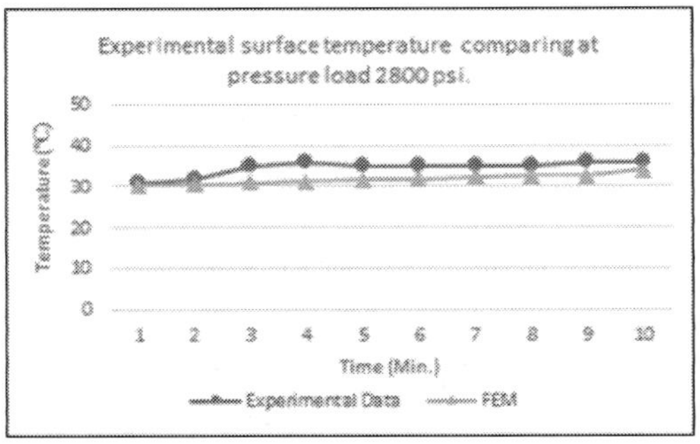

Fig.16. Comparison of temperature obtained from experimental data and FEM analysis at pressure load 2800 psi.

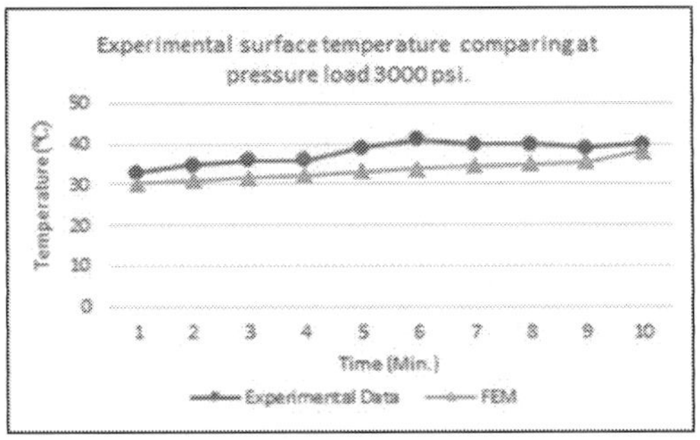

Fig.17. Comparison of temperature obtained from experimental data and FEM analysis at pressure load 3000 psi.

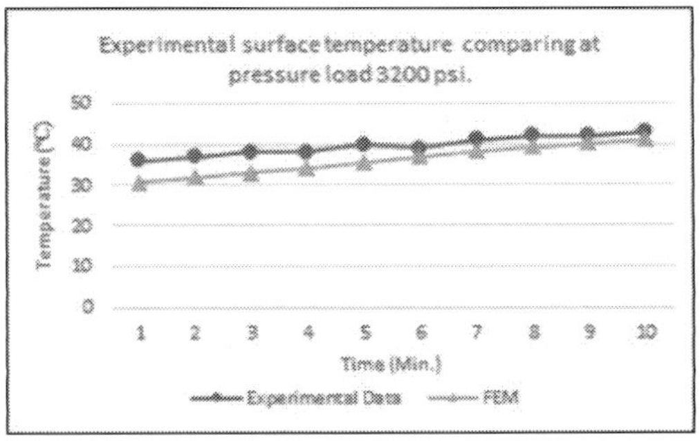

Fig.18. Comparison of temperature obtained from experimental data and FEM analysis at pressure load 3200 psi.

The results obtained from FEM were compared with the experiment results in order to validate the accuracy of the simulation results. It is observed that the simulated results were closely matched with that of the experiment results. The working temperature and the load of hydraulic top cap/oil pressure (psi) in Table 3 was plotted and shown in Fig.19.

Comparing with experiment and FEA results, the temperature difference are below 6% (~5.85%). Therefore, the FEA model in this study is valid and can be used to predict thermal stress and temperature of the journal bearing and shaft. Table 4 shows calculated temperature and Von-mises stress in various hydraulic pressure.

Table 3. The model analysis results.

The load of hydraulic top cap/Oil pressure (psi)	Working Temperature (°C)		Errors (%)
	Experimental data	FEM	
2600	34	32	5.88
2800	36	34	5.56
3000	41	38	7.32
3200	43	41	4.65

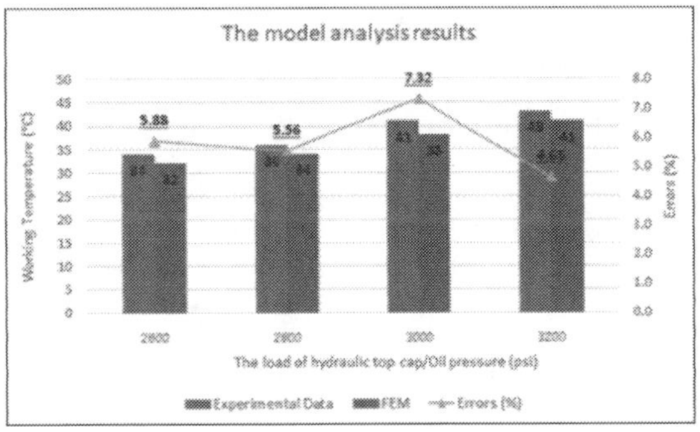

Fig. 19. Working temperature vs load of hydraulic top cap/oil pressure(psi) plot from FEM and experimental results.

Table 4. Results summary from finite element analysis.

The load of hydraulic top cap/Oil pressure (psi)	Temperature (°C)		Von-mises stress (MPa)	
	min	max	min	max
2600	29.55	375.36	145.87	490.28
2800	29.21	534.45	190.24	832.69
3000	28.66	827.63	247.50	1200.50
3200	27.94	1189.20	292.92	1535.60

The finite element analysis (FEA) result found that the maximum stress value that caused damage to the initial bearing of 3000 psi. Therefore, the value of the analysis will be applied in the next working year. The load of hydraulic top cap/oil pressure adjustment does not exceed 3000 psi., it was reduced journal bearing damage. which are shown in Table 5.

Table 5. Results of the failure on the top journal bearing.

No. of trial	Max. pressure used	No. of journal bearing		Defect (%)
		Total	Failure	
1st	>3000 psi	10	8	80
2nd	≤3000 psi	10	4	40
3rd	≤3000 psi	10	3	30

This research starts at 1st trial and applying the value of the analysis in 2nd and 3rd trials, which is reduced damage of journal bearing. This means the action force that affects to the damage of the shaft and bearings [8]. Finite element analysis can predict shaft and journal bearing damage from parameters that affects bearing operation such as speed, friction, clearance distance, and others [7]. These values can be used as indicator of damage through FEA analysis.

4 Conclusion

According to the research, the parameters affecting temperature and damage of the shaft and bearings was pressing oil from the hydraulic top cap, with a safe working value of less than 3000 (psi). This proves that the results of the experimental result are relatively close to the FEA analysis result and can be used in the other manufacturing processes. As a result of applying FEA analysis values to the device, it was found to reduce journal bearing damage and reduce waste of energy.

The FEA method has a few advantages, the results can be easily extracted from specific points and analysed with the tools available in the software. This reduces the time spent on data extraction and increased the overall project time efficiency. As a whole, the usage of computation simulation is an efficient and effective method of predicting the results in a thermal simulation.

Acknowledgement

This research was supported by M.W.A. Welding (Thailand) Company Limited.

References

1. S.M. Muzakkir, H. Hirani, G.D. Thakre, M.R. Tyagi, Tribological Failure Analysis of Journal Bearings Used in Sugar Mills, Engineering Failure Analysis, **18**, (2011): 2093–2
2. A.S. Lee, Catalytic hydrogenation of liquid ketones with emphasis on gas liquid mass transfer, Ph.D. Dissertation, Texas A&M University, (1999)
3. W.S. Rathod, C.T. Rathod, Design and Analysis of Two Roller Sugar mill using FEA Technique, International Journal of Scientific Engineering and Technology, **1**, 3, (2012):148-152
4. N. Nao, M. Keiji, Stress Concentration Factors for Shoulder Fillets in Round and Flat Bars under Various Loads, International Journal of Fatigue, **19**, 1, (1997): 75-84
5. E. Hugot, Handbook of cane sugar engineering third edition, Elsevier science publishers, (1986)
6. G.B. Richard, J.K. Nisbett, Mechanical Engineering Design. Eleventh Edition, McGraw-Hill Book Company, (2020)
7. S.M. Muzakkir, H. Hirani, K.P. Lijesh, Tribological Failure Analysis of Journal Bearings Used in Sugar Mills, Engineering Failure Analysis, **16**, (2015): 36843-36850.
8. N.P. Narayan, Theoretical and Experimental Investigation on Tribological Failure of Journal Bearing of Cane Mills Used in Sugar Industry, (2019): 1-20

https://doi.org/10.1051/e3sconf/202130201012

Performance analysis of two-stage pressure retarded osmosis

Suksun Amornraksa[1], *Kanokporn* Youyen[2], and *Lida* Simasatitkul[2,*]

[1]The Sirindhorn International Thai-German Graduate School of Engineering (TGGS) King Mongkut's University of Technology North Bangkok, 10800 Thailand
[2]Department of Industrial Chemistry, Faculty of Applied Science, King Mongkut's University of Technology North Bangkok, 10800 Thailand

Abstract. The performance of a two-stage pressure retarded osmosis (PRO) for power generation with a total membrane length of 1 meter was investigated and analyzed in this work. Two feed configurations of freshwater and seawater were studied: one with the freshwater entering at the first stage only and the other with freshwater entering at both stages. The effect of membrane length and flow ratio between freshwater and seawater on the PRO performance were also examined. The results revealed that the performances of both feed configurations were quite similar. The membrane with a shorter length offered a higher average power density than that of a longer length. It was also revealed that the flow ratio had a strong influence on the average power density produced. The maximum average power density of 10.15 W/m² was obtained at the applied hydraulic pressure of 12 bar, the flow ratio of 5, and the membrane length of both stages of 0.5 meter. The best water utilization was achieved at 65%.

Keyword. Two-stage pressure retarded osmosis, power density, water utilization, performance analysis, modelling

1 Introduction

Electricity demand is continually increasing because it is essential for household and industrial activities. The undesirable electricity from fossil fuels like coal leads to a search for renewable and clean energy such as wind, solar, and hydropower. Pressure retarded osmosis (PRO) is one of the latest renewable energy that has been developed. Electricity is generated from the difference between the salinity concentration of two fluids. In the PRO, a low salinity fluid, e.g., freshwater, permeates across a selective membrane to the other side where a high pressurized and high salinity fluid, e.g., seawater, flows to generate electricity [1].

Many studies tried to improve the efficiency of PRO through membrane characteristics by developing new membrane materials [2,3]. Other studies tried to investigate operating conditions, water resources, and membrane module design [4]. It is well-accepted that the optimal applied hydraulic pressure should be about half of the osmotic pressure. Although a large amount of energy can be harvested when mixing between freshwater and seawater, the actual harvested energy is limited due to internal concentration polarization (ICP) and external concentration polarization (ECP). So, thin-film composite membranes have been developed to reduce the effect of ICP and ECP. They can provide high water permeability and produce a power density of 6 W/m² or higher [5].

Generally, a design of the PRO module is performed by using a single-stage operation. Naguib et al. [6] found that the power density is lost along the membrane length since thin-film of draw concentration is covered on the support layer of the membrane. He et al. [7] also proposed two-stage PRO configurations and compared with the single-stage PRO. The distribution of feed flow rate and membrane area were important factors to average power density. Multistage PRO (10-stage PRO) increased net power density around 9% higher than a single-stage PRO with the same membrane area [8]. Thus, a two-stage PRO is proposed to reduce this effect and improve the overall performance in terms of power density. In addition, there is no need for an extra pump for the second-stage PRO. Two-stage PRO can also reduce the water requirement in the fresh feed stream. Thus, this work proposes two configurations of two-stage PRO. The effect of parameters such as flow ratio between draw solution and feed solution, applied hydraulic pressure, the fresh feed ratio to feed each stage, the flow ratio of draw solution and feed solution (Q_D/Q_F), and membrane length in each stage was investigated in terms of power density and water utilization.

2 Methodology

Mathematical modeling of two-stage PRO consisting of osmotic pressure, water flux, salt flux, power density, and concentration was performed using MATLAB. Because water flux and salt flux equations are differential equations, Euler's method was used to solve

* Corresponding author: lida.s@sci.kmutnb.ac.th

the differential equation. Osmotic pressure is determined by Van't Hoff equation as shown in eq (1).

$$\pi = iCRT \quad (1)$$

Power density is shown in eq. (2)

$$W = J_w \Delta P \quad (2)$$

Water flux is defined in eq. (3)

$$J_w = A(\Delta\pi - \Delta P) \quad (3)$$

Salt flux is defined in eq. (4)

$$J_s = B\Delta c_m \quad (4)$$

The effective concentration is defined in eq. (5)

$$\Delta c_m = \frac{C_{D,b}\exp\left(\frac{-J_w\delta_D}{D}\right) - C_{F,b}\exp\left(\frac{J_w(\delta_F + S)}{D}\right)}{1 - \frac{B}{J_w}\exp\left(\frac{-J_w\delta_D}{D}\right) + \frac{B}{J_w}\exp\left(\frac{J_w(\delta_F + S)}{D}\right)} \quad (5)$$

Film thickness and mass transfer coefficient is defined in eq. (6) – (9).

$$\delta_F = \frac{D}{K_F} \quad (6)$$

$$\delta_D = \frac{D}{K_D} \quad (7)$$

$$K_F = 1.62\left(\frac{u_F D^2}{d_{h,F} L}\right)^{1/3} \quad (8)$$

$$K_D = 1.62\left(\frac{u_D D^2}{d_{h,D} L}\right)^{1/3} \quad (9)$$

Mass balance of draw solution and feed solution is performed in eq. (10) – (13).

$$\frac{dQ_D}{dx} = J_w(A_r) \quad (10)$$

$$\frac{dQ_F}{dx} = -J_w(A_r) \quad (11)$$

$$\frac{d(Q_D C_D)}{dx} = J_s(A_r) \quad (12)$$

$$\frac{d(Q_F C_F)}{dx} = -J_s(A_r) \quad (13)$$

3 Process description

3.1 Configuration 1: two-stage PRO in series without new feed solution at inter-stage

Two-stage PRO is arranged in a series, as shown in Fig.1. The total membrane length of the two modules is 1 meter. Draw solution (seawater) is pressurized by a pressure exchanger and fed into the first membrane module. Feed solution enters the module at a pressure 1 bar. Because the draw solution has a high salinity, the feed solution permeates the membrane, resulting in seawater dilution. After that, the outlet streams from the first module enter the second module. A partial stream of outlet draw solution is sent to the pressure exchanger.

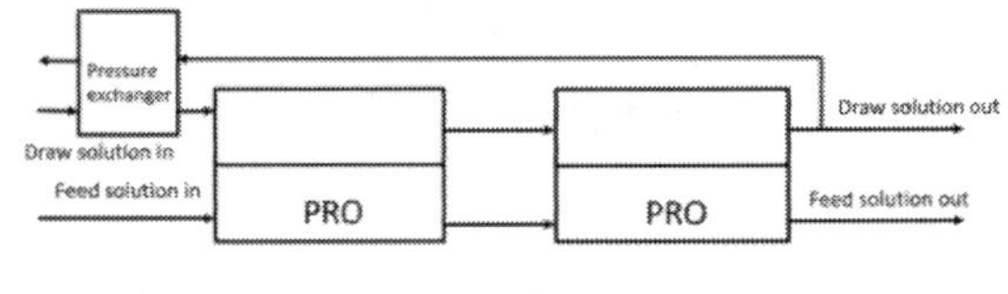

Fig. 1. Configuration 1 of two-stage PRO

3.2 Configuration 2: two-stage PRO in series with feed solution at inter-stage

The second configuration is very similar to the first configuration. However, the feed solution is split into two streams. The first stream is sent to the first module, while the second is sent to the second module. The input data used for both configurations is shown in Table 1.

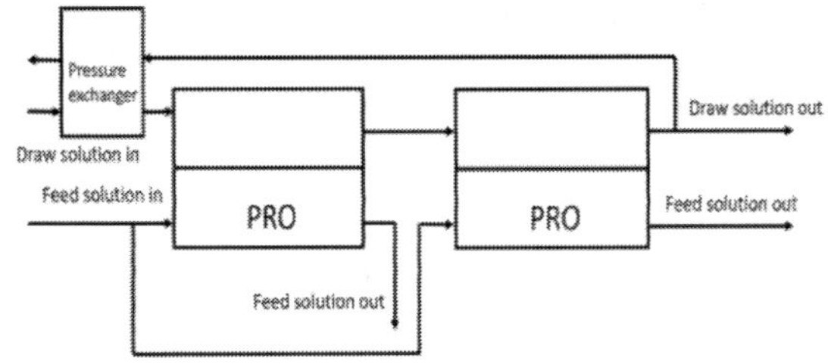

Fig. 2. Configuration 2 of two-stage PRO

Table 1. Input parameter of two-stage PRO [6]

Parameter	Condition
Feed concentration (g/l)	0
Draw concentration (g/l)	30
Temperature (°C)	25
Radius of hollow fiber (m)	0.1×10^{-3}
Radius of module casting (m)	0.15×10^{-3}
Water permeability ($\times 10^{-8}$ L m^{-2} h^{-1} bar^{-1})	14.2
Salt permeability ($\times 10^{-7}$ L m^{-2} h^{-1})	0.241
Structure parameter (m)	310×10^{-3}
Draw velocity (m/s)	0.25

The modeling of the configuration 1 and 2 are investigated in terms of membrane length of each stage and freshwater feed ratio. Precisely, configuration 1 consists of 3 sub-models while configuration 2 consists of 9 sub-models, as demonstrated in Table 2a and Table 2b, respectively.

Table 2a. Model of configuration 1

Model	Length of 1st stage (m)	Length of 2nd stage (m)
1	0.25	0.75
2	0.50	0.50
3	0.75	0.25

Table 2b. Model of configuration b

Model	Length of 1st stage (m)	Length of 2nd stage (m)	Feed ratio of the 1st stage
4	0.25	0.75	0.25
5	0.25	0.75	0.5
6	0.25	0.75	0.75
7	0.5	0.5	0.25
8	0.5	0.5	0.5
9	0.5	0.5	0.75
10	0.75	0.25	0.25
11	0.75	0.25	0.5
12	0.75	0.25	0.75

4 Results and Discussions

4.1 Influence of flow ratio (Q_D/Q_F) and applied hydraulic pressure in configuration 1

The effect of flow ratio on power density and water utilization of each stage is studied in a range of 1 to 7 and shown in Figure 3a – 3d. The results showed that the power density and water utilization increase with an increase in flow ratio because it provides a high salinity gradient. Also, a high flow ratio decreases the film thickness leading to increased effective concentration. Furthermore, osmotic pressure is high with respect to the high flow ratio. Therefore, the feed solution can be transferred across the membrane more easily. It was found that the flow ratio at 5 is suitable as it offers the optimal power density at the first stage and the second stage at 12.86 W/m^2 and 9.97 W/m^2, respectively. At the suitable flow ratio, water utilization of the first and second stages is 22.29% and 63.17%, respectively. It can be noticed that the power density of the first stage is greater than the second stage. This finding is caused by the decrease of salinity gradient along the membrane length, leading to reduced osmotic pressure. Oppositely, the water utilization of the second stage is higher than the first one. It implies that the second stage will improve the potential of freshwater utilization. The applied hydraulic pressure also has a strong effect on power density and water flux. When the applied hydraulic pressure is lower than the osmotic pressure, power density is increased while water utilization is decreased. However, water flux is decreased with an increase in applied hydraulic pressure. The suitable applied hydraulic pressure of both stages is 12 bar which is about half of the osmotic pressure.

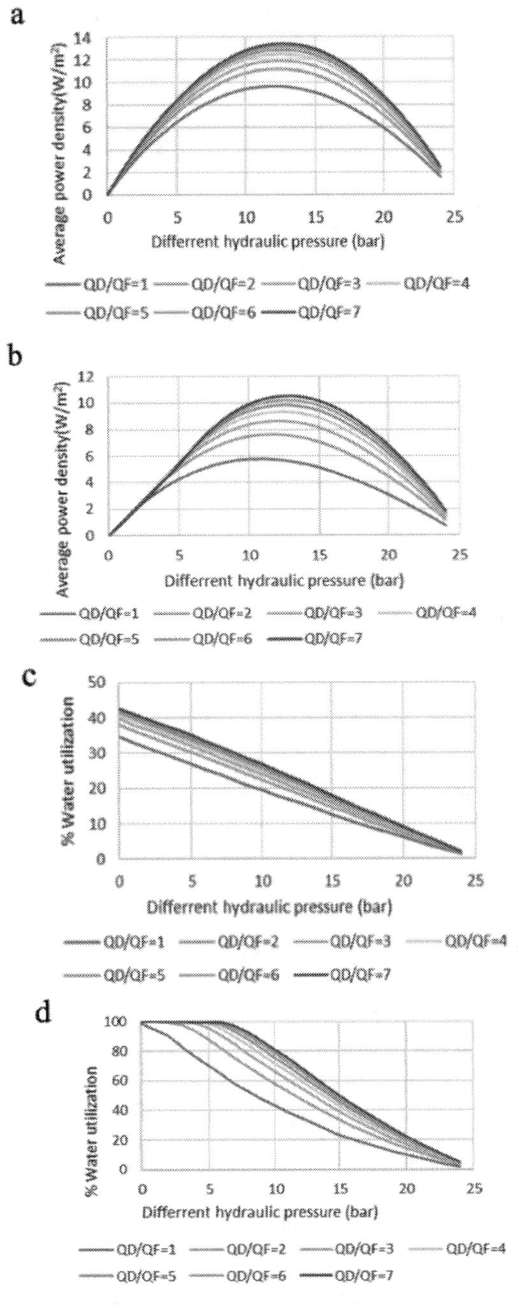

Fig. 3 Effect of flow ratio and hydraulic pressure of the configuration 1 (a) power density of 1st stage (b) power density of 2nd stage (c) water utilization of 1st stage (d) water utilization of 2nd stage

4.2 Influence of membrane length in configuration 1

The effect of membrane length is illustrated in Table 3. When the applied hydraulic pressure is fixed at 12 bar, and the flow ratio is fixed at 5, power density is decreased with an increase in membrane length. For example, when membrane length of the first stage and the second stage is 0.25 m and 0.75 m, respectively. The power density of the first stage is the highest at 12.86 W/m^2. This is because the ICP and ECP that affect the film thickness on the draw and feed sides are founded. They reduce the effective concentration along membrane length. The 1st stage of model 1 has low water utilization because of low permeation area. On the other hand, the 2nd stage of model 1 has the highest water utilization. According to the result, model 2 offered the highest total power density and total water utilization.

Table 3. Power density and water utilization of Model 1 - 3 in Configuration 1

Model	Power density (W/m^2)		% Water utilization	
	1st stage	2nd stage	1st stage	2nd stage
1	12.86	9.97	22.29	63.17
2	11.47	9.93	39.01	54.74
3	10.57	10.39	53.55	38.26

4.3 Influence of flow ratio, membrane length and applied hydraulic pressure in Configuration 2

The effect of membrane length in configuration 2 is similar to configuration 1. Figures 4-6 show the fresh feed ratio of 0.25 with different applied hydraulic pressures and flow ratios. At a low fresh feed ratio (model 1, 4, 7), increasing the membrane length of the 1st stage led to a lower power density but higher water utilization. Water utilization of the 1st stage (85%) and the 2nd stage (67%) of model 1 are carried out for the same reason as configuration 1. Complete water utilization of the 1st stage is obtained with an increase in membrane length and applied hydraulic pressure. It can be noticed that the result of power density and water utilization showed the opposite trend. This phenomenon leads to an increase in water flux resulting in high power density and water utilization. The volume of feed water that can permeate across the membrane is deficient with high hydraulic pressure, especially at the end of the membrane. However, the optimal applied hydraulic pressure of the 1st stage is higher than a half of osmosis pressure when the membrane length of the 1st stage is longer. For example, the applied hydraulic pressure of 17 bar offered the highest power density of 9.2 W/m^2 (model 4). Applied hydraulic pressure of 18 bars is optimal (model 7). The optimal applied hydraulic pressure of the 2nd stage is 12 bars.

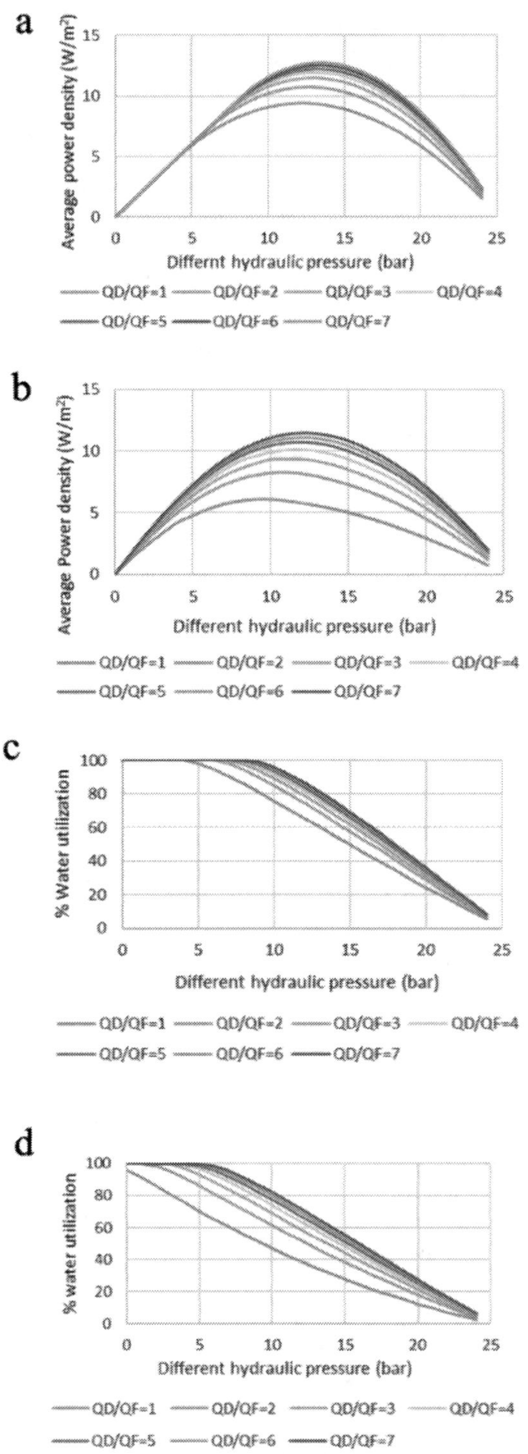

Fig. 4 Model 1 in configuration 2 at fresh feed ratio of 0.25 (a) power density of 1st stage (b) power density 2nd stage (c) water utilization of 1st stage (d) water utilization of 2nd stage

https://doi.org/10.1051/e3sconf/202130201012

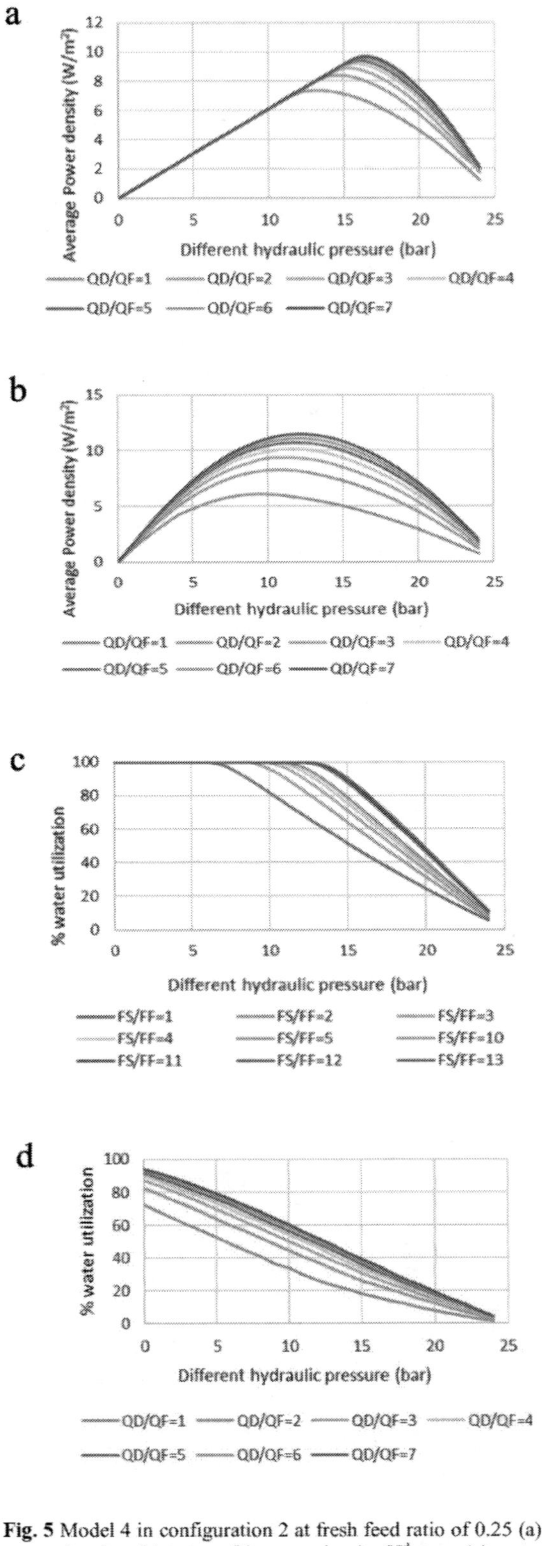

Fig. 5 Model 4 in configuration 2 at fresh feed ratio of 0.25 (a) power density of 1st stage (b) power density 2nd stage (c) water utilization of 1st stage (d) water utilization of 2nd stage

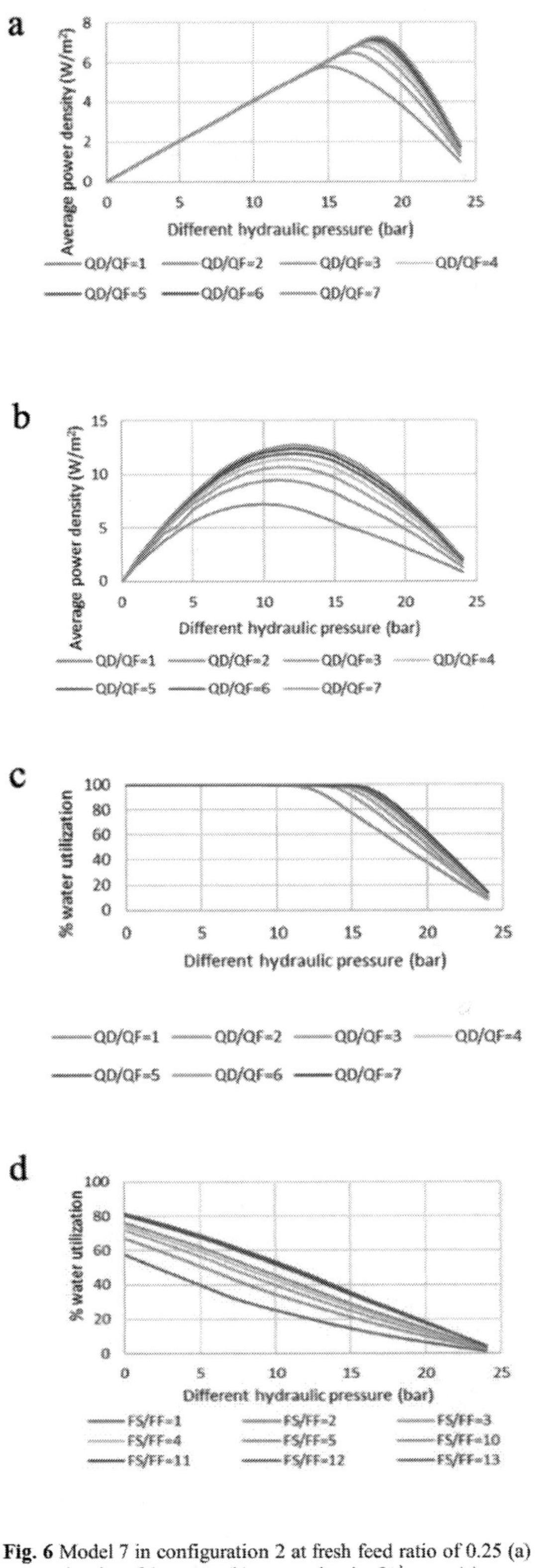

Fig. 6 Model 7 in configuration 2 at fresh feed ratio of 0.25 (a) power density of 1st stage (b) power density 2nd stage (c) water utilization of 1st stage (d) water utilization of 2nd stage

4.4 Influence of fresh feed ratio in configuration 2

Tables 4-6 show the power density and water utilization at a fixed flow ratio of 5 and applied hydraulic pressure of 12 bars. In this work, the fresh feed ratio is varied when draw velocity is fixed. An increase in fresh feed ratio resulted in high power density and water utilization. Because an increase in fresh feed ratio causes a high feed velocity, film thickness on the feed side is decreased, leading to high water flux. To consider models 4 to 6 in configuration 2, the power density of the 1st stage is greater than the 2nd stage. However, the power density of the 1st stage of models 7 to 12 is lower than those of the 2nd stage. It is because the 1st stage of models 4 to 6 has a shorter membrane length. The fresh feed ratio also influenced water utilization. Thus, the lower fresh feed ratio leads to higher water utilization of the 1st stage than the 2nd stage (model 4, 7, 10). However, membrane length substantially impacted water utilization, especially at medium and high fresh feed ratios. When comparing configuration 1 and configuration 2, it can be seen that the average power density of both configurations is similar. The insight behavior shows that film thickness and feeds velocity at the feed side slightly affect power density. The Configuration 1 and Configuration 2 can provide the average water utilization 43% and 78%, respectively. So The Configuration 2 offer more benefits than the Configuration 1. It is because the fresh feed ratio increases the salinity gradient, so water transport is driven. The best conditions of Configuration 2 are fresh feed ratio of 0.25, membrane length of 1st stage of 0.25 m, and membrane length of 2nd stage of 0.75 m. They offered the highest water utilization. To compared with the reported literature [9], the two – stage PRO can provide higher power density than commercial single stage PRO (6.5 W/m²). This is because two-stage PRO improve salinity gradient. The results of this work is similar to He et al. [7] which confirmed that two-stage PRO, divided feed solution performed the better performance in term of average power density. Although adding the second – stage does not improve power generation significantly when fresh feed water is used. The two-stage PRO will important when feed brine is used in the first-stage PRO and replacement the fresh solution in the second – stage PRO [10].

Table 4. Power density and water utilization of model 4- 5 (short membrane length of 1st stage) in configuration 2

Model	Power density (W/m²)		% water utilization	
	1st stage	2nd stage	1st stage	2nd stage
4	12.26	9.89	89	66.99
5	12.73	9.47	44.04	94.38
6	12.83	5.53	29.61	99.23

Table 5. Power density and water utilization of model 7- 9 (medium membrane length of 1st stage) in configuration 2

Model	Power density (W/m²)		% water utilization	
	1st stage	2nd stage	1st stage	2nd stage
7	7.92	10.58	97.37	48.39
8	10.13	10.13	75.64	69
9	11.39	7.91	51.62	99.36

Table 6. Power density and water utilization of model 10- 12 (high membrane length of 1st stage) in configuration 2

Model	Power density (W/m²)		% water utilization	
	1st stage	2nd stage	1st stage	2nd stage
10	5.7	11.88	98.35	27.66
11	9.85	11.05	97.38	38.56
12	10.43	10.69	70.4	73.98

5 Conclusion

The performance analysis of two-stage PRO (i.e., two-stage PRO in series with no feed interstage and two-stage PRO in series with feed interstage) is investigated in terms of power density and water utilization. The two-stage PRO in series with feed interstage is preferred because it can save freshwater requirements. The best condition is fresh feed ratio of 0.25, membrane length of 1st stage of 0.25 m, the flow ratio of 5, and applied hydraulic pressure of 12 bar (model 1).

Acknowledgment

The authors would like to acknowledge the support from King Mongkut's University of Technology North Bangkok fund (KMUTNB-60-GEN-050).

References

1. A. Altaee, A. Sharif. Desalination **356**, 31-46 (2015).
2. W. A. Phillip, J.S. Yong, M. Elimelech. Environmental Science & Technology Journals. **44**, 5170 – 5176 (2010).
3. I.L. Alsvik, M.-B. Hagg. Polymers, **5**, 303-327 (2013).
4. W.Y. Chia, S.R. Chia, K.S. Khoo, K.W. Chew, P. L. Show. Journal of Water Process Engineering. **39**, 101758 (2021).

5. A. Achili, J.L. Prantle, N.T. Hancock, E.B. Maxwell, A.E. Childress. Environmental Science & Technology Journals. **48**, 6437-6443 (2014).
6. M.F. Naguib, J. Maisonneuve, C.B. Laflamme, P. Pillay. Renew. Energy. **76**, 619 – 627 (2015).
7. W. He, Y. Wang, M. Shaheed. Desalination **359**, 186 – 199 (2015).
8. H. Chung, J. Saminathan, J. Lienhard. Desalination **476**, 114230 (2020).
9. N. Sawaki, C. Chen. Desalination **497**, 114767 (2021).
10. A. Alaneza, A. Altaee. Energy Procedia- Journals. **142**, 4182 – 4197 (2017).

The Mechanical Properties of Oil Palm by Using Semi-Damp Dryer

Sakultala Wannapakhe[1,*] and *Khridsadakhon* Booddachan [1]

[1]Department of Agricultural Engineering for Industry, Faculty of Industrial Technology and Management, King Mongkut's University of Technology North Bangkok, 129 Moo 21, Tumbol Noen-Hom, Amphur Muang, 25230, Prachinburi Province, Thailand

Abstract. Hot air drying of palm fruit is a simple method but results in low grade palm oil. The steaming of palm fruit results in high grade palm oil but it requires a wastewater treatment process. Both methods have both advantages and disadvantages, therefore if they were combined, it can result in the better grades of oil, which should be encouraged for farmers in Thailand to apply. This research aimed to determine mechanical characteristics of semi-damp dryer for oil palm. The results of a compression and relative humidity tests were used as the criteria for the tests. The oil palm semi-damp dryer was used in this research. The oil palm semi-damp dryer used the aerosolized technique with 100 °C hot water as a sprayer and a humidifier for the oil palm in the dryer. The purpose was to reduced palm fruit hardness. The study use tenera palm from Prachinburi province and neighborhood of Thailand. The drying time for testing was 1, 2, 3 and 4 hours. The temperature for testing was 70, 90 and more than 100 °C. The results of the compression force of oil palm by using a semi-damp dryer were lower than using a hot air drying. The compression was lower than using hot air drying. The best of mechanical characteristics palm was baking with semi-damp dry at 90 °C in 3 hours. Compression force was 0.675 kN and relative humidity value was 57.18%. The palm oil was tested for the %FFA and DOBI and showed that the grade of palm oil by using semi-damp drying was fair but still edible grade.

Keyword. Oil palm drying, Semi-damp dryer, Oil palm semi-damp drying, Oil palm mechanical properties

1 Introduction

The main process in palm oil production is drying or steaming of palm bunches to inhibit the enzyme reaction that will generate free fatty acids (FFA) and also help the palm fruits to be easily removed from the bunch. The drying operations is primarily influenced by the different drying parameters. Cosme-De Vera et al., (2021) studied and compiled research on drying methods under different drying conditions [1]. Jitwiriya et al., (2020) studied about heat loss analysis of continuous drying oven with outside conveyor chain [2]. The mathematical model can be designed and analyzed for the drying oven before building the hot air oven. For farmers in Thailand, the baking process is an easy-to-use process. It does not require a system for management with the wastewater generated by the palm steaming process. Although oil palm drying has the advantages of being easy-to-use and does not require wastewater management, the dehydrated palm fruit has the disadvantage of being unable to separate palm meat and palm seeds kernels. As a result, the palm oil obtained from the extraction process is lower grade than the palm oil obtained from the palm meat alone. If farmers are able to separate palm meat and palm seeds to enter the oil extraction process, they will get quality oil from palm meat which can be sold at a better price. Kanchanaworavanich et al., [3], National Institute for Materials Science, 2005, analyzed the advantages and disadvantages of steamed and non-steamed palm bunch drying. Ekarat Vaiyanitet et al., [4] developed the new system of palm bunch drying without using the steam. This helps removing the moisture from the palm, makes the palm off the shell more easily, and also inhibits the formation of free fatty acids. Therefore, the palm oil has good quality in A grade with low humidity and low fatty acids. In the last step, there is no need to chase away moisture from the oil again. It can save energy. And when not using water in the production process, it does not cause wastewater. Shehu et al., (2019) [5] studied in kinetics of thermal hydrolysis of crude palm oil with mass and heat transfer in a closed system. The deterioration of bleachability index (DOBI) and carotene content decreased by more than 90% and 95%, respectively, after 90 min of heating at 473 K, while at a temperature of 353 K, the DOBI and carotene content decreased by less than 10% for all samples with different initial water content. The increase in temperature and water content resulted in an increase in %FFA accumulation. There are also other studies that studies on the implementation of the process, from oil to analyze and create [6-12].

*Corresponding author: Sakultala.W@fitm.kmutnb.ac.th

From the study it was found that drying and steaming had different advantages and disadvantages. The advantages of each of these were designed to inhibit the enzyme reactions that will produce free fatty acids (FFA) and also help the palm fruits to be easily removed from the bunch and without the need for a wastewater treatment process. It was suitable for application to the farmers of Thailand. Therefore, to achieve a breakthrough in this regard, the researcher had an idea to learn about drying semi-damp palm oil for baking. The researchers had designed and built a semi-damp oil palm bunch dryer. After baking, the fruit is used to test the mechanical properties of palm oil after each drying temperature and humidity. Moreover, the semi-damp palm oil was studied for the basic properties of palm oil.

2. Materials and methods

2.1 Experimental setup and procedure

To test for mechanical properties of palm oil, the Wannapakhe et al., [13] researched on a hot air dryer with closed-loop oscillating heat pipe with check valves (CLOHP/CV) for reducing energy in drying as shown in Fig. 1. The heat generator was a 4,000 W heater. The hot air from the heater that had passed through the chamber before being released into the atmosphere was used as a source of heat to the CLOHP/CV. When the CLOHP/CV were heated, the heat was transferred to the air before it enters the heater, it was reducing the energy in the drying process. It was added to humidify the air in the hot air chamber. The size of the baking chamber can contain up to 10 kilograms of palm oil fruits, the maximum temperature was 120 °C. The ware forming machine from nature pulp by Juntralux and Watanasriyakul [14] as shown in Fig. 2, it was applied in the press of oil palm fruit to produce palm oil. The machine had the section of a vacuum for sucking water out of the mold, the liquid palm oil was sucked and released through the liquid outlet.

2.2 Operating conditions

For testing, the drying time was 1, 2, 3, and 4 hours. The temperature for testing was 70, 90, and 110 °C. The hot air velocity was 0.5 m/s. The palm fruits (2 kilograms per testing) used for the test were Tenera varieties from the area of Kabin-Buri District, Prachinburi Province and neighborhood. In increasing the moisture, the boiling water can be increased at 99.7 °C and the water pressure up to 1.2 bars.

Then, the palm fruits after drying were tested for finding the compression force by using Universal Testing machine model NRI – TSS500 – 2S. The palm oil was tested for finding the FFA and DOBI value. The FFA and DOBI of palm oils were tested as follows [15]. Grade A crude palm oil (CPO) is controlled with a chemical value of FFA (Free Fatty Acid) not more than 5% of the amount of CPO per unit and DOBI value not lower than 2.0. The appearance of the oil is yellow to red and thick and it is viscous. Grade B crude palm oil is controlled with a chemical value of FFA (Free Fatty Acid) at a level of not more than 8 - 10% of the amount of CPO per 1 unit. Its appearance is a color between yellow, red and dark brown. It depends on the production in the palm drying process of each producer that will affect the burnt palm fruit or the cleanliness of the palm fruit before the extraction process.

2.2.1 Finding FFA

Prepare Sodium Hydroxide (NaOH) into the burette by weighing 0.4g of Sodium Hydroxide into the 250 ml shampoo bottle and pour 100 ml of pure refining water into the 250 ml shampoo bottle. Use the stirring rod to stir the solution to be mixed well. After that, pour it into the burette. Bring the obtained palm liquid into the titrate process for finding FFA by putting the crude palm oil into the oven and setting the temperature at 115 °C to dehydrate for 2 hours. Weigh 5 g of palm oil to put into the 250 ml shampoo bottle. Pour 50 ml of ethanol 95 %v/v into the shampoo bottle containing the palm oil. Drop 1-2 drops of phenolphthalein solution into the bottle and perform titrate process until the sample turns constant pink. The volume of NaOH solution (V) was recorded to calculate the FFA as equation 1.

$$\text{FFA (wt.\%)} = \frac{V \times N \times 25.6}{M} \tag{1}$$

V is the volume of NaOH (ml)
N is the standard solution concentration of NaOH (m)
25.6 is the molecular mass weight of palmitic acid
m is the weight of palm oil (g)

2.2.2 Finding DOBI

Put the obtained palm oil into Spectrophotometer to find DOBI by putting the crude palm oil into the oven and setting the temperature at 115 °C to dehydrate for 2 hours. Put the palm oil into the glass tube and measure the value by using Spectrophotometer. The wavelengths for the set absorbance test were 446 and 269 nm. The absorbance was recorded to calculate the DOBI by using equation 2.

$$\text{DOBI} = \text{Abs}_{446} / \text{Abs}_{269} \tag{2}$$

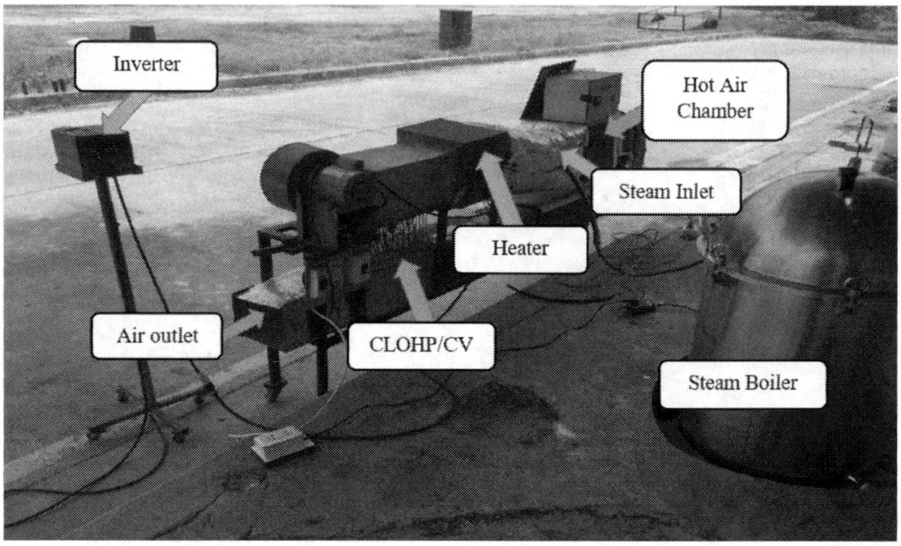

Fig. 1. The semi-damp dryer

Fig. 2. The ware forming machine from nature pulp [14]

3. Results and discussions

3.1 The results of the study on air conditions inside the drying chamber

Data from a study that tested a palm hot air and semi-dry. The relative humidity of both baking treatments was measured as shown in figure 6 – 8. If the air temperature was raised to more than 100 °C, the air in the palm fruit drying chamber cannot be semi-humidified. For measurements in the air temperature range of 70 and 90 °C, in the case of the hot air drying process, the hot air drying humidity was the percentage of about 5-20. In the case of the hot air semi-damp drying process, the hot air drying humidity was the percentage of about 37-60.

3.2 The results of the study on mechanical properties of palm fruits after a drying process

As shown in figure 3-4, the heat-treated palm fruit has a brown skin color and dry palm. The semi-dampened palm has a dark red skin color and the palm was moist enough with water. When the dried palm fruit was tested for mechanical properties by using the Universal Testing Machine model NRI – TSS500 – 2S as figure 5. The test results were shown in figure 9-12. As the baking temperature increases, the compression force required to press decreases. Due to the ripening of the palm fruit, it had an effect on the pulp of the palm fruit was soft. Compared with the drying process, it was found that the palm dried by using the hot air drying has a higher compression force value than the semi-dryer at all drying temperatures. in case of the drying temperature was more than 100 °C, it was found that the compression force used was similar between hot air drying and semi-damp drying. Because in the condition of drying at a temperature of more than 100 °C, the humidity of the air inside the drying chamber was not much higher than that of the general hot air dryer. It resulted in a semi-dry drying at temperatures exceeding 1 0 0 °C no different from hot air drying.

Fig. 3. The heat treated palms

Fig. 4. The semi-damp dried palms

Fig. 5. The compression force testing by Universal Testing Machine model NRI – TSS500 – 2S

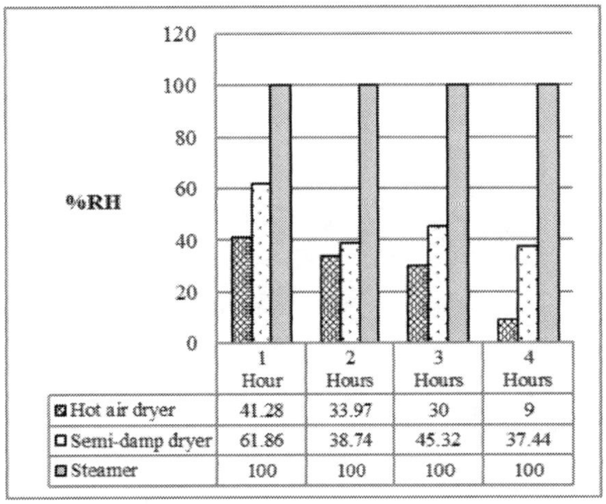

Fig. 6. Relative humidity in case of drying at 70 °C

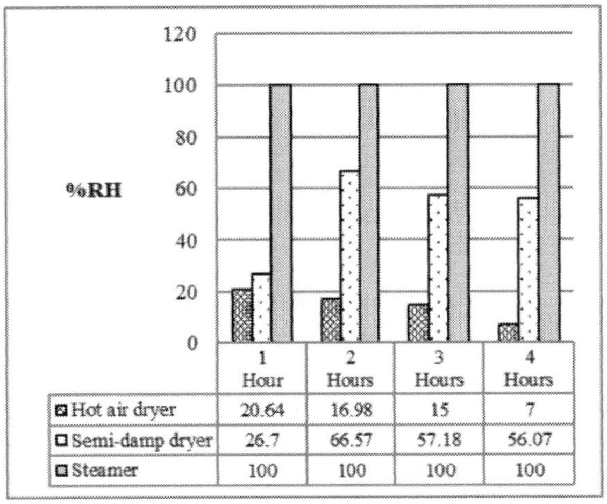

Fig. 7. Relative humidity in case of drying at 90 °C

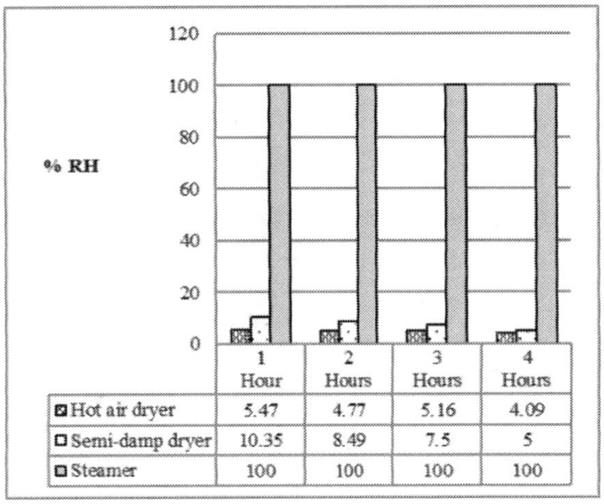

Fig. 8. Relative humidity in case of drying at 110 °C

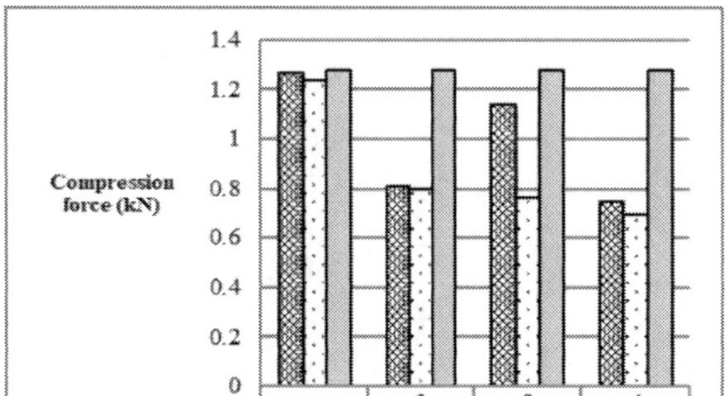

	1 Hour	2 Hours	3 Hours	4 Hours
Hot air dryer	1.269	0.81	1.139	0.743
Semi-damp dryer	1.238	0.8	0.76	0.694
Steamer	1.276	1.276	1.276	1.276

Fig. 9. Compression force in case of drying at 70 °C

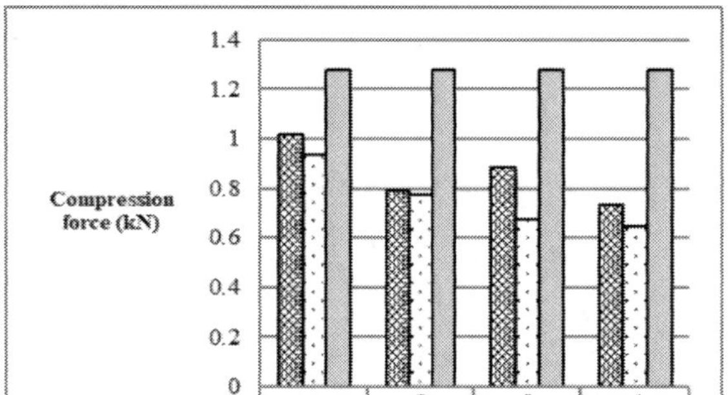

	1 Hour	2 Hours	3 Hours	4 Hours
Hot air dryer	1.019	0.792	0.882	0.733
Semi-damp dryer	0.938	0.773	0.675	0.645
Steamer	1.276	1.276	1.276	1.276

Fig. 10. Compression force in case of drying at 90 °C

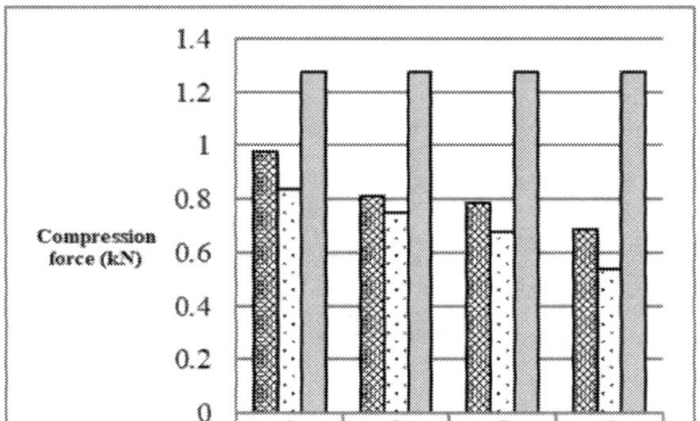

	1 Hour	2 Hours	3 Hours	4 Hours
Hot air dryer	0.976	0.809	0.784	0.685
Semi-damp dryer	0.835	0.752	0.677	0.537
Steamer	1.276	1.276	1.276	1.276

Fig. 11. Compression force in case of drying at 110 °C

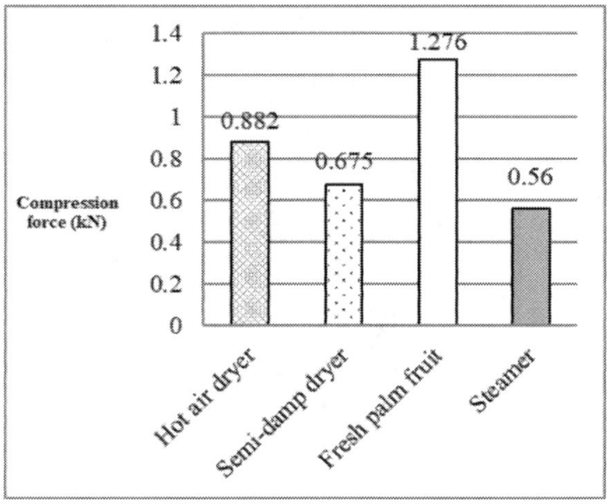

Fig. 12. Comparison of the compression force of 90 °C at drying time 3 hours

3.3 The results of testing properties of palm oil

From the test to study the condition of semi-damp drying, it was found that if the drying air temperature was increased more than 100 °C, semi-damp drying could not occur. The properties of palm oil were tested at temperatures of 70 and 90 °C only, and the drying time was 3 hours. From the test results as shown in figure 13, the results of the experiment showed that drying at 90°C for 3 hours, stored for 1 day in semi-damp drying, and was the best at % FFA. Because the semi-damp was helped to stop lipolysis better than hot air drying. The semi-damp drying, hot steam can penetrate into the palm fruit. As a result, the palm pulp obtained from the semi-damp drying process was better than the conventional hot air drying process which produced free fatty acids in the palm fruit. The free fatty acid content in oil is an indicator of the quality of the oil that will be used for biodiesel production or consumption. If the %FFA is high, it will easily oxidize in nature, causing a rancid smell. When the palm oil was analyzed for DOBI in palm oil by center of excellence-oil palm Kasetsart university, Thailand, it was found that the CPO from hot air drying at 90 °C, drying time was 3 hours, DOBI value was 0.63. The oil grade was based on the DOBI value of the freshness of crude palm oil. If the value is lower than 2.2, it means that the CPO was not fresh. The oil was not suitable for consumption. The CPO from semi-damp drying at 90 °C, drying time 3 hours, DOBI value was 2.35, CPO was in grade 2.31 - 2.92, which was usable quality. The oil was suitable for consumption. The results are similar to the research of Pootao and Kanjanapongkul [15]. Increasing voltage, temperature and time resulted in higher yield and lower FFA while quality of extracted palm oil still met the standard.

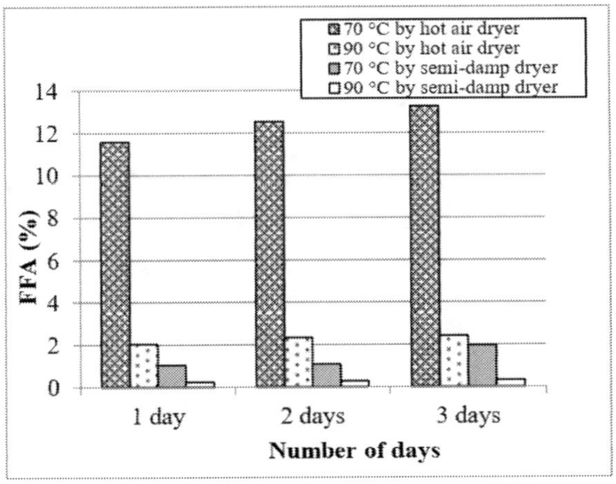

Fig. 13. The free fatty acid content (FFA) of hot air drying and semi-damp drying

4. Conclusion

Semi-dryer, if the air temperature was more than 100 °C, it was not possible to make the air in the semi-moisture drying chamber. The mechanical properties test by using Universal Testing Machine, it was found that when the drying temperature increased, the compression force used to press decreased. The palm drying using hot air dryer has higher compression force values than semi-damp dryer at all test temperatures. Except for the semi-damp dryer at the drying temperature of more than 100 °C, it was found that the compression force was similar between hot air drying and semi-damp drying. Therefore, it can be concluded that if semi-damp drying should be in the range of 70 - 90 °C. The properties of palm oil from the study data were conducted at 70°C and 90°C, both hot air and semi-damp drying for 3 hours, with the oil being stored for 1 day, 2 days and 3 days. It was found that the best value of CPO was semi-damp drying at 90 °C for 3 hours of drying time with 1 day of CPO left, %FFA was 0.25%. The best DOBI test was at Semi-damp drying at 90°C, drying time 3 hours, with DOBI value of 2.35. From the test results, it can be concluded that semi-damp drying should be done. Because the properties of CPO were a higher grade than CPO obtained from hot air drying.

Acknowledgements

The author would like to express their appreciation to King Mongkut's University of Technology North Bangkok for providing financial support for this study.

References

1. F.H. Cosme-De Vera, A.N. Soriano, N.P. Dugos, R.V.C. Rubi, A comprehensive review on the drying kinetics of common tubers, Applied Science and Engineering Progress, **14**, 2 (2021): 146-155
2. J. Weerayut, C. Tonkid, N. Udomkiat, Y. Pisit, Heat Loss Analysis of Continuous Drying Oven with Outside Conveyor Chain, Applied Science and Engineering Progress, (2020)
3. K. Boonyarak, Study and analyze the advantages and the disadvantages of non-steam and steam-based palm oil extraction systems. National Institute for Materials Science, (2015)
4. V. Ekarat, Non-steam palm oil extraction system. National Institute for Materials Science, (2015)
5. Shehu, E. Umar, Kinetics of thermal hydrolysis of crude palm oil with mass and heat transfer in a closed system, Food and Bioproducts Processing, **118** (2019): 187-197
6. S. Yusoff, Renewable energy from palm oil–innovation on effective utilization of waste, Journal of cleaner production, **14**, 1 (2006): 87-93
7. Y. Basiron, Palm oil production through sustainable plantations, European Journal of Lipid Science and Technology, **109**, 4 (2007): 289-295
8. P. Somporn, Gheewala, H. Shabbir, Full chain energy analysis of biodiesel production from palm oil in Thailand, Applied energy, **86** (2009): 209-214
9. M.J. Junaidah, A.R. Norizzah, O. Zaliha, Effect of sterilization process on deterioration of bleach ability Index (DOBI) of crude palm oil (CPO) extracted from different degree of oil palm ripens, **3**, 4 (2013): 322-327
10. B.A. Orhevba, O. Chukwu, V. Oguagwu, Z.D. Osunde, Effect of moisture content on some quality parameters of mechanically expressed neem seed kernel oil. The International Journal of Engineering and Science, **2**, 8 (2013): 01-07
11. I.M. Ogan, D. Marie-Josée, N. Michael, Palm oil: Processing, characterization and utilization in the food industry–A review, Food bioscience, **10** (2015): 26-41
12. B. Murad, S. Nadia, S. Jumat, D-Optimal Design Optimization for the Separation of Oleic Acid from Malaysian High Free Fatty Acid Crude Palm Oil Fatty Acids Mixture Using Urea Complex Fractionation, Applied Science and Engineering Progress, **14**, 2 (2021): 175-186
13. S. Wanapakhe, T. Chaiwong, M. Dandee, S. Prompakdee, Hot air dryer with closed-loop oscillating heat pipe with check valves for reducing energy in drying process, Procedia Engineering, **32** (2012): 77-82
14. J. Noppadol, W. Somnuk, Design and construction of ware forming machine from nature pulp, IE Network Conference, Phetchaburi, Thailand, (2012): 1770-1775
15. P. Sunisa, K. Kobsak, Pretreatment of oil palm fruit using ohmic heating, The 16th TSAE National Conference and the 8th TSAE International Conference, Bangkok, Thailand, (2015)

Comparison between pressure retarded osmosis model using batch and continuous water supply sources

Kingchat Potisa-ad[1], *Lida* Simasatitkul[2], and *Suksun* Amornraksa[1,*]

[1]The Sirindhorn International Thai-German Graduate School of Engineering (TGGS), King Monkut's University of Technology North Bangkok, Bangkok, 10800 Thailand
[2]Department of Industrial Chemistry, Faculty of Applied Science, King Monkut's University of Technology North Bangkk, Bangkok, 10800 Thailand

Abstract. Pressure retarded osmosis (PRO) is a novel renewable energy technology that generates electricity from two water sources. Due to the osmotic pressure difference, freshwater permeates across a membrane to the other side, where the high-pressure seawater flows and drives a turbine to generate power. Many mathematic models have been proposed to evaluate the performance of a PRO. However, it was found that most performance of the PRO that have been reported were performance by using freshwater with limited supply (batch) in the model. It is not accurate as, in practice, the supply of freshwater occurs in a continuous manner. In this work, the influence of batch and continuous supply of fresh water on the performance of PRO was demonstrated. The effect of flow direction, i.e., concurrent and counter-current flows, was also examined. The model simulation was performed by using MATLLAB program, and the performance of PRO is expressed in terms of average power density. The results revealed that the batch and continuous supplies of freshwater had a strong impact on the performance of the PRO. However, the performance of concurrent and counter-current flow were not significantly different.

Keyword. Pressure retarded osmosis, counter-current flow, Osmotic power

1 Introduction

The energy resources for human activities are supported by fossil combustion, which releases high emissions of greenhouse gases. It accelerates climate change and global warming. To reduce the reliance on fossil fuels, alternative sources, which are environmentally friendly, need to be developed and used. Among various renewable energy sources, pressure retarded osmosis is a new energy generation technology that has a good potential to harvest energy from nature. Salinity gradient energy, given by mixing two solutions with different salinity concentrations, is considered promising renewable energy [1]. In theory, the salinity gradient energy has the advantage of high energy density [2]. With a proper method, it is possible to generate electricity from the salinity gradient energy, i.e., by using pressure retarded osmosis (PRO) technology [3].

PRO is an attractive membrane technology for generating electricity by utilization a salinity gradient [4]. This technology operates based on different osmotic pressure between seawater (draw solution) and freshwater (feed solution). Fig. 1 inllustrates a schematic diagram of a typical PRO osmotic pressure process with a continuous and steady-state flow [5]. Both draw and feed solutions are fed into a membrane module with a selected semi-permeable membrane installed to separate both streams. The draw solution is pumped to high pressure then it passes through a pressure exchanger before entering the module, while the feed solution enters the module at atmospheric or low pressure. The osmotic pressure difference between both streams causes the freshwater to permeate across the membrane to the other side where the high salinity seawater flows. Freshwater passing through membrane to salty water increase the volume which is diviede into two streams. The first one is the permeated volume is then used to drive the turbine to produce energy [5, 6]. Other stream passes through pressure exchanger through pressure exchanger for energy recovery.

The performance of PRO is critical for its economic viability; therefore, it is very important to evaluate the performance accurately and precisely. There are two operation modes of the PRO process; concurrent and counter-current flow operations. Most simulation studies were performed based on the concurrent flow operation, even though some reports indicated that the counter-current flow is more efficient at extracting energy than concurrent flow [7]. It may be because the simulation steps for concurrent flow are much less complicated than counter-current flow. However, we have noticed that the PRO performance reported in many works of literature was performed based on mathematic models that used a limited supply of freshwater (batch volume) as feed [8-10]. We realized that they used the batch water supply when we tried to repeat and verify their simulation result. The results from those models that used the batch water supply may not represent the real performance as,

* Corresponding author: suksun.a@tggs.kmutnb.ac.th

in practice, the actual flow of freshwater is continuous (unlimited supply). Therefore, it is crucial to examine the effect of the batch and continuous freshwater supplies on the performance of PRO. In addition, the draw and feed solutions can flow into the PRO module in two different directions: concurrent and counter-current flows. The different directions may influence the mass transfer between the draw and feed solutions. In this regard, it is also essential to investigate the effect of flow direction on the performance of PRO.

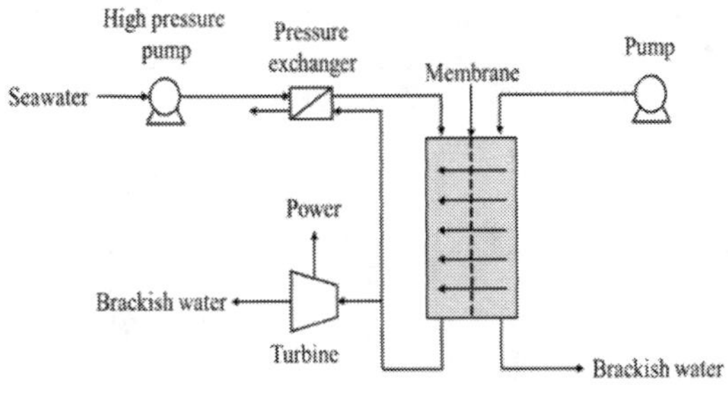

Fig. 1. Pressure retarded osmosis (PRO) presess [5].

2 Methodology

A mathematical model for the PRO process was developed and used to calculate the performance of the PRO process. The effect of concentration polarization, which consists of internal concentration polarization (ICP) and external concentration polarization (ECP), was also included in the model. A schematic diagram of the PRO module with concurrent flow and counter-current flow configurations is illustrated in Fig. 2.

(a)

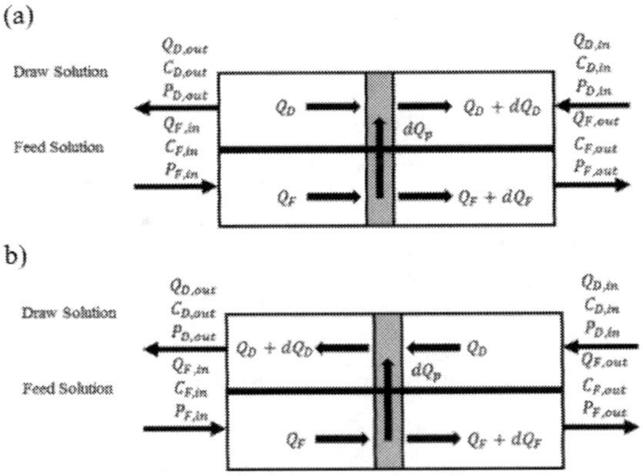

Fig. 2. (a) Schematic diagram of a PRO mass exchanger in concurrent flow configuration. (b) Schematic diagram of a PRO mass exchanger in counter-current flow configuration [11].

A semi-permeable membrane separates the module into two channels. A feed solution with a low salt concentration flows across the membrane to the other side, where a draw solution with high salt concentration flows. Both solutions flow in the same direction for the concurrent flow configuration (Fig. 1a) and in the opposite direction for the counter-current flow configuration (Fig. 1b). The inlet and outlet boundary conditions of both feed and draw streams are given as indicated in Fig. 2.

The performance of the PRO process can be defined as the power density (W), which can be calculated by,

$$W = A_W(\Delta\pi - \Delta P)\Delta P \quad (1)$$

where A_W is the membrane permeability coefficient, ΔP is the hydraulic pressure difference across the membrane, and $\Delta\pi$ is the osmotic pressure as a function of molar concentration following the van't Hoff equation:

$$\pi = \nu RTc \quad (2)$$

where ν, c, R, and T are the van't Hoff coefficient, the concentration difference across the membrane, the gas constant, and temperature. The mass transfer equations of water along the membrane channels in concurrent flow module for the draw and feed solution can be expressed as [12]:

$$\frac{dQ_D(S)}{dA_m} = J_W(C_D(S), C_F(S), \Delta P) \quad (3)$$

$$\frac{dQ_F(S)}{dA_m} = -J_W(C_D(S), C_F(S), \Delta P) \quad (4)$$

where Q_D and Q_F are the flow rates of the draw and feed solutions, respectively, C_D and C_F are the salt concentrations of the draw and feed solutions, respectively, A_W is defined as the total membrane area from the feed entrance, and J_W is the water flux through the membrane, which is a function of the salt concentration in the draw and feed solutions and the difference of hydraulic pressure. The salt concentrations distributions (J_S) in both draw and feed channels can be expressed as:

$$\frac{d(Q_D(S)C_D(S))}{dA_m} = -J_S(C_D(S), C_F(S), \Delta P) \quad (5)$$

$$\frac{d(Q_F(S)C_D(S))}{dA_m} = J_S(C_D(S), C_F(S), \Delta P) \quad (6)$$

In the counter-current flow operation, the mass transfer equations can be expressed as [12]:

$$\frac{dQ_D(S)}{dA_m} = J_W(C_D(S), C_F(S), \Delta P) \quad (7)$$

$$\frac{dQ_F(S)}{dA_m} = J_W(C_D(S), C_F(S), \Delta P) \quad (8)$$

$$\frac{d(Q_D(S)C_D(S))}{dA_m} = -J_S(C_D(S), C_F(S), \Delta P) \quad (9)$$

https://doi.org/10.1051/e3sconf/202130201014

$$\frac{d\left(Q_F(S)C_D(S)\right)}{dA_m} = -J_S\left(C_D(S), C_F(S), \Delta P\right) \tag{10}$$

With the effect of ICP and ECP, the water permeates flux, and salt permeate flux in PRO modules can be given in equations (11) and (12), respectively [13].

$$J_W = A_W\left\{\frac{\pi_{D,b}\exp\left(\frac{-J_W}{k}\right) - \pi_{F,b}\exp\left(J_W K\right)}{1+\frac{B}{J_W}\left[\exp\left(J_W K\right) - \exp\left(\frac{-J_W}{k}\right)\right]} - \Delta P\right\} \tag{11}$$

$$J_S = B\left\{\frac{\pi_{D,b}\exp\left(\frac{-J_W}{k}\right) - \pi_{F,b}\exp\left(J_W K\right)}{1+\frac{B}{J_W}\left[\exp\left(J_W K\right) - \exp\left(\frac{-J_W}{k}\right)\right]} - \Delta P\right\} \tag{12}$$

where $\pi_{D,b}$ and $\pi_{F,b}$ are the osmotic pressure of the bulk draw and feed solutions, respectively. B is the solute permeability coefficient, k is the mass transfer coefficient, and K is the solute resistivity for diffusion within the membrane porous support layer, which is calculated from the following equation [14]:

$$K = \frac{\delta}{D} \tag{13}$$

In equation (13), δ is a film thickness on both the feed and draw sides of the membrane, and D is a solute diffusion coefficient. On the other hand, k is calculated by equations (14) and (15), where dh is the hydraulic diameter of the flow channel, Re is Reynolds numbers, Sh is the Sherwood number, and Sc is the Schmidt number. Due to the mass transfer in the spacer-filled flow channel was turbulent at a relatively low Reynolds number (Re<50). In this case, the Sherwood number can be determined from equation (15) [15].

$$k = \frac{ShD}{dh} \tag{14}$$

$$Sh = 0.2\,\mathrm{Re}^{0.57}\,Sc^{0.4} \tag{15}$$

Other than the power density, the performance of the PRO process is also evaluated in terms of water utilization rate. The water utilization rate is the ratio of the permeate flow rate ($Q_{p,w}$) to the inlet flow rate ($Q_F(in)$) of the feed solution [10] as shown in equations (16) and (17). In implies how efficiently the feed water can be utilized. It is very useful information, especially when the freshwater is limited or not abundant. In the batch supply, the water utilization rate is less than 100%. However, in the continuous supply, the rate can be higher than 100% as water feed is always replenished along the membrane length.

$$\%Ut = \frac{\sum_{i=1}^{L} Q_{p,w}}{Q_F(in)} \mathrm{x}100 \tag{16}$$

$$Q_{p,w} = \int_a^b J_W \, \mathrm{x} d\, A_m \tag{17}$$

The model simulation in this work was performed by using the MATLAB program. The performance of the PRO process was evaluated in terms of average power density. It is assumed that the pressure drop through the flow channel on both sides of the membrane is negligible, and thus, the hydraulic pressure difference between the draw and feed side remains constant over the membrane length [11]. For simulation, the membrane is divided into n number of pieces. The water that flows into the channels at the starting point is denoted as i=0. The number continuously increases until it reaches n at the ending point. Water flow rates and salt concentration at membrane piece i+1 are calculated based on the permeate through membrane piece i. The flow chart for a mathematical model of the PRO power system is summarized in Fig. 3. Note that in the simulation, the feed flow rate decreases along the membrane length in the batch water supply. However, it is constant in the case of continuous supply.

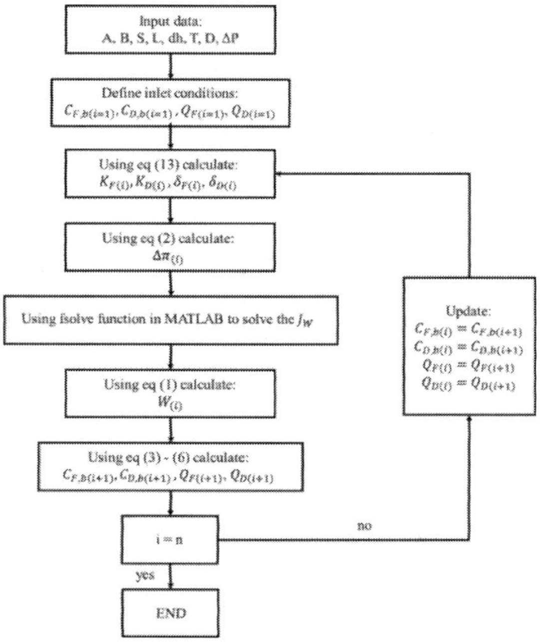

Fig. 3. Flow chart for mathematical model of concurrent PRO power system [8].

3 Results and discussions

The effect of batch and continuous water supply sources on the mass transfer of salt and water flux across the membrane were investigated and compared in this section. The configuration and operating conditions used for the simulations in this work were summarized and shown in Table 1. It is important to note that the membrane length at 10 meters was selected to demonstrate the effects more clearly. In practice, the membrane length is usually much shorter.

Table 1. Conditions for commercial-scale simulation [8].

Parameters	Value
Feed concentration, $C_{F,b\ (in)}$ (g/l)	0
Draw concentration, $C_{D,b\ (in)}$ (g/l)	30
Temperature, T (°C)	25
Membrane length, L (m)	10
The radius of hollow fiber, r (mm)	0.1
Radius of module casing, R (mm)	0.15
Feed cross-flow velocity, $u_{F\ (in)}$ (m/s)	0.25
Draw cross-flow velocity, $u_{D\ (in)}$ (m/s)	0.25
Salt diffusion coefficient, D (m^2/s)	1.5e-9

3.1 The effect of concurrent and counter-current flow operation

First of all, the effect of concurrent and counter-current flow operations was investigated to determine their impact on the PRO performance. The simulation was performed with the continuous water supply power over different applied hydraulic pressures. The power density obtained from both flow directions were illustrated in Fig. 4. It can be seen that the power density obtained from the concurrent and counter-current flow at the same apply hydraulic pressure were very similar. The counter-current offered slightly better values, especially at the pressure over 10 bars. However, the differences between them are generally less than 5%. Therefore, it can be concluded that the flow direction of the draw and feed solution did not significantly affect the performance of the PRO process. Consequently, to make the simulation much less complicated, all the simulations in the following sections were performed based the concurrent flow operation only.

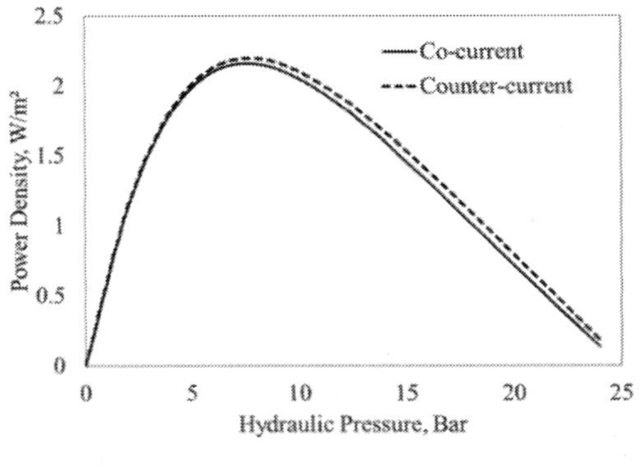

Fig. 4. The power density obtained from the concurrent flow and counter-current flow operations using the continuous supply of fresh water.

3.2 The effect of batch and continuous water supply sources

3.2.1 Salt concentration

The salt concentration distributions of feed and draw solutions along the membrane length are illustrated in Fig. 5. On the feed solution side, it can be seen that the salt concentration of the continuous feed solution is relatively constant. The reason is simply that the freshwater is continually replenished whenever it permeates across the membrane. So, the overall salt concentration inside the solution did not significantly change. However, on the other hand, the salt concentration of the batch feed solution is increased gradually and significantly along the membrane length. This is because once the freshwater permeates through the membrane, it is gone for good, with no water addition. So, as the salt flux comes from the other side, the salt concentration of the feed solution increases rapidly.

On the draw solution side, it can be seen that the salt concentration gradually decreases along the membrane length as a result of water permeate across the membrane. However, it can be noticed that the salt concentration in the case of batch water supply decreases slowly after passing half of the membrane range. It is likely because the salt concentration in the feed solution is relatively high. It resulted in less driving force for mass transfer (the difference between the osmotic pressure difference and the applied hydraulic pressure). So, lesser and lesser volume of freshwater could permeate across the membrane.

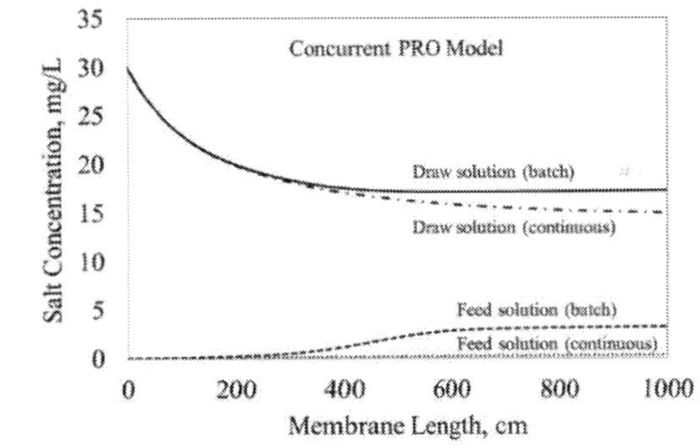

Fig. 5. Salt concentration distribution of draw and feed solutions along the membrane length.

It can be seen that if we operate the PRO process using a very long length of membrane or connecting the membranes in a series, the salt concentration, and thus the osmotic pressure difference, will significantly affect if we select the batch water supply. However, for a short length of membrane, i.e., less than 2 meters, the calculated performance of the PRO may still be acceptable even if the batch water supply is used.

3.2.2 Water flux

The water flux across the membrane length for batch and continuous water supply sources are presented in Fig. 6. It can be seen from the figure that the water flux for both cases had the same trends, i.e., continuously decreasing along the length of the membrane. However, the water flux of the batch water supply decreased more rapidly than that of the continuous water supply. It happened because there is no new fresh water in the case of the batch water supply; thus, the volume of remaining fresh water on the feed side is continuously decreasing along the membrane length. Consequently, when the salt permeates across the membrane, the salt concentration in the feed solution was increased rapidly. It is different for a continuous water supply because the volume of the remaining fresh water on the feed side is always constant due to fresh water makeup. So, the salt concentration in the feed solution was increased relatively slower than in the batch supply case. As the salt concentration in the feed solution is increased, the driving force, i.e., the osmotic pressure difference between the draw and feed solutions, is also deceased. A simple equation of water flux can explain this. $J_W = A(\Delta\pi - \Delta P)$ when the osmotic pressure difference between the draw and feed solutions ($\Delta\pi_{D,b} - \Delta\pi_{F,b}$), which as a function of salt concentration in the draw and feed solutions decreases, the permeate water flux J_W will be decreased as well. It can be noticed from Fig. 6 that the water flux in batch mode dropped to zero after approximately 6 meters of the length. This result explains why the salt concentration of the draw solution of the batch mode in Fig. 5 no longer decreased after 6 meters.

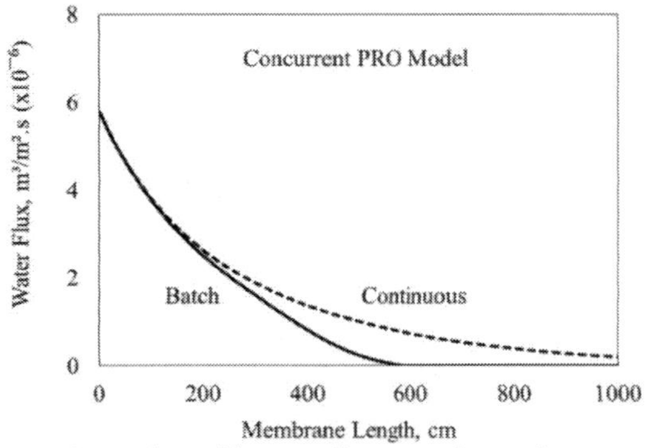

Fig. 6. Comparison of the water flux across the membrane length using batch continuous water supply sources.

In contrast, the water flux in the case of the continuous water supply was continuously penetrated across the membrane until the end of the length, although the salt concentration of the draw solution was continuously decreased. The increase in water flux is directly linked to the power density obtained by the PRO process (eq 1). Hence, it is expected that the power density of the continuous water supply will be higher than the batch water supply.

3.2.3 Power density

The power density produced from the PRO using batch and continuous water supply sources at different applied hydraulic pressure was determined and presented in Fig. 7. It can be seen that the power density of the batch and continuous water supply sources were significantly different, especially at the low range of applied hydraulic pressure. However, they are not much different when the applied hydraulic pressure was more than 15 bars. It was found that the maximum power density for the continuous water supply was 2.26 W/m² at 8 bars. For the same hydraulic pressure, the power density of the batch water supply as 0.99 W/m². Therefore, the percentage of error is 54.2 present. Suppose we use the data from batch water supply to do the economic calculation. Likely, the project will not be viable. Thus, it is essential to do the simulation correctly using the continuous water supply.

In the studies that use freshwater and seawater as the feed and draw solutions, the maximum power density for operation can be achieved when applied hydraulic pressure is equal to half of the osmotic pressure difference ($\Delta P = \Delta\pi/2$) is about 12 bars [13, 8]. This result is consistent with the result of batch water supply. However, if we use the continuous water supply model, the maximum power density could be obtained at lower applied hydraulic pressures. This result suggests that the PRO process may be suitable to operste at the applied hydraulic pressures lower than half of the osmotic pressure difference. This lower pressure condition is favorable as it can reduce the electricity cost of the pump and give more flexibility for the design and construction of the PRO system.

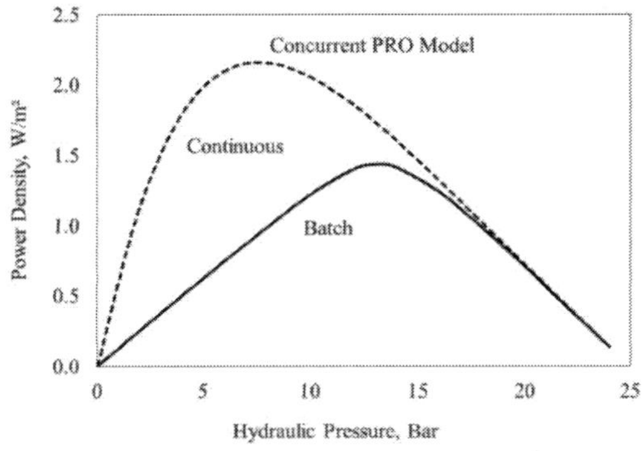

Fig. 7. Comparison of the power density of PRO obtained from batch and continuous water supply sources

3.2.4 Water utilization

The water utilization obtained from the batch and continuous water supply sources at different applied hydraulic pressure were calculated and shown in Fig. 8. In general, it can be seen that the maximum water utilization for batch water supply is 100 percent as all the water permeates through the membrane, and no additional water was added. On the other hand, at high pressure, the water hardly permeated through the membrane. However, some water remained in the water solution feed, and thus the water utilization was not 100 percent.

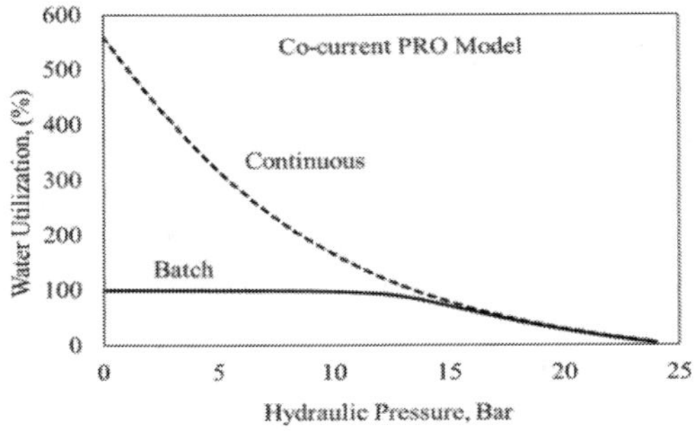

Fig. 8. Comparison of the water utilization of PRO obtained from batch and continuous water supply sources

In the continuous water supply, the water could easily permeate through the membrane, especially at low applied hydraulic pressure. As the water supply was constantly replenished, more volumes of water could permeate through the membrane. It made the percentage of water utilization to be higher than 100%. So, it depends on the availability of freshwater. We could formulate an operating strategy that utilizes the freshwater supply most efficiently while maximizing the power density.

4 Conclusion

For the results, it can be concluded that the effect of flow direction, i.e., concurrent and counter-current, did not significantly affect the performance of the PRO. On the other hand, the batch and continuous water supply sources have a strong influence on the PRO's performance. Using the continuous water supply can increase permeate water through the membrane, boosting the power density by 54.2%. Therefore, it is very important to select the condition of the water supply for the simulation carefully. Unless the designers have a specific reason to choose the batch operation, the continuous water supply should be used to simulate and estimate the project's cost.

Acknowledgement

This work was supported from King Mongkut's University of Technology North Bangkok found (KMUTNB-60-GEN-050) is gratefully acknowledged.

References

1. B. E. Logan, M. Elimelech, Membrane-based processes for sustainable power generation using water, nature, **488** (2012): 313–319.
2. N. Y. Yip, M. Elimelech, Thermodynamic and energy efficiency analysis of power generation from natural salinity gradients by pressure retarded osmosis, Environmental Science & Technology, **46** (2012): 5230-5239.
3. A.G.L. Borthwick, Marine renewable energy seascape, Engineering, **2** (2016): 69-78.
4. S. Loeb, R.S. Norman, Osmosis power plants, Science, **189** (1975): 654-655.
5. G. Han, S. Zhang, X. Li, and T.S. Chung, Progress in pressure retarded osmosis (PRO) membrane for osmotic power generation, Progress in polymer science, **51** (2015): 1-27.
6. S. Loeb, Large-scale power production by pressure-retarded osmosis using river water and seawater passing through spiral modules, Desalination, **143** (2002): 115-122.
7. S. Lin, A.P. Straub, M. Elimelech, Thermodynamic limits of extractable energy by pressure retarded osmosis, Energy & Environmental Science, **7** (2014): 2706–2714.
8. M. F. Naguib, J. Maisonneuve, C.B. Laflamme, and P. Pillay, Modeling pressure-retarded osmotic power in commercial length membrane, Renewable energy, **76** (2015): 619-627.
9. J. Maisonneuve, P. Pillay, C.B. Laflamme, Pressure-retarded osmotic power system model considering non-ideal effects, Renewable energy, **75** (2015): 416-424.
10. W. Yang, L. Song, J. Zhao, Y. Chen, and B. Hu, Numerical analysis of performance of ideal counter-current flow pressure retarded osmosis, Desalination, **433** (2018): 41-472.
11. M. H. Sharqawy, and L. D. Banchik, Effectiveness-mass transfer unit (Ɛ-MUT) model of an ideal pressure retarded osmosis membrane mass exchanger, Journal of membrane science, **445** (2013): 211-219.
12. A. P. Straub, S. Lin, and M. Elimelech, Module-scale analysis of pressure retarded osmosis: performance limitations and implications for full-scale operation, Environmental science & Technology, **445** (2013): 211-219.
13. A. Altaee, and N. Hilal, Dual stage PRO power generation from brackish water brine and wastewater effluent feeds, Desalination, **389** (2016): 68-77.
14. A. Achilli, T.Y. Cath, and A.E. Childress, Power generation with pressure retarded osmosis: an experimental and theoretical investigation, Journal of membrane science, **343** (2009): 42-52.
15. C.P. Koutsou, S.G. Yiantsios, A.J. Karabelas, A numerical and experimental study of mass transfer in spacer-filled channels: Effects of spacer geometrical characteristics and Schmidt number, Journal of Membrane Science, **326** (2009): 234–251.

Perspectives of anaerobic decomposition of biomass for sustainable biogas production: A Review

S. Indran[1,*], *D. Divya*[2], *Sanjay Mavinkere Rangappa*[3], *Suchart Siengchin*[3], *P. Merlin Christy*[4], *L.R.Gopinath*[4]

[1]Rohini College of Engineering and Technology, Palkulam-629401, Kanyakumari District, TN, India.
[2]Research and Development Department, Pinnacle Bio-Sciences, Kanyakumari-629701, TN, India.
[3] Natural Composites Research Group Lab, Department of Materials and Production Engineering, The Sirindhorn International Thai-German Graduate School of Engineering, King Mongkut's 13 University of Technology North Bangkok, Bangkok-10800, Thailand.
[4]Vivekanandha College of Arts and Sciences for Women, Tiruchengode-637205, Namakkal, TN, India.

Abstract. Biogas production from biomass is an eco-friendly approach that offers renewable energy generation, waste recycling, biofertilizer production along with maintaining environmental quality. Anaerobic decomposition is a familiar practice used for biogas production in worldwide, whereas only few substrates were convenient for attaining desired methane concentration in biogas. Hence, further advancements are necessary for the exploration and utilization of various complex organic materials for the purpose. This article gives a clear outlook on potential of various biomass for biogas production, necessity of pretreatment, applicability of microbial/enzyme addition, maintenance of various process parameters, formulation of suitable digester designs and future scope of this area. The livestock wastes and agricultural wastes possess high energy generation potential (71%) and sustainable utilization of such wastes are admirable to commercialize biogas production in future. Hence, selection of biomass through biochemical methane potential (BMP) analysis and biomass pretreatment prior to anaerobic decomposition is adequate to improve the quality and quantity of generated biogas. Addition of effective microorganisms or respective enzymes along with the employment of suitable bioreactors, are other perspectives to enhance decomposition. The single-stage and multi-stage systems possess much attention than other types of reactors since that offers accurate process management at four different stages of decomposition. Moreover, the maintenances of optimum pH, temperature, volatile fatty acids, carbon/nitrogen ratio, etc. are crucial to evade system unsteadiness during decomposition. Since comprehensive mathematical models are appropriate to make the anaerobic decomposition process economically feasible and advancement with these forecasts are adequate to commercialize this technology in the future.

Keyword. Biomass, Anaerobic decomposition, Biogas, Methane, Pretreatment

1 Introduction

Though many resources exist to meet energy demand, bioenergy draws more attention since it reduces the overreliance of fossil fuel, in addition to partake carbon neutral property [1]. Bioenergy is a form of renewable energy derived from various organic materials known as biomass and it can be used to make transportation fuels with appropriate treatments. According to Mercom India Solar Project Tracker (2020), the cumulative power capacity of renewable energy is 36.2% (Figure 1). Among various renewable energy sources, the power generated from bio-based materials accounts for 2.7% [2]. Table 1 demonstrates biofuels obtained from discrete feed stocks, specific conversion processes, and the energy yield from various sources. Among various biofuels, biogas produced through anaerobic digestion is a fascinating approach due to its certain criteria's such as waste management, skill less small-scale production, cost-effectiveness and multiple-benefits.

Anaerobic digestion is the historically recognized method for cleansing of various organic wastes. It was recognized in the 17th century by Robert Boyle and Stephen Hale from the sediments of lake streams. Later, the presence of methane in biogas was identified by Humphry Day in the year 1808 [3]. The first anaerobic digester in India was built in Bombay as a part of promoting renewable energy technology [4]. This technology has gained attention over other energy generation technologies due to its multiple benefits such as energy and heat generation, waste recycling, biofertilizer production, environmental protection from pollutants, greenhouse gas reduction, etc. [5]. Even though the improper management of environmental parameters, the complexity of substrates, the efficiency of microorganisms, and the unsteadiness of bioreactors were reported as the drawbacks that inclined the researchers to improve the biogas productivity through various prospects.

Biogas is composed of methane, carbon dioxide, nitrogen, hydrogen, hydrogen sulphide and oxygen, and the typical biogas concentration is represented in table 2 [6]. Where, methane is the flame source, and lower methane content impedes the usage of biogas for commercial purposes [7]. The typical methane

* Corresponding author: indransdesign@gmail.com

 https://doi.org/10.1051/e3sconf/202130201015

concentration was reported to be 50-73%, which is comparatively lower than that of LPG and CNG with a high methane concentration of around 95-99%. Hence, a deep understanding of the rate-limiting factors of biogas production is adequate and contemporary research and development of various prospects of anaerobic decomposition might significantly improve the methane concentration of biogas to accomplish its industrialization.

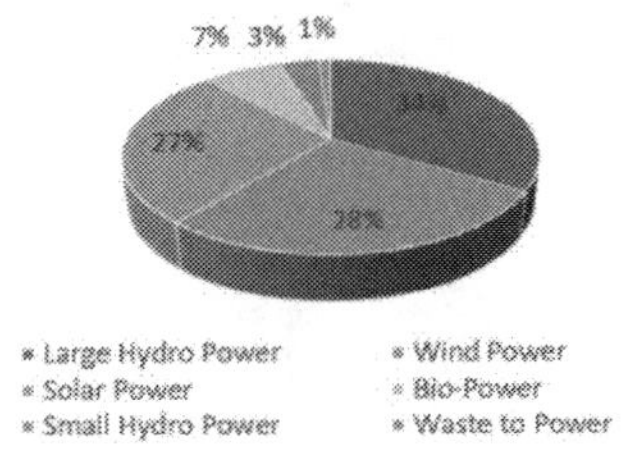

Fig. 1. Contribution of various renewable energy sources in overall production rate (36.2%) (2020)

Table 1. Energy production potential of various biofuels.

Biofuels	Feedstocks	Process	Energy yield
Biohydrogen	Synthetic waste water	Photo fermentation	3.6-7.9L/L/d
	Cellulose based biomass	Dark anaerobic fermentation	12.2 mol H_2 Kg^{-1}
	Macroalgal biomass	Microbial electrolysis	74.5ml/g COD
Biodiesel	Soybean	Transesterification	95% at 65^{0}C
	Rapeseed oil	Transesterification	95%
	Unrefined Palm Oil	Transesterification	90.4%
	Microalgae	Transesterification	12.03g m^{-2} day^{-1}
Bioethanol	Sugarcane bagasse	Sugar fermentation	6.83±0.07%
	Corn stalk juice	Sugar fermentation	6.01% (v/v)
Biogas	Cow dung	Anaerobic digestion	64%
	Agro-industrial wastes	Anaerobic digestion	6.55dm^3/kg
	Agricultural wastes	Anaerobic digestion	500cm^3
	Household waste	Anaerobic digestion	73%

The current understanding of biomass conversion into biogas through anaerobic decomposition is fascinated by four steps such as hydrolysis, acidogenesis, acetogenesis and methanogenesis. An outline of the anaerobic digestion process is depicted in figure 2.

Table 2. Composition of biogas [6]

Composition	Percentage (%)
Methane	50-75
Carbon dioxide	25-50
Nitrogen	0-10
Hydrogen sulfide	0-3
Hydrogen	0-2
Oxygen	0-2

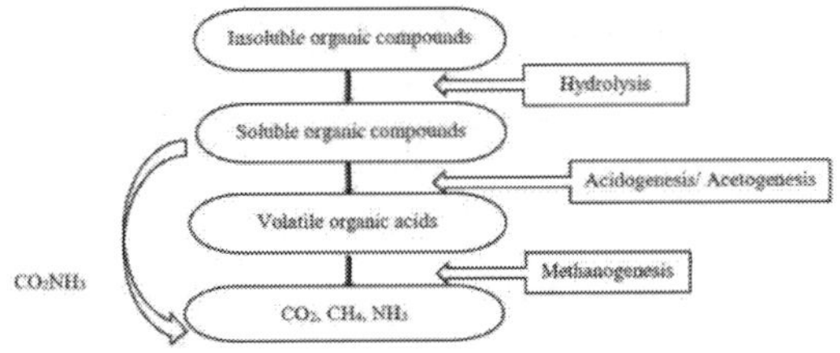

Fig. 2. Outline of the anaerobic decomposition process

The first rate-limiting stage is hydrolysis, where the complex organic materials (insoluble) are converted into simpler in line with this, acetogenesis plays a crucial role in converting long-chain fatty acids obtained by acidogenesis into acetate and CO_2. Finally, the acetate and CO_2 are converted to methane through methanogenesis [8]. The mechanism of bioconversion of organic materials during decomposition process is demonstrated in figure 3. There were several factors that influence the decomposition process with respect to different stages, biomass feedstock, type of bioreactors, etc. In this context, this article discusses the factors that remote the degradation of biomass and further prospects for enhancing biogas production with various aspects.

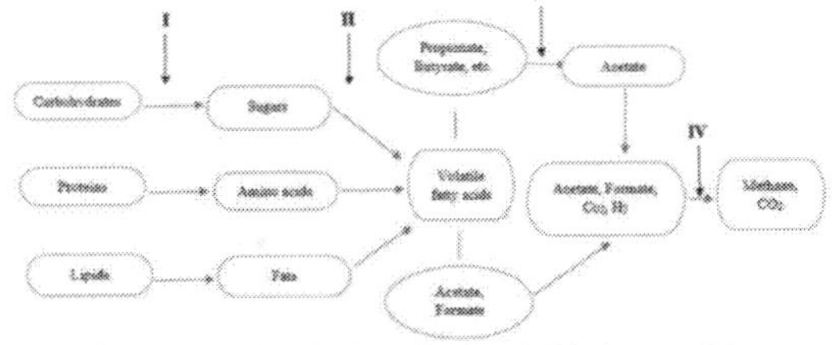

Fig. 3. Biochemical mechanism behind anaerobic decomposition process. I-Hydrolysis, II- Acidogenesis, III- Acetogenesis, IV- Methanogenesis

2 Potential of biomass for biogas production

Biomass is mainly composed of cellulose, hemicellulose, lignin, starch, proteins, lipids and some organic and inorganic components. Cellulose is a polysaccharide composed of D-glucose units with β-1,4-glycosidic bonds [2]. Biogas can be produced from various cellulosic/

https://doi.org/10.1051/e3sconf/202130201015

lignocellulosic biomass through the decomposition process. During the initial stage of degradation, the cellulose present in the feedstock is converted to D-glucose. Microbial enzymes can degrade cellulose, in fact, different microbial enzymes are required for diverse stages of anaerobic decomposition and their action is substrate specific [9]. Hemicellulose is a polymer of D-xylose, which is shorter than the cellulose chain. It can more easily break down than cellulose and soluble in alkaline solution. Lignin is a complex polysaccharide than other biomass constituents and the primary function of lignin is to give mechanical strength and protection to biomass in its native state. Microorganisms or chemicals cannot quickly degrade lignin due to their complex nature. Starch is another polysaccharide composed of D-glucose with α-glyosidic bonds, which are water-soluble in contrast to cellulose. Another type of organic compound is proteins, which are composed of amino acids. Proteins are the nutritive source in biomass along with other organic and inorganic components [10]. Besides, lipids also contribute a prominent part of biomass that possess high calorific value compared with carbohydrates and proteins present in it.

Since biomass is an abundant resource of energy deriving components that apposite for biogas production. The biomass had been considered as the primary source of energy in the form of firewood and charcoal in the 19th century. Consequently, it was replaced with coal and oil during the 20th century. However, the potential of biomass for biogas production is extensively established during the 21st century. Based on the source from where biomass obtained, it is characterized into three conventional biomass resources, biomass wastes and plantation biomass (Figure 4). The conventional biomass sources include agriculture, forestry, fishery and livestock. The biomass wastes or derivatives mainly include agricultural and livestock wastes. The aquatic, herbaceous and forestry biomass contributes to plantation biomass.

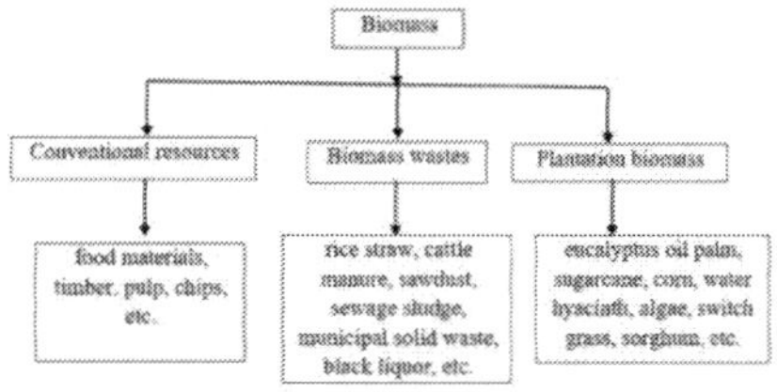

Fig. 4. Characterization of biomass

An assessment of biomass production showed that 1800 billion tons of total world biomass hosted on the ground and 4 billion tons presented in the ocean [10]. The ground biomass can provide 33,000 EJ, which can meet the total energy consumption of the world once it used in an accessible form. There are three significant sources of biomass energy that were established worldwide as livestock biomass, agricultural biomass and forestry biomass, and are illustrated in figure 5. In which, agricultural wastes (37%) contributes to prominent energy generation, compared to livestock wastes (34%) and forestry wastes (29%) [10]. Moreover, biomass-based energy production has been recognized as a thriving prospect to enhance energy generation at this energy demanding current scenario [11].

The biomass utilized for biogas production is limited as compared to the total amount of biomass available in the world. Among the agricultural biomass, rice contributes the principal source, followed by wheat and maize confer the massive production in world wide. The availability of bagasse was higher than the other agricultural biomass. Since the utilization of agricultural biomass for energy production is not yet industrialized due to some concerns. Among livestock biomass, cattle dung (22 EJ) contributes the most substantial part, which accounts that cattle dung is the primary source of biomass available in the world for bioenergy production. Followed by poultry waste and sheep waste are accessible substantially. Forestry biomass represents the most available industrial logs, fuel logs and wood wax. Additionally, the availability ratio for agricultural biomass and livestock were 25% and 12.5%. The highest availability ratio was noted for industrial logs were 75%. Overall, the utmost availability is reported for industrial logs (15 EJ) (Figure 6) [12]. Although, further advancements are necessary to utilize wide substrate and thereby to enhance biomass-based energy generation policies in the future.

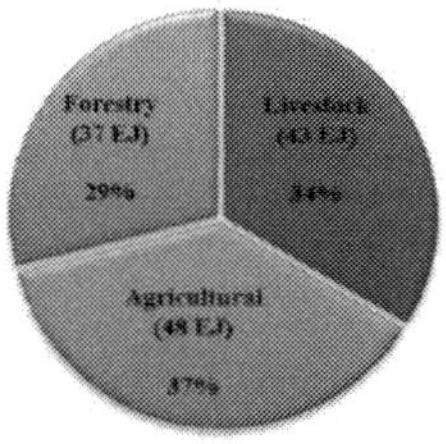

Fig. 5. Energy generation potential of biomass resources

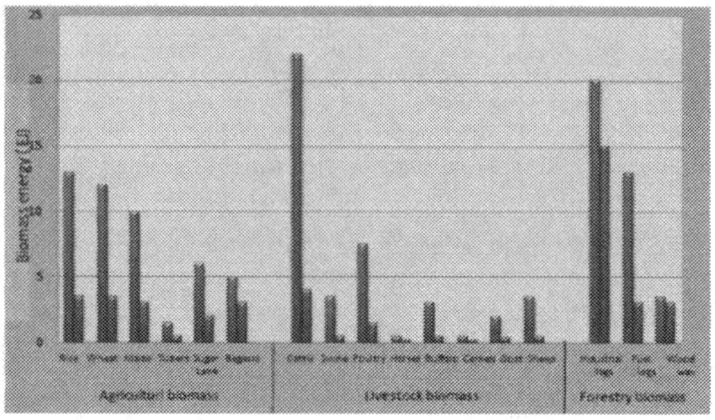

Fig. 6. Presence of biomass in the world and its availability [8]

3 Biomass selection criteria and pretreatment

The bio-methanation potential of each substrate is based on its carbohydrates, proteins, and lipids content. It is essential to delineate the chemical composition of the substrates prior to anaerobic decomposition in course to offer absolute decomposition. The biochemical methane potential (BMP) of the substrate to be statistically analysed in order to predict the potentiality of the substrate for the biogas production process. At the same time, the substrate availability, easy handling, low complexity, optimum carbon/nitrogen (C/N) ratio and moderate operational conditions are the other criteria to be taken in account for the selection of biomass for biogas production. Thus, the selection of materials based on biochemical methane potential of substrates and other relevant criteria could improve biogas production to a large extent [13].

The hydrolysis is the rate-limiting stage of biogas production, which largely influences the overall production process. Generally, particle size, age of the material, the composition of the feedstock, etc. are the factors that impact substrate degradation [14]. Most of the microbial enzymes cannot attack lignocellulose present in the complex organic materials. Lignocellulosic materials possess high nutrient content compared to simpler materials; which can be used as potential substrates for biogas production once it is pretreated. Pretreatment is necessary to accelerate the degradation potential of various complex solid substrates [15]. Pretreatment by hydrolytic bacteria made avail simpler organic compounds for other groups of bacteria. Physical, chemical, physico-chemical and biological treatment methods are available for pretreatment. Though, pretreatment of biomass before microbial degradation shows a significant effect on improving biogas production, as demonstrated in figure 7. Biological pretreatment paid more attraction than other methods due to its effectiveness and simple, effortless nature [16]. There were some disadvantages reported in the case of physical, chemical and physico-chemical pretreatments. The physical treatment is aligned with the particle size of the feedstock, while chemical treatment causes the excess release of volatile fatty acids that will lead to process inhibition. Furthermore, physical and chemical treatments are expensive and require high energy [17]. The cost-effectiveness, low energy consumption, higher substrate specificity and environmental friendliness are the outcomes of biological pretreatment [18]. Based on the nature of pretreatment, biological treatment can be categorized as aerobic microbial pretreatment and anaerobic microbial pretreatment. In aerobic pretreatment, naturally occurring mixed cultures that produce cellulase, xylanase and laccase are used to degrade cellulose, hemicelluloses, lignin, respectively.

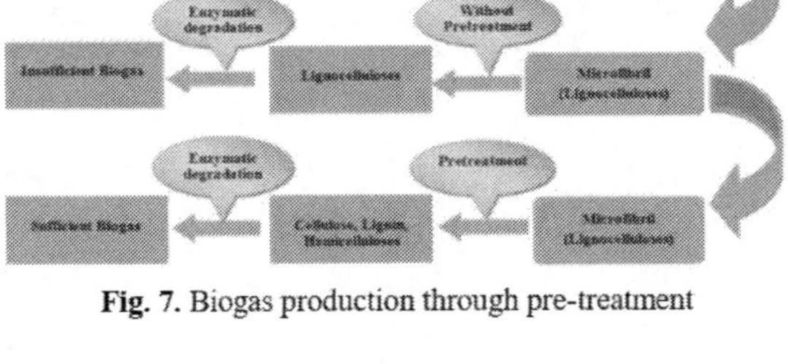

Fig. 7. Biogas production through pre-treatment

During anaerobic microbial treatment, a pre-acidification step offers an increased degradation rate by providing optimum conditions for hydrolytic enzymes [19]. Several fermentative bacteria (*Pseudomonas spp., Cellulomonas spp., Streptomyces spp., Bacillus spp. and Actinobacteria spp.*) and white-rot fungi (*Coriolus spp., Aspergillus spp., Trichoderma spp., Phanerochaete spp.* and *Penicillium spp.*) are preferred for productive pretreatment purposes [20,21]. Nevertheless, further advancement is admirable to find out potential microbes for pretreatment or absolute deployment of available microbes by providing optimum conditions for its enzyme production and activity.

4 Role of microbes in anaerobic decompositions

There are four different phases are involved in anaerobic digestion such as hydrolysis, acidogenesis, acetogenesis and methanogenesis. The hydrolytic, acidogenic, acetogenic and methanogenic bacteria are involved in the hydrolysis, acidogenesis, acetogenesis, methanogenesis stages, respectively. Different stages of anaerobic decomposition with probable microbes for each step are presented in table 3. The effectiveness of microbes involved in anaerobic decomposition potentially determine the rate of decomposition [22,1]. In general, microbes are ubiquitous in nature, while the effectiveness of bacteria depends on their habitat. The habitats with acidic/ anaerobic nature will be the chief carriers of acetogenic/ methanogenic bacteria, while fermentative bacteria are common in most of the habitats. The addition of microbial additives to the digester improves decomposition of the materials. In line with this, the usage of cow dung, agro-industrial and municipal solid wastes as microbial sources were the best practices for efficient biogas production. Consequently, rumen fluid was recognized as a potential source of anaerobic and fermentative bacteria that can be employed as a microbial additive for efficient biogas production.

Table 3. Major microbes and enzymes involved in four stages of anaerobic decomposition.

Stages	Microorganisms	Microbial Enzymes
Hydrolysis	*Bacillus, Streptococcus, Enterobacter*	Cellulase, Protease, Lipase, Amylase, Xylanase
Acidogenesis	*Micrococcus, Syntrophomonas, Pseudomonas, Flavobacterium, Clostridium, Acidobacterium*	Acetate kinase, Acetaldehyde dehydrogenase, Formate hydrogen lyase, Formate c-acetyl transferase
Acetogenesis	*Syntrophomonas, Syntrophobacter, Clostridium, Syntrophospora, Acetobacter*	Carbonic anhydrase, Hydrogenases, Carbon monoxide dehydrogenases
Methanogenesis	*Methanosarcina, Methanosprillum, Methanococcus, Methanobacteria, Methanobrevibacter, Methanothrix*	Methyl co-enzyme M-transferase, Methyl co-enzyme M- reductase, Formyl methanofuran dehydrogenases

The usage of microbial inoculums with different kinds of microbes that involves in respective stages of anaerobic decomposition revealed improvement in the production of biogas and methane. Thus, the exploration and expansion of significant microbial consortia for biogas production could serve as a prospect for the commercialization of biogas production technology [23].

The co-operation or synchronization of particular microorganisms of four stages is required to achieve complete decomposition. A synchronized action of each kind of microbes is essential since the product of one group of microbes is being the feedstock for the next group of bacteria. While a failure of microbes to utilize the products of the former stage will cause whole process failure. For instance, *Methanosarcina barkeri* capable of utilizing acetate, carbon dioxide, hydrogen and methanol, but *Methanosprillum hungaeti* does not utilize acetate but capable of utilize formate. Thus, the production will be lower when using microbes with narrow substrate specificity [24].

Substrate to inoculum ratio is another measure that affects entire decomposition process, hence it is to be maintained to accomplish desirable methane yield. The inoculums to substrate ratio of 1:1-1:3 showed improved methane yield, previously. While, few studies showed smaller inoculum size resulted in highest production [25]. Another prospect for improved biogas production is the identification and employment of potential bacteria for each stage of anaerobic decomposition and its synchronization. The mass production and addition of potential bacteria to respective stages of anaerobic decomposition will enhance the anaerobic decomposition process by reducing the lag period required for the action of a wide range of microbes in the system [26].

5 Enzymology of anaerobic decomposition

Enzymes are the biocatalysts involved in all kinds of metabolic reactions. Anaerobic decomposition is the result of complex reactions of various groups of microbial enzymes. However, the enzyme addition could not offer apposite decomposition in contrast to microbial addition. The microbes that has potential to produce diverse degradative enzymes in its utmost concentration are considered as promising candidates for anaerobic decomposition [27]. Diverse microbial enzymes responsible for various stages of anaerobic decomposition is also tabulated in table 3. The addition of amylase, lipase, protease, etc. showed increased fluidization of materials during anaerobic decomposition. Another investigation explored that the addition of xylanase, cellulase, protease and lipase were effective for solid waste digestion [28]. While enzyme addition to acetogenesis/methanogenesis phase were not accomplished process improvement [29].

Conversely, a mixture of enzymes increased methane production, where the concentration of enzymes applied was an important factor. It is also evidenced that the lowest concentration of 0.5ml/kg volatile solids resulted higher degree of decomposition [30]. There were two groups of enzymes involved in initial stage that are lytic enzymes those work cell wall disruption and hydrolytic enzymes that will cause breakdown of the macromolecules. In fact, the pretreatment with enzymes prior to anaerobic decomposition also resulted improved methane production. An increase production of 12-15% was achieved by enzyme addition to substrates in earlier. A mixture of cellulase and pronase-E showed an 80% reduction of solids compared to control and 97% COD removal was observed instead of 63% of control [31]. Enzymes can act at wide range of environmental conditions such as pH, temperature and salinity; while their utmost action observed at a stable optimum condition. The enzymes are easily accessible by feedstocks rather than the microorganisms. The use of microbes needs lag period for its acclimatization on feedstock, which can be reduced by the direct application of enzymes [32]. Few studies recognized that most of the organism involved in anaerobic decomposition were grown in the range of pH 5.5 to 8.5. While, the pH above or below the optimum range could reduce process efficiency. The pH tends to increase during hydrolysis, followed by decrease during acidogenesis/acetogenesis and again remains in the range of 7.2 to 8.2 when the production of methane stabilizes [33].

6 Prospects behind various parameters

Biogas production is influenced by a number of parameters such as feedstock's potential, digester design, inoculum size, substrate nature, pH, temperature, loading rate, hydraulic retention time (HRT), carbon/nitrogen ratio (C/N), volatile fatty acids (VFA), and so on [34,26]. The significant parameters required for digestion process are listed in table 4.

Table 4. Important parameters required for anaerobic digestion.

Parameters	Modes	Range
Temperature	Mesophilic	30-40°C
	Thermophilic	50-60°C
pH	Hydrolysis	6.5-7.5
	Acidogenesis	5.0-6.5
	Acetogenesis	4.0-5.5
	Methanogenesis	7.0-8.5
C/N ratio	-	10-35
Volatile fatty acids	-	30-300 mg/l
Loading rate/ Total solids	Small-scale system	<10%
	Large-scale system	>10%
Hydraulic retention time (HRT)	Small-scale system	30 days
	Large-scale system	30-45 days

6.1 Temperature

The psychrophilic (10-30°C), mesophilic (30-40°C) and thermophilic (50-60°C) temperature ranges were feasible for anaerobic digestion while the choice is based on the temperature tolerance of the microorganisms employed. The anaerobic decomposition under thermophilic temperature was more powerful than psychrophilic and mesophilic temperature ranges. Few studies revealed that an increase in temperature could significantly improve the performance of anaerobic decomposition system [35,36]. It is also established that mesophilic digestion often causes ammonia inhibition and the rate of inhibition was higher than that of thermophilic degradation [37]. However, a small fluctuation in thermophilic temperature would detrimentally affect the whole system and it implies that the maintenance of temperature is imperative to achieve stable digestion [38]. A sewage sludge mesophilic system with a total solid concentration of 10% showed desirable production whereas 7.5% total solid concentration was required for thermophilic system [39]. Thus, an increase in temperature (50-65°C) improves the rate of anaerobic decomposition and also higher metabolic rates were observed in thermophilic condition. However, the effortless management and ideal biogas production are the considerable features of mesophilic systems.

6.2 Volatile fatty acids and pH

The typical intermediates of anaerobic decomposition are long-chain/ short-chain volatile fatty acids that includes acetic acid, propionic acid, butyric acid, valeric acid, etc., where propionate and acetate take part the dominant portion. Since, the ratio of propionate to acetate can be used as the indicator of the process progress [40]. During acetogenesis, long-chain fatty acids are converted to short chain fatty acids or acetate by β- oxidation process. There is a probability of accumulation of long-chain fatty acids once it is not converted successfully or even due to the accumulation of acetate when it is formed excessively. The accumulation of volatile fatty acids can be evaded through certain approaches, either by maintaining the optimum solid content of the system nor by supplying effective inoculums with adequate ratio. Otherwise, the addition of adsorbents also showed admirable effect to neutralize volatile acids [30]. The ion-exclusion, HPLC and gas chromatography are usually employed methods for the analysis of volatile fatty acids concentration in the system.

The pH concentration of the system principally allied with the concentration of volatile fatty acids exhibited. Usually, a pH range of 4.0 to 7.5 is favourable for fermentative degradation while a pH of 7.0 to 8.5 is feasible for methanogenic bacteria [7]. The accumulation of volatile fatty acids especially butyrate and acetate lead to decrease in pH that causes adverse effect on microbial growth and process unsteadiness occurs. On the other hand, the excessive bioconversion of short-chain fatty acids leads to the accumulation of ammonia and as a result pH increases over 8.5, that inhibit methanogenesis. The excess or high concentrations of long chain fatty acids will inhibit the growth of methanogens by damaging the cellular membrane [41]. Though it is assigned that the bacterial population can be maintained by controlling pH of the system. The long chain fatty acid concentration of about 30-300 mg/l was found as appropriate for anaerobic decomposition [42]. The effluent and some additives were found as suitable to neutralize volatile fatty acids in many systems. Moreover, pH, temperature, volatile fatty acids, H_2 partial pressure, ammonia accumulation and heavy organic loading were the foremost parameters that causes process unsteadiness, is tabulated in table 5 [43,8].

6.3 Carbon/nitrogen (C/N) ratio

The Carbon/Nitrogen (C/N) ratio is another important factor that determines the decomposition rate and biogas production. Based on the carbon and nitrogen content, biomass can be distinguished in to carbon or nitrogen rich and carbon or nitrogen low materials. Usually microbes utilize carbon much faster than nitrogen. Hence the carbon content should be many folds higher than that of nitrogen content [44]. A carbon/nitrogen (C/N) ratio of 10-35 was established as optimum for maximum biomass degradation. An increase in C/N ratio causes increased uptake of carbon and thereby accumulation of volatile fatty acids occurs. The lower C/N ratio causes enhanced nitrogen uptake and thereby ammonia accumulation partakes. However, the optimization of carbon and nitrogen ratio of substrate reduces the accumulation of volatile fatty acids and ammonia.

Recent studies explored that co-digestion of two or more substrates assist to maintain C/N ratio and thereby reduce accumulation of inhibitors [45]. The advantages of co-digestion are it offers optimum C/N ratio, wide range substrate availability, better degradability, improved methane/biogas production, low energy requirement, utilization of complex substrates, highly nutrient biofertilizer, etc. Nowadays, co-digestion of lignocellulosic agricultural wastes or industrial wastes with animal manures become a new trend to enhance its degradability, instead of using alone. Co-digestion do not possess any positive impact on reducing hydraulic retention time, but increases the decomposition rate as well as biogas production [46].

6.4 Hydraulic retention time (HRT)

The hydraulic retention time (HRT) is the time taken to carry out the whole process. The small-scale systems generally work with low organic loading rate below 10%. For the industrial scale production, organic loading rate is the major measure and moderate/high organic rate is desirable to assure economical progression. Generally, high organic loading rate will cause process unsteadiness sometimes process failure, which needs long HRT. Hence further improvements are needed through microbial addition or other methods to improve organic loading rate and to reduce low HRT [47]. An effort has been made by Divya et al in 2019 through enzyme addition, that resulted significant reduction of HRT from 30 days to 20 days [9]. The HRT can be measured by using certain criteria with the aid of following equation (1). It is expressed in hours (d), where V is the volume of aeration tank (m3) and Q is the influent flow rate (m^3/d) [48].

$$HRT\ (d) = V/Q \qquad (1)$$

https://doi.org/10.1051/e3sconf/202130201015

6.5 Metal elements

In addition to nutrients, the light metal ions (Na, Mg, Al, Ca, K) and heavy metal ions (Co, Cr, Ni, Zn, Cu, etc.) are also necessitate for decomposition since which play a significant role in the production of degradative enzymes along with maintaining enzyme activities [49,50]. Even though some side effects caused by the presence of heavy metals in the feedstock are described here. Heavy metals are not easily degradable and would accumulate during decomposition that leads to process inhibition. The factors influencing metal accumulation are total concentration of metals, pH and redox potential. The accumulation of sodium and potassium was reported as the prime issue and co-digestion approaches are admirable to overcome these kinds of inhibition [51].

Many studies evidenced that the heavy metals Cr, Cd and Ni are highly toxic, that disrupt enzyme activity in large extent. Besides that, Zn, Cu, Pb and Fe showed inhibitory effect under certain conditions. The effect of heavy metals in decomposition is usually occur in the increasing order of Pb<Zn<Cu<Cd. The heavy metals possess chemical binding to the enzymes and makes changes in the orientation of enzyme. Although, the inhibitory effect of metals to microbes is based on its concentration, types of microbes and microbial resistance power over it. Since advanced research is necessary to identify the effect of various heavy metals on acetogenic and methanogenic stages of anaerobic decomposition [52].

7 Anaerobic bioreactor models

The anaerobic bioreactors are mainly classified in to three types based on the mode of treatments, reactor type and segregation of stages cast-off (Figure 8). The different reactor models frequently employed for biogas production process through oxygen free decomposition are presented below.

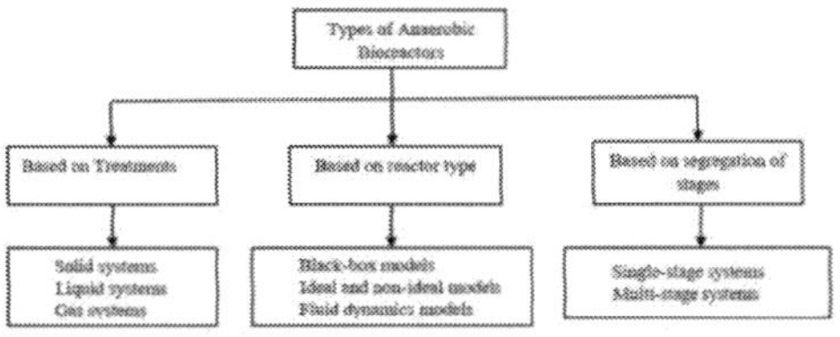

Fig. 8. Different types of anaerobic bioreactors for biogas production

7.1 Solid, liquid and gas phase systems

Based on the type of treatments used in bioreactor, they are classified in to solid phase system, liquid phase system and gas phase system. The solid phase system works with solid/sludge granules with a diameter of about 0.5-2.0 mm. The liquid phase systems are usually run with waste water, where the lower part contains sludge bed while the upper part contains solid, liquid and gas phases. Solid phase and liquid phase systems are generally used for treating organic wastes, which ensures biogas production once required. Nevertheless, gas phase systems are typically used for biogas production only, where organic materials such as dead plant and animal materials, animal feces, agro-industrial wastes, domestic wastes, etc. can be used for biomethanation.

7.2 Promising bioreactor models

Based on the type of reactors used, they are classified in to black-box models, phenomenological models (ideal and non-ideal models) and computational fluid dynamics models. The artificial neural networks are coming under black-box model that do not require knowledge concerning the key intermediates of anaerobic decomposition systems. The ideal models include continuous stirred tank reactors (CSTR) and plug flow reactor (PFR). CSTR can be used for treating complex materials in a continuous mode, which offers better decomposition too. Plug flow reactor and plug flow with dispersion reactor models were developed to treat organic wastes in stimulation mode [53]. Few digester models designed through comprehensive approach by CSTR and PFR configuration were found as non-ideal models that would assist improved anaerobic decomposition, which possess stirring device and optimization properties [54]. Computational fluid dynamic (CFD) model is a simulation approach to optimize parameters to improve the performance of the system. However, it is established that in-silico model simulation approaches are desirable to enhance the rate of decomposition through suitable mathematical modeling.

7.3 Single stage and multi-stage systems

The anaerobic digestion systems are distinguished into single stage and multi-stage systems based on the segregation of different stages. Single stage systems are usually used for past many decades where the decomposition carried out in a single system with batch or continuous mode. In single stage batch reactors, all the four steps of anaerobic digestion put together and as a result process unsteadiness or low biogas production occurs by the improper maintenance of optimum parameters of four stages. Multi-stage systems with the separation of acidogenesis and methanogenesis showed improved gas production in earlier. Lyberatos and Skiadas in 1999 proposed several mathematical models for improving biogas production [8]. They demonstrated that volatile fatty acids accumulation and pH changes are the major reasons that causes process unsteadiness. Moreover, it is presumed that separation of each of the four stages of anaerobic decomposition would improve substrate degradation as well as biogas/methane production [26]. The segregation of each of the four stages within a digester by specific microbial addition or enzyme addition along with optimum parameters, is a promising challenge to enhance methane production in future.

8 Future of biogas production

Biochemical methane potential is the measure to find out methane potential of diverse organic materials prior to anaerobic decomposition and according to that potent materials can be selected as best choice for the biogas production purpose. The substrate selection based on its availability throughout the year is another factor that offers continuous production and lead to commercialization of biogas as an alternative to non-renewable conventional energy sources. The socio-economic impacts such as biofertilizer production and materials recycling are the emerging challenges of this eco-friendly technology that can be potentially used for making a greener tomorrow by reducing environmental pollution. Co-digestion of different probable organic materials by maintaining C/N ratio propose improved substrate decomposition and biogas/methane production. Biological pretreatment of complex substrates prior to anaerobic decomposition is a promising prospect to increase the availability of nutrient rich substrates.

The optimization of specific parameters such as temperature, pH, organic loading rate, total solids, etc. would also augment biogas production. The analysis of pH, temperature, volatile solid content, chemical oxygen demand and volatile fatty acids during decomposition could benefit to trace out the rate of decomposition. Moreover, appropriate measures should be taken to maintain all imperative parameters like pH, temperature and volatile fatty acids concentration throughout the process. The digester design plays a significant role in maintaining the rate of decomposition. Digesters with all control measures to neutralize the fluctuations in parameters possess high economic impact. In line with this, the implementation of continuous stirred tank systems over batch systems with thermometer, digital pH analyser, pressure gauge, sampling unit, inlet, outlet, stirrer, gas jacket, etc. will be more convenient and significant for the effective process monitoring and controlling.

Conversely, multi-stage digesters are desirable for realizing maximum substrate decomposition. Segregation of anaerobic digestion in single stage system through microbial means by providing optimum conditions would be another task in this field. The microbial and enzyme addition also helps to improve substrate decomposition. Further exploration is needed to identify potential strains and to analyze enzyme stability in order to augment its utilization. The development of genetically modified anaerobic microbes is a great challenge in this area while that requires advanced techniques to overcome the highly sensitive nature of anaerobic bacteria. In this scenario, the implementation and usage of effective microbial consortia partaking high decomposition rate is another viable perspective [23]. Even though some techniques have been executed to purify the biogas for concentrating methane content. There is a need for cost-effective technologies for purification of biogas to enrich the methane concentration and thereby to increase flammability as well as its competence towards conventional energy sources.

9 Conclusion

Biogas production from biomass is an eco-friendly method that offers energy generation along with maintaining environmental standards. The exploration and utilization of various complex, potential biomass along with pretreatment, microbial/enzyme addition, parameters maintenance and process segregation are necessary for sustainable utilization of biomass for absolute biogas production. Moreover, cost-effective, small-scale anaerobic digestion systems with specific control measures are adequate at the moment to commercialize this technology. Hence, this study suggests that comprehensive approaches with such forecasts are admirable to extend the utility of diverse biomass to meet today's energy demand in sustainable means.

Acknowledgement

The authors acknowledge the Rohini College of Engineering and Technology, Vivekanandha College of Arts and Sciences for Women, and Pinnacle Bio-Sciences for providing research support to carry out this review work. The authors also acknowledge the Natural Composites Research Group Lab, Department of Materials and Production Engineering, The Sirindhorn International Thai-German Graduate School of Engineering, King Mongkut's 13 University of Technology North Bangkok, Bangkok-10800, Thailand. for their support to complete this review article.

References

1. P. Merlin Christy, L. R. Gopinath, D. Divya, A review on anaerobic decomposition and enhancement of biogas production through enzymes and microorganisms. Elsevier- Renewable and Sustainable Energy Reviews, **34** (2014): 167-173.
2. CEA, MNRE, Mercom India Solar Project Tracker, source: Mercom India Research, Installed Capacity and Capacity Utilization as on 31 Mar 2020, (2020).
3. S.Z.Bibi, M. Qaisar, P. Arshid, Characteristics and Performance of Anaerobic Wastewater Treatment (A Review), Journal of the Chemical Society of Pakistan., **35**, 1 (2013): 217-232
4. Y. Voegeli, C. Zurbrügg, Decentralised anaerobic digestion of kitchen and market waste in developing countries – "state-of-the-art" in south India, Proceedings Venice 2008, Second International Symposium on Energy from Biomass and Waste Venice, Environmental Sanitary Engineering Centre, Italy, 17-20 November 2008.
5. P. Valerio, P. Francesco, S. Marco, T. Laura, N. Naja, C. Angelo, Environmental impact of biogas: A short review of current knowledge, Journal of Environmental Science and Health, Part A, **53**, 10 (2018), 899–906
6. O.S. Joshua, G.J. Ejura, I.C. Bako, I.S. Gbaja, Y.I. Yusuf, Fundamental principles of biogas product,

International Journal of Scientific Engineering and Research, **2**, 8 (2014): 47-50.

7. D. Divya, L. R. Gopinath, P. Merlin Christy, A review on Trends, Issues and Prospects for biogas production in Developing Countries, International Research Journal of Environmental Sciences, **3**, 1 (2014): 62-69.
8. G. Lyberatos, I.V. Skiadas, Modeling of anaerobic digestion- a review. Global Nest: The international Journal, **1**, 2 (1999): 63-76.
9. D. Divya, L. R. Gopinath, S. Indran, Analysis of the effect of enzyme substitution on feedstock to enhance biogas production, International Journal of Research and Analytical Reviews, **6**, 2 (2019): 964-973.
10. Asian Biomass Handbook, Part 2, Biomass resources, University of Tokyo, Japan (2008).
11. D. Divya, L. R. Gopinath, S. Sreeremya, S. Indran, Enhancement of Substrate Decomposition through Potential Hydrolytic Bacteria for Cumulative Biogas Production, International Journal of Applied Science and Biotechnology, **6**, 4 (2018): 386-396.
12. D.O. Hall, F. Rosillo-Calle, R.H. Williams, J. Woods, Biomass for energy: supply prospects, in Renewable Energy: Sources for Fuels and Electricity, (ed.) Johansson, T.B., Kelly, H., Reddy, A.K.N., and Williams, R.H., Washington, DC, Island Press, (1998): 593- 651.
13. P.C. Suryawanshi, A.B. Chudhari, S. Bhardwaj, T.Y. Yeole, T.Y. Operating procedures for efficient anaerobic digester operation. Research Journal of Animal, Veterninary and Animal Sciences ; Fishery Sciences, **1**, 2 (2013): 12-15.
14. S. Aliyu and A. Zahangir, Pretreatment Methods of Organic Wastes for Biogas Production, Journal of Applied Sciences., **16** (3) (2016): 124-137
15. A. Javkhlan Ariunbaatar, P. Antonio, E. Giovanni, P. Francesco, N.L.L. Piet, Pretreatment methods to enhance anaerobic digestion of organic solid waste, Applied Energy, **123**, 15 (2014): 143-156
16. H.B. Gonzales, K. Takyu, H. Sakashita, Y. Nakano, W. Nishijima, M. Okada, Biological solubilization and mineralization as novel approach for the pretreatment of food waste, Chemosphere, **58** (2005): 57-63.
17. J.W. Lim, J.Y. Wang, Enhanced hydrolysis and methane yield by applying microaeration pretreatment to the anaerobic co-digestion of brown water and food water, Waste Management, **33** (2013): 813-819.
18. G.D. Saratale, S.D. Chen, Y.C. Lo, R.G. Saratale, J.S. Chang, Outlook of biohydrogen production from lignocellulosic feedstock using dark fermentation- A review, Journal of Scientific and Industrial Research, ,**67** (2008): 962-979.
19. F.R. Lucy, Pretreatment of feedstock for enhanced biogas production, IEA Bioenergy, First edition books, (2014).
20. J. Perez, J. Munoz-Dorado, T. Rubia, J. Martinez, Biodegradation and biological treatments of cellulose, hemicelluloses and lignin: An overview, International Journal of Microbiology, **5** (2002): 53-63.
21. U.G. Phutela, N. Sahni, S.S. Sooch, Fungal degradation of paddy straw for enhancing biogas production, Indian Journal of Science and Technology, **4**, 6 (2011): 660-665.
22. R. Manimegalai, L. R. Gopinath, P. Merlin Christy, D. Divya. Isolation and identification of acetogenic and methanogenic bacteria from anoxic black sediments and their role in biogas production, International Journal of Plant, Animal & Environmental Sciences, **4**, 3 (2014): 156-164.
23. L. R. Gopinath, P. Merlin Christy, K. Mahesh, R. Bhuvaneswari, D. Divya, Identification and evaluation of effective bacterial consortia for efficient biogas production, IOSR Journal of Environmental Science, Toxicology and Food Technology, **8**, 3 (2014): 82-86.
24. I.O. Okonko, O.T. Adeola, F.E. Aloysius, A.O. Damiola, O.A. Adewale, Utilization of food wastesfor sustainable development, Electronic Journal of Environmental Agricultural and Food Chemistry, **8**, 4 (2009): 263-286.
25. A.S. Fayyaz, M. Qaisar, M.H. Mohammad, P. Arshid, A.A. Saeed, Microbial ecology of anaerobic digesters: The key players of Anaerobiosis, The Scientific World Journal, (2014): 1-21.
26. D. Divya, L. R. Gopinath, P. Merlin Christy, A review on current aspects and diverse prospects for enhancing biogas production in sustainable means, Renewable and Sustainable Energy Reviews, **42** (2015a): 690-699.
27. W. Parawira, Enzyme research and applications in biotechnological intensification of biogas production, Critical Reviews in Biotechnology, (2011): 1-15.
28. J. Rintala, B. Ahring, Thermophilic anaerobic digestion of source-sorted households solid waste: the effects of enzyme additions, Applied Microbiology and Biotechnology, **40** (1994): 916-919.
29. J. Jeganathan, G. Nakhla, A. Bassi, Hydrolytic pretreatment of oily waste water by immobilized lipase, Journal of Hazardous Materials, **145** (2007): 127-135.
30. C. Zhang, H. Su, J. Baeyens, T. Tan, Reviewing the anaerobic digestion of food waste for biogas production, Renewable and Sustainable Energy Reviews, **38** (2014): 383-392.
31. H.J. Roman, J.E. Burgess, B.I. Pletschke, Enzyme treatment to decrease solids and improve digestion of primary sewage sludge, African Journal of. Biotechnology, **5** (2006): 963–967.

32. S.K. Ahuja, G.M. Ferreira, A.R. Moreira, Utilization of enzymes for environmental applications, Critical Reviews in Biotechnology, **24** (2004): 125-154.
33. T. Abbasi, S.M. Tauseef, A. Abbasi, Anaerobic digestion for global warming control and energy generation- An overview, Renewable and Journal of Plant, Animal and Environmental Sciences, **3**, 4 (2013): 65-70.
34. P. Merlin Christy, L. R. Gopinath, D. Divya. A review on decomposition as a technology for sustainable energy management, International Sustainable Energy Reviews, **16**, 5 (2012): 3228-3242.
35. H.M. Ei- Mashad, G. Zeeman, W.K.P. Van Loon, G.P.A. Bot, G. Lettinga, Effect of temperature fluctuation on thermophilic anaerobic digestion of cattle manure, Bioresource Technology, **95** (2004): 191-201.
36. E. Sanchez, R. Borja, P. Weiland, L. Travieso, A. Martin, Effect of substrate concentration and temperature on the anaerobic digestion of piggery waste in a tropical climate, Process Biochemistry, **37** (2001): 483-489.
37. C. Gallert, S. Bauer, J. Winter, Effect of ammonia on the anaerobic degradation of protein by mesophilic and thermophilic biowaste, Applied Microbiology and Biotechnology, **50** (1998): 495-501.
38. I. Apples, A.V. Assche, K. Willems, J. Degreve, J.V. Impe, R. Dewil, Peracetic acid oxidation as an alternative pre-treatment for the anaerobic digestion of waste activated sludge, Bioresource Technology, **102** (2011): 4124-4130.
39. T. Hidaka, F. Wang, T. Pascual, Comparative performance of mesophilic and thermophilic anaerobic digestion for high solid sewage sludge, Bioresource Technology, **149** (2013): 177-183.
40. U. Marchaim, C. Krause, Propionic to acetic acid ratios in overloaded anaerobic digestion. Bioresource Technology, **43** (1993): 195-203.
41. I. Angelidaki, B.K. Ahring, Effects of free long-chain fatty acids on thermophilic anaerobic digestion, Applied Microbiology and Biotechnology, **37** (1992): 808-812.
42. A. Fernández, A. Sánchez, X. Font, Anaerobic co-digestion of a simulated organic fraction of municipal solid wastes and fats of animal and vegetable origin, Biochemical Engineering Journal, **26**, 1 (2005): 22-28.
43. J. F. Andrews, Dynamic Model of the Anaerobic Digestion Process, Journal of the Sanitary Engineering Division, **1** (1969): 95-116.
44. P. Mahanta, U.K. Saha, A. Dewan, P. Kalita, B. Buragohain, Biogas digester: A discussion on factors affecting biogas production and field investigation of a novel duplex digester, Journal of the Solar Energy Society of India, **15**, 2 (2005): 1-12.
45. S. Park, Y. Li, Evaluation of methane production and macronutrient degradation in the anaerobic co-digestion of algae biomass residue and lipid waste, Bioresource Technology, **111** (2012): 42-48.
46. D. Divya, L. R. Gopinath, S. Indran, P. Merlin Christy, Enhancement of biogas production through sustainable feedstock utilization by co-digestion, International Journal of Plant, Animal and Environmental Sciences, **5**, 3 (2015b): 88-94.
47. A. Award, C. Xiaoguang, L. Jianshe, X. Xuehui, Z. Jian, Z. Kai, W. Heng, L. Na, Characteristics, process parameters, and inner components of anaerobic bioreactors, Biomed Research International, (2014): 1-10.
48. M.T. Kato, J.A. Field, R. Kleerebezem, G. Lettinga, Treatment of low strength soluble wastewaters in UASB reactors. Journ. Journal of Fermentation and Bioengineering., **77** (1994): 679.
49. I. Zhang, Y.W. Lee, D. Jhang, Anaerobic co-digestion of food waste and piggery waste water: focusing on the role of trace elements, Bioresource Technology, **102** (2011): 5048-5059.
50. A. Schattauer, E. Abdoun, P. Weiland, M. Plochl, M. Heiermann, Abundance of trace elements in demonstration biogas plants, Biosystem Engineering, **108** (2011): 57-65.
51. F. Cheng, K. Boe, A. Irini, Anaerobic co-digestion of desugared molasses with cow manure focusing on sodium and potassium inhibition, Bioresource Technology, **102** (2011): 1005-1011.
52. A. Mudhoo, S. Kumar, Effects of heavy metals as stress factors on anaerobic digestion processes and biogas production from biomass, International Journal of Environmental Science and Technology, **10** (2013): 1383-1398.
53. P. Buffiere, C. Fonade, R. Moletta, Mixing and phase hold-ups variations due to gas production in anaerobic fluidized-bed digesters: influence on reactor performance, Biotechnology and Bioengineering, **60** (1998): 36-43.
54. L. Yu, P.C. Wensel, J. Ma, S. Chen, Mathematical modeling in Anaerobic decomposition, Bioremediation and Biodegradation, **S4** (2013): 003.

Tensile Strength and Moisture Resistance Properties of Biocomposite Films Based on Polyvinyl Alcohol (PVA) with Cellulose as Reinforcement from Durian Peel Fibers

Melbi Mahardika[1,*], *Mochamad* Asrofi[2], *Devita* Amelia[3], *Edi* Syafri[4], *Sanjay* Mavinkere Rangappa[5] and *Suchart* Siengchin[5]

[1]Department of Biosystems Engineering, Institut Teknologi Sumatera, South Lampung, 35365, Indonesia
[2]Department of Mechanical Engineering, University of Jember, Jember, 68121, Indonesia
[3]Department of Chemical Engineering, Institut Teknologi Sumatera, South Lampung, 35365, Indonesia
[4]Department of Agricultural Technology, Agricultural Polytechnic, Payakumbuh, West Sumatra 26271, Indonesia
[5]Natural Composites Research Group Lab, Department of Materials and Production Engineering, The Sirindhorn International Thai-German Graduate School of Engineering, King Mongkut's University of Technology North Bangkok, Bangkok, 10800, Thailand

Abstract. Polyvinyl Alcohol (PVA) based biocomposite film with cellulose was successfully fabricated by the solution casting method. The cellulose fibers were obtained by extraction of durian peel using alkalization and bleaching treatments. These treated cellulose fibers were used for the fabrication of PVA-based biocomposites. The durian peel cellulose fibers were varied by 2%, 4%, 6%, and 8% in the PVA matrix. Tensile test and moisture resistance of biocomposites were evaluated. The 6% addition of cellulose fibers in biocomposites increases the tensile strength up to 54% (37 MPa) than pure PVA film (24 MPa). Conversely, it reduces the elongation at break of the biocomposite film. Meanwhile, the moisture resistance properties of the biocomposites increased with the addition of cellulose fibers. The tensile strength and moisture resistance of biocomposites have been increased due to the homogeneous dispersion of the cellulose fibers and PVA matrix. These biocomposites able to reduce the environmental impacts by utilizing residual lignocellulosic biomass.

Keyword. Polyvinyl Alcohol (PVA); Cellulose fiber; Durian peel fiber; Tensile strength; Moisture resistance

1 INTRODUCTION

Polyvinyl alcohol (PVA) is a potential material for the manufacturing of bioplastics because of its biodegradability properties and water solubility [1-3]. PVA is a general bio-plastic with good characterization including chemical resistance and has good gelatin formation. In addition, PVA is medical and packaging applications [4-6]. However, PVA has a high moisture absorption because of the large number of OH groups [7, 8]. An effort to overcome the weakness is by mixing PVA with a hydrophobic filler such as clay particles or natural fibers [9, 10]. Recent studies have reported that PVA based biocomposites show higher moisture resistance after the addition of filler such as cellulose nanofibers [11] and nanosilica [12]. The good distribution of the filler in the PVA matrix can be produced using an ultrasonic bath to optimize the tensile and physical properties of the biocomposite film. Previous studies reported the use of ultrasonic baths for the preparation of polyvinyl alcohol films blended with nanofiber [13], montmorillonite [14], halloysite nanoclay [15], and nanocrystalline cellulose [16] to improve the properties of the biocomposite film. This is due to ultrasonic bath treatment to filler leads to the homogeneous dispersion of the filler in the matrix [17, 18].

Ultrasonic bath can improve the polymer structure in matrix which causes changes in its characteristic [19, 20]. The effect of this treatment on the properties of the starch-based biocomposite film was reported in a previous study [21]. The results of this study explain that sonicated films have a more compact and homogeneous polymer structure resulting in higher tensile strength and higher moisture resistance compared to non-sonicated films [21, 22].

From the literature studies, it has been identified that very little work has been focused on the various concentration of durian peel fibers in PVA matrix to develop biocomposite film. Also, from the literature it is well known that, use of cellulose fibers as a reinforcement in biocomposites is one way for improving the environmentally friendly products [23-32].

* Corresponding author: melbi.mahardika@tbs.itera.ac.id

2 MATERIALS AND METHODS

2.1 Materials

Polyvinyl alcohol 2488 CAS number 9002-89-5 was purchased from Prima Global Chemical Ltd., Indonesia. PVA before using it and drying it in an oven at 45°C for 20 hours. Other chemicals were purchased from UD. Aneka Kimia (Jember). Durian peel fiber (DPF) was obtained from Durian seller in Jember. The cellulose content of durian peel fiber before extraction was 55-58% [33].

2.2 Extraction of Durian Peel Fiber

Initially, durian peel was cleaned with water until it was clean from dirt. Durian peel was dried for about 3 days until the constant weight under the sun. After that, the durian peel was cut into small pieces. The durian peel was blended at 25000 rpm for 20 minutes until it becomes a fiber powder as previous report [34]. The fiber powder was alkalized using 8% NaOH for 2 hours at 60°C. The fiber was neutralized with distilled water until the pH becomes 7. Then, it was dried under the sun for 3 hours and formed into sheets of paper fibers.

2.3 Preparation for PVA/Biocomposite Films

The 10 gr of PVA pellets were dissolved in 100 mL of distilled water using a hot plate stirrer (Daihan scientific, model MSH-20D) at 70°C and 800 rpm for 1.5 hours until to gel form. The PVA gel was treated ultrasonic bath (BAKU BK-2400) for 10 minutes under 50°C. After that, it was poured into a petri dish (diameter 150 mm) and dried in drying oven (Universal Oven KLAZ) at 60°C for 24 hours. Each PVA/biocomposite film was labeled with varying concentrations of durian peel fiber namely PVA-0 (0% fiber), PVA-2DPF (2% fiber), PVA-4DPF (4% fiber), PVA-6DPF (6% fiber) and PVA-8DPF (8% fiber). All of biocomposite films were storaged in Tupperware container size 15 cm x 20 cm with relative humidity 50% for one day before characterization.

2.4 Tensile Test

The tensile test was performed using a Tensile Machine (Wang Ta type SF093A) to measure modulus of elasticity, elongation at break, and tensile strength. All biocomposite films were cut according to ASTM D882-12 standards. Prior to the test, the PVA/biocomposite films were conditioned for 2 days under 50 ± 5% relative humidity at 28°C. The tensile test was repeated three times for each sample.

2.5 Moisture Absorption

Moisture absorption (MA) was measured by the same method as previous studies [35]. The PVA and biocomposite films were cut into 2 cm x 2.5 cm rectangular pieces and dried in drying oven (Universal Oven KLAZ) until constant weight. The 75% relative humidity (RH) in a closed chamber was used saturated NaCl solution. The dried PVA/biocomposite films were placed in a closed chamber at 28°C for 6 hours and weighed using a precision scale of 0.0001 g (Kenko KK-LAB) every 30 minutes. The MA was calculated using the Equation 1:

$$MA\ (\%) = [(w_t - w_0)/w_0] \times 100 \quad (1)$$

where w_t final weight sample dan w_0 initial weight sample.

2.6 Scanning Electron Microscopy (SEM)

The fracture surface of PVA and its biocomposite film after tensile test were observed using scanning electron microscopy (SEM) Hitachi TM3030 Plus. The test was carried out at 5 kV conditions and at room temperature. The fracture sample was formed with a size of 5 mm x 12 mm and coated with gold.

2.7 Statistical Analysis

The experimental data analysis was performed using IBM SPSS Statistics 25.0 (IBM Corporation, Chicago, USA). Analysis of variance (ANOVA) were performed to determine the effect of addition fiber in PVA matrix on the tensile and moisture absorption properties. The difference in the mean score of each film was based on Duncan multiple range test, 95% confidence level ($p \leq 0.05$).

3 RESULT AND DISCUSSION

3.1 Tensile Strength

Figure 1 shows the average values of tensile strength, modulus of elasticity, and elongation at break of pure PVA and its biocomposite films. After the addition of 6% durian peel fibers, the value of tensile strength was 37 MPa compared to pure PVA film (24 MPa), increased by 54%. The increasing in tensile strength for PVA-DPF was due to the addition of durian peel fibers in matrix as shown in Figure 1. Meanwhile, the elongation at break was improved by 79% after addition of 6% DPF compared to pure PVA film. From the ANOVA statistical analysis showed a significant increase in tensile strength along with an increase in the volume of fiber fraction. This is due to the enhancement degree of crystal orientation of the durian peel fibers [36]. This result supported by previous studies the biocomposite PVA/cellulose was higher than PVA film due to the high crystalline structure of PVA [37, 38]. The mechanical strength of cellulose composites for bioplastic applications was important to study because these materials are able to withstand high loads.

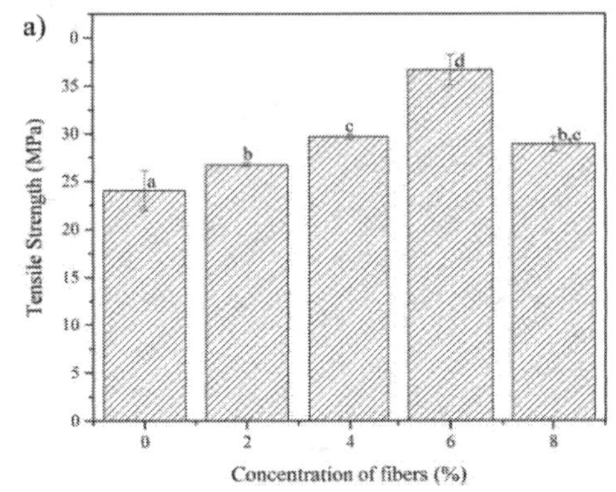

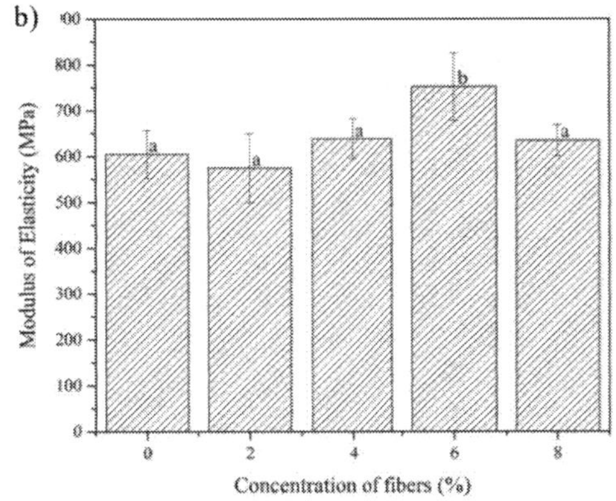

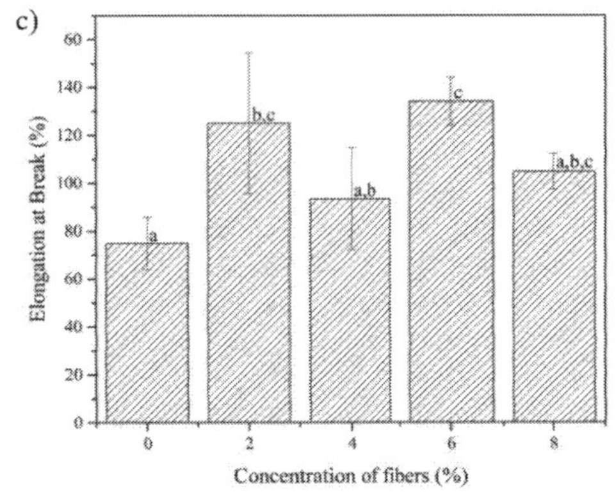

Fig. 1. Tensile Strength (a), Modulus of Elasticity (b), and Elongation at Break (c) for PVA/biocomposite films.

3.2 Moisture Absorption

Figure 2 shows the moisture absorption (MA) value of the PVA/biocomposite film. PVA film has the highest MA (14.2% after 6 h) due to the much amount of free O-H group [28]. After the addition of the DPF by 2% (6 h), the moisture resistance of the film increased. MA for PVA-4DPF (11.3%) is lower than pure PVA film (14%) after 6h. After the fiber fraction increased by 6%, MA decreased again associated with the hydrophilic functional group. For example, after 6 hours in a humid room, the MA for PVA-8DPF was 9.7%, absorbing 24.8% less moisture than PVA-2DPF. This is due to the strong intermolecular interactions of the PVA chain via the intermolecular hydrogen bonds. Moisture absorption can diffuse through a hydrophilic film. From the ANOVA statistical analysis data showed the significance of improving moisture resistance along with the increase of the volume fraction DPF. These results are in accordance with previous studies which showed a decrease in moisture absorption in the biocomposite film after the addition of cellulose fibers [8, 11, 38, 39]. Moisture absorption testing was important to ensure that no moisture enters the bioplastic in order to keep the product in good condition.

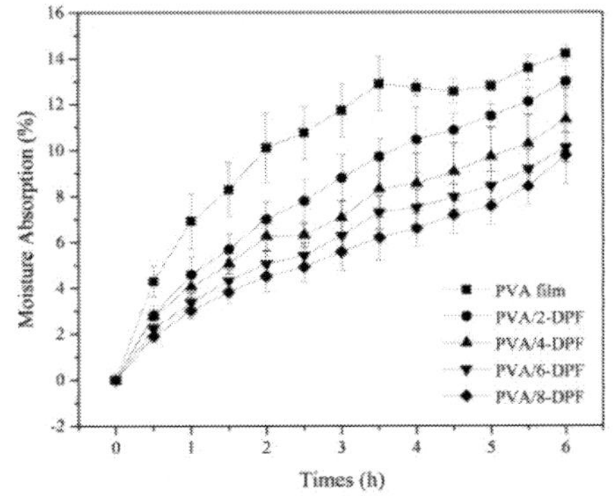

Fig. 2. Moisture Absorption of the PVA/biocomposite films

3.3 SEM

Scanning electron microscopy image of fracture surface of the PVA/biocomposite films are shown in Figure 3. Figure 3a showing the smooth surface fracture of pure PVA film. After the addition of 4% durian peel fiber, the surface of the film becomes rough and wavy like waves. This is due to the contribution of the durian peel fibers that bind to the PVA matrix which makes the polymer chain structure compact as shown in Figure 3b (indicated with yellow arrow). Figure 3c depicts the surface of the PVA/biocomposite films with the addition of 8% durian skin fibers. In this figure, it can be seen that the surface is getting rougher compared to Figure 3b (indicated with yellow arrow) due to the contribution of the increasing concentration of durian skin fibers in the PVA matrix. These results are supported by previous studies which show the same results [8, 12].

4 CONCLUSION

This study showed that the addition of DPF significantly increaased the tensile strength and moisture resistance of the PVA/biocomposite film. The results showed that the addition of DPF increased the compatibility of the polymer chain structure as evidenced by the fracture surface morphology of SEM. This phenomenon leads to higher tensile strength, and improve moisture resistance properties than pure PVA film. Also shows that the durian peel fibers contribute to improve some of the properties required for food packaging materials. This research needs to be continued further, especially in the characterization of biodegradation and its thermal stability.

a)
b)
c)

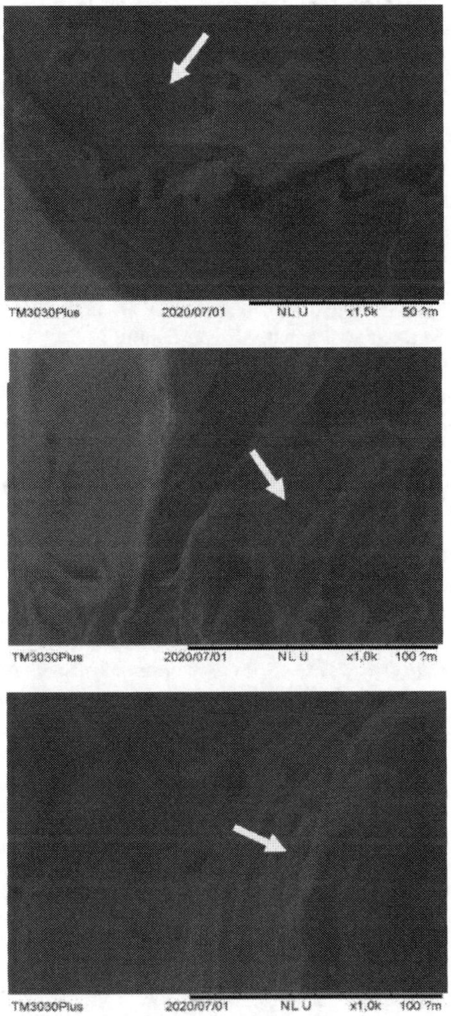

Fig. 3. SEM images of a fracture surface of PVA (a), PVA/biocomposites reinforced with Durian peel fiber (6%) (b), PVA/biocomposites reinforced with Durian peel fiber (8%) (c).

Acknowledgements

This research was supported by Institut Teknologi Sumatera. In addition, researchers are also grateful to the laboratory of material testing, Faculty of Engineering, University of Jember for facilitating research, especially in the manufacture of PVA biocomposite and durian peel fibers.

References

1. A.I. Cano, M. Cháfer, A. Chiralt, C. González-Martínez, Physical and microstructural properties of biodegradable films based on pea starch and PVA, Journal of Food Engineering, **167**, (2015): 59–64
2. D. Domene-López, M.M. Guillén, I. Martin-Gullon, J.C. García-Quesada, M.G. Montalbán, Study of the behavior of biodegradable starch/polyvinyl alcohol/rosin blends, Carbohydrate Polymers, **202**, (2018): 299–305
3. Z.W. Abdullah, Y. Dong, Biodegradable and water resistant poly (vinyl) alcohol (PVA)/starch (ST)/glycerol (GL)/halloysite nanotube (HNT) nanocomposite films for sustainable food packaging, Frontiers in Materials, **6**, 2019: 58
4. Y. Dou, B. Zhang, M. He, G. Yin, Y. Cui, I. N. Savina, Keratin/polyvinyl alcohol blend films cross-linked by dialdehyde starch and their potential application for drug release, Polymers (Basel), **7**, (2015): 580–591
5. J. Zhang, W.R. Xu, Y.C. Zhang, X.D. Han, C. Chen, A. Chen, In situ generated silica reinforced polyvinyl alcohol/liquefied chitin biodegradable films for food packaging, Carbohydr. Polymer, **238**, (2020): 116182
6. S. Mathew, A. Jayakumar, V.P. Kumar, J. Mathew, E.K. Radhakrishnan, One-step synthesis of eco-friendly boiled rice starch blended polyvinyl alcohol bionanocomposite films decorated with in situ generated silver nanoparticles for food packaging purpose, International Journal of Biological Macromolecules, **139**, (2019): 475–485
7. H. Abral, A. Atmajaya, M. Mahardika, Fadli Hafizulhaq, Kadriadi, D. Handayani, S. M. Sapuan, R. A. Ilyas, Effect of ultrasonication duration of polyvinyl alcohol (PVA) gel on characterizations of PVA film, Journal of Materials Research and Technology, **9**, (2020): 2477–2486
8. T. Song, S. Tanpichai, K. Oksman, Cross-linked polyvinyl alcohol (PVA) foams reinforced with cellulose nanocrystals (CNCs), Cellulose, **23**, (2016): 1925–1938
9. Z.W. Abdullah, Y. Dong, I.J. Davies, S. Barbhuiya, PVA, PVA blends, and their nanocomposites for biodegradable packaging application, Polymer-Plastics Technology and Engineering, **56**, (2017): 1307–1344
10. K.E. Bourakadi, N. Merghoub, M. Fardioui, M.E. M. Mekhzoum, I.M. Kadmiri, E.M. Essassi, A.E. K. Qaiss, R. Bouhfid, Chitosan/polyvinyl alcohol/thiabendazoluim-montmorillonite bio-nanocomposite films: Mechanical, morphological and antimicrobial properties, Composites Part B: Engineering, **172**, (2019): 103–110
11. A.N. Frone, C.A. Nicolae, R.A. Gabor, D.M. Panaitescu, Thermal properties of water-resistant starch–polyvinyl alcohol films modified with

cellulose nanofibers, Polymer Degradation and Stability, **121**, (2015): 385–397

12. Y.C. Ching, A. Rahman, K.Y. Ching, N.L. Sukiman, H.C. Cheng, Preparation and characterization of polyvinyl alcohol-based composite reinforced with nanocellulose and nanosilica. BioResources, **10**, (2015): 3364–3377
13. Y. Huang, T. Zhou, S. He, H. Xiao, H. Dai, B. Yuan, X. Chen, X. Yang, Flame-retardant polyvinyl alcohol/cellulose nanofibers hybrid carbon aerogel by freeze drying with ultra-low phosphorus, Applied Surface Science, **497**, (2019): 143775
14. J. Li, M. Zhou, G. Cheng, F. Cheng, Y. Lin, P.X. Zhu, Fabrication and characterization of starch-based nanocomposites reinforced with montmorillonite and cellulose nanofibers, Carbohydr. Polymer, **210**, (2019): 429–436
15. S. Radoor, J. Karayil, J. Parameswaranpillai, S. Siengchin, Adsorption of methylene blue dye from aqueous solution by a novel PVA/CMC/halloysite nanoclay bio composite: Characterization, kinetics, isotherm and antibacterial properties, Journal of Environmental Health Science and Engineering, **18**, (2020): 1311–1327
16. N. Noshirvani, W. Hong, B. Ghanbarzadeh, H. Fasihi, R. Montazami, Study of cellulose nanocrystal doped starch-polyvinyl alcohol bionanocomposite films, International Journal of Biological Macromolecules, **107**, (2018): 2065–2074
17. M.J. Dunlop, B. Acharya, and R. Bissessur, Study of plant and tunicate based nanocrystalline cellulose in hybrid polymeric nanocomposites, Cellulose, **27**, (2020): 249–261
18. K. Vaezi, G. Asadpour, S.H. Sharifi, Bio nanocomposites based on cationic starch reinforced with montmorillonite and cellulose nanocrystals: Fundamental properties and biodegradability study, International Journal of Biological Macromolecules, **146**, (2020): 374–386
19. N.E. Kochkina, O.A. Butikova, Effect of fibrous TiO2 filler on the structural, mechanical, barrier and optical characteristics of biodegradable maize starch/PVA composite films, International Journal of Biological Macromolecules, **139**, (2019): 431–439
20. Z. Wu, Y. Huang, L. Xiao, D. Lin, Y. Yang, H, Wang, Y, Yang, D. Wu, H. Chen, Q, Zhang, W. Qin, S. Pu, Physical properties and structural characterization of starch/polyvinyl alcohol/graphene oxide composite films, International Journal of Biological Macromolecules, **123**, (2019): 569–575
21. M. Asrofi, H. Abral, Y.K. Putra, S.M. Sapuan, H.J. Kim, Effect of duration of sonication during gelatinization on properties of tapioca starch water hyacinth fiber biocomposite, International Journal of Biological Macromolecules, **108**, (2018): 167–176
22. A. Jokar, M.H. Azizi, Z. Hamidi Esfehani, Effects of ultrasound time on the properties of polyvinyl alcohol-based nanocomposite films, Nutrition and Food Sciences Research, **2**, (2015): 29–38
23. S.M. Rangappa, S. Siengchin, H.N. Dhakal., Green-composites: Ecofriendly and sustainability. Applied Science and Engineering Progress, **13**, (2020) 183-184
24. M.R. Sanjay, S. Siengchin, J. Parameswaranpillai, M. Jawaid, C. I. Pruncu, A. Khan. A comprehensive review of techniques for natural fibers as reinforcement in composites: Preparation, processing and characterization, Carbohydrate Polymers, **207**, (2019): 108-121
25. G. Rajeshkumar, G.L. Devnani, J. Prakash Maran, M.R. Sanjay, Suchart Siengchin, Naif Abdullah Al-Dhabi, K. Ponmurugan. Characterization of novel natural cellulosic fibers from purple bauhinia for potential reinforcement in polymer composites, Cellulose, **28**, (2021): 5373-5385
26. M.R. Sanjay, P. Madhu, Mohammad Jawaid, P. Senthamaraikannan, S. Senthil, S. Pradeep. Characterization and properties of natural fiber polymer composites: A comprehensive review. Journal of Cleaner Production, **172**, (2018): 566-581
27. A. Vinod, M.R. Sanjay, S. Suchart, P. Jyotishkumar. Renewable and sustainable biobased materials: An assessment on biofibers, biofilms, biopolymers and biocomposites, Journal of Cleaner Production - Journal, **258**, (2020): 120978
28. M. Ramesh, C. Deepa, L. Rajesh Kumar, M.R. Sanjay, S. Siengchin. Life-cycle and environmental impact assessments on processing of plant fibres and its bio-composites: A critical review, Journal of Industrial Textiles, (2020): 1528083720924730
29. P. Madhu, P.M.R. Sanjay, P. Senthamaraikannan, S. Pradeep, S.S. Saravanakumar, B. Yogesha. A review on synthesis and characterization of commercially available natural fibers: Part II. Journal of Natural Fibers, **16**, (2019): 25-36
30. G. Rajeshkumar, S. Arvindh Seshadri, G.L. Devnani, M.R. Sanjay, S. Siengchin, J.P. Maran, N.A. Al-Dhabi, P. Karuppiah, V.A. Mariadhas, N. Sivarajasekar, A.R. Anuf, Environment friendly, renewable and sustainable poly lactic acid (PLA) based natural fiber reinforced composites–A comprehensive reviewJournal of Cleaner Production, (2021): 127483
31. J. Praveenkumara, P. Madhu, T.G. Yashas Gowda, M. R. Sanjay, S Siengchin, A comprehensive review on the effect of synthetic filler materials on fiber-reinforced hybrid polymer composites, The Journal of The Textile Institute, (2021): 1-9
32. M.R. Sanjay, S. Siengchin. Editorial corner-a personal view exploring the applicability of natural fibers for the development of biocomposites, Express Polymer Letters, (2021): 193-193

33. R. Lubis, S.W. Saragih, B. Wirjosentono, E. Eddyanto, Characterization of durian rinds fiber (Durio zubinthinus, murr) from North Sumatera, In AIP Conference Proceedings, **2049**, (2018): 1
34. M. Asrofi, H. Abral, A. Kasim, A. Pratoto, Characterization of the microfibrillated cellulose from water hyacinth pulp after alkali treatment and wet blending, In IOP Conference Series: Materials Science and Engineering, **204**, (2017): 1
35. M. Mahardika, H. Abral, A. Kasim, S. Arief, F. Hafizulhaq, M. Asrofi, Properties of cellulose nanofiber/bengkoang starch bionanocomposites: Effect of fiber loading, LWT, **116**, (2019): 108554.
36. S. Shrestha, F. Montes, G.T. Schueneman, J. F. Snyder, J. P. Youngblood, Effects of aspect ratio and crystal orientation of cellulose nanocrystals on properties of poly (vinyl alcohol) composite fibers, Compos. Science and Technology, **167**, (2018): 482–488
37. M. Asrofi, D. Dwilaksana, H. Abral, R. Fajrul, Tensile, thermal, and moisture absorption properties of polyvinyl alcohol (PVA)/bengkuang (pachyrhizuserosus) starch blend films, Material Science Research India, **16**, (2019): 70–75
38. H. Abral, J. Ariksa, M, Mahardika, D. Handayani, I. Aminah, N. Sandrawati, S. M. Sapuan, R. A. Ilyas, Highly transparent and antimicrobial PVA based bionanocomposites reinforced by ginger nanofiber, Polymer Testing, **81**, (2020): 106186
39. E. Syafri, Sudirman, Mashadi, E. Yulianti, Deswita, M. Asrofi, H. Abral, S.M. Sapuane, R.A. Ilyas, A. Fudholi, Effect of sonication time on the thermal stability, moisture absorption, and biodegradation of water hyacinth (Eichhornia crassipes) nanocellulose-filled bengkuang (Pachyrhizus erosus) starch biocomposites, Journal of Materials Research and Technology, **8**, (2019): 6223–6231

Value Added Gummy Jelly from Palmyra Palm (*Borassus flabellifer* Linn.)

Nutsuda Sumonsiri*, *Pornthip* Phalaithong, *Amornrat* Mukprasirt, and *Rattakorn* Jumnongpon

Department of Agro-Industrial, Food, and Environmental Technology, Faculty of Applied Science, King Mongkut's University of Technology North Bangkok, Bangkok, Thailand

Abstract. The palmyra palm (*Borassus flabellifer* Linn.) is a multipurpose tree. It is commonly found along the coasts of Africa, South Asia and Southeast Asia due to its strong tap root system's ability to store a large volume of water and raise the water table level in the surrounding area. Several uses of this plant include beverages, food, medicine, fiber and timber. Unfortunately, due to the short shelf-life of palmyra palm fruit, more than 60% of the annual fruit yield is lost within 10 days of harvesting, and there is limited commercial use of ripe palm fruit. As a result, the physical properties, physicochemical properties, and proximate composition of value-added gummy jelly from palmyra palm were evaluated. The results showed that up to 26% (w/w) of fruit juice (prepared by mixing the mesocarp of ripe palmyra fruit with water (at a 1:1 w/w ratio) and then removing the insoluble pulp) can be added to the gummy jelly recipe with significant effects on textural and color characteristics, as well as changes in moisture, protein, carbohydrate and energy content.

Keyword. Palmyra palm, gummy jelly, value added product

1 Introduction

The palmyra palm (*Borassus flabellifer* Linn.) is a sugar palm that is commonly grown in Africa, South Asian (including India, Bangladesh and Sri Lanka), and Southeast Asia (for example, Thailand, Myanmar, Vietnam, Cambodia, Indonesia and Malaysia) [1, 2]. This plant is also called by various names depending on different languages, such as, Talgachh and Tarkajhar in Hindi and Bangali, Lulu or Tadi in Telugu, Karimpana in Malayalam, Fan palm, Toddy palm, Tala palm and Brab tree in English [3]. This perennial plant has a maximum life span of 100 years. It can reach a height of 30 meters and bear fruit in 15 and 25 years in areas with adequate water and in arid areas, respectively [4]. Palmyra trees are typically planted near waterbodies such as lakes, rivers, and wells since their strong tap root system can penetrate up to 5 ft radially round the topsoil and 5-10 ft deep in the soil, bind the sandy soil around them, absorb moisture and store water. As a result, water resources are protected [4, 5], storm or hurricane speeds are reduced, and soil erosion is prevented [3].

Palmyra trees have nearly 800 applications, including beverages, food, medicine, fiber, and timber [6]. Palmyra products are claimed to have a variety of pharmacological properties, including anti-inflammatory, antiarthritic, cytotoxic, antibacterial, analgesic, antioxidant, antipyretic and hypoglycemic activities [7]. The pharmacological properties of each part of the palmyra palm are listed in Table 1.

Table 1. Pharmacological properties of each part of palmyra palm.

Part of the palmyra palm	Pharmacological properties	Ref.
Seed and seed coat extract	Antimicrobial activity	[3, 8]
Male inflorescence (flower)	Anti-inflammatory activity	[3]
Young roots	Diuretic agents, anthelmintic, coolant, and treatment of respiratory diseases	[9]
Palm fruit	Anti-inflammatory, anti-oxidant and hydration properties, prevention of malnutrition on children and adults	[10, 11]
Fruit pulp	Reduction of skin inflammations, treatment of nausea and vomiting	[3]
Spadix (blossom)	Treatment of heartburn and enlarged spleen and liver	[3]
Flower sap	Tonic, laxative, diuretic agents, stimulant, and amebicide	[3, 12]
Palm sugar	Treatment of liver disorder	[12]
Toddy	Ulcer treatment	[9]
Neera (palm nectar, juice)	Controlling gastric problems and ulcer treatment	[3]

* Corresponding author: nutsuda.s@sci.kmutnb.ac.th

According to the palm sugar global market trajectory and analysis report [13], the palm sugar market was valued at USD 1.7 billion in 2020 and is expected to reach USD 2 billion by 2027, with a compound annual growth rate (CAGR) of 2.1% from 2020 to 2027. Aside from palm sugar, all parts of the palmyra tree provide several useful products, as presented in Table 2. The total income from one palmyra palm is approximately USD 230-256. In one acre of land, 400-500 palmyra trees can be planted [14]. An acre of land with palmyra palms could generate a total annual income of USD 92,000-128,000. As a result, this plant can provide a year-round source of income for agricultural families.

Table 2. Products from a single palmyra palm per year.

Products	Quantity	Approximately value (USD)
Neera (juice)	150 L	55
Jaggery (palm sugar)	20-25 kg	50-75
Palm candy	16 kg	120
Matured leaves	10 kg	1.38
Coir	11 kg	2.75
Leaf ribs	2.25 kg	0.96
Palm fiber	16-20 kg	0.96
Firewood	10 kg	0.69

Sources: adapted from [3, 4, 14]

The most valuable part of the palmyra palm is the fruit. Like all *Borassus* species, the palmyra palm has male and female flowers on separate plants. Fruits with a black to brown husk and a diameter of 10-18 cm develop after pollination. Each fruit contains 1-3 seeds [8], as shown in Fig. 1. The seeds (endosperm) of an unripe fruit can be eaten as a fruit [9]. The unripe fruit of the palmyra palm is extremely difficult to consume since the hard rind and fibrous portion must be removed before consumption [15]. However, when the fruit is ripe, the fibrous outer layer of the fruit can also be eaten raw, roasted, or boiled [9].

The edible pulp of the ripe palmyra palm is a soft orange-yellow mesocarp that accounts for approximately 51.07% of the total fruit (Fig. 1). This sugary part is slightly bitter and rich in vitamin C and β-carotenoids at 16.9 mg per 100 g of fruit pulp and 3 ppm per g of fruit pulp, respectively [1, 2, 16-18]. Moreover, it is considered to be a good source of pectin, iron, phosphorus, calcium, magnesium and antioxidants [19].

The pulp can be extracted manually or with a fruit pulp extractor using a pulp-to-water ratio of 1:1 or 1:2 [16, 20, 21]. It can be processed into soft beverages, toffee, cakes, jam, jelly, ice cream, sweets, cordials and other delicious food products [19, 22-24]. Unfortunately, over 60% of the annual fruit yield of the palmyra palm is lost within 10 days after harvesting due to its high moisture content, resulting in a short shelf-life [1, 2, 16] and limited commercial use of ripe palm fruit [16]. It is estimated that approximately 10,000 tons of fruit pulp are wasted each year [18].

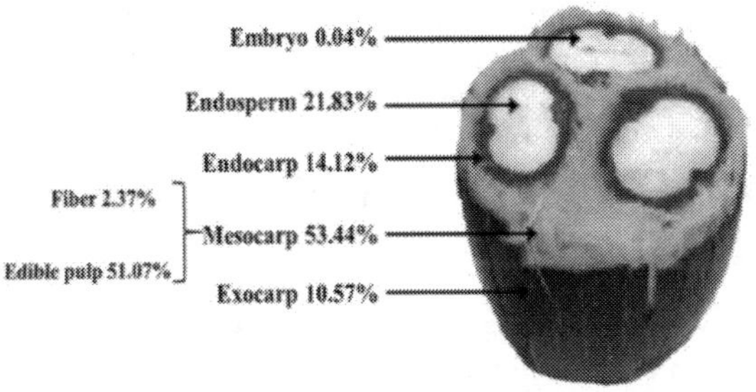

Fig. 1. Components of ripe palmyra palm fruit (adapted from [1, 25]).

The global market for jellies and gummies was valued at USD 13.9 billion in 2018, and it is expected to expand at a CAGR of 3.5% between 2019 and 2025. This market's size has grown as a result of the growing candy industry, the vegan population, and increased consumer spending. The main factors affecting consumers' purchasing decision are usually distinctive flavors, shape, textures, product design, packaging and sugar content [26].

Gummy jelly is one of the confectionery products, often containing high sugar (especially sucrose and glucose syrup) as well as one or more gelling components, flavorings, coloring and food acid [27, 28]. This product is usually produced by heating a sugar solution with gelling components, pouring it into various shapes of molds, and then removing it with powder after hardening [29]. In confectionery gels, sucrose is used as a sweetener while glucose syrup is often added more than sucrose to slow down sucrose crystallization and sucrose molecule migration [29, 30]. Moreover, glucose syrup is used to reduce water activity, which prevents microbial growth [31].

The gelling components used in confectionery gels are polysaccharides, such as thin-boiling starch, pectin, amylopectin, agar, and gum arabic [29]. Pectin, a linear chain of galacturonic acids, is usually extracted from apple pomace or citrus peel [32]. In confectionery gels, high methoxyl pectin or pectin with a degree of esterification greater than 50% is combined with sugar and food acid in order to produce a firm gel [29]. Some of the most common food acids are malic acid, citric acid, fumaric acid, adipic acid and tartaric acid. Food

https://doi.org/10.1051/e3sconf/202130202002

acids can also lower pH and improve product shelf life. In confectionery production, a 50% citric acid solution is usually added at the end of gel processing [29, 31].

Several studies have recently developed gummy jelly containing natural plant extracts and fruit juice, such as *Psidium guajava* leaf extract [33], pineapple and papaya peel powders [34], strawberry syrup [35], honey and propolis [36], *Thunbergia laurifolia* L. extract [37], rosemary (*Rosmarinus officinalis* L.) extract [38] and Babbler's Bill leaf [39]. The purpose of this study was to investigate the physical properties, proximate composition, and energy content of a value-added gummy jelly product made from palmyra palm fruit pulp.

2 Materials and methods

2.1 Preparation of palmyra palm fruit juice

Ripe palmyra palm fruits (*Borassus flabellifer* Linn.) were harvested approximately 2-3 months after pollination in Baan Lard district, Phetchaburi province in Thailand. Each fruit weighed 1-2 kg. The fruits were washed under running tap water and the exocarp (Fig.1) was peeled. The seeds were removed, and the fruit pulp (mesocarp in Fig.1) was then manually extracted into water (pulp:water ratio = 1:1) [16]. After that, the extract was sieved to remove the insoluble pulp and obtain palmyra palm fruit juice. The juice was pasteurized at 65°C for 5 min [1, 40] and stored at -18°C until used.

2.2. Sample preparation

The ingredients used in sample preparation were glucose syrup (Capital Glucose Co., Ltd., Nakorn Pathom, Thailand), sucrose (Baanrai Sugar Industry Co., Ltd., Uthai Thani, Thailand), high methoxyl pectin (Union Chemical 1986 Co., Ltd., Nonthaburi, Thailand), sodium citrate (Chemipan Corporation Co., Ltd., Bangkok, Thailand), fructose (Naturbiotic Co., Ltd., Nakorn Pathom, Thailand), and 50% (w/v) citric acid (Chemipan Corporation Co., Ltd., Bangkok, Thailand). According to the preliminary experiments, the formulation of gummy jelly products in this research is presented in Table 3 and the procedures of sample preparation are shown in Fig.2. The preparation processes were independently repeated on 3 separate days as replications.

2.3 Determination of pH and a_w

In order to determine pH, 5±1 g of samples were cut into thin slices and mixed with distilled water (sample:water = 1:3, w/w) [41]. The mixture was heated at 60±2 °C and constantly stirred until completely dissolved. It was cooled at 25°C before being measured with a pH meter (pH510, Eutech Instruments, Klang Selangor D.E., Malaysia), which had previously been calibrated with buffer solutions of pH = 7 and 4 [42].

The water activity of gummy jelly samples was determined at 25±2°C using a water activity meter (Aqualab, CX-2, Decagon Device, Inc., Pullman, WA, USA), which had previously been calibrated with standard solutions of a_w = 0.250 (13.4 mol/kg LiCl), a_w = 0.5 (8.57 mol/kg LiCl), and a_w = 0.984 (0.5 mol/kg KCl) [42]. 2±0.5 g of food product was placed in the sample holder of the water activity meter. All analyses were performed in triplicate.

Table 3. Formulation of gummy jelly from palmyra palm fruit juice.

Ingredients	Sample			
	No fruit juice added	**20% (w/w) fruit juice***	**23% (w/w) fruit juice***	**26% (w/w) fruit juice***
Glucose syrup (%)	42.94	42.92	41.28	39.67
Sucrose (%)	28.87	28.86	27.75	26.67
High methoxyl pectin (%)	2.25	2.25	2.17	2.08
Sodium citrate (%)	0.22	0.19	0.17	0.17
Fructose (%)	4.51	4.51	4.34	4.17
Water (%)	20.00	-	-	-
Palmyra palm fruit juice* (%)	-	20.00	23.00	26.00
50% (w/v) Citric acid (%)	1.21	1.27	1.28	1.24

*The fruit juice was prepared by mixing mesocarp the of ripe palmyra fruit with water (at a 1:1 w/w ratio) and the insoluble pulp was then removed.

Source: adapted from [34]

2.4 Texture profile analysis

Texture profile analysis (TPA) of gummy jelly was measured using a texture analyzer TA.XT2i (Stable Micro Systems Ltd., UK) as described by Amjadi *et al.* [43]. Food samples were compressed twice without being penetrated. Three replicates of the analysis were performed at 25±2°C. The following texture parameters of the samples were reported: hardness, adhesiveness, springiness, cohesiveness and chewiness [41].

2.5 Color analysis

Color values of samples were measured at room temperature using a Hunter colorimeter (Color Quest 45/0, Hunter Associates Laboratory, Inc., Reston, VA) as described by Charoen *et al.* [44]. The instrument was calibrated prior to measurement using standard white and black reflector plates [33]. Four readings per sample were conducted by changing the position of the sample. Three samples of each formulation were measured in each replicate. CIE L*a*b* color parameters (L*: lightness, a*: a positive a* represents redness and a negative a* represents greenness, and b*: a positive b* represents yellowness and a negative b* represents blueness) for 10° vision angle and D65 illuminant were measured.

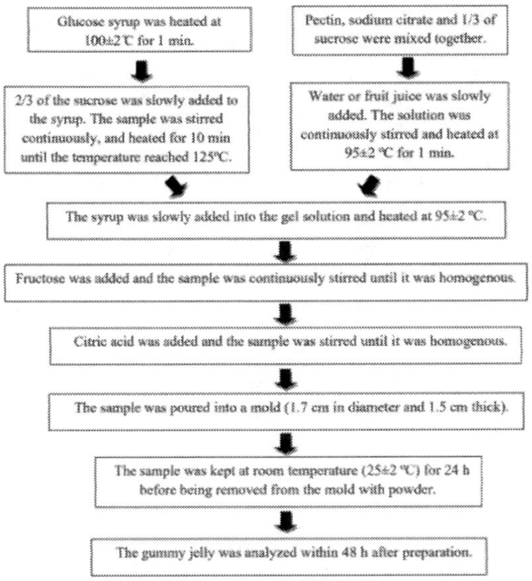

Fig. 2. Workflow for sample preparation.

2.6 Proximate analysis

Proximate analysis, including moisture, protein, lipid, ash, crude fiber and carbohydrate contents, was performed according to the Association of Official Analytical Chemists' procedures [45]. Carbohydrate (%) was calculated using equation (1) [35].

$$Carbohydrate(\%) = 100 - (moisture + ash + lipid + protein) \quad (1)$$

The energy content of each sample was calculated by multiplying the percentages of lipid, protein and carbohydrate by 9, 4 and 4, respectively, and then adding all of the multiplied products together. The energy content in calories was then multiplied by a factor of 4.185 to convert it into kJ [46].

2.7 Statistical analysis

Every analysis was carried out in triplicate, with three subsamples. Significant differences among averages were analyzed in a completely randomized design (CRD) with one-way analysis of variance (one-way ANOVA) and Duncan's multiple range test (DMRT) at a 95% confidence level using IBM SPSS Statistics 26 (IBM Corporation, Armonk, NY).

3 Results and discussion

3.1 pH and a_w

Table 4 presents the pH and a_w of gummy jelly with different concentrations of palmyra palm fruit juice. In this research, citric acid was added to every sample to obtain a pH of ≈ 2.0-3.7, which is the recommended pH range for gelation of high methoxyl pectin [47]. In acidic conditions, pectin chains have lower electrostatic repulsive forces since there is less dissociation of carboxyl groups of galacturonic acid residues [48]. Gummy jelly samples with the addition of palmyra palm fruit juice had a slightly higher pH than the sample without fruit juice ($p<0.05$). Since the pH of palmyra fruit juice was 4.52-4.87, the addition of fruit juice could lead to a slight increase in the pH of the final product. As concentrations of palmyra palm fruit juice increased, a_w of the samples tended to slightly increase since palmyra palm fruit pulp had a high moisture content of 82.13±0.42%. Some gummy jelly products supplemented with natural extracts, such as *Psidium guajava* leaf extract, Babbler's Bill leaf, and juçara and passion fruits, also have a_w of 0.7-0.8 [44, 39, 49].

Table 4. pH and a_w of gummy jelly from palmyra palm fruit juice.

Parameters	Sample			
	No fruit juice added	**20% (w/w) fruit juice***	**23% (w/w) fruit juice***	**26% (w/w) fruit juice***
pH	3.12^{b} ±0.01	3.22^{a} ±0.03	3.23^{a} ±0.02	3.25^{a} ±0.03
a_w	0.704^{c} ±0.003	0.685^{d} ±0.018	0.751^{b} ±0.003	0.779^{a} ±0.004

*The fruit juice was prepared by mixing the mesocarp of ripe palmyra fruit with water (at a 1:1 w/w ratio) and the insoluble pulp was then removed.

Data are presented as mean ± SD (n = 3).

Different superscripts within the same row indicate significant difference between the samples in the Duncan's multiple range test ($p < 0.05$).

3.2 Texture profile analysis

Texture profiles are one of the important factors affecting consumer acceptability. The texture profiles of gummy jelly with and without the addition of palmyra palm fruit juice are shown in Table 5. The texture profiles studied in this research included hardness, adhesiveness, springiness, cohesiveness and chewiness. The force required to compress a food product by a given amount is referred as its hardness [41].

Adhesiveness is defined as the energy required to overcome the attractive forces between the food surface and the surface of other materials contacting the food [50]. Springiness is elastic recovery when the compressive force is removed. Cohesiveness is the strength of the internal bonds in the food sample. Chewiness is the amount of energy required to chew a solid food product until it is ready for swallowing [41].

The samples with 23-26% (w/w) fruit juice tended to have higher hardness, springiness, cohesiveness, and chewiness as compared to the gummy jelly without the addition of fruit juice. Since there is approximately 11.6-13.3% sugar found in palmyra fruit pulp [3], the addition of fruit juice to the product can also increase the sugar concentration of the product. Sugar concentration in confectionery products can reduce the hydration of pectin molecules [51]. The hydrophobic interactions of methoxyl groups could also stabilize the gel structure [48]. Furthermore, because palmyra palm fruit is a good source of pectin [19, 52, 53], adding the fruit juice can increase pectin in the food system, resulting in a slight increase in gel strength and, as a result, increasing the hardness, springiness, cohesiveness, and chewiness of the product. A previous study on gummy candies enriched with pineapple and papaya peel powders also found that the modified candies were firmer and more elastic than the control samples [34]. Unlike other texture profiles, the adhesiveness of samples decreased as concentrations of palmyra palm fruit juice increased. In gummy jelly, adhesiveness is usually undesirable [53], since high adhesiveness indicates that the food product would stick to other non-food surfaces, such as the teeth, palate and tongue [50]. Therefore, the product with 26% (w/w) fruit juice, which provided the lowest adhesiveness, was highly recommended over the gummy jelly without fruit juice.

Table 5. Texture profiles of gummy jelly from palmyra palm fruit juice.

Texture	Sample			
	No fruit juice added	**20% (w/w) fruit juice***	**23% (w/w) fruit juice***	**26% (w/w) fruit juice***
Hardness (g.force)	953.55^{b} ±19.88	881.51^{c} ±57.49	$1,067.95^{a}$ ±32.81	$1,088.23^{a}$ ±32.03
Adhesiveness)g.force(	221.47^{b} ±7.87	325.08^{a} ±9.22	115.42^{c} ±2.83	60.96^{d} ±7.43
Springiness (g.force.sec)	0.968^{b} ±0.003	0.964^{b} ±0.001	0.952^{c} ±0.002	0.979^{a} ±0.003
Cohesiveness	0.639^{a} ±0.006	0.602^{c} ±0.004	0.631^{b} ±0.003	0.637^{ab} ±0.003
Chewiness	563.62^{b} ±15.28	457.95^{c} ±6.06	609.01^{a} ±9.76	571.69^{b} ±19.34

*The fruit juice was prepared by mixing the mesocarp of ripe palmyra fruit with water (at a 1:1 w/w ratio) and the insoluble pulp was then removed.

Data are presented as mean ± SD (n = 3).

Different superscripts within the same row indicate significant difference between the samples in the Duncan's multiple range test ($p < 0.05$).

3.3 Color

The color parameters (L*, a*, b*) of gummy jelly with various concentrations of palmyra palm fruit juice are presented in Table 6. As concentrations of fruit juice increased, all parameters significantly increased ($p<0.05$). The products with palmyra palm fruit juice were orange-yellow (Fig.3) since the fruit pulp is bright yellow or orange (Fig.1) according to the high amount of pro-vitamin A, carotenoids [1, 18, 54]. Srithuvaragan and Anuluxshy could extract pigments from palmyra palm fruit pulp and the yield of β-carotene determined in the literature was 3 ppm per g of fruit pulp [18]. As a result, the palmyra fruit pulp can also be used as a natural food colorant.

Table 6. Color of gummy jelly from palmyra palm fruit juice.

Color	Sample			
	No fruit juice added	**20% (w/w) fruit juice***	**23% (w/w) fruit juice***	**26% (w/w) fruit juice***
L*	20.08^{c} ±0.42	23.81^{a} ±0.32	24.21^{a} ±0.50	21.60^{b} ±0.48
a*	1.75^{d} ±0.10	4.31^{c} ±0.06	5.43^{b} ±0.08	5.79^{a} ±0.15
b*	4.46^{d} ±0.11	5.42^{c} ±0.02	9.21^{b} ±0.14	10.74^{a} ±0.46

*The fruit juice was prepared by mixing the mesocarp of ripe palmyra fruit with water (at a 1:1 w/w ratio) and the insoluble pulp was then removed.

Data are presented as mean ± SD (n = 3).

Different superscripts within the same row indicate significant difference between the samples in the Duncan's multiple range test ($p < 0.05$).

Fig. 3. Gummy jelly with 26% palmyra palm fruit juice.

3.4 Proximate composition

Table 7 presents the proximate composition of the gummy jelly with different concentrations of palmyra palm fruit juice. There was no significant difference in lipid, ash and crude fiber between samples with and without the addition of fruit juice (p≥0.05) since there were low levels of these compositions in the fruit pulp (approximately 2.03-2.06% of lipid, 1.23-1.27% of ash and 0.49-0.52% of fiber) [2]. The addition of fruit juice increased the moisture and protein content of the product while significantly reducing the carbohydrates, resulting in a lower energy content. Although the palmyra palm fruit is high in sugar and carbohydrates [2, 3], the pulp must be extracted using water [16, 20, 21], resulting in the diluted fruit juice and a reduction of carbohydrates in the final product.

Table 7. Proximate composition of gummy jelly from palmyra palm fruit juice.

Composition	Sample			
	No fruit juice added	**20% (w/w) fruit juice***	**23% (w/w) fruit juice***	**26% (w/w) fruit juice***
Moisture (%)	11.60^{d} ±0.13	14.22^{c} ±0.19	14.84^{b} ±0.36	15.79^{a} ±0.31
Protein (%)	0.26^{b} ±0.03	0.43^{a} ±0.04	0.46^{a} ±0.02	0.47^{a} ±0.01
lipid (%)^{ns}	0.16^{b} ±0.03	0.19^{a} ±0.05	0.17^{a} ±0.03	0.14^{a} ±0.03
Ash (%)^{ns}	0.39 ±0.09	0.45 ±0.09	0.42 ±0.07	0.48 ±0.07
Crude fiber^{ns} (%)	0.21 ±0.01	0.21 ±0.01	0.20 ±0.01	0.20 ±0.01
Carbohydrate (%)	87.59^{a} ±0.19	84.70^{b} ±0.20	84.11^{c} ±0.45	83.12^{d} ±0.27
Energy content (kJ)	$1,476.4^{a}$ ±3.6	$1,432.4^{a}$ ±3.1	$1,422.2^{a}$ ±6.6	$1,404.5^{a}$ ±4.8

*The fruit juice was prepared by mixing the mesocarp of ripe palmyra fruit with water (at a 1:1 w/w ratio) and the insoluble pulp was then removed.

Data are presented as mean ± SD (n = 3).

Different superscripts within the same row indicate significant difference between the samples in the Duncan's multiple range test ($p < 0.05$).

^{ns}The samples were not significantly different (p≥0.05).

4 Conclusions

Ripe palmyra fruit pulp can be used to produce gummy jelly with significant effects on the textural and color characteristics of the product, as well as a slight increase in moisture and protein content, but a decrease in carbohydrate and energy content. Furthermore, the addition of fruit juice could also reduce the undesired adhesiveness of the final product. The palmyra palm fruit juice concentration in the product is recommended to be up to 26% (w/w) of fruit juice (prepared by mixing the mesocarp of ripe palmyra fruit with water (at a 1:1 w/w ratio) and then removing the insoluble pulp). This product's production could be potentially profitable for agricultural families.

Acknowledgements

This research was funded by Faculty of Applied Science, King Mongkut's University of Technology North Bangkok (KMUTNB), Contract no. 642053.

References

1. S. Artnarong, P. Masniyom, J. Maneesri, Isolation of yeast and acetic acid bacteria from palmyra palm fruit pulp (*Borassus flabellifer* Linn.), International Food Research Journal, **23**, 3 (2016): 1308-1314
2. M.K. Golly, A.S. Amponsah, V. Mintah-Prempeh, E. Agbamakah, M.A. Akari, L. Adu-Poku, G.F. Paokuaa, V. Gandaa, B. Agodey, Development of food products from palmyra palm (*Borassus flabellifer* L.) fruit pulp for possible commercialization, STU International Journal of Technology, **1**, 4 (2017): 89-102
3. T.R. Sridevi Krishnaveni, R. Arunachalam, M. Chandrakumar, G. Parthasarathi, R. Nisha, Potential review on palmyra (*Borassus flabellifer* L.), Advances in Research, **21**, 9 (2020): 29-40
4. P. Veilmuthu, *Palmyra – nature's perennial gift in the face of climate crisis* (Climate South Asia Network, Bangalore, 2020)
5. B. Kolappan, *Still standing: how palmyra trees withstood the force of the cyclone* (The Hindu, THG Publishing Private Limited, Chennai, 2018)
6. S. Arulraj, B.J. Augustine, Underutilized palms. In: K.V. Peter (ed), *Underutilized and underexploited horticultural crops*, (New India Publishing Agency, New Delhi, 2008)
7. V.P. Gummadi, G.R. Battu, K. Diyya, K. Manda, A review on palmyra palm (*Borassus flabellifer*), International Journal of Current Pharmaceutical Research, **8**, 2 (2016): 17-20
8. H. Jana, S. Jana, Palmyra palm: Importance in Indian agriculture, Rashtriya Krishi, **12**, 2 (2017): 35-40
9. G. Shirisha, R. Sarshaik, J. Nagasowjanya, *Borassus flabellifer* fruit versatile pharmaceutical application: An overview, International Journal of Advanced Research in Medical and Pharmaceutical Sciences, **3**, 4 (2018): 12-16
10. H.J. Pramod, A.V. Yadav, V.N. Raje, M. Mohite, G. Wadkar, Antioxidant activity of *Borassus flabellifer* (Linn.) fruits, Asian Journal of Pharmacy and Technology, **3**,1 (2013): 16-19
11. A. Jerry, A Comprehensive review on the medicinal properties of *Borassus flabellifer*, Journal of

Academia and Industrial Research, **7**, 7 (2018): 93-97

12. P. Saranya, *Characterization of palmyra fruit Borassus flabellifer Linn. pulp and development of ready to serve beverage from palmyra fruit pulp* (Periyar University, Salem, 2016)
13. Global Industry Analysis, Inc., *Palm sugar global market trajectory and analysis (2020-2027)* (Global Industry Analysis, Inc., San Jose, 2021)
14. S. Karthigayini, S. Sahana, V. Sinthujan, Analyzing and recommending marketing strategies for palmyra based productions in Sri Lanka, Journal of Academic and Industrial Research, **8**, 10 (2020): 166-169
15. A. Mani, B. Krishna, P. Dutta, An array of processed and value-added products from palmyra palm, Multilogic in Science, **8**, Special Issue A (2018): 154-156
16. A.K. Chaurasiya, I. Chakraborty, J. Saha, Value addition of palmyra palm and studies on the storage life, Journal of Food Science and Technology, **51**, 14 (2014): 768-773
17. V. Kumari, P. Kiranmayi, P.C. Vengaiah, Estimation of vitamins, minerals and amino acids in palmyra palm (*Borassus flabellifer* L.) fruit pulp, International Research Journal of Pharmacy, **7**, 12 (2016): 70-73
18. R. Srithuvaragan, B. Anuluxshy, Extracting pigments from palmyrah fruit pulp (*Borassus flabellifer* L.) for the production of natural colorants for food, International Journal of Science: Basic and Applied Research, **48**, 4 (2019): 145-151
19. D.D. Ariyasena, E.R. Jansz, A.M. Abeysekera, Some studies directed at increasing the potential use of palmyrah (*Borassus flabellifer* L.) fruit pulp, Journal of the Science of Food and Agriculture, **81**, 14 (2001): 1347-1352
20. D.D. Ariyasena, E.R. Jansz, S. Jayesekera, A.M. Abeysekara, Inhibitory effect of bitter principle of palmyrah (*Borassus flabellifer* L) fruit pulp on the growth of mice: evidence using bitter and non bitter fruit pulp, Journal of the Science of Food and Agriculture, **80**, 12 (2000): 1763-1766
21. E.R. Jansz, N.T. Wickremasekara, K.A.V. Sumuduni, A review of the chemistry and biochemistry of seed shoot flour and fruit pulp of the palmyrah palm (*Borassus flabellifer* L), Journal of the National Science Foundation of Sri Lanka, **30**, 1&2 (2002): 61-87
22. B.C. Das, S.N. Das, *Cultivation of minor fruits* (Kalyani Publishers, New Delhi, 2003)
23. P.J. Rao, M. Das, S. Das, Thermophysical properties of sugarcane, palmyra palm, and date-palm granular jaggery, International Journal of Food Properties, **11**, 4 (2008): 876-886
24. I. Chakraborty, A.K. Chaurasiya, J. Saha, Quality of diversified value addition from some minor fruits, Journal of Food Science and Technology, **48**, 6 (2011): 750–754
25. T.A. Davis, D.V. Johnson, Current utilization and further development of the palmyra palm (*Borassus flabellifer* L., Arecaceae) in Tamil Nadu State, India, Economic Botany, **41**, 2 (1987): 247-266
26. Grand View Research, Inc., *Jellies & Gummies Market Size, Share & Trends Analysis Report By Flavor (Cherry, Berries), By Distribution Channel (Store Based, Non-store Based), By Region, And Segment Forecasts, 2019 - 2025* (Grand View Research, Inc., San Francisco, 2019)
27. P. Burey, B.R. Bhandari, R.P.G. Rutgers, P.J. Halley, P.J. Torley, Confectionery gels: a review on formulation, rheological and structural aspects, International Journal of Food Properties, **12**, 1: 176-210
28. P. Marfil, A. Anhe, V. Telis, Texture and microstructure of gelatin/corn starch based gummy confections, Food Biophysics, **7**, 3 (2012): 236-243
29. H.-D. Belitz, W. Grosch, P. Schieberle, *Food Chemistry (4th edition)* (Springer-Verlag, Berlin, 2009)
30. W.P. Edwards, *The Science of Sugar Confectionery (2nd edition)* (The Royal Society of Chemistry, Cambridge, 2018)
31. E.B. Jackson, *Sugar Confectionery Manufacture (2nd edition)* (Blackie Academic and Professional, London, 1995)
32. A.C. Hoefler, Hydrocolloids: Practical Guides for the Food Industry (Eagen Press, St Paul, 2004)
33. R. Charoen, W. Savedboworn, S. Phuditcharnchnakun, T. Khuntaweetap, Development of antioxidant gummy jelly candy supplemented with *Psidium guajava* leaf extract, KMUTNB International Journal of Applied Science and Technology, **8**, 2 (2015): 145-151
34. K.F. Romo-Zamarrón, L.E. Pérez-Cabreral, A. Tecante, Physicochemical and sensory properties of gummy candies enriched with pineapple and papaya peel powders, Food and Nutrition Sciences, **10**, 11 (2019): 1300-1312
35. S. Takeungwongtrakul, P. Thavarang, S. Sai-Ut, Development of Strawberry gummy jelly with reduced sugar content from strawberry syrup, International Journal of Agricultural Technology, **16**, 5 (2020): 1267-1276
36. R. Rivero, D. Archaina, N. Sosa, G. Leiva, B. Baldi Coronel, C. Schebor, Development of healthy gummy jellies containing honey and propolis, Journal of the Science of Food and Agriculture, **100**, 3 (2020): 1030-1037.
37. T. Kitpot, Development of gummy jelly mixed with *Thunbergia laurifolia* Linn. extract for reducing breath alcohol concentration, Food and Applied Bioscience Journal, **8**, 3 (2020): 29-37.
38. C. Cedeño-Pinos, M. Martínez-Tomé, M.A. Murcia, M.J. Jordán, S. Bañón, Assessment of rosemary

(*Rosmarinus officinalis* L.) extract as antioxidant in jelly candies made with fructan fibres and stevia, Antioxidants, **9** (2020), 1289.

39. T. Suwan, W. Pradutprom, W. Ngamroop, N. Choosuk, C. Phungamngoen, Development of Babbler's Bill leaf gummy jelly, Burapha Science Journal, **22**, 1 (2017): 189-201
40. S. Artnarong, P. Masniyom, J. Maneesri, Preparation the substrate from palmyra palm fruit by *Candida stellimalicola* fermentation for acetic acid production, KKU Research Journal, **21**, 2 (2016): 397-403
41. P. Delgado, S. Bañón, Determining the minimum drying time of gummy confections based on their mechanical properties, CyTA - Journal of Food, **13**, 3 (2015): 329-335
42. Herbstreith & Fox GmbH & Co. KG, Confectionery Gum and Jelly Products (Herbstreith & Fox GmbH & Co. KG, Neuenbürg/Württ, 2004)
43. S. Amjadi, M. Ghorbani, H. Hamishehkar, L. Roufegarinejad, Improvement in the stability of betanin by liposomal nanocarriers: Its application in gummy candy as a food model, Food Chemistry, **256** (2018): 156-162
44. N. Sumonsiri, N., Kundacha, N. Pom-iam, Effect of microcrystalline cellulose on physical characteristics and sensory acceptance of chocolate flavored milk, Current Research in Nutrition and Food Science, **6**, 3 (2018): 781-788
45. AOAC. *Official Methods of Analysis of AOAC International (19th edition)* (AOAC International, Arling, 2012)
46. M.K. Golly, B. Amadotor, Nutritional composition of the seed of *Icacina senegalensis/oliviformis* (flase yam), Pakistan Journal of Nutrition, **12**, 1 (2013): 80-84
47. A.M. Stephen, G.O. Philips, P.A. Williams, *Food Polysaccharides and Their Applications (2nd edition)* (CRC Press, Boca Raton, 2006)
48. D. Gawkowska, J. Cybulska, A. Zdunek, Structure-related gelling of pectins and linking with other natural compounds: a review, Polymers, **10**, 7 (2018): 762
49. J.S. Miranda, B.V. Costa, I.V. de Oliveira, D.C.N. de Lima, E.M.F. Martins, B.R. de Castro Leite Júnior, W.C.A. do Nascimento Benevenuto, I.C. de Queiroz, R.R. da Silva, M.L. Martins, Probiotic jelly candies enriched with native Atlantic Forest fruits and *Bacillus coagulans* GBI-30 6086, LWT Food Science and Technology, **126** (2020), 109275
50. K. Nishinari, Y. Fang, A. Rosenthal, Human oral processing and texture profile analysis parameters: Bridging the gap between the sensory evaluation and the instrumental measurements, Journal of Texture Studies, **50**, 5 (2019): 369-380
51. P. Sriamornsak, Chemistry of pectin and its pharmaceutical uses: a review. Silpakorn University International Journal, **3** (2003): 206–228
52. S. Assoi, K. Konan, L.T. Walker, R. Holser, G.N. Agbo, H. Dodo, L. Wicker, Functionality and yield of pectin extracted from Palmyra palm (*Borassus aethiopum* Mart) fruit, LWT Food Science and Technology, **58** (2014):214-221
53. S. Assoi, K. Konan, G.N. Agbo, H. Dodo, R. Holser, L. Wicker, Palmyra palm (*Borassus aethiopum* Mart.) fruits: novel raw materials for the pectin industry, Journal of the Science of Food and Agriculture, **97**, 7 (2016): 2057-2067
54. M.H. Rodiah, B. Jamilah, B, S.M. Sharifah Kharidah, A.R. Russly, Physico-chemical and antioxidant properties of mesocarp and exocarp from *Borassus flabellifer*, International Food Research Journal **26**, 5 (2019): 1469-1476

Utilization of RS-3 Rice Starch into Development of Food for Elderly: A Ready-to-mix Beverage Case

Atittaya Tandhanskul[1,*], *Patchanee* Yasurin[2], *Praewpan* Chavanon[2], *Pattarawadee* Watanakijcharoenman[2], *Malinee* Sriariyanun[3], *Supawan* Rattanakom[4], and *Ir.* Lindayani[5]

[1]Department of Food Technology, Theophane Venard School of Biotechnology, Assumption University, Bangkok, 10240, Thailand
[2]Department of Food Biotechnology, Theophane Venard School of Biotechnology, Assumption University, Bangkok, 10240, Thailand
[3]Department of Chemical and Process Engineering, The Sirindhorn International Thai-German Graduate School of Engineering (TGGS), King Mongkut's University of Technology North Bangkok, Bangkok, 10800, Thailand
[4]Department of Mathematics, Martin de Tours School of Management and Economics, Assumption University, Bangkok, 10240, Thailand
[5]Food Technology Program, Soegijapranata Catholic University (SCU), Semarang, Indonesia

Abstract. As population aging is becoming a global phenomenon, the search for food for elderly continues. Since elderly population is associated with non-communicable diseases, elderly food development must provide nutritious value and physiological needs. The objective of this study was to develop a ready-to-mix beverage with resistant rice starch type 3 (RS-3) as one of the ingredients. Five Thai rice varieties were modified into resistant-starch type 3 (RS-3). The dietary quality was conducted by measuring rapidly available glucose (RAG) and slowly available glucose (SAG) which was used for primary study of the glycemic index (GI). Riceberry variety showed the highest potential and was used to formulate. Market survey with elderly population (n=400, age>50) revealed additional protein and vegetables that should be included in the beverage. Mixture design was used to initiate the formulation before fine-tuning with sensory analysis. Sucralose was utilized as an alternative sweetener and xanthan gum was added to improve the texture and viscosity to assist population with dysphagia. Using a 7-point hedonic face scale, panelists liked the product slightly to moderately with 0.7% xanthan gum (5.7±1.09) and 0.1% sucralose (5.5±0.89). The viscosity of water and beverage mix at 150:12 (w/w) were found appropriate for patient with dysphagia (700±12.50 mPa.s). The appropriate formulation of RS3-rice starch into elderly beverage were developed, however, the consumer acceptance test is suggested to be continued.

Keyword. Elderly beverage; RS-3 rice starch; elderly food development; dysphagia beverage; beverage for diabetes

1 Introduction

Aging population is on the rise globally. This requires the continuous development of a healthy meal for elderly that should reduce the burden of chronic diseases and avoid loss of function and disability. The role of nutrients, foods, and diet patterns became the subject of study for identifying the best food intake recommendations to achieve a better health and wellbeing [1]. With the decline in digestive, olfactory, and salivary functions, as well as in hormonal profile affect the nutritional status of elderly population. The design for a better meal replacement is a challenge as elderly population is often associated with the presence of chronic diseases, multiple medications, cognitive impairment, and depression. [2] In Thailand, an elderly population is expected to meet 25% of total Thai population, or approximately 17.5 million people in 2028 [3].

Consumption of well-balanced meals is a key to maintaining a healthy lifestyle. Carbohydrates are required for elderly nutrition as they provide human body a primary source of fuel for body and brain functioning. Rice is one of the global staples and a good option for developing elderly food. It is easy to cook, digest, and comparatively less expensive than other food items offering similar nutritional value. Rice is cholesterol free, low in sodium and helps in managing blood pressure [4]. However, ingesting rice can become problematic in elderly since rice can be digested too quickly which increases body blood sugar level and raises risk of diabetes.

To solve this challenge, starch is modified into resistant starch, fractions that are indigestible by body enzymes [5]. Although there are 4 types of resistant starch, resistant starch type 3 (RS3) is starch which is made resistant to hydrolysis by exhaustive α-amylase and pullulanase treatment in vitro [6]. The benefits of resistant starch involve in the prevention and control of chronic human diseases, including diabetes, colon cancer, and obesity [7]. Designing an elderly food should concern the prevention of insufficient nutritional status. Selection of ingredients to be added into the prototype is subsequently performed based on consumer need.

* Corresponding author: atittayatnd@au.edu

Since swallowing difficulty, known as dysphagia, and diabetes mellitus are commonly found in the elderly, there is still a need for development of a texture-modified foods suitable for blood sugar control. This study aims to use RS-3 rice starch as an ingredient to help control blood sugar level and modify the beverage texture to suit those suffered from dysphagia.

2 Materials and Method

Initially, 5 Thai rice varieties were selected and purchased from local supermarket (red, riceberry, black glutinous, Sao Hai, and brown). Each rice variety was modified enzymatically and physically as described by [8-9]. Vegetables used in the formula were decided from the results of the survey (carrot, corn, and pumpkin) and processed into dry powder form. Pea protein isolate was used as a protein source.

2.1. Determination of Rapidly available Glucose (RAG) and Slowly Available Glucose (SAG)

Five ml pepsin was added into 0.1 grams sample. Sample was heated at 37°C in water bath for 30 minutes. After that, 10 ml of 0.1 M sodium acetate was added into sample. Five glass beads with 1.5 cm diameter were added into sample and followed by addition of 2.5 ml enzyme mixtures which include pancreatin, invertase and amyloglucosidase enzymes. The sample was shaken by shaker with the control temperature at 37°C for 20 minutes and the sample was collected (G20). The shaking continued for another 100 minutes before the sample was collected (G120). Amount of glucose was determined by using blood glucose monitoring system (ACCU-chek instant S blood glucose meter) [10-11].

2.2 Market Survey of Elderly Need towards a new developed product

A survey was conducted with 400 elderly people in Bangkok Metropolitan region by the participants whose age were more than 50 years old. A questionnaire was used to screen the consumer need in order to formulate and develop the product. In the questionnaire, there were 4 parts including (1) the consumer behavior, (2) the factors that influence of purchasing rice beverage, (3) the consumer opinion on rice beverage ingredients and (4) demographic data.

2.3 Designing a prototype

2.3.1 Texture adjustment

Resistance starch type 3 (RS3) rice powder was mixed with carrot powder, corn powder, pumpkin powder and pea protein isolate. Xanthan gum was added at 0% as a control, 0.3%, 0.5%, 0.7% and 0.9% to study the effect of texture modification. The hot water (150 ml) was added into samples and stirred until mixed before serving.

2.3.2 Sweetness adjustment

Resistance starch type 3 (RS3) rice powder was mixed with carrot powder, corn powder, pumpkin powder and pea protein isolate. Sucralose was added at 0% as a control, 0.1%, 0.2% and 0.3% to study the effect of sweetness modification. The hot water (150 ml) was added into samples and stirred until mixed before serving.

2.3.3 Viscosity measurement

The formulas as described in 2.3.1 were additionally determined for their viscosity. The NDJ-1 Rotational Viscometer at rotor no.2 at 12 rpm was used to measure viscosity.

2.3.4 Sensory evaluation

Sensory analysis of formulas at different concentration of xanthan gum and sucralose were evaluated. The 30 panelists, whose age were over 50 years old, were included for sensory testing. The sensory form template was designed to use a 7-point hedonic face scale which measured the sensory properties of appearance, color, rice flavor, vegetable flavor, saltiness, sweetness, viscosity and overall acceptability to find the best formula of the RS3-rice beverage (Fig. 1). Three-digit random numbers were used to code the samples and the serving order was counterbalanced.

Fig. 1. The 7-point hedonic face scale.

2.3 Statistical Analysis

The data collected from was statistically analyzed using SAS software (version 9.4). The results were reported as mean ± standard deviations. One-way ANOVA for Randomized Complete Block Design (RCBD) was used to analyze the result and Duncan Multiple Range Test was further analyzed when the significantly difference was detected. Significance of difference were defined as $p<0.05$.

3 Results and Discussion

3.1 Determination of Rapidly available Glucose (RAG) and Slowly Available Glucose (SAG)

Selection of Thai rice varieties for developing a prototype for RS-3 rice beverage was performed by determination of RAG and SAG of the rice samples. All rice varieties exhibited lower in level of glucose detected. Modified riceberry starch showed the lowest RAG or the glucose level measured after 20-min digestion (26.13±0.53) and a slightly higher level of glucose after 120-min digestion (27.71±0.61). This result indicated that riceberry was slowly digested to release the glucose into the system (Table 1). Riceberry is also known as a medium GI rice variety that has been used to develop riceberry rice pudding for people with dysphagia and those need to control blood sugar level [12].

Table 1. RAG and SAG of native rice starches and modified rice starches.

Varieties of rice	RAG (g/100g sample)		SAG (g/100g sample)	
	Native	Modified	Native	Modified
Red rice	40.89 ± 0.56^{b}	31.09 ± 0.62^{b}	38.44 ± 0.35^{b}	26.95 ± 0.53^{a}
Riceberry	38.85 ± 0.46^{c}	26.13 ± 0.53^{c}	38.68 ± 0.53^{b}	27.71 ± 0.61^{a}
Black glutinous	47.78 ± 0.46^{a}	40.66 ± 0.53^{a}	52.04 ± 0.62^{a}	19.83 ± 0.53^{b}
Sao Hai	48.28 ± 0.59^{a}	41.26 ± 0.62^{a}	55.11 ± 0.48^{a}	22.79 ± 0.27^{b}
Brown rice	41.15 ± 0.64^{b}	33.97 ± 0.47^{b}	38.05 ± 0.38^{b}	27.94 ± 0.71^{a}

3.2 Market Survey of Elderly Need towards a new developed product

After a survey with 400 elderly participants, those who aged more than 50 years old, using questionnaire, the consumer behaviors, factors affecting buying decision, and consumer's opinion towards a beverage product were analyzed.

3.2.1 Consumer behaviors

From Table 2, most consumers had meals 3 meal per day (58.8%) and decided to buy food by themselves (67.25%) which the food cost per meal was in the range of 50 THB to 100 THB, or USD 2-4 (data not shown). Moreover, food with fortified vitamins and minerals interested the elderly the most (16.2%). The survey revealed that the cost of buying the healthy food in 1 month for elderly was in the range of 501 THB to 1,500 THB, or USD 15-50. Most of the participant population bought food from the supermarket every week and they have never bought food online. In addition, elderly participants preferred to eat or drink food that contained no preservative in product (data not shown).

3.2.2 Factors influencing buying decision

The important factors for the selection of purchasing ready-to-drink rice beverage or soup from consumer survey (1 refers to "least important" and 5 refers to "most important") were investigated (Table 3). According to the survey, 'safety' of the product was rated as the highest factor is fairly important to the elderly (4.34 ± 0.87) while 'nutrition' and 'freshness' factors were in the range of "fairly important" but scores were lower, 4.23 and 4.09, respectively. For 'source of raw material', 'taste', 'price', 'net volume', 'packaging', 'brand' and 'promotion' provided score in the range of "important". On the other hand, the 'popularity' of the product was the only slightly important factor, 2.96, that the elderly people focused.

Table 2. Elderly consumer behaviors.

Criteria	Percentage
Number of meal per day	
3 meals	58.8
2 meals	35.5
1 meal	1.0
Other	5.0
Decision Making	
Self	67.25
Self-decision but other buys	20.0
Other	12.75
Characteristics elderly interested	
Vitamin fortified	16.2
Nourishing	15.9
Weight control	12.3
Disease prevention	12.2
Create fullness	9.7
Snack	9.4
Meal substitute	9.0
Rice substitute	8.4
Dinner substitute	3.8
Other	0.1

Table 3. Factors affecting buying decision.

Factors	Importance score
Safety	4.34 ± 0.87
Nutrition	4.23 ± 0.86
Freshness	4.09 ± 0.92
Source of raw material	3.90 ± 0.91
Taste	3.80 ± 0.98
Price	3.58 ± 1.05
Net volume	3.31 ± 0.97
Packaging	3.28 ± 1.06
Brand	3.16 ± 0.89
Promotion	3.11 ± 1.19
Popularity	2.96 ± 0.95

3.2.3 Consumer opinions

According to Table 4, it can be described that the elderly participants preferred the characteristic of the ready-to-drink rice beverage in form of beverage powder. The type of vegetables which most of elderly choose to add in product including carrot (11.78%), corn (10.77%) and pumpkin (10.20%), respectively (data not shown). The characteristic of vegetables added was to be in form of blended with the beverage for smooth texture (27.79%), followed by adding as small pieces (23.35%). From figure 2, the elderly population who were interested in a new product made from modified rice and a variety of vegetables, processed to have a low GI, suitable for the elderly and people who want to control their blood sugar levels was 74%.

Table 4. Factors affecting buying decision.

Criteria	Percentage
Characteristic of beverage	
Beverage powder	23.96
Ready-to-drink	19.82
Ready-to-eat soup	18.43
Soup powder	14.86
Smooth texture	14.75
Contain meat for chewing	8.18
Characteristic of vegetables added	
Blended with beverage	27.79
Small piece	23.35
Small dice for chewing	21.95
Dehydrated	19.74
Original size	7.27

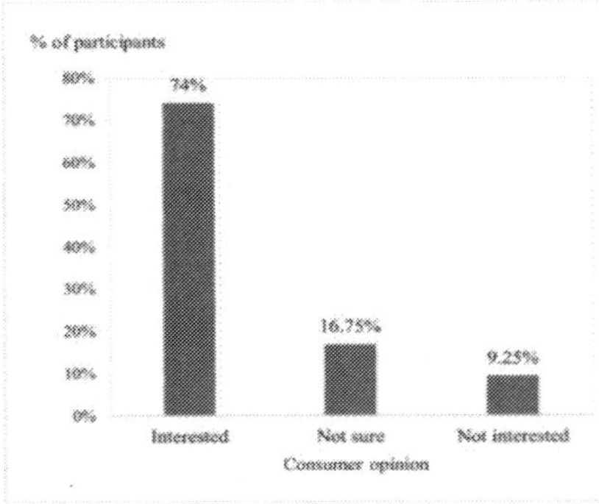

Fig. 2. The percentage of elderly participants that were interested in purchasing the new product.

3.3 Designing a prototype

3.3.1 Texture adjustment

The sensory test was done using 7-point hedonic face scale. The liking scores of sensory acceptability of formulas with five different concentration xanthan gum in attributes including appearance, color, rice flavor, vegetable flavor, saltiness, sweetness, viscosity and overall liking of the samples were shown in Table 5. There was no significant difference ($p \leq 0.05$) in color, rice flavor, vegetable flavor and saltiness when compared to the control (0% of xanthan gum). However, the result of 0.7% of xanthan gum showed the significant difference in sweetness, viscosity and overall liking (4.4 ± 1.42, 5.2 ± 1.28 and 5.7 ± 1.09, respectively). The 0.7% of xanthan gum was selected to develop the further formula due to the overall liking score was highest.

According to the previous study [13], the flavor and appearance influenced to the liking of the food samples based on a 3-point hedonic scale in texture-modified foods (TMFs). Table 5 showed that the appearance affected to the overall liking which corresponded to the research of Okkels et al. [13].

3.3.2 Sweetness adjustment

Barragán et al. [14] reported the preference of sweet foods was consistent although increase in age affected perceived taste sensitivity including salty, sour and bitter tastes were influenced. However, a study by Wiriyawattana et al. [15] showed that the detection and recognition in sweet taste in elderly needed higher sucralose concentration than the younger ones. According to Schiffman [16], the decline in taste sensitivity with aging related to the sweetener molecules by large sweetener molecules had more possible sites for hydrogen bonding with the receptors than small sweetener molecules which the molecular weight of sucralose was 397.64 g/mol.

In this study, the effect of difference concentration of sucralose found no significant difference ($p \leq 0.05$) in appearance, color, rice flavor, vegetable flavor and saltiness of the samples (Table 6). However, the result of 0.1% of sucralose showed the significant difference in sweetness (4.6 ± 1.43) and overall liking (5.5 ± 0.89) when compared to the control (0% of sucralose).

Table 5. Sensory evaluation of different amount of xanthan gum used in development of RS3-rice beverage.

Attributes	Amount of xanthan gum (%)				
	0%	0.3%	0.5%	0.7%	0.9%
Appearance	4.6 ± 1.23^{ab}	4.7 ± 1.09^{ab}	5.2 ± 1.24^{a}	4.9 ± 1.32^{ab}	4.4 ± 1.49^{b}
Color^{ns}	4.5 ± 1.43	4.5 ± 1.10	4.8 ± 1.47	5.2 ± 1.31	4.9 ± 1.43
Rice flavor^{ns}	4.5 ± 1.23	4.1 ± 1.71	4.8 ± 1.39	5.0 ± 1.31	4.7 ± 1.52
Vegetable flavor^{ns}	4.7 ± 1.47	4.8 ± 1.03	4.8 ± 1.35	4.6 ± 0.97	4.9 ± 1.47
Saltiness^{ns}	4.9 ± 1.28	4.8 ± 0.95	4.9 ± 1.17	5.4 ± 1.18	4.7 ± 1.17
Sweetness	4.5 ± 1.47^{ab}	4.4 ± 1.42^{b}	5.2 ± 0.93^{a}	4.4 ± 1.42^{b}	4.5 ± 1.54^{ab}
Viscosity	4.2 ± 1.44^{b}	4.8 ± 1.11^{ab}	5.2 ± 1.06^{a}	5.2 ± 1.28^{a}	3.9 ± 1.77^{b}
Overall liking	4.2 ± 1.35^{c}	4.5 ± 1.32^{cb}	5.1 ± 1.12^{ab}	5.7 ± 1.09^{a}	3.9 ± 1.54^{c}

Note: Different letters in the same row represent significant difference ($p \leq 0.05$)

Table 6. Sensory evaluation of different amount of sucralose used in development of RS3-rice beverage.

Attributes	Amount of sucralose (%)			
	0%	0.1%	0.2%	0.3%
Appearance[ns]	4.6 ± 1.23	5.4 ± 1.27	5.3 ± 1.07	4.7 ± 1.81
Color[ns]	4.6 ± 1.12	5.4 ± 1.38	5.3 ± 1.25	4.7 ± 1.23
Rice flavor[ns]	4.0 ± 1.92	4.7 ± 1.42	4.4 ± 1.54	4.8 ± 1.37
Vegetable flavor[ns]	4.6 ± 1.15	4.8 ± 1.36	5.4 ± 1.23	5.0 ± 1.03
Saltiness[ns]	4.7 ± 1.13	4.9 ± 1.27	5.1 ± 1.07	4.9 ± 1.35
Sweetness	3.6 ± 1.67^{b}	4.6 ± 1.43^{a}	5.1 ± 1.43^{a}	2.4 ± 1.14^{c}
Viscosity	4.5 ± 1.43^{c}	4.7 ± 1.42^{cb}	5.4 ± 0.99^{ba}	5.7 ± 0.98^{a}
Overall liking	3.9 ± 1.52^{b}	5.5 ± 0.89^{a}	5.3 ± 1.13^{a}	3.6 ± 1.57^{b}

Note: Different letters in the same row represent significant difference ($p \leq 0.05$)

Moreover, the increase in amount of sucralose affected the overall liking. The score of overall liking related to the score of sweetness by the score was constant at the 0.1% and 0.2% of sucralose when compared with the sucralose at 0% and 0.3%. These findings indicate that the elderly could detect and recognize the sweet taste of different amount of sucralose which in agreement with the result from the previous research [15] in term of the detection and recognition sucralose concentration in sweet taste in elderly. Hence, the 0.1% of sucralose was chosen to develop formulas since it received the highest overall liking score.

3.3.3 Viscosity measurement

Table 7. The viscosity of RS3- rice beverage with different amount of xanthan gum.

Amount of xanthan gum (%)	Viscosity (Mean ± SD) (mPa.s)
0%	285 ± 8.66^{c}
0.3%	283.33 ± 10.41^{c}
0.5%	301.67 ± 18.76^{c}
0.7%	426.67 ± 50.64^{b}
0.9%	733.33 ± 87.79^{a}

Note: Different letters in the same row represent significant difference ($p \leq 0.05$)

The different amounts of xanthan gum incorporated into RS-3 rice beverage were measured for their viscosity and the result was presented in the Table 7. Compared with the 0% of xanthan gum, the result showed that the increasing amount of xanthan gum affected the viscosity. The xanthan gum was hydrocolloid and thickening agent which used to modify the viscosity [17] and improved consistency of food to ease swallowing [18]. The 0.7% of xanthan gum showed significant difference in viscosity that it was not too thin nor too thick (426.67 ± 50.64 mPa.s). The viscosity ranges for thickened liquids in the National Dysphagia Diet (NDD) stated that the viscosity of honey thick liquids was between 351 – 1,750cP which all the results fell in the range of honey thick liquids except the 0% of xanthan gum was in the range of nectar thick liquids [19]. In addition, the result of sensory evaluation on effect of different concentration of xanthan gum (Table 5) was found that the viscosity at 0.7% of xanthan gum received 5.2 ± 1.28 liking score on 7-point hedonic scale. With appropriate viscosity for patient with dysphagia, 0.7% of xanthan gum was selected as the suitable concentration to incorporate in the formula.

4 Conclusion

The aging population is increasing inevitably around the world. Development of food for elderly should be well designed since there is a close relationship between nutritional status and health in the elderly. The utilization of RS3-rice starch incorporated with plant-based ingredients into a ready-to-mix beverage is just one of the health claimed food products for elderly which the number of this type of products is still limited in the current market. This product is designed for people coping with dysphagia and diabetic mellitus by modifying texture and usage of alternative sweetener, sucralose. Even though the results from this study provided the appropriate formulation of RS3-rice with plant-based on ingredient beverage, the consumer acceptance test for this formulation is suggested to be continued. Furthermore, random controlled test in a specific target group will be conducted in the next phase of research.

References

1. B. Roman-Viñas, L. Serra-Majem, Diet and Healthy Patterns in the Elderly, Current Nutrition Report **3** (2014):69–87
2. L.M. Donini, P. Scardella, L. Piombo, B. Neri, R. Asprino, A.R. Proietti, Malnutrition in elderly: social and economic determinants. Journal of Nutritional Health Aging, **17**, 1 (2013): 9-15.
3. SCB Intelligence Center, The business of aging and opportunities in care for the elderly (2015) https://www.scbeic.com/en/detail/product/1676
4. A. Rohman, S. Helmiyati, M. Hapsari, D.L. Setyaningrum, Rice in health and nutrition, International Food Research Journal, **21**, 1 (2014): 13-24

5. M.G. Sajilata, R.S. Singhal, P.R. Kulkarni, Resistant starch – A review, Comprehensive Reviews in Food Science and Food Safety, **5** (2006) 1-17
6. H. Englyst, H.S. Wiggins, J.H. Cummings, Determination of the non-starch polysaccharides in plant foods by gas-liquid chromatography of constituent sugars as alditol acetates. Analyst **107** (1982): 307– 318
7. D.F. Birt, T. Boylston, S. Hendrich, J.L. Jane, J. Hollis, L. Li, K. Schalinske, Resistant starch: promise for improving human health. Advances in Nutrition, **4,** 6 (2013): 587-601
8. M. Miao, B. Jiang, and T. Zhang, Effect of pullulanase debranching and recrystallization on structure and digestibility of waxy maize starch. Carbohydrate Polymers, **76** (2009): 214-221
9. J. Pongjanta, A. Utaipattanaceep, O. Naivikul, and K. Piyachomkwan, Debranching enzyme concentration effected on physicochemical properties and α-amylase hydrolysis rate of resistant starch type III from amylose rice starch. Carbohydrate Polymers. **78** (2009): 5-9
10. H. Englyst, Classification and measurement of nutritionally important starch fractions. European journal of clinical nutrition, **46**, (1992): 33-50
11. A. Wasusun, Rice Variety RD43: Medium Glycemic Index Rice for Niche Market. Division of rice research and development, (2017): 309-322
12. P. Suttireung, T. Winuprasith, W. Srichamnong, W. Paemuang, T. Phonyiam, D. Trachootham, Riceberry rice puddings: rice-based low glycemic dysphagia diets. Asia Pacific Journal of Clinical Nutrition, **28,** 3 (2019): 467-475
13. S.L. Okkels, M. Saxosen, S. Bügel, A. Olsen, T. W. Klausen, A. M. Beck, Acceptance of texture-modified in-between-meals among old adults with dysphagia, Clinical nutrition ESPEN, **25**, (2018): 126-132
14. R. Barragán, O. Coltell, O. Portolés, E.M. Asensio, J.V. Sorlí, C. Ortega-Azorín, D. Corella, Bitter, sweet, salty, sour and umami taste perception decreases with age: Sex-specific analysis, modulation by genetic variants and taste-preference associations in 18 to 80 year-old subjects, Nutrients, **10,** 10, (2018): 1539
15. P. Wiriyawattana, S. Suwonsichon, T. Suwonsichon, T. 2018. Effects of aging on taste thresholds: A case of Asian people, Journal of Sensory Studies, **33**, 4, (2018): e12436
16. S.S. Schiffman, 2008. The aging gustatory system, *The senses: A comprehensive reference* (2008): 479–498. San Diego, CA: Academic Press.
17. R. Newman, N. Vilardell, P. Clave, R. Speyer, Effect of bolus viscosity on the safety and efficacy of swallowing and the kinematics of the swallow response in patients with oropharyngeal dysphagia: White paper by the European society for swallowing disorders (ESSD), Dysphagia, **31**, (2016): 232–249
18. M. Sharma, L.J.F. Duizer, Characterizing the Dynamic Textural Properties of Hydrocolloids in Pureed Foods—A Comparison Between TDS and TCATA., **8**, (2019): 184
19. R. J. Leonard, C. White, S. McKenzie, P.C. Belafsky, Effects of bolus rheology on aspiration in patients with dysphagia. Journal of the Academy of Nutrition and Dietetics, **114**, 4, (2014): 590-594

Preservations, packaging materials and temperatures affected on quality changes of ready-to-eat bamboo shoot during storage

Ratchanee Charoen[1,*], *Plengsuree* Thiengnoi[2], *Regine* Schoenlechner[3] and *Sutee* Wangtueai[4]

[1]Faculty of Agro-Industry, King Mongkut's University of Technology North Bangkok, Muang, Prachinburi 25230, Thailand.
[2]Academic and Curriculum unit, Mahidol University, Nakhonsawan campus, Nakhonsawan 60130, Thailand.
[3]Department of Food Science and Technology, University of Natural Resources and Life Sciences, Vienna, 1180, Austria.
[4]College of Maritime Studies and Management, Chiang Mai University, Samut Sakhon 74000, Thailand.

Abstract. The influences of preservatives types (sodium benzoate and citric acid), storage temperatures (-10, 4 and 25 degrees Celsius) and container types (low-density polyethylene; LDPE and laminate nylon) on quality changes (physical, microbiological and sensory) of ready-to-eat trimmed and sliced bamboo shoot (slice shape) during storage were investigated. The ready-to-eat bamboo shoot treated with citric acid had good color and microbial stabilities than the control (without the preservatives) and the treated with sodium benzoate at 4th week of storage period. However, the weight loss percentage of the sodium benzoate treatment was lower than the citric acid treatment and the control. At 4 degrees Celsius of storage temperature, the bamboo shoot had the lowest of weight loss percentage compared to other treatments, which the qualities of bamboo shoot at 14 days of storage in this temperature showed an acceptable of weight loss, color, microbial and sensory qualities and could be stored up to 21 days before spoilage. While the storage at 25 degrees Celsius, the bamboo shoot had a shelf life for 7 days. The storage of bamboo shoot in 60 micron LDPE bag was comparable to the laminate nylon bag in term of all quality aspects.

Keyword. Bamboo shoot, Fresh cut, Ready-to-eat, Preservative

1. Introduction

Bamboo is a very useful plant. The traditional wicker is one of bamboo benefit in the past to present. Bamboo is more valuable. It is used in cosmetics, fabric industry, renewable energy such as biodiesel and bio gas, and activated charcoal production. In the present, Young bamboo shoots are used as a common food material in many countries like China, Japan, Philippines, India, US, Nepal, Bhutan, Korea, Australia, New Zealand, Malaysia, Indonesia, and Thailand [1-2] which the commercially major export countries are China, Taiwan, and Thailand [3].

In Thailand, bamboo shoot production is about 300,518 tonnes a year with 67,800 hectare of producing area [2]. Recently, one of important cultivation area is Prachinburi province as Prachantakham, Amphoe Muang, and Nadi district, which produces a bamboo shoots namely Pai-tong)*Dendrocalamus asper*(. The plantation areas and yield are approximately 8,600 hectare and 16 tonnes per hectare [2, 4]. Thus, the bamboo shoot is not only economy crop, but also it is one of reputation product of Prachinburi province, especially during August to September. In during the season, the amount of fresh Pai-tong bamboo shoot is over demand for fresh consumption and the price dramatically decline to about 0.2-0.5 USD/kg. To solvent this problem, the bamboo shoot has been produced to variety products such as dried, fermented, canned bamboo shoot, and boiled bamboo shoot in bucket [5-6]. Most of the consumption of bamboo shoots in season is consumed as fresh while bamboo shoots consumed as dried, fermented or pickled style for off-season. Fresh bamboo shoots are produced by chopping inner soft portion of shoots into thin slices after peeling off and treating them in water for 2-3 hours. Another fresh bamboo species is boiled for 1 hour and shocked in water to remove bitterness of shoot and able to keep until consumed for 2 day and 5-7 day in the fridge. For the shelf stable, fresh bamboo are available for limited period and have a short shelf life. Therefore, it must be processed and used immediately after harvesting [7-8].

Generally, the storage temperature, time, preprocessing methods and packaging influence the shelf life of fresh bamboo shoot, which the deterioration is characterized by unusual increase in color, firmness and stiffness, and fungal infection in the edible parts of shoots [1, 9]. In present, the market of the processed bamboo shoots accounts in a large proportion, while consumer prefer a fresh bamboo shoots because it has much better taste and higher nutrition. In addition, the various factors during postharvest handling was affected

* Corresponding author: ratchanee.c@agro.kmutnb.ac.th

the shoots qualities leading to the subsequent rapid deterioration [1, 3].

In Thailand, the bamboo shoot harvesting is done at early morning and transported to preprocessing plant for trimming, peeling, packing, and transporting to wholesale market. Some chemicals such as sulfur dioxide)SO_2(are treated in the whole peeled bamboo shoot for marketing reason. After peeling, the whole shoots are soaked in the 10-30%)w/v(SO_2 solution for 3- 5 second and then packed into polypropylene bags about 10 kg/bag [5, 9-10]. Other preservatives sodium benzoate citric acid, the food additives that are generally recognized as safe by the Food and Drug Administration. Both preservatives are highly effective in food and commonly used in the food industry. In addition, these additives are economically and the possibility of production costs [7, 10]. The comparison of the packaging type is challenging. Packaging material, thickness and permeability properties of the packaging influence the decision on the selection of suitable package for consumer distribution. LDPD plastic type is safe for food, more transparent while Laminate/nylon plastic is stiffer and more opaque, including the strengths of this package is a good barrier to oxygen and water permeability [3, 8, 10].

Therefore, the processed bamboo shoot is not only one of the many approaches to increase the value of the products expanding marketing channels but also the consumers could easily use for cooking when it had cutting and trimming into appropriated shapes and sizes [11]. According to previous reports, bamboo shoot required some heat treatments, for example, blanching for at least 3 min in order to prevent discoloration of the fresh bamboo shoot. Due to the heating cloud be used for inactivation of related enzymes)such as polyphenol oxidase, peroxidase, and phenylalanine ammonialyase(, preventing pigment instabilities and browning reactions during storage [12-15]. The heat treatment could be a potential tool to decrease disease incidence, respiration and ethylene production rate and lead to a consistent with the reduction in shoot firmness during storage at 20°C [15]. Moreover, the storage conditions and the food additive using have been reported to prevent a spoilage of bamboo shoots for example sodium nitroprusside treatment [16], salicylic acid treatment [17], 1-methylcyclopropene)1-MCP(treating [18]. The effect of packaging and storage temperature to fresh bamboo shoots were studied [3] and the coating of bamboo shoot with antimicrobial and enzymatic antibrowning film resulted in lower weight loss, color changes and reduction in surface microbial count [19]. As a guideline, this study aimed to improve the overall quality of ready-to-eat bamboo shoot products, which is common trade barrier. The combination effect of preservatives, storage temperatures, and container materials on some physicochemical, microbiological quality, and the preference of ready-to-eat bamboo shoots were investigated during 1 month storage.

2. Materials and Methods

2.1 Preparation of bamboo shoot

Fresh Pai-tong bamboo shoots)*Dendrocalamus asper*(were locally purchased from farmers in Noen Hom district, Prachinburi, Thailand during August-October. The bamboo shoots were kept at 4±1°C until use)no longer than three days(. Fresh bamboo shoots were washed to remove other residuals, and peeled the skin off. The peeled bamboo shoots were then trimmed and sliced with 3.0 mm thickness using slicer machine)Savioli, Food EQ Co., LTD, Thailand(. The sliced bamboo shoot was blanched by using boiling water)95±5°C(for 10 min, cooled down in cold water)4±1°C(, and drained in plastic basket for 5 min. Food grade of sodium benzoate and citric acid was obtained from Amata food & Chemical, CO., LTD, Thailand. The low-density polyethylene)LDPE( )60µm of thickness(and laminate/nylon bags)80µm of thickness(were purchased from T.S.T Plaspack Co., Ltd, Thailand and Qingdao RTGYH Packaging Co., LTD, China, respectively.

2.2 Chemical treatments, packaging, and storage conditions

Pretreated bamboo shoots as described in the previous section were treated with and without food preservatives such as citric acid)0.1% w/v(and sodium benzoate)1.0% w/v(by soaking for 10 min and drained in plastic basket for 1 min. The pretreated bamboo shoots)110±7.8 g/bag(were transferred into commercial low-density polyethylene)LDPE(bags)60µm of thickness(, and laminate/nylon bags. Then, the products were stored at shelf storage temperature)25±2°C(, cold storage temperature)4±2°C(, and freezer temperature)-10±2°C(for 30 days. Changes in weight loss, color)L* and b*(, microbiological, and sensory of the products were investigated.

2.3 Physical determination

2.3.1 Weight loss

Weight loss was determined using the method of Badwaik et al [19]. The weight loss during the storage was performed as the percentage of the initial fresh shoot weight remaining after storage. The weight of water condensed in packages was calculated as the difference between fresh weight loss of bamboo shoots and weight loss from packages.

2.3.2 Color measurement

The surface color of bamboo shoot slide was carried out using a Minolta model CM-3500d)Minolta, Osaka, Japan(. Five measurement were taken for each treatment. Colors were expressed as CIELAB coordinates. The L* represents the color lightness on a 0–100 points scale from black to white and b* is the position between yellow)+(and blue)-(.

2.4 Microbiological analysis

Total plate count)CFU/g(and coliform bacteria)MPN/g(were investigated according to FDA-Bam)2001(. Briefly, the 50 g of bamboo shoot sample was blended with 450 ml of 0.85% sterilized normal saline using a stomacher. From this dilution, additional 10-fold dilutions were prepared with sterile normal saline. Pour-plate method using plate count agar)PCA(media)Merck, Germany(was used for total plate counts. Most probable number)MPN(method was used for coliform bacteria determination. The inoculated plates and tubes were incubated at 35±1°C for 48±2 h. All analyses were performed in triplicate.

2.5 Sensory evaluation

The preference test in terms of odor and color was performed by 30 untrained panelists. Bamboo shoot samples were packed into plastic tray covered with wrap film and subjected to panelists for evaluation. The 7-hedonic scale was used that included 1 = dislike extremely, 2 = dislike very much, 3 = dislike moderately, 4 = neither like nor dislike, 5 = like moderately, 6 = like very much, and 7 = like extremely [20].

2.6 Statistical analysis

The means and standard variation)SD(of weight loss and color changes were calculated from 3 and 5 replicates, respectively. For experimental design in physical properties, a 2x3x2 factorial completely randomized design)CRD(was performed. A 2x3x2 randomized complete block design)RCBD(was carried out for sensory evaluation. The analysis of variance)ANOVA(was computed at 95% confident intervals for the population means.

3. Results and discussions

3.1 Physiochemical changes of the ready-to-eat bamboo shoot product

The processed bamboo shoots treated with different preservatives and container materials were kept at different storage temperatures for 1 month. The bamboo shoots were evaluated for changes in weight loss, color and microbes.

3.1.1 Changes in weight loss

As shown in Table 1, without adding any preservatives as the control sample, the samples exhibited increases in the averaged weight loss during the storage from 1 to 4 weeks. It was noted that the weight loss was dramatically increased from the 1st to 2nd week, and the trend was slowed down after the 3rd and 4th week in all treatments. After 4 weeks of the storage, the bamboo shoot stored at -10 °C in both LDPE and laminate/nylon bags exhibited the highest weight loss; however, the bamboo shoots stored at 4 °C in both types of bags showed the lowest weight loss. It should be noted that the bamboo shoots stored at 25 °C could not be investigated since they were spoiled after the 3rd week of the storage.

For the bamboo shoots treated with citric acid and sodium benzoate, results exhibited that after storage at -10 °C for 4 weeks, the bamboo shoots treated with citric acid and sodium benzoate had the highest total weight loss of the products. Results suggested that the storage temperature at 4 °C was considered as the best condition to maintain water release from the product in comparison with other storage temperatures. Nevertheless, the study on the effect of LDPE and laminate/nylon bags on the weight loss of the products at the same temperature revealed that using different container materials did not exhibit significant differences in weight loss. From statistical analysis, preservatives and storage temperatures were significantly influenced on the weight loss of the products)P<0.05(. In addition, the maturity of sliced bamboo shoots affecting the initial moisture content of the bamboo shoots in the range of 83.6-92.5% so the weight loss was highly variable.

Kleinhenz and et al. [3] reported the effect of storage temperatures (1, 2, 8, 11, 20 and 25 °C) and container types (heat-sealed PVC, LDPE, LDPE film, micro-perforated LDPE and macro-perforated LDPE) on the total weight loss of the whole bamboo shoot. Results indicated that the bamboo shoot stored in the macro-perforated LDPE bag at high temperatures exhibited higher rate of the weight loss than other conditions. When considering at 5% weight loss as the critical value, the shelf life of the bamboo shoots stored at 1, 8, 11, and 25 °C was 17, 10, 5 and 1 days, respectively. In addition, the weight loss of the bamboo also related to moisture content and humidity of the products. Overall results suggested that the bamboo shoots stored in macro-perforated LDPE exhibited superior ability to prevent the weight loss than other bag types. Macro-perforated LDPE could prolong the shelf life of the product up to 4 weeks in comparison with LDPE (21 days) and heat-sealed PVC (14 days).

Table 1. The percentage of weight loss of bamboo shoots with and without preservatives a. control; b. citric acid (0.1% w/v); c. sodium benzoate (1.0% w/v) in different container materials and storage temperatures.

Sample	Storage	-10°C		4°C		25°C	
	Time (week)	LDPE	Laminate	LDPE	Laminate	LDPE	Laminate
a. Control	1	26.0 ± 3.1^{a}	25.0 ± 4.3^{a}	17.0 ± 4.8^{c}	12.7 ± 3.8^{d}	26.0 ± 3.8^{a}	20.7 ± 3.1^{b}
	2	42.0 ± 3.1^{a}	30.0 ± 2.8^{b}	18.0 ± 4.1^{d}	20.3 ± 2.2^{d}	34.0 ± 4.7^{ab}	26.4 ± 4.2^{c}
	3	32.0 ± 4.5^{a}	30.0 ± 3.0^{a}	21.0 ± 2.6^{b}	17.8 ± 4.4^{b}	22.0 ± 3.4^{b}	30.2 ± 5.0^{a}
	4	31.0 ± 2.9^{a}	33.0 ± 2.7^{a}	20.0 ± 4.0^{b}	19.0 ± 2.6^{b}	N/A	N/A
b. Citric acid	1	28.0 ± 2.5^{b}	33.7 ± 3.3^{a}	28.0 ± 2.8^{b}	23.6 ± 2.8^{c}	33.5 ± 4.8^{a}	30.8 ± 4.2^{a}
	2	45.0 ± 3.0^{a}	44.2 ± 3.8^{a}	25.0 ± 3.1^{c}	28.7 ± 3.2^{c}	41.0 ± 3.7^{ab}	33.4 ± 4.4^{b}
	3	36.0 ± 3.5^{a}	35.4 ± 2.7^{a}	22.0 ± 3.6^{c}	22.9 ± 3.4^{c}	36.0 ± 4.4^{a}	31.2 ± 3.3^{b}
	4	41.5 ± 2.0^{a}	37.4 ± 1.9^{b}	22.0 ± 3.2^{c}	23.6 ± 3.6^{c}	N/A	N/A
c.Sodium benzoate	1	35.0 ± 1.9^{a}	33.5 ± 2.2^{a}	26.0 ± 3.4^{b}	23.0 ± 3.9^{bc}	$19.5\pm4.4c$	24.0 ± 3.4^{b}
	2	41.0 ± 2.3^{a}	44.3 ± 2.7^{a}	27.0 ± 2.1^{b}	25.8 ± 4.2^{b}	25.5 ± 4.2^{bc}	22.2 ± 2.8^{cd}
	3	36.5 ± 3.5^{a}	39.7 ± 3.4^{a}	22.0 ± 3.5^{c}	21.4 ± 1.9^{c}	26.5 ± 4.3^{b}	28.2 ± 3.7^{b}
	4	46.0 ± 3.1^{a}	48.7 ± 3.1^{a}	24.0 ± 2.8^{b}	22.3 ± 4.0^{c}	N/A	N/A

Volues are means ± SD (n=3 and $p<0.05$), Mean value within each row followed by different letters differ significantly at $p<0.05$

3.1.2 Changes in the apparent color

Results suggested that preservative types, container materials, and storage temperatures were strongly influenced on both lightness (L*) and yellowness (b*). After 4 weeks of storage, the bamboo shoot treated with citric acid exhibited the highest L* value (69.63) followed by the control sample (67.79) and the sample treated with sodium benzoate (66.27), respectively. In addition, considering the result of container materials on the lightness of products, the bamboo shoots treated with citric acid and the control sample in LDPE bags had higher lightness value than the fresh bamboo shoots stored in the laminate/nylon bags.

The bamboo shoots treated with citric acid in LDPE bags exhibited the highest L* value (70.32), while the bamboo shoot treated with sodium benzoate in LDPE bags showed the lowest L* value (66.03). Changing storage temperatures strongly influenced on the L* value of the products. Overall, the bamboo shoots stored at 4 °C showed the lowest L* value in comparison with the products stored at -10 and 25 °C. Sodium benzoate-treated bamboo shoot stored in LDPE at 25 °C had the highest L* value (72.44). The bamboo without preservatives in laminate/nylon bag at -10 °C showed the lowest L* value (63.50) as Table 2.

For the yellowness)b* value(in Table 3, after 4 weeks of storage, sodium benzoate-treated bamboo shoot exhibited the lowest b* value)22.84(followed by the bamboo shoot treated with citric acid)23.54(and the bamboo shoot without addition of any preservatives)25.48(. For the container materials, the bamboo shoots stored in the LDPE bags had higher b* value than the bamboo shoot stored in the laminate/nylon bags. The bamboo shoot without preservatives in LDPE bags showed the highest b* value at 26.11, while the bamboo shoot treated with citric acid in laminate/nylon exhibited the highest b* value at 22.10. Regarding the effect of storage temperatures, the bamboo shoot without any preservatives stored at -10 °C had the highest b* value at 29.06, but the bamboo treated with citric acid in LPDE stored at 0-4 °C had the lowest b* value at 21.05.

The statistical analysis)ANOVA(of L* and b* value indicated that major factors influenced the color changes)L* and b*(were preservatives and storage temperatures at 95% confident intervals. However, the container types did not show significant influence on L* value at 95% confident intervals. In addition, we also found that combined effects of 2 or 3 factors could influence on the color changes.

In addition to the extrinsic factors influencing the color changes as demonstrated in this report, other the intrinsic factors influencing on lipid peroxidation in plants are related three enzymes such as polyphenol oxidase)PPO(, polyphenol peroxidase)POD(and phenylalanine ammonialyase)PAL(. These enzymes can cause enzymatic browning reaction under oxygen environment from polyphenolic compounds such as chlorogenic acid. PAL in the phenyl propanoid pathway also relates to the formation of cinnamic acid, coumaric acid and chlorogenic acid. Furthermore, it is well known that PPO can cause browning formation during harvesting and food processing. In the case of POD in fruits and vegetables can affect to oxidation of phenolic compounds via reduction of hydrogen peroxide to water [14]. Thus, in this study, the blanching process of the bamboo shoots for 10 min should be able to inactivate these mentioned enzymes. Moreover, application of citric acid or sodium benzoate could increase the acidity of the products causing partially inactivation of the enzymes. However, these enzymatic inactivation processes should be further investigated.

Table 2. Changes in lightness (L*) of bamboo shoots with and without preservatives a. control; b. citric acid (0.1% w/v); c. sodium benzoate (1.0% w/v) in different container materials and storage temperatures.

Sample	Storage Time (week)	-10°C		4°C		25°C	
		LDPE	Laminate	LDPE	Laminate	LDPE	Laminate
a.Control	1	65.0±3.7	66.0±4.3	66.0±6.9	65.0±3.8	68.0±5.0	68.0±3.9
	2	68.0±3.7^{ab}	70.0±4.8^{a}	65.0±4.1^{b}	67.0±5.7^{ab}	72.0±3.6^{a}	72.0±3.6^{a}
	3	66.0±4.6^{b}	70.5±4.4^{ab}	67.0±4.1^{b}	67.5±5.1^{b}	75.0±5.2^{a}	75.0±4.7^{a}
	4	69.0±3.7^{ab}	72.0±4.5^{a}	65.0±3.5^{b}	71.0±5.3^{a}	N/A	N/A
b.Citric acid	1	71.0±4.4	71.0±6.5	70.0±5.7	72.0±5.2	71.8±5.8	72.8±4.2
	2	72.0±4.5^{ab}	71.1±6.6^{b}	73.0±6.5^{a}	73.0±5.2^{a}	73.0±5.5^{a}	74.0±6.1^{a}
	3	74.0±5.9^{a}	73.0±4.9^{ab}	74.0±4.3^{a}	72.9±3.7^{ab}	75.0±5.1^{a}	76.0±4.6^{a}
	4	75.0±6.2	74.1±3.4	73.0±4.0	74.0±3.9	N/A	N/A
c.Sodium benzoate	1	63.0±5.8	62.0±3.9	65.0±4.3	64.0±3.4	64.0±4.7	67.0±4.3
	2	64.0±4.7^{b}	62.8±3.1^{b}	68.0±4.5^{a}	69.7±5.2^{a}	67.0±5.1^{ab}	71.0±2.7^{a}
	3	66.4±3.8^{ab}	65.0±3.9^{b}	67.0±4.3^{a}	69.0±4.5^{a}	71.0±2.6^{a}	72.0±5.5^{a}
	4	66.5±3.5^{ab}	64.4±4.4^{b}	70.0±3.7^{a}	71.6±3.9^{a}	N/A	N/A

Volues are means ± SD (n=3 and $p<0.05$), Mean value within each row followed by different letters differ significantly at $p<0.05$

Table 3. Changes in yellowness (b*) of bamboo shoots with and without preservatives a. control; b. citric acid (0.1% w/v); c. sodium benzoate (1.0% w/v) in different container materials and storage temperatures.

Sample	Storage Time (week)	-10°C		4°C		25°C	
		LDPE	Laminate	LDPE	Laminate	LDPE	Laminate
a.Control	1	21.0±2.6	21.4±3.3	20.0±1.9	21.0±3.8	20.0±2.0	20.7±1.9
	2	23.0±2.7^{b}	27.1±2.8^{a}	23.6±2.1^{b}	23.1±1.7^{b}	23.8±3.6^{b}	24.6±1.6^{ab}
	3	22.4±2.6^{b}	26.8±2.4^{a}	23.0±2.1^{b}	22.7±3.1^{b}	25.0±2.2^{ab}	26.0±1.7^{a}
	4	23.4±2.5^{b}	28.8±2.5^{a}	23.2±3.5^{b}	20.8±2.3^{b}	N/A	N/A
b.Citric acid^{ns}	1	20.0±2.4	20.7±2.5	20.0±3.7	21.0±2.2	21.5±1.8	21.6±2.5
	2	22.5±3.5	20.1±3.6	20.5±1.5	22.3±2.2	22.7±2.5	22.3±2.2
	3	21.4±1.9	21.0±2.9	20.4±2.3	21.5±1.7	23.0±3.1	22.0±2.8
	4	21.6±2.2	22.2±3.2	19.5±2.4	21.1±3.2	N/A	N/A
c.Sodium benzoate^{ns}	1	21.0±2.8	21.4±2.9	20.0±2.3	21.0±1.9	20.0±1.7	20.7±2.3
	2	21.4±2.7	21.9±2.8	22.4±2.5	21.2±3.2	22.1±3.1	21.0±1.7
	3	22.0±2.8	22.0±3.4	22.8±2.3	20.7±2.4	22.0±2.6	22.4±2.5
	4	23.8±3.5	22.4±2.1	23.7±3.1	21.3±1.8	N/A	N/A

Volues are means ± SD (n=3 and $p<0.05$), Mean value within each row followed by different letters differ significantly at $p<0.05$

3.2 Changes in microorganisms

Changes in total plate count (TPC) and total coliform bacteria of the bamboo shoots during 4 weeks of storage were investigated. Results showed that the bamboo shoot treated with citric acid had the longest shelf life in terms of microbial stability as indicated by the lowest TPC and total coliform bacteria followed by the bamboo shoots

 https://doi.org/10.1051/e3sconf/202130202004

without any preservatives (control) and the treatment with sodium benzoate. After stored for 4 weeks, TPC value and coliform bacteria of the bamboo shoot kept at -10 °C in the LDPE bags were 54 CFU/g and <3 MPN/g, respectively. In case of the control sample, TPC value and coliform bacteria were 80 CFU/g and < 3 MPN/g, respectively. In addition, in case of the bamboo shoots treated with sodium benzoate, TPC value and coliform bacteria were 1.4×10^3 CFU/g and 9.2 MPN/g, respectively (Table 4). This was showed a higher storage temperature affected to a higher microorganism in the bamboo shoot. Therefore, the deterioration of physical characteristic was obviously observed such as fermentation or spoilage smells and unacceptable soft texture. It was found that the bamboo shoots treated with citric acid and the control sample stored at 4°C in LDPE and laminate/nylon bags had longer shelf life based on texture properties (<4 weeks) than the bamboo shoot treated with sodium benzoate (<2 weeks) (Table 4, Table 5). The same was true for the bamboo shoot kept at 25 °C. In general, sodium benzoate exhibited poor antimicrobial property since the bamboo shoot treated with sodium benzoate could be stored less than 2 weeks.

For the effect of storage temperature, it was clear that the lower temperature, the lower microbial growth was observed. The bamboo shoots stored at -10 °C exhibited the highest microbial stability in comparison with other temperatures. After 2 weeks of storage, total microbial plate counts of the products stored at -10, 4 and 25 °C were <10)none(, 1.9×10^3 and 1.2×10^6 CFU/g, respectively. Nevertheless, with different container materials, it was found that the container materials did not show significant differences on microbial growth during 4 weeks of storage. Badwaik et al.)2014(reported that the surface microbial in bamboo shoot was increased from 4.60 to 6.20 log CFU/cm^2 during 5days storage at room temperature)30±2°C(and relative humidity of 64 ± 3%.

In general, total plate and coliform bacteria counts are used as the indicator for shelf life estimation. These plate count techniques determine the number of aerobic bacteria at normal temperature, at which they are not specific to types of bacteria. Environmental factors such as temperature, acidity, and oxygen content can influence on the growth of microorganisms. Moreover, coliform bacteria count is used as a sanitation indicator that could be contaminated from hands, containers, soil, air, and water during processes into food products, since coliform bacteria are commonly found in the digestive tracts of animals. However, coliform bacteria can be easily destroyed by heating at 60-70 °C.

Besides the storage temperature, the acidity of the bamboo shoots was also important parameters In this study, two types of acidify agents)citric acid and sodium benzoate(were used to lower the pH value of products. Both acidify agents could decrease the pH value of the bamboo shoots to 3.27±0.15 for citric acid and sodium benzoate. However, sodium benzoate can decrease the pH value of product to 5.53±0.38. In general, at the pH lower than 4.5, most anaerobic pathogenic bacteria such as *Clostridium* sp.are effectively inhibited. Products treated with sodium benzoate may carry a high risk of causing food poisoning from thermophilic and mesophilic flat-sour organisms, and thermophilic anaerobes. Thus, it was necessary to determine these microorganisms for a human risk assessment from consumption of bamboo shoots.

Table 4. Total Plate Count (cfu/g) and Coliform bacteria (MPN/g) in bamboo shoot under LDPE packaging during storage for 4 weeks at different temperature.

Sample	Storage Time (week)	-10°C		4°C		25°C	
		TPC	coliform	TPC	coliform	TPC	coliform
Control	1	<10	<3	$1.2x10^2$	<3	$3.2x10^2$	<3
	2	<10	<3	$1.9x10^3$	<3	$1.2x10^6$	<3
	3	$1.0x10^2$	<3	$5.4x10^7$	23	N/A	N/A
	4	80	<3	N/A	N/A	N/A	N/A
Citric acid	1	<10	<3	<10	<3	90	<3
	2	<10	<3	$1.1x10^2$	<3	$3.4x10^5$	<3
	3	35	<3	$3.4x10^7$	460	N/A	N/A
	4	54	<3	N/A	N/A	N/A	N/A
Sodium benzoate	1	<10	<3	$1.2x10^3$	9.2	$3.6x10^6$	460
	2	<10	<3	$1.3x10^7$	240	N/A	N/A
	3	$1.0x10^2$	<3	N/A	N/A	N/A	N/A
	4	$1.4x10^3$	9.2	N/A	N/A	N/A	N/A

Volues are means ± SD (n=3 and p<0.05)

Table 5. Total Plate Count (cfu/g) and Coliform bacteria (MPN/g) in bamboo shoot under laminate /nylon packaging during storage for 4 weeks at different temperature.

Sample	Storage Time (week)	-10°C		4°C		25°C	
		TPC	coliform	TPC	coliform	TPC	coliform
Control	1	<10	<3	45	<3	$1.2x10^3$	<3
	2	<10	<3	$1.7x10^2$	<3	$4.5x10^4$	<3
	3	$3.0x10^2$	<3	$5.4x10^7$	9.2	N/A	N/A
	4	$1.2x10^2$	<3	N/A	N/A	N/A	N/A
Citric acid	1	<10	<3	85	<3	$3.1x10^4$	<3
	2	<10	<3	<10	<3	$2.0x10^7$	<3
	3	<10	<3	$1.5x10^7$	<3	N/A	N/A
	4	20	<3	N/A	N/A	N/A	N/A
Sodium benzoate	1	<10	<3	$3.4x10^3$	460	$1.1x10^7$	460
	2	<10	<3	$1.3x10^6$	240	N/A	N/A
	3	<10	<3	N/A	N/A	N/A	N/A
	4	$1.4x10^3$	9.2	N/A	N/A	N/A	N/A

Volues are means ± SD (n=3 and p<0.05)

3.3 Changes in sensory qualities

Preference test in terms of odor, color, and overall likeness of bamboo shoots during 1 to 4 weeks of storage was evaluated (Table 6, Table 7). Results showed that for the bamboo shoots stored at 4°C for 4 weeks, the odor and color liking scores of the control sample and the citric acid-treated bamboo shoot obtained were approximately 4 to 5 (between neither like nor dislike and like moderately). However, these linking scores were lower in the sodium benzoate-treated bamboo shoot, which were around 2 to 3 (between dislike moderately and dislike very much). Overall, the obtained color linking score of the products was higher than the obtained odor linking score. It was found that some bamboo shoots were spoiled, so they could not be performed sensory evaluation. For instance, the sodium benzoate-treated bamboo shoots stored at 25 °C in both container materials had spoiled at 2 weeks of storage. In addition, after 4 weeks of storage at 4 and 25 °C, all bamboo shoot samples had spoiled. Only the bamboo shoot samples stored at -10 °C could be stored up to 4 weeks without noticeable microbial spoilage.

Table 6. Sensory evaluation of bamboo shoot under LDPE packaging during 4 weeks storage at different temperatures.

Sample	Storage	-10°C		4°C		25°C	
	Time (week)	Flavor	Color	Flavor	Color	Flavor	Color
Control	1	5.33±0.52	4.83±0.75	6.33±0.58	5.50±0.85	4.33±0.53	4.33±0.78
	2	5.21±1.07	4.13±1.02	5.03±0.83	5.06±0.72	2.63±0.91	2.14±1.17
	3	5.00±1.32	4.22±0.64	4.89±1.03	4.78±0.67	N/A	N/A
	4	4.17±1.14	4.06±0.73	4.18±0.73	4.01±1.06	N/A	N/A
Citric acid	1	4.67±1.02	4.67±0.57	4.17±0.41	4.67±0.93	4.00±0.63	3.17±0.41
	2	3.92±1.31	2.89±0.94	1.72±1.05	3.16±0.78	2.82±0.95	2.42±0.78
	3	3.17±1.06	3.04±0.93	1.11±1.12	3.89±0.83	N/A	N/A
	4	3.76±1.14	3.09±1.09	1.73±1.05	3.92±0.91	N/A	N/A
Sodium benzoate	1	5.00±1.06	4.17±0.82	4.33±0.52	3.33±0.92	3.67±0.72	2.17±0.41
	2	3.17±0.74	3.07±0.74	2.72±0.48	3.33±0.52	N/A	N/A
	3	2.89±1.45	2.56±0.88	3.06±0.73	3.02±0.67	N/A	N/A
	4	2.74±1.06	2.94±0.97	N/A	N/A	N/A	N/A

Volues are means ± SD (n=3 and p<0.05)

Table 7. Sensory evaluation of bamboo shoot laminate /nylon packaging during 4 weeks storage at different temperatures.

Sample	Storage	-10°C		4°C		25°C	
	Time (week)	Flavor	Color	Flavor	Color	Flavor	Color
Control	1	5.83±0.78	5.50±0.57	6.13±0.55	5.17±0.41	4.83±0.98	4.00±0.63
	2	4.83±0.82	4.33±1.02	4.36±1.09	4.41±1.32	3.78±0.55	3.23±1.14
	3	4.17±0.84	4.07±0.87	4.47±0.41	4.17±0.52	N/A	N/A
	4	4.06±0.92	4.14±0.91	3.91±0.67	4.02±0.57	N/A	N/A
Citric acid	1	4.17±0.41	4.83±0.72	5.33±0.52	5.17±0.41	4.50±1.05	4.17±0.41
	2	4.33±0.71	4.11±0.52	4.02±0.94	4.22±0.67	2.73±0.59	2.78±0.67
	3	4.42±0.57	4.23±1.04	3.89±0.78	3.04±1.04	N/A	N/A
	4	4.33±0.95	4.78±1.20	3.92±0.83	3.36±0.95	N/A	N/A
Sodium benzoate	1	4.67±0.82	4.33±0.82	4.17±0.64	4.17±0.48	4.50±0.55	5.00±1.41
	2	3.97±0.91	3.72±0.97	4.02±0.59	3.74±0.48	N/A	N/A
	3	3.10±0.53	3.02±0.46	3.27±1.07	3.00±0.87	N/A	N/A
	4	2.83±0.82	2.11±1.04	N/A	N/A	N/A	N/A

Volues are means ± SD (n=3 and p<0.05)

4 Conclusions

Overall quality of ready-to-eat trimmed and sliced phai-tong bamboo shoot products is strongly influenced by container, storage temperature and preservatives. Sodium benzoate could prevent discoloration of bamboo shoot products; however, the bamboo shoot had shorter shelf life in terms of microbial stability in comparison with other preservatives. Controlling storage temperatures had dramatically influenced on all tested parameters including weight loss, discoloration, shelf life, and sensory quality. In general, the bamboo products tested for long-term storage (4 weeks) at lower temperature had higher qualities in terms of sensory quality, color, and microbial stability than the product stored at higher temperature. Optimum storage temperature was around at 4 °C since the bamboo shoot product had the lowest weight loss; however, increases in the storage temperature beyond this point, the weight loss of the product also increased. Using improved bamboo processing and optimal packaging contributes to the fresh bamboo shoot quality which is safe and prolong shelf stable until consumption. The fresh cut bamboo shoots market for perspective consumers who expect exotic flavor is very special and large market. Therefore, effective marketing strategies need to be created to reach both customers domestically and internationally. Fresh cut bamboo shoots with advance technology (High Pressure) are further studies on the satety and nutritional value. It should be an opportunity for product to make consumers confident in their consumption.

References

1. F. Zeng, Z. Luo, J. Xie, S. Feng, Gamma radiation control quality and lignification of bamboo shoots)*Phyllostachys praecox* f. *prevernalis*.(stored at low temperature, Postharvest Biology and Technology, **102** (2015):17–24
2. R. Pattanavibool, Bamboo research and development in Thailand (2000) *in* Bamboo-Conservation, Diversity, Ecogeography, Germplasm, Resource Utilization and Taxonomy. https://www.bioversityinternational.org/fileadmin/bioversity/publications/Web_version/572/ch32.htm [available online 29 April, 2021]
3. V. Kleinhenz, M. Gosbee, T.W. Lyall, K. Blackburn, K. Harrower, D.J. Midmore, Storage methods for extending shelf life of fresh, edible bamboo shoots]*Bambusa oldhamii*)Munro([, Postharvest Biology and Technology, **19** (2000): 253-264
4. Prachinburi Provincial Agriculture Office, Statistical of gross domestic product of Prachinburi province, 2554/2555.]Available on http://klang.cgd.go.th/pri/menu3.html.)June, 2015([
5. C. Techapun, Feasibility study on establishment of bamboo shoot products plant in Chiang Mai, Royal Forest Department, International Tropical Timber Organization, Bangkok, (2004): 46
6. K. Chiangthong, L. Chayawat, Bamboo shoot stream processing from Tha Sao community, Sai Yoke district, Kanchanaburi Province. Kanchanaburi 71000: Department of Science and Technology, Kanchanaburi Rajabhat University, Thailand. www.brt. biotec.or.th/upload/Eng%208-bambo.doc. (2009)
7. P.Nongdam and L. Tikendra, The nutritional facts of bamboo shoots and their usage as important traditional foods of northeast india, Interanational scholarly research notices, (2014): 17
8. A.K. Pandey,V. Ojha, S.K. Choubey, Development and shelf-life evaluation of value added edible products from bamboo shoots, American Journal of Food Technology, 7 (2012): 363-371

9. J. Tipmontian, R.K. Suryaprasa, M. Rajmohan, P. Limpapawich, Assessment and selection of post-harvest treatment in a fresh bamboo shoot supply chain", 15th Food Innovation Asia Conference 2013, June 2013, Bangkok, Thailand

10. B.P. Bhatt, L.B. Singha, K. Singh, M.S. Sachan, Some commercial edible bamboo species of North East India: production, indigenous uses, cost-benefit and management strategies, The Journal of the American Bamboo Society, **17**, 1 (2003(: 4-20

11. Y. Qing, B.D. Zhu, L.W. Zheng, H.H. Kai, X.S. Qi, H.P. Zhen, Bamboo resources, utilization and ex-situ conservation in Xishuangbanna, South-eastern China, Journal of Forestry Research, **19**, 1)2008(: 79-83

12. Y. Xu, M. Zhang, D. Tu, J. Sun, L. Zhou, A.S. Mujumdar, A two-stage convective air and vacuum freeze drying technique for bamboo shoots, International Journal of Food Science and Technology, **40,** 6 (2005(: 589-595

13. T. Wongsakpaiod, Bamboo shoot drying using superheated steam [Meng. Thesis] Bangkok, Thailand : King Mongkut's University of Technology (2000)

14. Q. Shen, F. Kong, Q. Wang, Effect of modified atmosphere packaging on the browning and lignification of bamboo shoots, Journal of Food Engineering, **77** (2006): 348-354

15. Z. Luo, S. Feng, J. Pang, M. Mao, H. Shou, J. Xie, Effect of heat treatment on lignification of postharvest bamboo shoots)*Phyllostachys praecox* f. *prevernalis*(, Food Chemistry, **135** (2012a): 2182–2187

16. H. Yang, C. Zhou, F. Wu, J. Cheng, Effect of nitric oxide on browning and lignification of peeled bamboo shoots, Postharvest Biology and Technology, **57** (2010): 72–76

17. Z. Luo, X. Wu, Y. Xie, C. Chen, Alleviation of chilling injury and browning of postharvest bamboo shoot by salicylic acid treatment. Food Chemistry, **131** (2012b): 456–461

18. Z. Luo, X. Xu, Z. Cai, B. Yan, Effects of ethylene and 1-methylcyclopropene)1-MCP(on lignification of postharvest bamboo shoots, Food Chemistry, **105** (2007) : 521–527

19. L.S. Badwaik, P.K. Borah, S.C. Deka, Antimicrobial and enzymatic antibrowning film used as coating for bamboo shoot quality improvement, Carbohydrate Polymers, **103** (2014): 213-220

20. H.T. Lawless, H. Heymann, Sensory Evaluation of Food: Principles and Practices, Aspen Publishers, Inc., New York (1999)

Pretreatment enhanced the physical and antioxidant stability of dried *Cordyceps militaris* by different drying conditions

Dechapon Chaiya[1], *Chanthima* Phungamngoen[1]*, *Sunee* Eadmusik[1], *Wichien* Sriwichai[2] *and Ekkasit* Subcharoen[3]

[1]Department of Agro-industry and Management, Faculty of Agro-industry, King Mongkut's University of Technology North Bangkok, Prachinburi 25230, Thailand
[2]Department of Innovation and Product Development Technology, Faculty of Agro-industry, King Mongkut's University of Technology North Bangkok, Prachinburi 25230, Thailand
[3]ACX Herb Co., Ltd., Prachinburi 25230, Thailand

Abstract. *Cordyceps militaris* has been treated as a functional food and supplementary food for several years. In recent studies, active components in *Cordyceps militaris* have received great attention and promoted health. Drying technology is a common method and it was applied to *Cordyceps spp.* many times for extending their shelf life and preserving their nutrients. A combination of pretreatment with drying technology could promote processing efficiency. The aim of this study is to study the effect of pretreatments (Soaking in 0.5% citric acid, boiling, steaming and desiccating) combined with drying method by vacuum dryer (VD) at 50, 60 and 70 °C on physical quality and antioxidant capacity in dried *Cordyceps militaris*. The result showed pretreatment by soaking in acid 0.5% citric acid, boiling and steaming took drying time more than those desiccating and untreated dried sample to desire the final moisture content (approximately 8% db.). Increasing of drying temperature in VD could reduce drying time. Desiccating prior to drying samples exhibited the highest L^* value and the lowest of a^* and b^* values when compared with other pretreatments. High drying temperature tend to decrease L^* value while a^* and b^* values are increasing. It was due to the occurring of browning reaction during drying process. Desiccating samples showed color value closest to fresh samples. These results were consistent with all photographs from digital camera. The glass transition temperature of all samples was about 39.43-45.82 °C and the highest value was found in the case of desiccating sample. Antioxidant properties, pretreatment by soaking in acid combined with drying at 70 °C by VD showed the highest content of phenolic compound, DPPH and superoxide radical scavenging activity. It was concluded that pretreatment could preserve important substances, for example, cordycepin and adenosine contents of *Cordyceps militaris* during drying.

Keyword. Antioxidant, Cordyceps, Drying, Pretreatment

1. Introduction

Cordyceps militaris is a member of the fungus. It has been being interested by many countries in Asia and used to be functional food and supplementary foods for several years [1]. Pharmacological properties in *C. militaris* have showed antioxidant, anti-inflammatory and antidiabetic activities. *C. militaris* consists of nucleosides and active compounds such as adenine, cordycepin, polysaccharide and phenolic compound which could be promoted antioxidant capacity [2]. *C. militaris* has a short shelf life and deteriorate immediately in post-harvest [3]. Drying is common method use for extend shelf life of *Cordyceps spp.*

Normally, dried *Cordyceps spp.* was applied by hot air dryer at 50 °C [4]. Hot air drying has been showed low energy consumption and low cost. The increasing of drying temperature resulted in increase of drying rate and drying time decrease. The shorter of drying time led to lower energy consumption [5]. However, hot air drying showed long drying time leading to quality degradation [6]. Vacuum drying is operating under oxygen-free condition and low temperature. It has been showed shorter drying time and improve the quality of dried product compared with traditional drying process [7]. Applied low temperature under low pressure of vacuum drying exhibited the high levels of total phenolic compound and antioxidant capacity in Robusta wet coffee pulp [8]. ABTS radical scavenging activity also showed higher level of vacuum dried *Paramignya trimera* root than those form hot air drying [9]. For the novel drying technologies were used in *C. militaris* for preserving the degradation of bioactive compounds. For example, Wu et al. [10] studied effect of infrared drying on quality of *C. militaris* reported that infrared drying temperature at 60 °C was the optimal condition to

* Corresponding author: chanthima.p@agro.kmutnb.ac.th

maintain the adenosine, cordycepin and total phenolics. Wu et al. [11] studied the combination of drying technologies by microwave assisted pulse fluidized bed freeze-drying (MPFFD) on *C. militaris*. It was found that MPFFD could preserve bioactive compounds of product. However, industries usually use a common drying methods; hot air drying and vacuum drying for producing dried products. Pretreatment was used prior to drying. It showed many advantages such as reducing the initial load of microorganism and inactivate enzymatic activity in vegetables [12]. Pretreatment by blanching is usually combine with drying process for inhibit enzymatic browning reaction. It could promote color retention of product [13] and reduce the pathogens on product surface [14, 15]. The ß-carotene retentions of acid pretreated carrot exhibited unchanged during drying while untreated showed continuously decreased [16].

As mentioned above, there are many researches study on the effect of drying methods on quality of *C. militaris*. However, the combinations effect of pretreatment and drying is still lacking. The aim of this work is to study the physical quality and antioxidant stability of dried *C. militaris* by the combine effect of pretreatments and vacuum drying at different drying temperature. The application of this study could be use to guideline for the consumer to produce dried *C. militaris*.

2. Materials and methods

2.1 Raw material

Fresh *C. militaris* (cultivated indoor) size is 7-10 cm. length and 3 mm. diameter. Samples were obtained from ACX herb Co., Ltd. (Prachinburi province, Thailand). They were cultured in rice. Samples were washed, drained and stored at 4 °C before conducted the experiments.

2.2 Pretreatments

Samples were pretreated with different methods prior to drying. (1) Soaking in 0.5 % (v/v) of citric acid at the sample to acid solution ratio of 1: 10 for 3 min. (2) Boiling sample in hot water at temperature of ~97 °C. The ratio between the sample and boiled water was 1: 5 for 3 min. (3) Steam blanching in water bath (WNE 29, Memmert, Germany) at 95 °C for 3 min. (4) Desiccating in the room (25 °C) for 12 h.

2.3 Drying conditions

Pretreated samples were then dried by vacuum dryer (VO500, Memmert, Germany). The drying temperatures were 50, 60 and 70 °C under pressure of 60 mbar. 100 g of *C. militaris* was placed in a single layer on the tray. The final moisture content was below 8% (db). Drying time was obtained from the drying curves.

2.4 Moisture content and water activity analysis

The moisture content was evaluated follow AOAC [17]. Sample was dried at 105 °C in an oven until constant weight and then calculated moisture content of product using a standard gravimetric method. A water activity (a_w) of *C. militaris* was analyzed by a water activity meter (Series 3TE, AquaLab, Washington).

2.5 Color measurement

Color of *C. militaris* was determined using a colorimeter (ColorFlex, HunterLab, Virginia). Ground sample was put in a cuvette. After the calibration a sample was placed inside the chamber and then analyzed the color. The lightness (L^*), redness or greenness (a^*), yellowness or blueness (b^*) and ΔE were used to describe the total color difference between fresh and dried samples.

2.6 Micrograph of dried *C. militaris* by Field Emission Scanning Electron Microscope (FE-SEM)

The appearance of different dried *C. militaris* was evaluated follow methods of Voda et al. [18] with some modifications. Dried *C. militaris* was broke into two parts for observing a cross-section. The surface of sample was coated with platinum before measuring. The sample was put on to a chamber of field emission scanning electron microscope (Auriga, Zeiss, Germany) at 30× magnification to study the characteristics of dried *C. militaris*.

2.7 Glass transition temperature (T_g) measurement

The transition temperature (T_g) of different dried *C. militaris* was evaluated follow methods of Cao et al. [19] with some modifications. A differential scanning calorimeter (DSC3+, Mettler-Toledo, Switzerland) was applied to analysis. About 4 mg of dried samples were filled and encapsulated into a DSC aluminum pan (40 µL). An empty aluminum pan was used as a reference. Samples were cooled from room temperature to −20 °C, held for 2 min and heated from −20 to 180 °C at 10 °C/min. Then they were cooled from 180 to −20 °C at 50 °C/min and reheated again to 180 °C at 10 °C/min.

2.8 Total phenolic compound analysis

Total polyphenol content (TPC) of different dried *C. militaris* was measured by the colorimetric Folin-Ciocalteu method according to Chen et al. [20] with minor modifications. Firstly, 1 mL of *C. militaris* extracts was mixed with 5 mL of Folin-Ciocalteu's reagent (1:10, v/v). Then the solution was blend with 4 mL of sodium carbonate solution (7.5%). The mixture was heated at 45 °C for 15 min and the absorbance was determined at 765 nm.

https://doi.org/10.1051/e3sconf/202130202005

2.9 Antioxidant activity

Dried sample (1 g) was extracted with 100 mL of boiling water for 3 min. The solution was filtrated by No.1 filter paper for determine the antioxidant activity.

DPPH radical scavenging activity was evaluated following Mao et al. [21] with a minor modification. The extraction (200μL) was mixed with 600 μL of 0.2 mM 2,2-diphenyl-1-picrylhydrazyl (DPPH) and 5200 μL of 85% ethanol. Then the mixture was hold in the dark room for 30 min. The absorbance was measure at 515 nm. The calculation of percentage DPPH scavenging effect was calculated as follow:

$$\text{Scavenging activity (\%)} = \left[1 - \left(\frac{A_S - A_{S0}}{A_0}\right)\right] \times 100 \quad (1)$$

Where A_0 was the absorbance of DPPH, A_s was the absorbance of sample and A_{s0} was the absorbance of sample without DPPH.

Superoxide anion radical scavenging activity follow the 1,2,3-trihydroxybenzene method [22] with minor modifications. The extraction and 0.4 mL of 1,2,3-trihydroxybenzene were added to the Tris-HCl buffer (4.5 mL, 0.05 M, pH 8.2). After incubating at 25 °C for 5 min, the reaction was terminated by the addition of 0.1 mL of HCl (8 M). The mixture (1 mL) diluted with 2 mL of distilled water. Absorbance of the mixture was measured using a UV-Vis Spectrophotometer (Libra S22 Biochrom, United Kingdom) at 299 nm. The superoxide anion radical scavenging activity was expressed as:

$$\text{Scavenging activity (\%)} = \left(1 - \frac{A_2 - A_1}{A_3}\right) \times 100\% \quad (2)$$

Where A_2 is the absorbance of the sample, A_1 is the absorbance of the control and A_3 is the absorbance of the blank sample.

2.10 Statistical analysis

Analysis of variance (ANOVA) was used to obtain the mean value of the results and the standard deviation using the IBM Statistical Package for the Social Sciences (SPSS, IBM Inc., Chicago, Illinois) software, version 21.0. The significant difference between two mean values was calculated using Duncan's multiple range tests at the 5% significance level. Three replicates were used for each treatment.

3. Results and discussion

3.1 Moisture content

The initial moisture content in fresh *C. militaris* was about 532.59 ± 6.14 % db. Evolution of moisture content during drying are showed in Fig. 1. Moisture content in the cases of desiccating pretreatment and high drying temperature decreased at much higher rates than others; this observation agrees well with the previously reported results of Thirugnanasambandham and Sivakumar [7]. The final moisture content in dried samples was controlled below 8% db., the drying times are shown in Table 1. Drying time decreased as the drying temperature increased. This is because higher drying temperature could evaporate the moisture from food matrix easily [8]. A desiccating sample resulted lowest of drying time compared with other pretreatments. The shortest drying time could explain the lowest energy consumption [5]. In the case of steam blanching showed the highest drying time. This might be due to during the desiccating pretreatment moisture removal from plant cell while soaking in acid, boiled water blanching and steam blanching leading to high initial moisture content before drying. Nonthermal pretreatment (soakind in acid and desiccating pretreatment) exhibited lower of energy consumption than thermal pretreatment (boiling and steaming pretreatment) in the combination between pretreatments and drying process of fruits and vegetables [23].

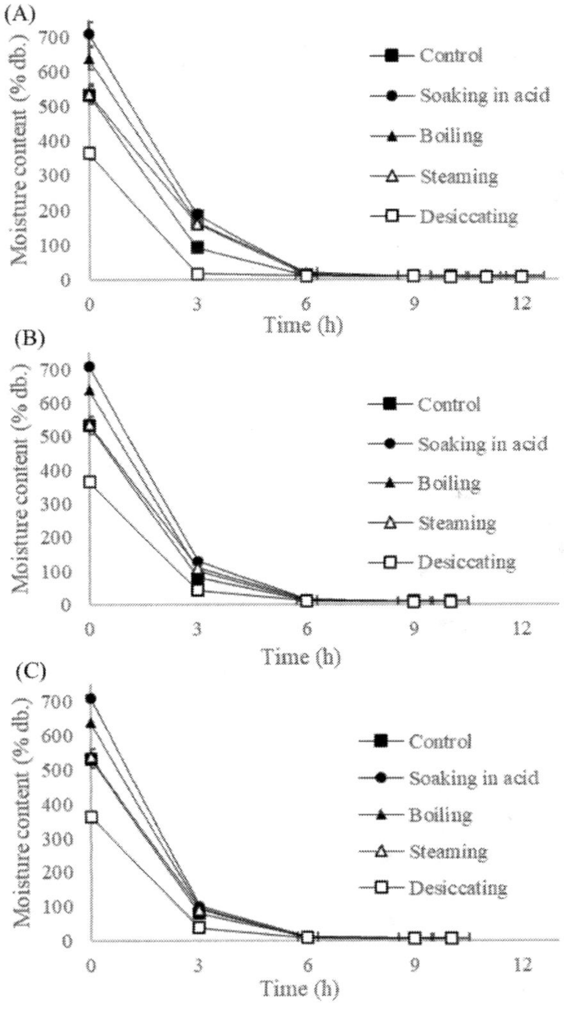

Fig. 1. The drying curves of pretreated *C. militaris* at different drying temperature : (A) 50 °C, (B) 60 °C and (C) 70°C.

Table 1. The drying time, final moisture content, water activity and glass transition temperature of dried *C. militaris*.

Temperature (°C)	Pretreatment	Drying time (hr)	Final moisture content (%)	Water activity (a_w)	Glass transition temperature (°C)
50	Untreated	10	7.20 ± 0.47^{d}	0.323 ± 0.005^{e}	40.54 ± 0.02^{h}
	Soaking in acid	10	7.15 ± 0.24^{de}	0.309 ± 0.002^{f}	41.34 ± 0.03^{g}
	Boiling	10	7.10 ± 0.06^{e}	0.314 ± 0.004^{f}	39.43 ± 0.07^{k}
	Steaming	10	7.21 ± 0.81^{d}	0.344 ± 0.003^{d}	40.21 ± 0.23^{ij}
	Desiccating	10	7.45 ± 0.26^{c}	0.373 ± 0.003^{a}	42.96 ± 0.07^{c}
60	Untreated	8	7.78 ± 0.40^{a}	0.314 ± 0.002^{f}	42.08 ± 0.02^{e}
	Soaking in acid	8	7.39 ± 0.37^{c}	0.308 ± 0.001^{f}	42.16 ± 0.01^{e}
	Boiling	8	7.37 ± 0.43^{c}	0.371 ± 0.002^{a}	39.98 ± 0.08^{j}
	Steaming	8	7.19 ± 0.65^{d}	0.322 ± 0.002^{e}	41.39 ± 0.12^{g}
	Desiccating	8	7.08 ± 0.44^{a}	0.359 ± 0.001^{b}	44.74 ± 0.08^{b}
70	Untreated	7	7.17 ± 0.87^{d}	0.285 ± 0.002^{h}	42.54 ± 0.02^{d}
	Soaking in acid	7	7.14 ± 0.10^{de}	0.343 ± 0.003^{d}	43.07 ± 0.04^{c}
	Boiling	7	7.58 ± 0.04^{b}	0.353 ± 0.002^{c}	40.43 ± 0.31^{hi}
	Steaming	7	7.61 ± 0.34^{b}	0.290 ± 0.003^{g}	41.75 ± 0.06^{f}
	Desiccating	7	7.42 ± 0.65^{c}	0.362 ± 0.002^{b}	45.82 ± 0.14^{a}

Different letters (a-k) within the same column indicate a significant different ($P < 0.05$).

3.2 The water activity and glass transition temperature

The water activity (a_w) and glass transition temperature (T_g) of dried *C. militaris* are shown (Table 1). The initial water activity of *C. militaris* was approximately 0.835 ± 0.030 while the dried samples is shown in a narrow range of 0.284 - 0.376 due to the final moisture content was controlled. It was observed that glass transition temperature was in range of 39.41 - 45.63 °C. Increasing drying temperature resulted in T_g was increased. The results consistent with Wu et al. [9] reported that the T_g of dried *C. militaris* by infrared dryer at 40, 50, 60 and 70 °C at 7 % (d.b) final content of product resulted in the T_g was range of 38.45 - 48.40 °C. On the other hand, Santhalakshmy *et al.* [24] reported that decrease T_g while inlet temperature increased in spray drying of jamun fruit juice powder. Normally, at low water activity and high T_g could control the storage condition and extend shelf life of *C. militaris*.

3.3 The color and appearance of *C. militaris*

The color of dried *C. militaris* are shown in Table 2. Increase in drying temperature leading to $L*$ decreased due to the biochemical reaction in food such as enzymatic reaction [10, 25]. Pretreated sample by boiled water blanching and steaming showed the brown color and darker than other samples due to the browning reaction by thermal stress. Boiled water blanching showed the lowest of $a*$ and $b*$ values due to affected heating and soaking in water simultaneous. This is because after applying heat pretreatment the naturally enzyme population was reduced, plant tissue was softer and more flexible. The appearance of *C. militaris* at different pretreatment is shown in Fig. 2.

These results correlated well with Hiranvarachat *et al.* [16]. They reported that blanching in acid had slightly darker red color comparing with fresh carrot and had no significantly difference with blanching in water due to the breaking down of pigments in samples. It was observed that the pretreated sample by desiccating combined with drying at 70 °C exhibited the lowest ΔE value, this might be due to some moisture content was evaporated during desiccating pretreatment would result in the shortest drying time, reduce the exposure to thermal stress.

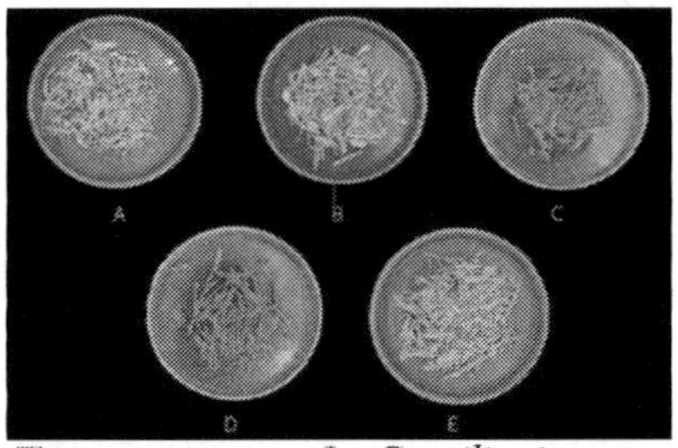

Fig. 2. The appearance of *C. militaris* at different pretreatment: (A) untreated sample, (B) soaking in acid, (C) boiled water blanching, (D) steaming and (E) desiccating.

Fig. 3 shows the micrograph of dried *C. militaris*. The micrograph could describe the shrinkage of *C. militaris* during drying. Pretreated samples with desiccating resulted the minimal of shrinkage. This might be the desiccating occur the evaporation of moisture in food matrix. Boiled water blanching samples showed the highest of shrinkage cause the detrimental of cell structure by thermal stress. Hiranvarachat et al. [16] reported that the pretreated carrot by water blanching showed the thinner of cell wall. This resulted could describe some solubilization and degradation of cell components of sample.

Table 2. The color parameters of dried *C. militaris* by different conditions.

Temperature (°C)	Pretreatment	Color			
		L^*	a^*	b^*	ΔE
50	Untreated	46.43 ± 0.06^{e}	22.43 ± 0.21^{d}	36.27 ± 0.04^{e}	34.95
	Soaking in acid	44.36 ± 0.24^{g}	24.35 ± 0.09^{b}	35.48 ± 0.09^{f}	56.10
	Boiling	33.49 ± 0.07^{m}	21.61 ± 0.02^{f}	19.90 ± 0.11^{k}	426.85
	Steaming	34.17 ± 0.05^{k}	22.08 ± 0.04^{e}	23.56 ± 0.04^{j}	338.32
	Desiccating	47.28 ± 0.02^{d}	22.54 ± 0.05^{d}	36.38 ± 0.03^{de}	29.93
60	Untreated	47.55 ± 0.23^{d}	24.36 ± 0.07^{b}	37.10 ± 0.07^{b}	26.81
	Soaking in acid	45.77 ± 0.04^{f}	24.33 ± 0.04^{b}	34.96 ± 0.07^{g}	50.08
	Boiling	33.88 ± 0.09^{l}	24.64 ± 0.07^{b}	26.66 ± 0.05^{i}	292.54
	Steaming	35.21 ± 0.02^{j}	24.61 ± 0.03^{b}	29.07 ± 0.10^{h}	234.09
	Desiccating	48.29 ± 0.04^{c}	24.40 ± 0.41^{b}	37.89 ± 0.03^{a}	19.92
70	Untreated	49.12 ± 0.03^{b}	25.02 ± 0.11^{a}	36.82 ± 0.02^{c}	24.26
	Soaking in acid	44.61 ± 0.05^{g}	24.33 ± 0.04^{b}	36.56 ± 0.05^{d}	47.27
	Boiling	35.55 ± 0.39^{i}	24.64 ± 0.07^{b}	26.66 ± 0.05^{i}	263.72
	Steaming	37.21 ± 0.02^{h}	24.61 ± 0.03^{b}	29.07 ± 0.10^{h}	202.51
	Desiccating	49.56 ± 0.07^{a}	23.96 ± 0.05^{c}	38.01 ± 0.27^{a}	14.60

Different letters (a-m) within the same column indicate a significant different ($P < 0.05$).

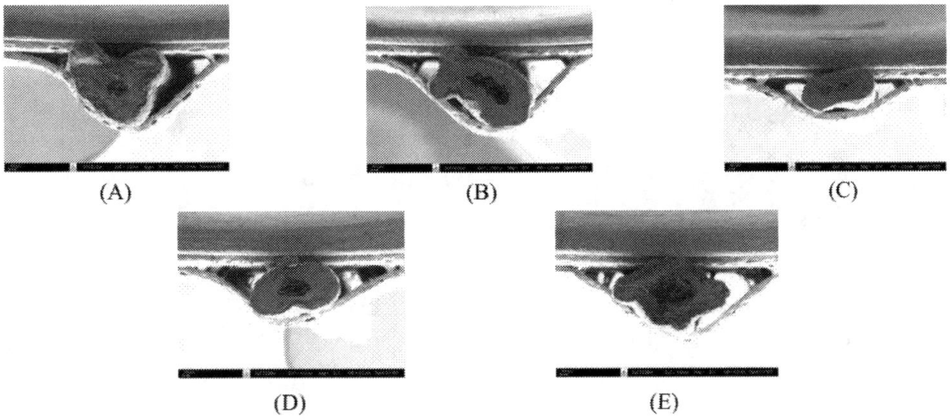

Fig. 3. The micrograph of dried *C. militaris* and different pretreatment: (A) untreated sample, (B) soaking in acid, (C) boiled water blanching, (D) steaming and (E) desiccating.

3.4 Total phenolic compounds

Fig. 4 shows the total phenolic content (TPC) of dried *C. militaris* by combined effect of pretreatments and drying conditions. The TPC in fresh *C. militaris* was 54.48 ± 2.25 mg GAE/g (db.). As drying temperature increase, the TPC increased. This might be because the reduction of drying time could preserve TPC in *C. militaris*. The reduction of TPC caused by oxidation reaction and changed in chemical structure of dried sample [26]. Therefore the prolong drying time would deteriorate the phenolic compound. This results consistent with Wu et al. [10]. They reported that the highest TPC in sample was found in the case of drying temperature at 70 °C when compared with drying temperature 40-60 °C. Soaking in acid sample showed the highest TPC. In contrast boiled water blanching sample showed the lowest amount of TPC.

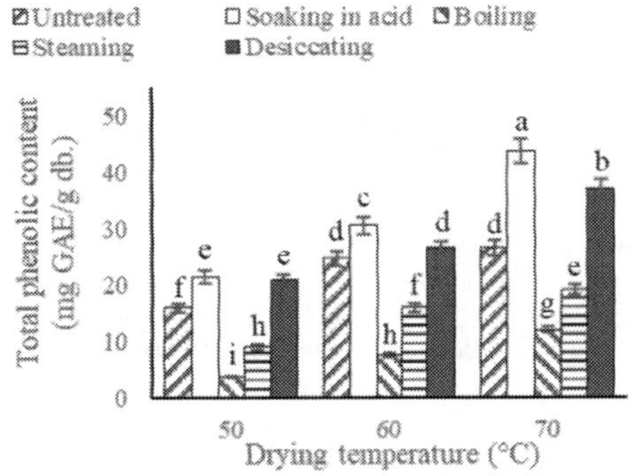

Fig. 4. The total phenolic content of dried *C. militaris* by different conditions.

https://doi.org/10.1051/e3sconf/202130202005

3.5 Antioxidant activities

The DPPH scavenging activity and superoxide anion radical scavenging activity of dried *C. militaris* are shown in Fig. 5A and 5B. Increasing drying temperature showed the higher DPPH scavenging activity. Boiled water blanching sample showed the lowest of antioxidant activity. This is because decomposition of bioactive compound by thermal stress. These results correlated well with the total phenolic content of dried *C. militaris*.

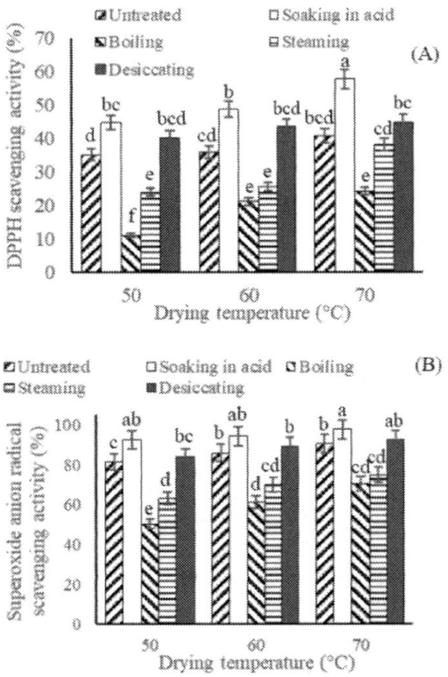

Fig. 5. The Scavenging activity of DPPH (A) and superoxide anion radicals (B) of dried *C. militaris* pretreated by different method.

In terms of superoxide anion radical scavenging activity of dried *C. militaris* under combination of different pretreatments and drying conditions showed similarity trend with DPPH scavenging activity and total phenolic compound. The vacuum drying temperature at 70 °C exhibited the superoxide anion radical scavenging activity higher than drying temperature at 50 and 60 °C. The long thermal exposure could degrade the compound that linked the antioxidant activity. Wu et al. [11] reported the superoxide anion radical scavenging activity of dried *C. militaris* by microwave assisted pulse fluidized bed freeze dryer and freeze dryer had no significant difference about 85.7-90.4% but in hot air drying showed the value very low. The antioxidant activity had correlation with polysaccharide, phenolics and flavonoids [25; 27]. The antioxidant mechanism of these compounds was donating a hydrogen atom or single electron to the hydroxyl radical [28]. The hydroxyl radical scavenging was described the inhibition of hydroxyl radical generating and removal hydroxyl radical. The highest superoxide anion radical scavenging activity was found in the case of pretreated by soaking in acid combined with vacuum drying at 70 °C.

4. Conclusion

Effects of pretreatment combined with vacuum drying at 50-70 °C of *C. militaris* were evaluated. Desiccating prior drying process had shortest of drying time and closet of color value with fresh sample. The boiled water blanching prior drying process exhibited the most difference of color, lowest of total phenolic retention and antioxidant activity due to high temperature during pretreatment affecting quality of *C. militaris*. The increasing of drying temperature exhibited the higher glass transition temperature, total phenolic compound and antioxidant activity than low temperature caused from shorter drying time. The recommended condition of dried *C. militaris* was pretreated by soaking in 0.5 % citric acid and then vacuum drying at 70 °C under pressure of 60 mbar. It showed the high levels of glass transition temperature, good appearance, the highest total phenolic compound and antioxidant activity.

Acknowledgement

We would like to thank you the Agricultural Research Development Agency (ARDA), Thailand, ACX Herb Co., Ltd., Thailand and King Mongkut's University of Technology North Bangkok for support my project.

References

1. S.S. Chiang, Z.C. Liang, Y.C. Wang, C. H. Liang, Effect of light-emitting diodes on the production of cordycepin, mannitol and adenosine in solid-state fermented rice by *Cordyceps militaris*, Journal of Food Composition and Analysis, **60** (2017): 51–56
2. C. Chen, X.P. Liu, W. Jiang, B. Zeng, W. Meng, L.P. Huang, Anti-effects of cordycepin to hypoxia-induced membrane depolarization on hippocampal CA1 pyramidal neuron, European, Journal of Pharmacological, **796** (2017): 1–6
3. Y. Xiao, G. Xing, X. Rui, W. Li, X. Chen, M. Jiang, M. Dong, Enhancement of the antioxidant capacity of chickpeas by solid state fermentation with *Cordyceps militaris* SN-18, Journal of Functional Foods, **10** (2014): 210–222
4. F.S.H. Hsiao, Y.H. Cheng, S.K. Wang, Y.H. Yu, *Cordyceps militaris* hot water extract inhibits lipopolysaccharide-induced inflammatory response in porcine alveolar macrophages by the regulation of mitogen-activated protein kinase signaling pathway, Canadian Journal of Animal Science, **98** (2017): 44–52
5. H. Flordeliza, A.N. Soriano, N.P. Dugos, R.V.C. Rubi, A comprehensive review on the drying kinetics of common tubers, Applied Science and Engineering Progress, **14**, 2 (2021): 146-155
6. W. Yan, M. Zhang, L. Huang, J. Tang, A.S. Mujumdar, J. Sun, Studies on different combined microwave drying of carrot pieces, International

Journal of Food Science and Technology, **45**, 10 (2010): 2141–2148
7. K. Thirugnanasambandham, V. Sivakumar, Enhancement of shelf life of *Coriandrum sativum leaves* using vacuum drying process : Modeling and optimization, Journal of the Saudi Society of Agricultural Sciences, **15** (2016): 195-201
8. T.K. Tran, T. Kirkman, M. Nguyen, Q.V. Vuong, Effects of drying on physical properties, phenolic compounds and antioxidant capacity of Robusta wet coffee pulp (Coffea canephora), Heliyon, **6** (2020): e04498
9. V.T. Nguyen, N.M.Q. Pham, Q.V. Vuong, M.C. Bowyer, I.A. van Altena, C.J. Scarlett, Phytochemical retention and antioxidant capacity of xao tam phan (*Paramignya trimera*) root as prepared by different drying methods, Drying Technology, **34** (2016): 324–334
10. X.F. Wu, M. Zhanga, Z. Li, Influence of infrared drying on the drying kinetics, bioactive compounds and flavor of *Cordyceps militaris*, LWT - Food Science and Technology, **111** (2019): 790–798
11. X.F. Wu, M. Zhanga, B. Bhandarid, Z. Lie, Effects of microwave assisted pulse fluidized bed freeze-drying (MPFFD) on quality attributes of *Cordyceps militaris*, Food Bioscience, **28** (2019): 7–14
12. C. Phungamngoen, N. Chiewchan, S. Devahastin, Effects of various pretreatments and drying methods on *Salmonella* resistance and physical properties of cabbage, Journal of Food Engineering, **115** (2013): 237–244
13. J.I. Maté, M. Zwietering, K. van't Riet, The effect of blanching on the mechanical and rehydration properties of dried potato slices, European Food Research and Technology **209** (1999): 343–347
14. M.S. McCann, J.J. Sheridan, D.A. McDowell, I.S. Blair, Effects of steam pasteurization on *Salmonella* Typhimurium DT 104 and *Escherichia coli* O157:H7 surface inoculated onto beef, pork and chicken, Journal of Food Engineering **76** (2006): 32–40
15. D.O. Ukuku, Effect of sanitizing treatments on removal of bacteria from cantaloupe surface, and re-contamination with *Salmonella*, Food Microbiology, **23** (2006): 289–293
16. B. Hiranvarachat, S. Devahastin, N. Chiewchan, Effects of acid pretreatments on some physicochemical properties of carrot undergoing hot air drying, Food and Bioproducts Processing **89** (2011): 116–127
17. AOAC. Official Method of Analysis 2008.06 (39.1.39) *Moisture and Fat in Meats Final Action*., (2013).
18. A. Voda, N. Homan, M. Witek, A. Duijster, G.V. Dalen, R.V.D. Sman, J. Nijsse, L.V. Vilet, H.V. As, J.V, Duynhoven, The impact of freeze-drying on microstructure and rehydration properties of carrot, Food Research International, **49** (2012): 687-693
19. X. Cao, M. Zhang, A.S. Mujumdar, Q. Zhong, Z. Wang, Effects of ultrasonic pretreatments on quality, energy consumption and sterilization of barley grass in freeze drying, Ultrasonics Sonochemistry, **40** (2018): 333–340
20. L. Chen, R. Wang, X. Wang, Y. Luo, J. Li, Phenolic contents, cellular antioxidant activity and antiproliferative capacity of different varieties of oats, Food Chemistry **239** (2018): 260–267
21. G. Mao, Y. Zou, W. Feng, W. Wang, T. Zhao, C. Ye, Extraction, preliminary characterization and antioxidant activity of se-enriched Maitake polysaccharide, Carbohydrate Polymers, **101** (2014): 213-219.
22. H.F. Gu, C.M. Li, Y.J. Xu, W.F. Hu, M.H. Chen, Q.H. Wan, Structural features and antioxidant activity of tannin from persimmon pulp, Food Research International, **41,** 2 (2008): 208–217
23. E.J. Bassey, J.H. Cheng, D.W. Sun, Novel nonthermal and thermal pretreatments for enhancing drying performance and improving quality of fruits and vegetables, Trends in Food Science & Technology, **112** (2021): 137-148.
24. S. Santhalakshmy, S.J.D. Bosco, S. Francis, M. Sabeena, Effect of inlet temperature on physicochemical properties of spray-dried jamun fruit juice powder, Powder Technology, **274** (2015): 37–43
25. C.O. Perera, Selected quality attributes of dried foods, Drying Technology, **23**, 4 (2005): 717–730
26. L. Valadez-Carmona, C.P. Plazola-Jacinto, M. Hernández-Ortega, M.D. Hernández-Navarro, F. Villarreal, H. Necoechea-Mondragón, G. Ceballos-Reyes, Effects of microwaves, hot air and freeze-drying on the phenolic compounds, antioxidant capacity, enzyme activity and microstructure of cacao pod husks (*Theobroma cacao* L.), Innovative Food Science & Emerging Technologies, **41** (2017): 378–386
27. S. Wu, F. Li, S. Jia, H. Ren, G. Gong, Y. Wang, Z. Lv, Drying effects on the antioxidant properties of polysaccharides obtained from Agaricus blazei Murrill, Carbohydrate Polymers, **103** (2014): 414–417
28. N. Russo, The molecular basis of working mechanism of natural polyphenolic antioxidant, Food Chemistry, **125** (2010): 288–306

Effect of protein and prebiotic on the survival of encapsulated probiotic during storage

Patchanok Apiwattanasiri [1], *Ratchanee* Charoen [2], *Kriangkrai* Phattayakorn [3], *Sakwiboon* Jantrasee [4] *and Wanticha* Savedboworn [1*]

[1] Department of Agro-Industry Technology and Management, Faculty of Agro-Industry, King Mongkut's University of Technology North Bangkok, Prachinburi, Thailand
[2] Department of Innovation and Product Development Technology, Faculty of Agro-Industry, King Mongkut's University of Technology North Bangkok, Prachinburi, Thailand
[3] Department of Food Technology and Nutrition, Faculty of Natural Resources and Agro-Industry, Kasetsart University, Chalermphrakiat Sakon Nakhon Province Campus, Sakon Nakhon, Thailand
[4] Department of Applied Physics, Faculty of Engineering, Rajamangala University of Technology Isan, Khon Kaen Campus, Khon Kaen, Thailand

Abstract. The aim of this research was to investigate the impact of encapsulation system on its protective capability during freeze drying of *Lactobacillus casei* TISTR 1463 and its storage at 4°C and 30°C for 90 and 60 days, respectively. Sericin extract is a kind of protein with gel-like property that could aid in binding and also contains important amino acids. The combinations of sericin and prebiotics, including isomalto-oligosaccharide and modified starch were evaluated during microencapsulation of probiotic *L. casei* TISTR 1463 in alginate beads. The highest encapsulation efficiency was 97.61% with the addition of sericin-isomalto-oligosaccharide. Encapsulated probiotic with sericin-modified starch had the lowest specific rate of degradation (k) of 3.74×10^{-2} day^{-1} and 1.09×10^{-1} day^{-1} at both storage temperatures of 4°C and 30°C, respectively. Probiotic property of cell surface hydrophobicity (CSH) was also carried out.

Keyword. Encapsulation; sericin; probiotic; prebiotic.

1 Introduction

Currently, the consumers are more interested in personal health. Therefore, consuming foods that help promote or prevent various diseases and dietary supplements are on the rise [1]. Probiotics are the term for life and are defined as "Living microorganisms that positively affect the health of the host by balancing the microbial communities in intestine" [2]. Probiotic is a group of microorganisms that are beneficial to the human body and animals that can be found in the intestine called Gastrointestinal (GI) Tract. The survival of probiotic strain is a great significance to ensure a positive result on human health. Several reports have been found indicating low survivability of probiotic strain in products [3,4]. Therefore, the use of techniques or methods to maintain probiotic survival is becoming interest. Microencapsulation is the technology that is effective in helping to preserve and protect bacterial cells and it has been continuously generate to use in food industry [5]. Microencapsulation has been performed as an efficient way to protect probiotic cells from detriment in improper conditions before reaching the colon [6,7,8]. Prebiotics are food components that are not digestible in the stomach and small intestine but stimulate the development of beneficial bacteria in the gut. Good prebiotics should be able to withstand unchanged digestion of stomach acid and go down into the large intestine [9]. Sericin is a water-soluble protein obtained as a by-product in the silk industry. Sericin is very useful as biomaterial in tissue engineering, drug delivery and cell delivery [10]. This study aimed to determine the efficiency of alginate, sericin protein and prebiotics on the survival of encapsulated *Lactobacillus casei* TISTR 1463 during storage. Moreover, the probiotic properties of cell surface hydrophobicity was also evaluated.

2 Material and methods destroyed

2.1 Bacterial culture

Probiotic *Lactobacillus casei* TISTR 1463 was purchased from TISTR Culture collection, Thailand. Probiotic microorganism was subcultured twice in MRS broth (Merck, Darmstadt, Germany) and incubated at 37 °C for 36 h. Probiotic culture was centrifuged at 4000 rpm for 10 min. Cell pellets were washed with 0.85% NaCl and prepared for encapsulation.

2.2 Freeze drying of encapsulated probiotic

Encapsulated materials including alginate (Chemipan, Chemipan Corporation Co., Ltd., Thailand), sericin and each prebiotic consisted of modified starch (phyboplus;

* Corresponding author: wanticha.s@agro.kmutnb.ac.th

 https://doi.org/10.1051/e3sconf/202130202006

SMS Corporation, Thailand) and isomalto-oligosaccharide (Isomalt; Brenntag Ingredients, Thailand) were combined with *L. casei* TISTR 1463 at a ratio of 0.5% : 0.5% : 15% (w/v) (sericin : prebiotic : probiotics) or 1%: 15% (w/v) (sericin : probiotics). Then, 60 mL of 2% (w / v) sodium alginate were mixed with the suspension and then added to 18 G syringe (Nipro Corporation, Belgium). The suspensions were dripped down into 0.2 M $CaCl_2$ and mixed for 30 min. Gel beads were washes with 0.85% (w / v) NaCl and viable cell number were enumerated by drop plate technique. Encapsulated probiotics were applied to a plate freezer at -44 °C for 24 hr. The freeze-drying process was carried out at 0.12 mbar, -50°C for 19 h with a freeze dryer (Alpha LSC1-4, Christ, Osterode amHarz, Germany). 0.5 g of freeze-dried encapsulated beads were packed into sealed polyethylene aluminum foil bags. The freeze-dried samples were kept at 4°C and 30°C for 90 days and 60 days, respectively. Each sample was analysed at the time interval to evaluate viable cell number.

2.3 Encapsulation efficiency

1.0 g of fresh encapsulated probiotics were added to 9.0 mL of 2% sodium citrate at pH 8.3 and vortex for 10 min. Ten-fold serial dilution with 0.85% NaCl was evaluated. Drop plate technique was performed to investigate the viable cell number of probiotic. Encapsulation efficiency was measured as this equation:

$$\text{Encapsulation efficiency (\%)} = \frac{X_t}{X_0} \times 100 \qquad (1)$$

where X_t is the number of viable probiotic released from the microparticles bacteria and X_0 is the initial amount of bacteria [11].

2.4 The stability of freeze-dried encapsulated probiotic

1.0 g of freeze-dried encapsulated probiotics were added to 9.0 mL of 2% sodium citrate at pH 8.3 and vortex for 10 min. Ten-fold serial dilution with 0.85% NaCl was evaluated. Drop plate technique was used to enumerate viable cell count on the MRS agar with the addition of 0.5% $CaCO_3$. Survival rate was calculated as the following equation:

$$\text{Survival rate (\%)} = \frac{N}{N_0} \times 100 \qquad (2)$$

where N is probiotic viability in term of (log CFU/g) at a particular time and N_0 defined as the initial viable cell in term of (log CFU/g).

The specific rate of degradation (k, day^{-1}) of dried encapsulated microparticles was investigated as a first order kinetic from the following equation:

$$\log\frac{N}{N_0} = -k.t \qquad (3)$$

where N is the cell number at a particular time (CFU/g), N_0 is the viable cell count at the initial time (CFU/g) and t is the storage time (days) [12].

2.5 Cell surface hydrophobicity

Cell surface hydrophobicity was evaluated as the methods of Savedboworn [13] with minor modifications. 1.0 g of freeze-dried encapsulated probiotics were released with 9 mL of 2% sodium citrate at pH 8.3 and vortex for 10 min. The solution was regulated to OD_{600} of 0.5. Then, cell suspension (3 mL) was transferred to test tube and xylene (1 mL) (Ajax finechem, Australia) was applied. The mixture was mixed for 90 sec and left to set for 15 min at room temperature. The bottom transparent of mixture was taken and analysed by a UV-Vis Spectrophotometer (Libra S22, Biochrom Ltd., UK) at 600 nm [12] and calculated as a percentage of cell surface hydrophobicity as this equation:

$$\text{Cell surface hydrophobicity} = \left[1 - \left(\frac{A_1}{A_0}\right)\right] \times 100 \qquad (4)$$

Where A_0 is the absorbance at 600 nm before and A_1 is the absorbance at 600 nm after mixing with xylene.

2.6 Size measurement

The 20 microcapsules of each treatment were measured using a vernier caliper to measure the diameter. The measuring unit in term of mm. was expressed.

2.7 Microcapsules morphology

The structure of freeze-dried encapsulated cells was studied by a Scanning electron microscope (SEM) (Auriga, Zeiss, Germany). Dried encapsuled probiotic beads with gold plating were mounted on a double carbon tape on a stud. The images were observed and photographed at 3000X magnification.

2.8 Statistical analysis

The mean value and standard variation of at least triplicate were calculated. Data was analysed by SPSS statistical program version 26. The analysis of variance (ANOVA) was performed at 95% confidence intervals. A significant difference among mean value was determined with DMRT.

3 Results and discussion

3.1 Encapsulation efficiency of probiotics with various wall materials

The number of probiotic encapsulated with alginate, sericin and different prebiotics including Phyboplus and Isomalt were determined. A significant difference ($p<0.05$) in encapsulation efficiency after encapsulation was obtained. The encapsulation of 2% Sodium Alginate

+ 0.5% Sericin + 0.5% Isomalt had the highest encapsulation efficiency of 97.60 % followed by 2% Sodium Alginate (control), 2% Sodium Alginate + 0.5% Sericin, 2% Sodium Alginate + 1% Sericin and 2% Sodium Alginate + 0.5% Sericin + 0.5% Phyboplus, respectively (Fig. 1).

Encapsulation of probiotic cells in the combination with prebiotics increased cell survival due to encapsulation techniques performed in mild conditions [14]. No significant difference ($p \geq 0.05$) in diameter of encapsulated microgel was achieved. The size of gel beads was in the range of 2.320-2.403 mm (Table 1). Chandramoulia et al. [15] suggested that probiotic survival increased with gel bead size. Moreover, Mortazavian, et al. [16] revealed a correlation between size of microcapsule and solution viscosity. A higher solution viscosity resulted in a larger size of microcapsule particles. Large microcapsule particles resulted in longer exposure to the unsuitable environment than the smaller particles obtained by the use of low concentrations of encapsulation agents.

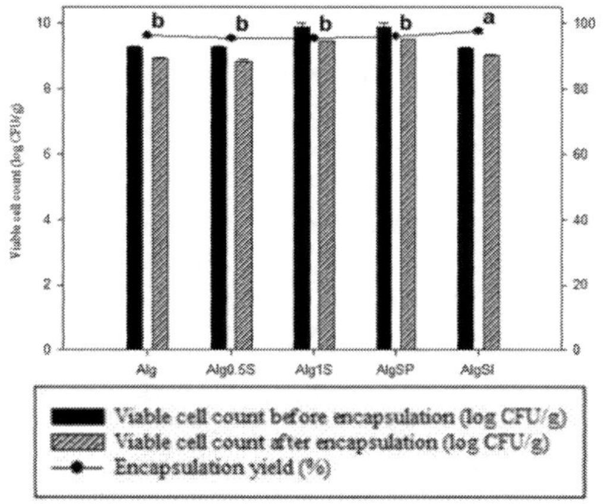

Fig. 1. Encapsulation efficiency of probiotic with various wall materials. The values of encapsulation efficiency expressed with different lowercase letters are significant differences ($p<0.05$). Alg = 2% Sodium Alginate, Alg0.5S= 2% Sodium Alginate + 0.5% Sericin, Alg1S = 2% Sodium Alginate + 1% Sericin, AlgSP = 2% Sodium Alginate + 0.5% Sericin + 0.5% Phyboplus, AlgSI = 2% Sodium Alginate + 0.5% Sericin + 0.5% Isomalt.

Table 1 The particle size of microcapsules with different wall materials. ns showed no significant difference ($p \geq 0.05$).

Wall materials	Size (mm) ^{ns}
2% Alginate	2.403±0.13
2% Alginate + 0.5% Sericin	2.320±0.14
2% Alginate + 1% Sericin	2.363±0.10
2% Alginate + 0.5% Sericin + 0.5% Phyboplus	2.357±0.13
2% Alginate + 0.5% Sericin +0.5% Isomalt	2.372±0.15

3.2 The survival rate of encapsulated cells with various prebiotics after freeze-drying

The survival rate of encapsulated probiotic strain was performed after freeze drying. The survival of encapsulated microparticles with alginate, sericin and different prebiotics and subjected to freeze-drying process were statistically differences ($p<0.05$). The survival rate of encapsulated probiotic with 2% Sodium alginate + 0.5% Sericin + 0.5% Isomalt had the highest survival rate of 99.78% (Fig. 2). Adding protective agent appears to enhance cell viability of probiotic. The ability of prebiotics to protect probiotic cells is related to its ability to replace water during dehydration [17].

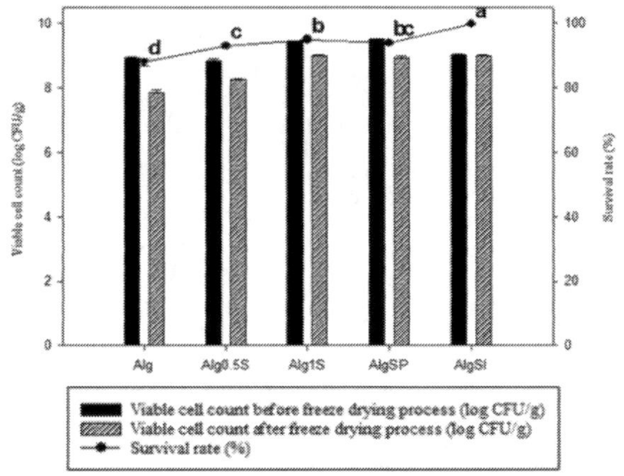

Fig. 2. The survival rates of encapsulated probiotic with different prebiotics after freeze drying process. The values of survival rate expressed with different lowercase letters are significant differences ($p<0.05$). Alg = 2% Sodium Alginate, Alg0.5S= 2% Sodium Alginate + 0.5% Sericin, Alg1S = 2% Sodium Alginate + 1% Sericin, AlgSP = 2% Sodium Alginate + 0.5% Sericin + 0.5% Phyboplus, AlgSI = 2% Sodium Alginate + 0.5% Sericin + 0.5% Isomalt.

3.3 The stability of freeze-dried encapsulated probiotic during storage

Specific rate of degradation (k) was investigated in the correlation of microorganism reduction rate over the storage time. Through the encapsulating cells with different wall materials, lower storage temperature at 4°C resulted in higher survival rate with lower k value and the co-encapsulation with prebiotics exhibited a great storage stability.

Encapsulated probiotic with 2% Sodium Alginate + 0.5% Sericin + 0.5% Phyboplus demonstrated the lowest k value of 3.74×10^{-2} day^{-1} and 1.09×10^{-1} day^{-1} at 4°C and 30°C, respectively (Table 2). The temperature of 4°C provided higher probiotic survival than at 30°C storage. This results suggested that co-encapsulation of alginate, sericin and prebiotic constitutes an alternative for maintaining probiotic viability during storage in refrigeration for up to 90 days.

 https://doi.org/10.1051/e3sconf/202130202006

Table 2 Viability of dried encapsulated probiotic kept at 4°C and 30°C

Protectants	Storage temperature 4°C	Storage temperature 30°C
	$k_{4°C}$ (day^{-1})	$k_{30°C}$ (day^{-1})
2% Alginate	5.27x10^{-2}	2.27x10^{-1}
2% Alginate + 0.5% sericin	4.10x10^{-2}	2.19x10^{-1}
2% Alginate + 1% sericin	4.91x10^{-2}	4.62x10^{-1}
2% Alginate + 0.5% sericin + phyboplus	3.74x10^{-2}	1.09x10^{-1}
2% Alginate + 0.5% sericin + isomalt	4.88x10^{-2}	1.61x10^{-1}

3.4 Microstructure of capsules

Characterization of encapsulated probiotic gel beads with alginate, sericin extract and different prebiotics were carried out by stereo microscope and Scanning Electron Microscope (SEM) were obtained as shown in (Figure 3). The morphology of encapsulated cells after freeze-drying by stereo microscope showed that the gel beads contracted and formed a porous appearance, resulting in non-spherical shape. The constricted and porous structure of encapsulated microbeads resulted from the water loss during dehydration [18]. The morphology of encapsulated cells after freeze-drying by SEM showed that the probiotic cells were homogeneous with the protective agents on the surface

3.5 Cell surface hydrophobicity of encapsulated probiotic

The hydrophobic properties of encapsulated probiotic after freeze drying were determined. Encapsulated probiotic with 2% sodium alginate had the highest cell surface hydrophobicity value of 71.65%, while 2% Sodium Alginate + 0.5% Sericin + 0.5% Phyboplus has the lowest adhesion value of 21.16% (Table 3). The hydrophobicity of probiotic cells is associated with their ability to bind in the gut, which is considered a property of probiotics. Hydrophobia related with the amount of surface protein in probiotic strain *Lactobacillus*. Different levels of hydrophobic cells resulted in different adhesion variations [19]. The encapsulated probiotic with only 2% sodium alginate demonstrated the highest cell surface hydrophobic which link to the adhesion to the epithelial wall which is considered a hallmark of being a probiotic.

4 Conclusion

In conclusion, the encapsulation of probiotic *L. casei* TISTR 1463 with 2% sodium alginate + 0.5% sericin +0.5% Isomalt exhibited the highest level of encapsulation efficiency and survival rate after freeze drying compared with other conditions. Encapsulated probiotic by sericin combined with different prebiotics promoted survival rate better than free cells. While, 2% sodium alginate + 0.5% sericin + 0.5% phyboplus showed the highest storage stability related to the lowest k values at 4°C and 30°C. This research could be used to develop new functions to increase the protection of probiotics to be more effective in the future.

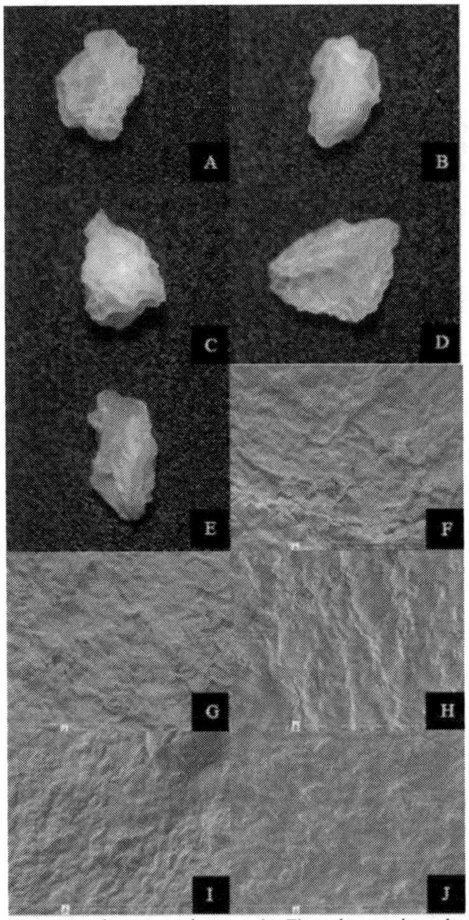

Fig.3 Stereo microscope images (A-E) and scanning electron micrographs images (F-J) of freeze-dried encapsulated probiotic with various wall materials. A,F = 2% Sodium Alginate, B,G = 2% Sodium Alginate + 0.5% Sericin, C,H = 2% Sodium Alginate + 1% Sericin, D,I = 2% Sodium Alginate + 0.5% Sericin + 0.5% Phyboplus, E,J = 2% Sodium Alginate + 0.5% Sericin + 0.5% Isomalt.

Table 3 Cell surface hydrophobicity probiotics with different prebiotics

Wall materials	Cell surface hydrophobicity (%) ±SD
2% Sodium Alginate	71.65±1.81[a]
2% Sodium Alginate + 0.5% Sericin	54.07±4.26[bc]
2% Sodium Alginate + 1% Sericin	44.41±4.69[c]
2% Sodium Alginate + 0.5% Sericin + 0.5% Phyboplus	21.16±2.89[d]
2% Sodium Alginate + 0.5% Sericin +0.5% Isomalt	47.71±4.09[bc]

The values of cell surface hydrophobicity expressed with different lowercase letters are significant differences by DMRT ($p<0.05$)

Acknowledgement

I would like to extend my thanks to my scholarship provider, the Agricultural Research Development Agency (ARDA), Thailand, who gave me an unforgettable opportunity to come to explore the scientific world at the King Mongkut's University of Technology North Bangkok to fulfil my master degree study.

References

1. K. Kailasapathy, Encapsulation technologies for functional foods and nutraceutical product development, CAB Reviews: Perspectives in Agriculture, (2009).
2. R. Fuller, Probiotics in man and animals, Journal of Applied Microbiology, **66** (1989): 365–378
3. P. De Vos, M.M. Faas, M. Spasojevic, J. Sikkema, Encapsulation for preservation of functionality and targeted delivery of bioactive food components, International Dairy Journal, **20** (2010): 292–302,
4. M.J. Chen, K.N. Chen, Applications of probiotic encapsulation in dairy products, In: Lakkis, M. Jamileh, Encapsulation and Controlled Release Technologies in Food Systems, Wiley-Blackwell USA, (2007): 83–107
5. M. Borgogna, B. Bellich, L. Zorzin, R. Lapasin, A. Cesàro, Food microencapsulation of bioactive compounds: rheological and thermal characterisation of non-conventional gelling system, Food Chemistry, **122** (2010): 416–423
6. M.T. Cook, G. Tzortzis, D. Charalampopulos, V.V. Khutoryanskiy, Microencapsulation of probiotics for gastrointestinal delivery, Journal of Controlled Release, **162** (2012): 56-67
7. M.D. Piano, S. Carmagnola, S. Andorno, M. Pagliarulo, R. Tari, L. Mogna, L. Capurso, Evaluation of the intestinal colonization by microencapsulated probiotic bacteria in comparison with the same uncoated strains, Journal of Clinical Gastroenterology, **44**, (2010): 42-46
8. A. Sipailiene, S. Petraityte, Encapsulation of probiotics: Proper selection of the probiotic strain and the influence of encapsulation technology and materials on the viability of encapsulated microorganisms, Probiotics and Antimicrobial Probiotic, **10** (2018): 1-10,
9. L.K. Sarao, M. Arora, Probiotics, prebiotics, and microencapsulation: A review. Critical Reviews in Food Science and Nutrition, **57** (2017): 344-371
10. L. Lallepak, G. Mario, Y. Guang, W. Qun, Silk sericin: A versatile material for tissue engineering and drug delivery, Biotechnology Advances, **33** (2015): 1855-1867
11. S.B. Doherty, V.L. Gee, R.P. Ross, C. Stanton, G.F. Fitzgerald, A. Brodkorb, Development and characterisation of whey protein micro-beads as potential matrices for probiotic protection, Food Hydrocolloids, **25** (2011): 1604–1617
12. W. Savedboworn, K. Teawsomboonkit, S. Surichay, W. Riansa-ngawong, S. Rittisak, R. Charoen, Impact of protectants on the storage stability of freeze-dried probiotic Lactobacillus plantarum, Food Science and Biotechnology, **28** (2019): 795–805
13. W. Savedboworn, W. Riansa-ngawong, W. Sinlapacharoen, S. Pajakang, B. Patcharajarukit, Assessment of Probiotic Properties in Lactic Acid Bacteria Isolated from Fermented Vegetables, Applied Science and Engineering Progress, **7** (2014) : 53-65
14. W. Krasaekoopt, B. Bhandari, H. Deeth, Evaluation of encapsulation techniques of probiotic for youhurt, International Dairy Journal, **13** (2003): 3-13
15. V. Chandramoulia, K. Kailasapathya, P. Peirisb, M. Jones, An improved method of microencapsulation and its evaluation to protect Lactobacillus spp. In simulated gastric conditions, The Journal of Microbiological Methods, **56** (2004): 27-35
16. A. Mortazavian, S.H. Razavi, M.R. Ehsani, S. Sohrabvandi, Principles and of microencapsulation of probiotic microorganisms, Iranian Journal of Biotechnology, **5** (2007): 1-18
17. J.H. Crowe, J.F. Carpenter, L.M. Crowe, The role of vitrification in anhy-drobiosis. Annual Review of Physiology, **60** (1998): 73–103
18. S. Ambros, A.H. Vollmer, N.N. Youssef, U. Kulozik, Structural basis of the impact of microwave drying on survival and shelf life of *Lactobacillus paracasei*, Lebensmittel-Wissenschaft & Technologie, **98** (2018): 291–298
19. T.P. Singh, R.K. Malik, G. Kaur, Cell surface proteins play an important role in probiotic activities of *Lactobacillus reuteri*. Nutrire, **41** (2016): 1–10

Optimization and Characterization Freeze Dried Fish Protein Hydrolysate Production

Sasithorn Kongruang[1,*], *Nicharee* Wisuthiphaet[1]

[1]Bioprocess Engineering and Biotechnology Center, Deperatment of Biotechnology, Faculty of Applied Science, King Mongkut's University of Technology North Bangkok, Bangkok, 10800, Thailand

Abstract. The production of fish protein hydrolysate from underutilized fish species is attracting the industrial interest for increasing the rich protein values. The central composite design was used to optimize the degree of enzymatic hydrolysis of freeze-dried fish protein hydrolysate (FPH) production by the application of commercial Alcalase on the recovery mixed small fish protein. The effects of time, temperature, pH and enzyme concentration on the degree of hydrolysis (DH) of five strains of fish as *Rastrelliger brachysoma* (short-bodied mackerel), *Rastrelliger kanagurta* (Indian mackerel), *Leiognathidae* (Ponyfish), *Amblygaster leiogaster* (Smooth belly sardinella) and *Selaroides leptolepis* (yellow-stripe scad) were experimented. Result showed that the FSH production was optimized at 2.85%v/w enzyme concentration at 61℃, pH 8.50 for 27 min with 89.42% DH. Mathematical model was proposed and validated under the optimum condition. The high proportion (46.43%) of smaller molecular weight <1 kDa was found in hydrolysate. Freeze-dried fish protein hydrolysate was produced and revealed that three predominant amino acids as glutamine, lysine and alanine. Based on amino acid compositions, the waste fish hydrolysate showed nutritional value and high potential for the applications of feed supplementation.

Keyword. Fish hydrolysate, Enzyme hydrolysis, Alcalase, Response surface methodology

1 Introduction

The fishery industries have been substantially increasing in catching because of the continuously increasing demand of the protein derived product to 463.8 million tons by 2050 [1]. To secure the sustainable economic impact and meet the competition of the world markets, there is a requirement to derive the fishery waste into valuable protein source of products through fish protein hydrolysate (FPH) production. As a great of the market size, the discard of rich dark muscle of small fishes has a potential to add up this demand. Hence, utilization by transformation of these rich proteins into protein hydrolysate produces highly valuable products. Several fish species mostly from North Atlantic and Pacific Ocean have studied for protein hydrolysates [2-4]; however, the related work in discard fish hydrolysate rarely reported in the fish species in tropical waters of Indo-West Pacific region. FPH plays an important role in the human and animal diet as a nutritional quality and functional food with highly digestible peptide and protein. This functional protein has used in many food applications such as emulsion properties, water holding capacity, fat absorption and foaming ability [2-8]. Moreover, bioactive properties of FSH also well known as antioxidant, antiproliferate activity and antihypertensive activity [9]. The choice of using these enzymes either endoproteinase, exoproteinase or the mixture of both is the crucial factor as the main activity with specificity toward the substrate resulting in the amino acid sequence of peptides or amino acid residues that lead to different bioactive, functional and nutritional properties in the hydrolysates [9-14]. Among these hydrolyzing enzymes, Alcalase (EC 3.4.21.62) is an approved food grade endopeptidase and a highly efficient bacterial protease from *Bacillus licheniformis*. This enzyme has developed for the hydrolysis of all kinds of proteins with the optimum temperature range between 55 and 70°C, depending on the specific substrate and pH values between 6.5 and 8.5 [15]. The process parameters for production FPH involve on several factors including enzyme to substrate ratio, pH, time and temperature, while the response variables were degree of hydrolysis and protein yield [16-22]. Since production optimization using a one factor at a time is not effective and not include the interactions among effecting parameters, experimental design as statistical completely central composite by response surface methodology has to apply to the degree of hydrolysis in order to find the appropriate enzyme concentration, hydrolysis time, hydrolysis temperature and pH. The final aim was to provide the alternative processes to reduce the production cost depend upon the availability of operation. Comparison of amino acid content from amino acid profile of FSH was also determined.

2 Materials and methods

2.1 Enzyme

The enzymatic hydrolysis from bacterial protease from *Bacillus licheniformis* was carried out using a commercial protease, Alcalase®2.4 L (Activity =2.59

* Corresponding author: wanticha.s@agro.kmutnb.ac.th

U.mL^{-1}; EC 3.4.21.14) was obtained from EMD Millipore, USA and stored at 4^{0}C until used. HPLC grade acetonitrile, chloroform and acetone were supplied by Fisher Scientific. All other chemicals used analytical grade.

2.2 Substrate

Five fish strains as *Rastrelliger brachysoma* (short-bodied mackerel), *Rastrelliger kanagurta* (Indian mackerel), *Leiognathidae* (Ponyfish), *Amblygaster leiogaster* (Smooth belly sardinella) and *Selaroides leptolepis* (yellow-stripe scad) were provided by animal feed company, INTEQC Group., Co Ltd, Thailand. Equal proportion of fish was washed, minced to homogenous paste (< 3 mm in size) and then placed in plastic bags immediately after process. These 100 g of samples was stored in -20 ^{0}C individually until further use.

2.3 Proximate chemical analysis

The chemical compositions of mixed fish samples and FPH were determined according to Association of Official Analytical Chemists (AOAC, 1990). All experiments performed in triplicate.

2.4 Enzymatic hydrolysis

Each experiment run was carried out 10 g of samples were thawed at room temperature and adjusted pH 4-12 according the experiment plan in 250 mL Erlenmeyer flask. Samples were then adjusted the final volume to 30 mL and incubated in shaking incubator at desired temperature from 40-80 ^{0}C at 200 rpm for the reaction times from 0-40 min. Different concentrations of Alcalase were added at 1-5 %v/w during the control hydrolysis reaction temperature. To stop the reaction, the mixtures were stopped at 85 ^{0}C for 5 minutes. The samples were then transferred to microcentrifuge tube after centrifuged at 12,000 rpm for 15 min for analytical analysis.

2.5 Central composite design

The chosen parameters on FPH were evaluated using response surface methodology (RSM). The Central Composite Statistical Design (CCD) was carried out the statistical model for the individual and interactive the effects of time (0-40 min), temperature (40-80°C), pH (4-12) and enzyme concentration (1-5% v/w) on the degree of hydrolysis. Levels of these factors were optimized for maximum degree of hydrolysis (the response) according to Table 1. All the 30-trial experiments were investigated in triplicate. Degree of hydrolysis was determined as the response for the combination of the reaction effects given in Table 2. The regression analysis of data obtained on degree of hydrolysis was chosen as quadratic model by equations of the following form:

$$y_1 = \beta_0 + \sum_{i=1}^{4}\beta_i X_i + \sum_{i=1}^{4}\sum_{j=1}^{4}\beta_{ij}X_iX_j + \sum_{i=1}^{4}\beta_{ii}X_i^2 \quad (1)$$

Here, X_1, X_2, X_3, X_4, are the parameter values for the independent variables (time, temperature, pH, enzyme concentration) as described in Table 2. The constants β_0, β_i, and β_{ij} (i, j = 1, 2, 3, 4) are coefficient estimates for degree of hydrolysis (y_1), where β_0 is an intercept term, β_i are linear terms, β_{ii} are quadratic terms, and β_{ij} are interaction terms. The accuracy and general suitability of the above polynomial model was evaluated by the coefficient of determination (R^2). The experimental data was analyzed using the statistical software, Design-Expert software version 8.0.6 (STAT-EASE Inc., Minneapolis, MN, USA)

Table1. Levels of factor for fish protein hydrolysis production.

Factor	Level				
	1	2	3	4	5
A: Time (min)	0	10	20	30	40
B: Temp (^{0}C)	40	50	60	70	80
C: pH	4	6	8	10	12
D: Concentration (%v/w)	1	2	3	4	5

2.6 Degree of hydrolysis

DH of FPH was analyzed for the content of α-amino acid from the supernatant of each hydrolysis run according to Adler-Nissan, 1979 [23]. The degree of hydrolysis calculation reported according to equation 2.

$$DH= [(L_t - L_0) / (L_{max} - L_0)] \times 100 \quad (2)$$

2.7 Molecular distribution of FPH

The supernatant of FPH was subjected to sequentially ultramembrane filtration by using a series of membranes (Centrifugal Devices, PALL Corporation, USA) from 100, 30, 10, 5, 3 and 1 kDa, respectively. Fractions of hydrolysate separation were centrifuged at 2,000 g-force for 5 min. to obtain both retentate and permeate. Both fractions were collected and measured its protein concentration by Nanodrop (Optizen NANOQ, Mecasys Co., Ltd, Korea).

2.8 Amino acid composition analysis

Chromatographic conditions of the HPLC system consisted of 626 pump, a 727 autosampler, 600S controller (Waters, MA), and a scanning fluorescence detector (Hewlett Packard), Millinnium™ 2010 were used to determine the amino acid profile. The amino acid composition and chemical score of amino acid composition were previously calculated [24].

2.9 Statistical analysis

All data were expressed as mean ± SD for all experiment. ANOVA was used to determine the significance of each term in the fitted equation. A *p*-value less than 0.05 was considered statistically significant.

3 Results and discussion

3.1 Degree of hydrolysis

The influential parameters including time, temperature, pH and enzyme concentration on the Alcalase hydrolysis to DH were determined using CCD as described on the previous section. The design matrix of all the independent variables and the result of 30 runs in triplicate on DH are exhibited in Table 2. The observed results found DH in the range of 0% to 95.38%. The highest observed condition for the DH of 95.38% was found at 60 ^{0}C, pH 8.0 with 3% v/w enzyme concentration over 40 min. The quadratic term for hydrolysis temperature was found highly significant affecting DH ($p < 0.0001$) followed by pH with p-value of 0.0019 as shown in the ANOVA (Table 3).

Table 2. Actual levels of Alcalase hydrolysis conditions along with the observed values for the response variable, degree of hydrolysis.

Run	Time (min)	Temp. (^{0}C)	pH	Conc. (% v/w)	DH
1	30	70	6	2	64.04 ± 4.62
2	30	70	10	2	69.62 ± 6.40
3	20	60	4	3	32.5 ± 1.90
4	30	50	6	2	35.96 ± 1.28
5	20	60	8	3	82.88 ± 1.17
6	20	80	8	3	8.85 ± 4.89
7	0	60	8	3	0
8	20	60	8	1	86.92 ± 0.74
9	20	60	8	5	84.81 ± 0.09
10	40	60	8	3	95.38 ± 1.68
11	20	60	8	3	82.12 ± 5.40
12	20	60	8	3	83.46 ± 2.14
13	10	70	6	2	46.15 ± 6.04
14	30	70	10	4	66.54 ± 2.06
15	20	60	8	3	82.69 ± 2.03
16	30	50	6	4	49.23 ± 2.69
17	10	50	10	4	53.46 ± 1.54
18	10	50	6	2	33.46 ± 0.62
19	10	70	6	4	56.15 ± 3.08
20	30	70	6	4	65.38 ± 3.81
21	20	60	8	3	81.92 ± 0.78
22	10	70	10	4	51.54 ± 0.99
23	30	50	10	2	56.15 ± 0.38
24	20	60	8	3	82.31 ± 0.54
25	10	50	6	4	25.00 ± 0.59
26	20	40	8	3	25.00 ± 4.35
27	30	50	10	4	62.69 ± 0.40
28	10	50	10	2	59.62 ± 0.44
29	10	70	10	2	51.54 ± 2.31
30	20	60	12	3	52.69 ± 2.17

However, p-values of hydrolysis time and enzyme concentration were insignificant (p=0.2569 and 0.7740). Results from the interaction terms between hydrolysis time and temperature, time and pH, time and concentration were not significant with the p-values, 0.6065, 0.7796 and 0.6728, respectively as well as other two interactions with temperature (Table 3). For the DH, the high coefficient of determination value (R^2= 0.8465) indicated that 84.65% of the variability in the response could be explained by the model. The independent and dependent variables were re-analyzed to obtain a real effect parameter model from regression equation that could predict the response within the given range. The regression equation for DH was reevaluated by the regression coefficient of linear, quadratic and interaction terms to fit a response surface model.

Table 3. ANOVA of enzymatic hydrolysis of fish protein affected by condition parameters for full quadratic equation.

Source	Sum of Squares	df	Mean Square	F Value	p-value Prob >F
Model	14344.12	14	1024.58	5.91	0.0008
A-Time	240.93	1	240.93	1.39	0.2569
B-Temp	5997.38	1	5997.38	34.58	< 0.0001
C-pH	2437.11	1	2437.11	14.05	0.0019
D-concentration	14.82	1	14.82	0.085	0.7740
AB	47.99	1	47.99	0.28	0.6065
AC	14.08	1	14.08	0.081	0.7796
AD	32.18	1	32.18	0.19	0.6728
BC	407.54	1	407.54	2.35	0.1461
BD	0.59	1	0.59	3.397E-003	0.9543
CD	22.21	1	22.21	0.13	0.7254
A^2	1731.42	1	1731.42	9.98	0.0065
B^2	6706.16	1	6706.16	38.67	< 0.0001
C^2	2331.08	1	2331.08	13.44	0.0023
D^2	70.10	1	70.10	0.40	0.5345
Residual	2601.20	15	173.41		
Cor Total	16945.31	29			

The resulting ANOVA for the reduced quadratic model (Table 4) summarizes the analysis of variance of each response and shows the significant model terms. The F-value of 11.57 implies that the model is significant (at $p < 0.0001$). Values of "Prob > F" (p values) less than 0.05 indicated that the model terms were significant. There is only a 0.01% chance that a "Model F Value" this large could occur due to noise. In this case hydrolysis temperature and pH were highly significant model terms with p-values of < 0.0001 and 0.0003. For the reduced model, we found R^2=0.8726 showing that the model gives a satisfactory fit to the experimental data. The following reduced quadratic model in terms of tested fermentation parameters was achieved as shown in equation (5).

$$DH = -840.53 + (3.79*Time) + (20.97*Temp) + (56.70*pH) + (0.02*Time*Temp) - (0.05*Time*pH) - (0.25*Temp*pH) - (0.08*Time^2) - (0.16*Temp^2) - (2.36*pH^2) \quad (5)$$

Table 4. ANOVA of enzymatic hydrolysis of fish protein affected by condition parameters for reduce quadratic equation.

Source	Sum of Squares	df	Mean Square	F Value	*p*-value Prob > F
Model	1425.49	9	1579.5	11.57	< 0.0001
A-Time	371.15	1	371.15	2.72	0.1148
B-Temp	6836.00	1	6836.00	50.08	< 0.0001
C-pH	2664.31	1	2664.31	19.52	0.0003
AB	47.99	1	47.99	0.35	0.5599
AC	14.08	1	14.08	0.10	0.7514
BC	407.54	1	407.54	2.99	0.0994
A^2	1870.56	1	1870.56	13.70	0.0014
B^2	7047.31	1	7047.31	51.63	< 0.0001
C^2	2499.01	1	2499.01	18.31	0.0004
Residual	2729.82	20	136.49		
Cor Total	16945.31	29			

Results found that DH of FSH production was significantly influenced by temperature and pH. Our relationship of fish waste hydrolysis is well in agreement with the previous studies reported a quadratic model for production of protein hydrolysate [18, 25].

3.2 Optimization of FSH condition

From the optimization analysis of Alcalase hydrolysis to produce FPH, the suggested optimum levels of all the variables determined by the quadratic model of CCD in this study to obtained 89.42% DH were 2.85%v/w Alcalase concentration at pH 8.48 for 27.18 minutes of hydrolysis time as 61°C. In terms of pH and temperature, the results in this present study are in accordance with the results of study of Yuan et al., 2008 [18] showing Alcalase 2.4 L exhibited the highest values in terms of DH when optimum temperature at 60 °C, pH 8.0 with high protein yield up to 80.33%. Figure 1 shows the respective response surface plots (3-D). These plots illustrate the significant interaction of the two independent variables to determine the optimum conditions for enzymatic hydrolysis in this study. Figure1A shows the effect of temperature and reaction time on DH. Results showed DH increased as time and temperature increased. When the reaction temperature increased from 50°C to 60 °C, the DH significantly increased from 20% to 89%.

The reaction temperature has a significantly impact on the DH of this mixed fish species. However, by further increasing the reaction temperature to 70°C, a decline in the DH observed. The reduction in the DH possibly resulted from the thermal denaturation of enzyme increased since the reaction shifted out from the optimum temperature, thus leading to a decrease in the DH. In this production, 61°C was chosen as the optimum reaction temperature once all other influential parameters were kept in the ranges. Our results agree with Vannabun et al., 2014 [26] who characterized alkaline protease with optimal temperature at 60°C for hydrolysis the viscera of farmed giant catfish. From the result presented in Figure1B, the DH gradually increased with the increasing pH from 6 to 8.5. The obtained results agreed with the nature of enzyme activity of commercial Alcalase® which normally active between pH 6.5 and 8.5. As a further increase in pH up to 10, the DH gradually decreased within the time range of 10-30 min. However, the DH gradually increased to a plateau between 25 to 30 min reaction time resulting in the DH raised from 60% to 85% as in Fig.2A. The longer reaction time would allow Alcalase to act more extensively on the fish muscle, thus resulting in an increment in the DH as previously reported by Haslaniza et al., 2010 [27].

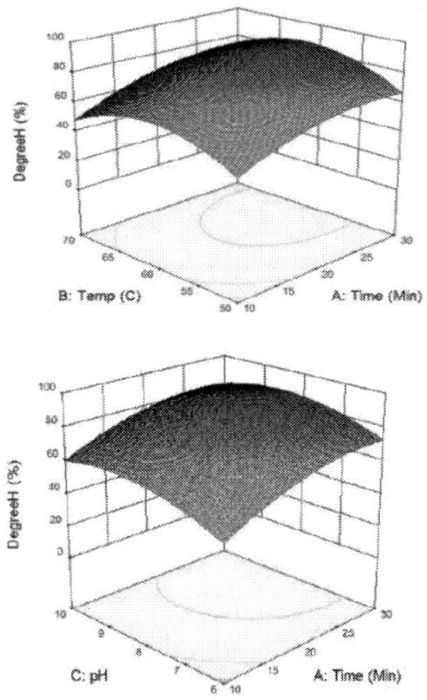

(A)

(B)

Fig.1. The 3-D surface plot of interaction between the parameters on degree of hydrolysis, (A) temperature and time, (B) pH and time.

However, the prolonging the reaction time beyond 25 min, no significant increase in DH occurred. Based on the obtained results, 27 min was selected as the suitable reaction time. The effect of pH and hydrolysis temperature on DH was shown in Fig. 2B, the DH was increased 25% to 89% as the pH increased from 6 to 8.5. However, by further increasing pH to 10, an obvious decrease in the DH was detected once temperature exceed to 70 °C. The effect of enzyme concentration on the DH displayed in Fig.2C. The DH of hydrolysates was insignificantly (p=0.7740) increased when enzyme concentration was increased from 2.0% to 4.0% at a fixed temperature. As temperature gradually increased from 50 to 60 °C at 2% enzyme concentration, pH 6.0 resulting in the DH raised from 48% to 73 %. However, by extending the reaction temperature to 70 °C, a decline in the DH was observed. The influence of enzyme concentration and pH change on the DH accessed within the same range as previously discussed and the results displayed in Fig 2D.Results showed that the DH was slightly increased from 70 to 83% when pH gradually increased from 6 to 8.5 and slightly decreased to 79% when the pH reach to 10.

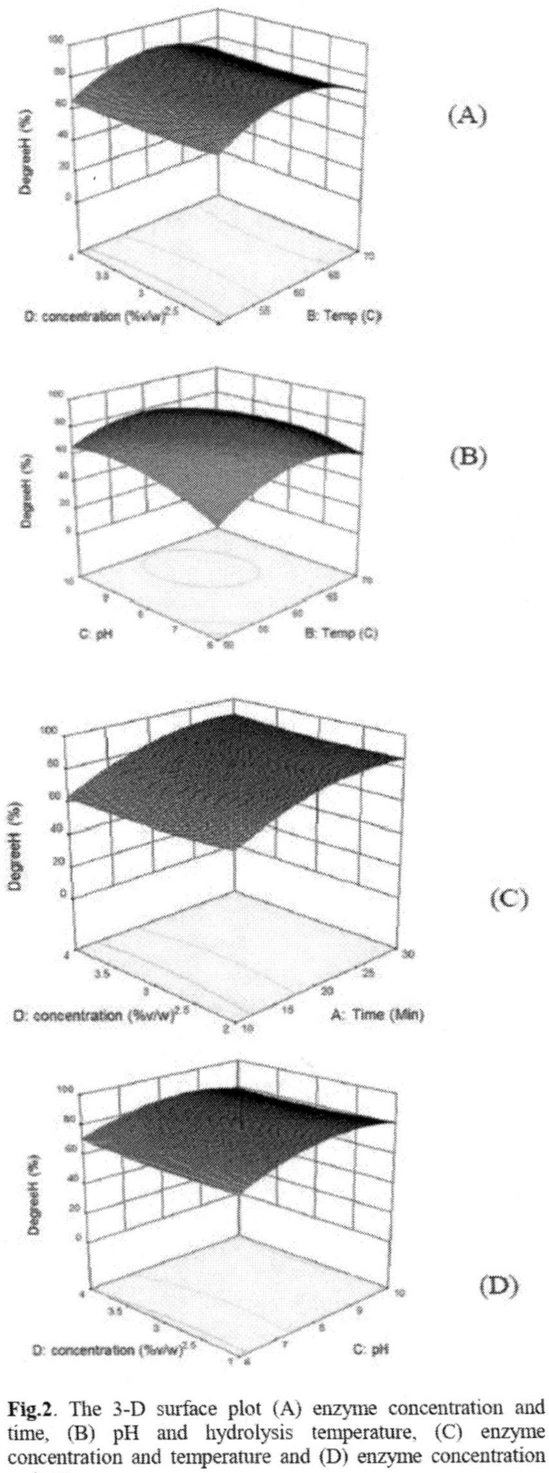

Fig.2. The 3-D surface plot (A) enzyme concentration and time, (B) pH and hydrolysis temperature, (C) enzyme concentration and temperature and (D) enzyme concentration and pH.

3.3 Validation the CCD model of FPH

In order to validate the statistical optimization of hydrolysis conditions of Alcalase on mixed fish towards degree of hydrolysis using response surface design. The validation experiments for confirming the adequacy of model was attempted for the accuracy of the model. The validation of the model performed under the optimum conditions in triplicate and found good agreement of DH as 78.27% obtained from the model prediction value of 89.42%. The difference between the predicted and experimental value was 11.16%. The difference of DH was a little higher than 10% enough to justify the validity of the response model with the high degree of accuracy leading to the hydrolysis yield.

3.4 Criteria of bioprocess operation

FSH production in this study can be manipulated the condition according to the criteria of process in terms of reaction speed, operation temperature, pH and enzyme concentration in order to obtain the desired product under the constrain condition or lowest operation cost [28]. We would suggest the following conditions to produce the FPH to make the hydrolysis process more controllable as the following four conditions. Firstly, in order to minimize the hydrolysis time and maximize degree of hydrolysis while temperature, pH and enzyme concentration kept in range of this experiment, the FPH production could be carried out at 65.58% DH using 3.74% v/w of enzyme concentration at 59.72^{0}C, pH 8.72 for 10 min. Secondly, an alternative process can be done if the consideration focus on minimizing the hydrolysis temperature and maximization of degree of hydrolysis while pH, enzyme concentration and time were kept in range of this study. FSH can be produced with 71.73% DH by using 2.38% v/w of enzyme concentration at 50 ^{0}C, pH 9.08 for 25.86 min. Thirdly, to minimize the pH and maximize DH with maintaining time, temperature and enzyme concentration in range of this study, the FPH production could run to obtain 75.53% DH using 2.90% v/w of enzyme at 62.88 ^{0}C, pH 6.00 for 28.09 min. For final option of operation process, it required 3.15% v/w of enzyme loading at 60.83 ^{0}C, pH 8.49 for 27.18 min of operation time. This option would resulting in the DH of 89.42 %.

3.5 Proximate composition

Result of proximate analysis revealed that protein content was relatively the same between FSH (87.11%) and dry weight powder of mixed fish raw material (87.11%) as depicted in Table 5. The freezed-dried FSH produced from this experiment was a brownish powder with fishy smell. The hydrolytic process of minced fish raw material by enzyme could generate molecules ranging from smaller proteins, peptides and individual amino acid. This FSH was then characterized molecular distribution through the sequential membrane filtration. Result showed the smaller size proteins in range of 30–100, 10–30, 5–10, 3–5, 1–3 and <1 kDa of 7.35, 10.36, 10.24, 25.62 and 46.43%, respectively. Fractionation of

FSH especially <1 kDa has a great potential to explore for bioactive peptides such as antioxidant activity, Angiotensin-Converting Enzyme inhibitor activity and hypocholesterolemic activity. However, the higher sizes of FSH are also attract the interest to find the capacity of food and feed application in terms of emulsifying properties, water-holding capacity and foaming capacity.

Table 5. Proximate chemical composition of mixed fish raw material and freezed-dried fish protein hydrolysate.

Components (%)	Wet weight	Dry weight	FPH
Moisture	77.92 ± 0.95	-	5.03+0.78
Protein	15.70 ± 0.47	87.11±2.28	88.61+2.81
Lipid	2.56 ± 0.07	8.83±0.65	1.30+0.05
Ash	2.74 ± 0.11	3.06±0.15	5.06+0.15

3.5 Amino acid composition

The composition of amino acid in freezed dried FPH is presented in Table 6. The major amino acids in FPH were glutamine, asparagine and lysine, respectively. To evaluate the nutritional properties of FPH, the chemical score was calculated in comparision with FAO/WHO and NRC standard protein [29,30]. The results indicated that the nutritional value of FPH are less than both standard almost all amino acid except histidine and lysine in reference protein 1. However, a large proportion of total amino acid normally occurs in short chain peptides proven to benefit for FPH production in aqafeed. The essential amino acid makes up 30.84 % of all amino acids. The hydrolysates contain enough content of the flavor enhancers, glutamic acid, asparagine, alanine by 24.56% of the total amino acids, which account for the good taste.

For nutritional applications, it is important to gain a high DH in protein hydrolysates to boost the absorption in digestive tract of animal digestion. The nutritional quality in terms of amino acid composition comparison between the experiments and the report values of fish meal and soy bean meal as previously investigation by Swanepoel and Goosen [31] shown in Table 7. Results showed the amount of EAA in the FSH was 292.6 g/kg total amino acid which was lower than those of fish and soybean meal by 1.62 and 1.48 times. All EAA were present in FPH with the leading amount amino acids as lysine, leucine and threonine except for tryptophan. Tryptophan is an acid labile amino acid destroyed during acid hydrolysis for sample preparation [31]. For non-essential amino acid (NEAA) results was found that glutamic acid was the most abundant amino acid present in freeze died FSH followed by aspartic acid as shown in Table 7.

Results showed the amount of EAA in the FSH was 292.6 g/kg total amino acid which was lower than those of fish and soybean meal by 1.62 and 1.48 times. All EAA were present in FPH with the leading amount amino acids as lysine, leucine and threonine except for tryptophan. Tryptophan is an acid labile amino acid destroyed during acid hydrolysis for sample preparation [32].

Table 6. Composition of amino acids and chemical score.

Amino acids	Amount (g/100g proteins)	Ref. Protein 1[a]	Ref. Protein 2[b]	Chemical score	
				RP-1	RP-2
Essential amino acids					
Histidine	2.25	2.00	2.10	1.13	1.07
Threonine	3.15	4.00	3.90	0.79	0.81
Arginine	4.27	5.00	1.31	0.85	3.26
Tyrosine	1.58	-	-	-	-
Valine	2.66	5.42	3.60	0.49	0.74
Methionine	1.95	3.50	3.10	0.56	0.63
Phenylalanine	2.47	4.29	6.50	0.58	0.38
Isoleucine	2.24	4.00	2.50	0.56	0.90
Leucine	4.72	7.00	3.30	0.67	1.43
Lysine	5.55	5.50	5.70	1.01	0.97
Non-Essential amino acids					
Asparagine	6.21				
Glutamine	10.26				
Serine	2.71				
Glycine	3.74				
Alanine	4.35				
Cysteine	0.41				
Proline	2.56				
ΣAAs	61.08				
ΣEAAs	30.84				
ΣN-ΣAAs	30.24				
ΣEAAs/ΣAAs (%)	50.49				
ΣEAAs/ΣN-ΣAAs	1.02				

[a]Reference protein 1: Essential amino acid reference protein according to FAO/WHO (1985) [28]
[b]Reference protein 2: Essential amino acid requirement of common carp according to NRC (1993) [29]

For non-essential amino acid (NEAA) results was found that glutamic acid was the most abundant amino acid present in freeze died FSH followed by aspartic acid as shown in Table 7. This results of high levels of nonessential amino acid as glutamic and aspartic acid was similar to several reports from previous study [4, 33-37]. Chemical composition of FPH is important in both nutrition perspective of human health. Our amino acid profile results are in the same range of several reports on the chemical composition of FPH that are prepared from various fish protein sources [33-37]. Chemical composition of FPH is important in both nutrition perspective of human health. Our amino acid profile results are in the same range of several research documents have been reported the chemical composition of FPH that are prepared from various fish protein sources. These reports included catfish (*Clarias gariepinus*) [31], bluewing searobin (*Prionotus punctatus*) [36] and *Misgurnus anguilliacaudatus* [37]. According to Watford report [38], some non-essential amino acids (like glutamic acid), presented an important role in the maintaining optimal health and growth levels.

4 Conclusion

Alcalase concentration at 2.85% (v/w), pH 8.45 for 27 min. incubation time as 61°C were the potential optimum conditions that gave the highest DH of 78.27% to produce FSH. The experimental data obtained based on the optimized conditions was in close agreement with the RSM model prediction. Four alternative conditions through RSM approach of optimization have a promising for FPH production in the future. In addition, the potential of this improved approach in the present study is affording a method for the efficiency of production process, which may be beneficial for reducing enzymatic loading and hydrolysis temperature leading to reduction of raw material for production and energy costs. The implementation of optimization report is appropriate for FPH in pilot scale and further study of bioactive compounds in feed and nutraceutical application.

Table 7. Amino acid composition comparison

Amino acid	Feed ingredients (g/kg total amino acids)		
	Fish protein hydrolysate	Fish meal[a]	Soybean meal[b]
Arginine	42.7	65.4	81.3
Histidine	22.5	19.5	33.7
Isoleucine	22.4	34.6	39.7
Leucine	47.2	77.7	75.9
Lysine	55.5	101.7	30.5
Methionine	19.5	44.6	8.6
Phenylalanine	24.7	39.8	79.2
Threonine	31.5	44.4	40.4
Tryptophan[1]	ND	ND	ND
Valine	26.6	46.6	43.6
∑EAA	292.6	474.3	432.9
Alanine	43.5	65.2	43.6
Aspartic acid	62.1	103.9	109.7
Glutamic acid	102.6	154.2	182.5
Glycine	37.4	66	57.9
Proline	0	50.4	64
Serine	27.1	48.5	52.4
Tyrosine	0	37.6	57
∑NEAA	15.8	525.8	567.1

[a,b] Amino acid composition of principle protein ingredients in the diets from Swanepoel and Goosen, 2011[30]

Acknowledgement

Author are indebted to a financial grant from KMUTNB-GOV-59-35.The author also would like to thank INTEQC Group., Co Ltd, Thailand for raw material and initiation to implement the fish protein hydrolysate production.

References

1. I. Richter, J. Thøgersen, C.A. Klöckner, Sustainable seafood consumption in action : relevant behaviors and their predictors, Sustainability, **9** (2017): 1-19
2. S. I. Aspmo, S. J. Horn, V. G.H. Eijsink, Hydrolysates from Atlantic cod (Gadus morhua L.) viscera as components of microbial growth media, Process Biochemistry, **40** (2005): 3714-3722
3. O. Martínez-Alvarez, S. Chamorro, A. Brenes, Protein hydrolysates from animal processing by-products as a source of bioactive molecules with interest in animal feeding: A review, Food Research International, **73** (2015): 204-212
4. M. Chalamaiah, B. Dinesh Kumar, R. Hemalatha, T. Jyothirmayi, Fish protein hydrolysates: proximate composition, amino acid composition, antioxidant activities and applications: a review, Food Chemistry, **135** (2012): 3020–3038
5. M. Chalamaiah, W. Yu, J. Wu, Immunomodulatory and anticancer protein hydrolysates (peptides) from food proteins: A review, Food Chemistry, **245** (2017): 205–222
6. P.J. García-Moreno, I. Batista, C. Pires, N. M. Bandarra, F. J. Espejo-Carpio, A. Guadix, and E.M. Guadix, Antioxidant activity of protein hydrolysates obtained from discarded Mediterranean fish species, Food Research International, **65** (2014): 469-476
7. N.A. Zakaria, N.M. Sarbon, Physicochemical properties and oxidative stability of fish emulsion sausage as influenced by snakehead (Channa striata) protein hydrolysate, Lebensmittel-Wissenschaft & Technologie, **94** (2018): 13-19
8. O. Villamil, H. Váquiro, J. F. Solanilla, Fish viscera protein hydrolysates: Production, potential applications on functional and bioactive properties, Food Chemistry, **224** (2017):160-171
9. A. Noman, Y. Xu, W. Q., AL-Bukhaiti, S. M., Abed, A. H., Ali, A. H., Ramadhan, W. Xia, Influence of enzymatic hydrolysis conditions on the degree of hydrolysis and functional properties of protein hydrolysate obtained from Chinese sturgeon (Acipenser sinensis) by using papain enzyme, Process Biochemistry, **67** (2018): 19-28
10. F. Alavi, M. Jamshidian, K. Rezaei, Applying native proteases from melon to hydrolyze kilka fish proteins (Clupeonella cultriventris caspia) compared to commercial enzyme Alcalase, Food Chemistry, **277** (2019): 314-322
11. M.S. Benhabiles, N. Abdi, N. Drouiche, H. Lounici, A. Pauss, M.F.A. Goosen, N. Mameri, Fish protein hydrolysate production from sardine solid waste by crude pepsin enzymatic hydrolysis in a bioreactor coupled to an ultrafiltration unit, Materials Science and Engineering: C, **32** (2012): 922-928
12. O. Villamil, H. Váquiro, J. F. Solanilla, Fish viscera protein hydrolysates: Production, potential applications and functional and bioactive properties, Food Chemistry, **24** (2017): 160-171
13. N.R.A. Halim, H.M. Yusof, N.M. Sarbon, Functional and bioactive properties of fish protein hydolysates and peptides: A comprehensive review, Trends in Food Science and Technology, **51** (2016): 24-33
14. H.C. Tran, H.A.T. Le, T.T. Le, V.M. Phan, Effects of enzyme types and extraction conditions on protein recovery and antioxidant properties of hydrolysed proteins derived from defatted Lemna minor, Applied Science and Engineering Progress, (2021) doi: 10.14416/j.asep.2021.05.003
15. T. Aspevik, H. Egede-Nissen, Ĺ. Oterhals, A systematic approach to the comparison of cost efficiency of endopeptidases for the hydrolysis of Atlantic salmon (Salmo salar) by-products. Faculty of Food Technology and Biotechnology. **54** (2016): 421-431

16. A.M. Liceaga-Gesualdo, E.C.Y Li-Chan, Functional properties of fish protein hydrolysate from Herring (Clupea harengus). Journal of Food Science, **64** (1999): 1000-1004
17. H.G. Kristinsson, B.A. Rasco, Fish protein hydrolysates: Production, biochemical and functional properties. Critical Reviews in Food Science and Nutrition, **40** (2000): 43-81
18. X. Yuan, X. Gu, J. Tang, Optimization of the production of Momordica charantia L. Var. abbreviata Ser. protein hydrolysates with hypoglycemic effect using Alcalase, Food Chemistry, **111** (2008): 340-344
19. T. P. Singh, R. A. Siddiqi, D. S. Sogi, Statistical optimization of enzymatic hydrolysis of rice bran protein concentrate for enhanced hydrolysate production by papain, Lebensmittel-Wissenschaft & Technologie, **99** (2019): 77-83
20. D. Betancur-Ancona, R. Martinez-Rosado, A. Corona-Cruz, A. Castellanos-Ruelas, M.E. Jaramillo-Flores, L. Chel-Guerrero, Functional properties of hydrolysates from Phaseolus lunatus seeds, International Journal of Food Science and Technology, **44** (2009): 128-137
21. N. Bhaskar, T. Benila, C. Radha, R.G. Lalitha, Optimization of enzymatic hydrolysis of visceral waste proteins of Catla (Catla catla) for preparing protein hydrolysate using a commercial protease, Bioresource Technology, **99** (2008): 335-343
22. K. Zhang, B. Zhang, B. Chen, L. Jing, Z. Zhu, K. Kazemi, Modeling and optimization of Newfoundland shrimp waste hydrolysis for microbial growth using response surface methodology and artificial neural networks, Marine Pollution Bulletin, **109** (2016): 245-252
23. J. Adler-Nissan, Determination of the degree of hydrolysis of food proteins hydrolysates by Trinitrobenzenesulfonic acid, Journal of Agricultural and Food Chemistry, **27** (1979):1256-1262
24. N. Wisuthiphaet, S. Klinchan, S. Kongruang, Fish Protein Hydrolysate Production by Acid and Enzymatic Hydrolysis, Applied Science and Engineering Progress, **9** (2016): 261–270
25. M.A. Amiza, S. Nurul Ashikin, A.L. Faazaz, Optimization of enzymatic protein hydrolysis from silver catfish (Pangasius sp.) Frame, International Food Research Journal, **18** (2011): 775-781
26. A. Vannabun, S. Ketnawa, S. Phongthai, S. Benjakul, S. Rawdkuen, Characterization of acid and alkaline proteases from viscera of farmed giant catfish, Food Bioscience, **6** (2014): 9-16
27. H. Haslaniza, M.Y. Maskat, W.M. Wan Aida, S. Mamot, The effects of enzyme concentration, temperature and incubation time on nitrogen content and degree of hydrolysis of protein precipitate from cockle (Anadara granosa) meat wash water, International Food Research Journal, International Food Research Journal, **17** (2010): 147-152
28. V.Venugopal, Enzymes from seafood processing waste and their applications in seafood processing, in Advances in Food and Nutrition Research, S.-K. Kim and F. Toldrá, Academic Press, **78** (2016): 47-69
29. FAO/WHO, Energy and protein requirements, Report of joint FAO/WHO/UNU Expert Consultation Technical Report. FAO/WHO and United Nations University, Geneva, **724** (1985): 116-129
30. NRC National Research Council, Nutrient Requirement of Fish, National Academy of Science Washington, (1993): 124.
31. J.C. Swanepoel, N.J. Goosen, Evaluation of fish protein hydrolysates in juvenile African catfish (Clarias gariepinus) diets, Aquaculture, **496** (2018): 262–269
32. S. MacKenzie, Recent developments in amino acid analysis by gas-liquid chromatography, in Methods of biochemical analysis, D. Glick, Ed. John Wiley and Sons, **27** (2009): 1-88
33. S. Benjakul, M.T. Morrissey, Protein hydrolysates from pacific whiting solid wastes, Agricultural and Food Chemistry, **45** (1997): 3423-3430
34. S.F. See, L.L. Hoo, A.S. Babji, Optimization of enzymatic hydrolysis of Salmon (Salmo salar) skin by Alcalase, International Food research, **18** (2011): 1359-1365
35. H. Yin, J. Pu, Y. Wan, B. Xiang, P.J. Bechtel, S. Sathivel, Rheological and functional properties of Catfish skin protein hydrolysates, Journal of Food Science, **75**, 1 (2010): E11-E17
36. S.A., Santos, V.G. Martins, M. Salas-Mellado, C. Prentice, Evaluation of functional properties in protein hydrolysates from Bluewing Searobin (Prionotus punctatus) obtained with different microbial enzymes, Food and Bioprocess Technology, **4** (2011): 1399–1406
37. L. You, M. Zhao, J. M. Regenstein, and J. Ren, In vitro antioxidant activity and in vivo anti-fatigue effect of loach (Misgurnus anguillicaudatus) peptides prepared by papain digestion, Food Chemistry, **124** (2011): 188-194
38. M. Watford, Glutamine and glutamate: Nonessential or essential amino acids?" Animal Nutrition, **1** (2015): 119-122

Production and Quality Evaluation of Young Thai Jasmine Rice Flake Using Drum Dryer

Kullamethee Piyarach[1*], *Phumpakdee* Sutham[1], *Khong-im* Pongsiri[1], *Alqurashi* Randah[2] and *Suwan* Teerawan [1]

[1]Department of Innovation and Product development Technology, Faculty of Agro-industry, King Mongkut's University of Technology North Bangkok, 129 Moo21 Nernhom, Prachinburi, 25230, Thailand

[2]Department of Food and Nutrition, College of Agriculture and Food Sciences, King Faisal University, Al-Ahsa 31982, Saudi Arabia

Abstract. Immature rice grains contain high bioactive compounds. Drying has been used for the food preservation and reduction of agricultural waste. This study aimed to produce the rice flake from young Thai Jasmine rice grain (light green and soft texture) using double drum dryer at 120 °C, drum rotation speed 0.5, 0.7, and 0.9 rpm, and the ratios of grain to water of 1:3 1:5 and 1:7w/w. The nutritional properties of rice flake were also investigated. The results showed that increasing rotation speed gives a significant effect on the higher moisture content and water activity. Using lower water ratio with slowing down rotation speed resulted in higher brightness of the flake products including quicker dehydration. The selected optimal condition of preparing rice flake was 1:5 w/w young grain to water at 0.5 rpm drum speed. The flake contains fat, protein, carbohydrate, ash, approximately 0.004%, 6.140%, 88.686% and 0.92%, respectively. Phenolic content and antioxidant capacity of rice flake were decreased after drying. Total phenolic content of the flake was 280.931 mg GAE/g and antioxidant activities; DPPH, FRAP and ABTS were 254.158, 123.172 and 372.298 mg TE/g, respectively. Young rice flake remains a great nutritional value which has potential in the food industry.

Keyword. Young rice grain, Thai Jasmine rice, Double drum dryer, Rice flake, Total phenolic, Antioxidant activity

1 Introduction

Rice has been the main staple food of choice for more than half of the world population and important to Thailand's economy especially Thai Jasmine rice (Hom Mali 105) variety (*Oryzae sativa.* L). Thailand is moving toward industrialization and production of agricultural products has declined including rice production sector; accordingly, the gross domestic product (GDP) has been decreasing. Therefore, an aim to increase value added of rice and nutritional value becomes a priority. At the moment, utilization of rice in the industries as a composition in processed foods and healthy foods products tends to increase. Among Asian countries where rice is mainly consumed as carbohydrate source, Thai's jasmine rice is very desirable due to its aromas and is strongly demanded for an import. Development of rice products is an alternative to increase an income for Thai farmers, yet marketing strategy and further research is essential [1].

Interestingly, immature kernel is traditionally used as nourishing food in most Asian countries. Thailand and Vietnam are commonly used for producing as a snack food (flatted milky rice mixed with sugar and scraped coconut) on the special occasion. The milky stage of the young rice was reported with high numbers of bioactive compounds than its mature stage, which observed that when developed the starch contained in rice is hence to degrade the bioactive compounds in rice [2-4]. A previous study by S. Butsat *et al.* [5] and S. Jiamyangyuen *et al.* [3] was described the 5 stages of rice grain development which include the "flowering" as the first stage (age 0-7 days) that has small and flat endosperm with the green colour of the seed coat. The second stage "milk grain or immature grain or young grain" (age 8-14 days after post-anthesis) (Figure 1(a)). In the second stage the endosperm begins forming milky liquid and increased the size of the rice and in this stage, the colour of the coat starts to change dependent on the cultivars of the rice. The third stage is "dough grain" (age 15-21 days after post-anthesis) which is the milky liquid of the endosperm that starts to become solidify into sticky white type. Also, in this stage, the endosperm continues to increase and become bigger and the colour of the seed coat becomes clearer than in the previous stage. The Fourth stage called "mature" (age 22-28 days after post-anthesis) which the grain become mature and hard also the endosperm is increased their size more. In the final stage called "fully ripe" (age 29-35 days after post-anthesis) which the grain becomes fully mature and hard, and the endosperm size increased to the largest size. The colour of that seed coat strongly presenting the colour of the other types of rice cultivars. However, the stage of immature milky of rice (second stage of rice grain development) has not the previous study well.

The study of bioactive compounds in rice during grain development was found that two developing rice grains (waxy red rice and non-waxy rice) contain higher bioactive compounds, total phenolics, and total flavonoids than its mature grain. The high content of free and soluble ferulic acids, α-Tocopherol, α- and β-tocotrienol, dietary fiber, are abundant in the early development stage of brown grains at day 15-18 after anthesis [6]. Rice grains at the stage of immature contain

* Corresponding author: piyarach.k@agro.kmutnb.ac.th

a high number of bioactive compounds and antioxidant activities which have potential health benefits and can apply in the food industry [7]. At present, people's lifestyle has changed due to the speed of life and they prefer to prepare and consumed food quickly especially in a breakfast meal. Therefore, in the food industry, the demand of cereals and food that contained fiber, proteins, and antioxidant activity is increasing dramatically. Some of cereals are available in the markets such as legumes, buckwheat, semolina, barley, rye, oats, and rice [8].

Nowadays, rice processing is the trend which commonly using as a breakfast cereals dish mixed with milk or water and added flavorings such as honey, dried fruit, sugar, or syrup. Alternatively, mixing rice products with vegetable and species is an option to create savory dishes. To be able to fulfill the demand of some people who focus on healthy lifestyle, rice is processed into the form of porridge for the ease to use and lower calories in the meals [9-10]. The rice porridges usually recooking and made as dried flakes (Figure 1(b)) and powder before mixing with other ingredients. These flakes of the rice swell after mix with a liquid such as milk or warm water [10]. Drying cereals and rice is one of the main important processes which have affected the quality of the final product. It can be affected the concentration of bioactive compounds and other nutrients also the textures of the product. Drum drying is a technique (short-time with high-temperature) widely uses in the food industry particularly for heat-sensitive products that uses the drying of a thin film of solution or suspension on a heated drum and subsequent removal of the film of dry solids [11]. Double drum dryer is generally used for preservation in the food industries because this equipment is suitable for a wide range of materials. This method is one of the most appropriate techniques for using liquid feedstock form (solution, slurry, and pastes) and providing the final product such as powders or flakes when compared with other drying technology. Moreover, this process is effortless and economical method during the operation for pre-gelatinization [12-14]. Energy consumption of liquid type of feedstock such as drum dryer, fluid bed dryer, and spray dryer is in the range between 3,200-6,500, 4,000-6,000, and 4,500-11,500 (kJ/kg of H_2O evaporated), respectively in the common industrial dryers [14]. C. Córdova *et al* [15] demonstrated that drum drying has great potential to dehydrate broccoli pulp and maintained functional compound such as sulforaphane, phenolic compounds, and antioxidant activities. A study by K. Ruttarattanamongkol *et al.* [16] reported that after decreasing drum speed and increasing drum drying temperature, the moisture content of the sweet potato flour was declined. On the contrary higher temperature of drum drying significantly resulted in the lower total phenolic content and antioxidant activity. This was also the observation of P. Wiriyawattana *et al.* [17] that addressed the effects of drum drying on the physical and antioxidant properties of pregelatinized rice berry flour. The finding has shown that drying resulted in reduction of both total phenolic content and antioxidant activity.

The developing rice grains have a relatively high amount of phytochemical sources, especially young rice grain at the milky period that would be suitable for healthy food applications such as porridge, cereal, and soup. Early harvesting rice grain at the immature stage instead of the fully ripe would be alternative process for resources conservation such as water consumption, land use, soil organic matter and fertilizer [18, 19]. In addition to high nourishment of immature rice grain that causes short shelf life, colour changing, microbial contamination affects its marketing value. Therefore, dehydration process has been widely used in order to preserve raw materials and reduce food waste and loss of agricultural resources [14, 15, 19]. The optimal condition to produce Thai Jasmine rice (milky period) flake using a double drum dryer is therefore interested. We determined the ratios of young brown rice grain to water and drum rotation speed on rice flake qualities. The physical, chemical, and microbial properties of young rice flakes were also estimated. This production would be useful for the rice farm owner who would like to value-added their agricultural product as well as the food industry that desires to produce the instant products from immature kernel.

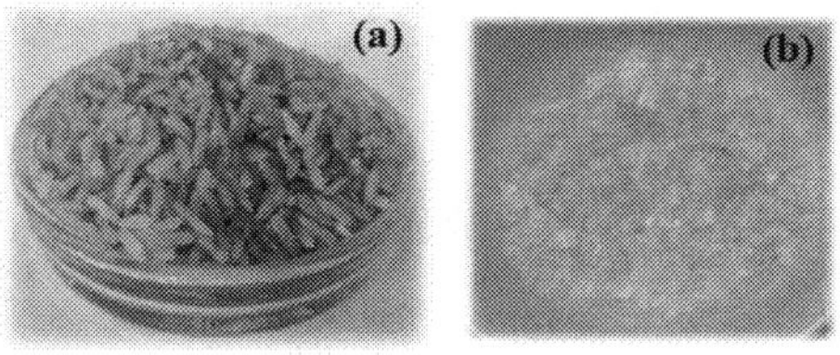

Fig. 1. Images of (a) immature brown rice grain and (b) the thin film of rice flake.

2 Materials and Methods

2.1. Materials

Immature stage of Thai Jasmine rice 105 or Khao Dawk Mali 105 (*Oryza sativa* L.) (Figure 1(a)), was obtained from Ban-nakham organic rice field (community enterprise network) in Ban-nakham, Kabinburi, Prachinburi province, Thailand. The stage of young grain development was collected at 8-14 days after the flowering day. Brown young rice grain was separated from hull using machine and kept in polyethylene vacuum bag covering with aluminum foil at room temperature until use.

2.2 Chemicals

Reagents for antioxidant assays and phenolic standards such as Gallic acid, Folin-Ciocalteu, 2,2'azinobis (3-ethylbenzothaiazoline-6 -sulfonic acid) (ABTS) were purchased form Sigma-Aldrichz, USA. 2,2-diphenyl-1picrylhydrazyl (DPPH) and Trolox were obtained from Sigma-Aldrichz, Germany and Russia, respectively. Other unmarked reagents were used analytical grade.

2.3 Drying experiment

Before drying process, young brown rice was prepared by soaking with water for about 4 hrs. Then, draining off water and preparation the ratios of young rice to water at 1:3 1:5 and 1:7 w/w and each treatment was blended for 5 mins using food processor (Sharp, EM-ICE2, Thailand) at high-speed level (17,000 rpm). After that, the slurry was pre-heat using electric hot plate (400 W) (Master Cooking, MC-1, France) at middle heat level for about 30 mins. (90 ± 2°C). The pregelatinized slurry was feed in double drum dryer (electric heating, DD-D6L12-HT-INV, Thailand) at 112 ± 5 g feed rate with the different of drum rotation speed (0.5, 0.7 and 0.9 rpm), 0.25 mm. of the gap between drums and 120 °C of drying temperature. The dried flake of young brown rice (Figure 1(b)) was collected as a thin film from the drum surface. Procedure of the drying experiment presented in Figure 2.

2.4 Proximate composition analysis, physical and microbial properties

Proximate composition analysis (moisture content, protein, fiber, ash and fat contents) of the brown immature rice flake were performed following the official methods described by Association of Official Analytical Chemistry (AOAC 2005) with three replications. Water activity meter (AQUA 3TE, USA) was used for water activity evaluation. The colour of the sample was measured using Hunterlab MiniScan (ColorFlex, USA) and reported as CIE system (International Commis-sion on Illumination). L^* (0 = black, 100 = white), a^* (-a* = greenness, +a* = redness), and b^* (-b* = blueness, +b* = yellowness) parameters indicated lightness, redness/greenness, and yellowness /blueness, respectively. Color measurement was carried out in three replicates for each treatment.

Total viable count and yeast and mold test were used as the microbiological assessment to monitor the young rice flake product. 1 g of sample was collected aseptically in a stomacher bag and 9 volumes of peptone solution (0.01%, w/v) were added. After homogenization in a Stomacher blender for 2 mins, a series of tenfold dilutions were made using the same solution. According to Thai Industrial Standard No. 315-2548 of the instant rice porridge (congee), the optimal conditions were expected to reach the final moisture content less than 7%. Moreover, the microbial standard of the product was not exceed the level of total microbial count $< 1\times10^5$ CFU/g, yeast and mold < 100 CFU/g. Thai Community product Standard No.1068/2558 of the instant drink brown rice was used for indication water activity less than 0.6 in order to maintain the desirable quality of food products and their shelf-life.

2.5 Water absorption index (WAI) and water soluble index (WSI)

The water absorption index (WAI) and water soluble index (WSI) properties of the young rice flake was determined as described by R. A. Anderson, *et al* [20] with some modification. Briefly, 2.5 g rice flake (weight of rice flake; Wrf) was suspended in distilled water at temperature 30 °C for 30 mins., gently stirred and centrifuged at 2,200 rpm for 15 mins. The supernatant was decanted into an evaporating dish. This supernatant was then dried at 105 °C until constant weight and the weight of dissolved solids (Wds) in the supernatant was recorded. The remaining sediment in the centrifuge tube was weighted (weight of remaining sediment; Wrs). WAI and WSI values were calculated according to Equations (1) and (2), respectively.

$$\mathrm{WAI} = \frac{\mathrm{Wrs\,(g)}}{\mathrm{Wrf\,(g)}} \quad (1)$$

$$\mathrm{WSI\,(\%)} = \frac{\mathrm{Wds}}{\mathrm{Wrf}} \times 100 \quad (2)$$

2.6 Extraction of bioactive compounds

Rice samples (young brown rice grain and the dried rice flake) was extracted for determination the bioactive compound as described by S. Jiamyangyuen *et al.* [3]. The rice samples were freeze dried at -40 °C for 4 hrs and ground into powder. The powder of rice samples was kept at -40 °C before use. 0.5 g of each sample (rice grain and young rice flake) was extracted with 20 ml. acidified methanol/1N HCl in water (85:15 by volume) in centrifuge tube for 1 hour and vortexed for 2 mins. The soluble extract was centrifuged at 5,000 rpm for 10 mins and the supernatant was collected. The residue was re-extracted twice using the same method and all supernatants were pooled. The supernatant was evaporated under rotary evaporator at 40 °C for 5 hrs. and made up completely dry using hot air oven in order to concentrate the extract and eliminate the solvent. The concentrated extract was redissolved to 15 mg/ml by adding methanol. The solution was then stored at -40°C for further analysis. The rice extract was used for determination total phenolic content and antioxidant activity (2,2-diphenyl-1picrylhydrazyl (DPPH), Ferric ion reducing antioxidant power (FRAP) and 2,2-azinobis 3-ethylbenzthiazoline-6-sulphonic acid (ABTS) radical scavenging assay)

2.7 Determination of total phenolic content (TPC)

The TPC was analyzed through the reducing power against Folin-Ciocalteu using the method of O. Folin and V. Ciocalteu [21] and P. Wiriyawattana *et al.* [17] with minor modification. 0.4 ml of rice sample extracts were mixed with 2 ml of freshly prepared diluted Folin–Ciocalteu reagent (10% v/v) and incubated at room temperature for 5 mins. After mixing, 1.6 µl of sodium carbonate aqueous solution (5% w/v) was added. The reagent was then stored under dark condition at room temperature for 30 mins to allow the completion of the reaction. The intensity of the developed blue colour was measured at 765 nm using a microplate reader (Accu Reader, M965/M965+, Taiwan). Different concentration of gallic acid dissolved in methanol were used to prepare the standard curve ($R^2 = 0.9929$). The result of TPC was reported as milligram of gallic acid equivalent (mg GAE/100 g dry weight (dw)). Estimation of the TPC was carried out in triplicate.

2.8 Determination of DPPH scavenging activity

The scavenging capacity of the rice sample extracts against 2,2-diphenyl-1picrylhydrazyl (DPPH) radical was measured according to W. Daiponmak *et al.* [22]. with slight modification. Briefly, 100 µl of extract solution was mixed with 2,900 µl DPPH solution (100 µM). The mixture was shaken vigorously and incubated in dark condition at room temperature for 30 mins. After the reaction, the absorbance (Abs) was measured at 517 nm using a microplate reader. Trolox was used as a standard antioxidant and the result was expressed as milligram Trolox equivalent (mg TE/g dw). Estimation of the DPPH activity was carried out in triplicate. The scavenging activity was calculated using the formula.

$$\text{scavenging activity (\%)} = \frac{Abs_{control} - Abs_{sample}}{Abs_{control}} \times 100$$

2.9 Determination of FRAP scavenging activity

Ferric ion reducing antioxidant power (FRAP) of the rice extract was analyzed based on Benzie and strain [23] and S. Jiamyangyuen *et al.* [3] with some modification. The reagent of FRAP was freshly prepared as following reagents: 300 mM acetate buffer (pH 3.6), 10 mM of 2,4,6-Tripyridyl-s-triazine (TPTZ) solution in 40 mM HCL, and 20 mM ferric chloride ($FeCl_3$) in methanol, at ratio of 10:1:1 by volume. 0.1 ml extract solution and 2 ml FRAP reagent were mixed and incubated for 30 mins. The developed blue solution was measured at 595 nm using microplate reader. Trolox was used as a standard reference and the result was converted as milligram trolox equivalent (mg TE/100 g dw). The scavenging activity was then calculated using the formula.

$$\text{scavenging activity (\%)} = \frac{Abs_{control} - Abs_{sample}}{Abs_{control}} \times 100$$

2.10 Determination of ABTS scavenging activity

2,2-azinobis 3-ethylbenzthiazoline-6-sulphonic acid (ABTS) radical scavenging assay of rice extracts was determined using a modification of the procedure described by Re *et al.* [24] and W. Daiponmark, *et al.* [22]. The stock solution was prepared with 2:1 volumes of 7 mM ABTS solution and 2.45 mM potassium persulfate solution. The mixture was stand in the dark condition at room temperature for 12 hrs. The working solution was freshly prepared for each assay by diluting the stock solution with 95% ethanol until the absorbance at 734 nm was measured 0.670 - 0.730. 1 ml of the rice extract was added into 2 ml of the ABTS working solution and mixed thoroughly. The decrease in absorbance at 734 nm was then recorded and expressed as mg Trolox/gram extract dry weight. The scavenging activity was estimated based on the percentage of ABTS radicals scavenged according to Equation.

$$\text{scavenging activity (\%)} = \frac{Abs_{control} - Abs_{sample}}{Abs_{control}} \times 100$$

2.11 Statistic

The experiment was conducted in a complete randomized design with three replications. The results were subjected to analysis of variance and mean comparison by One-Way ANOVA followed by Duncan's multiple range tests at 95% confidence level ($p \leq 0.05$), in order to determine any differences among treatments The paired Student's t-test were also used in order to determine any significance of the differences in the physical, chemical, and microbial properties between rice grain and rice flake. The data was performed using SPSS.

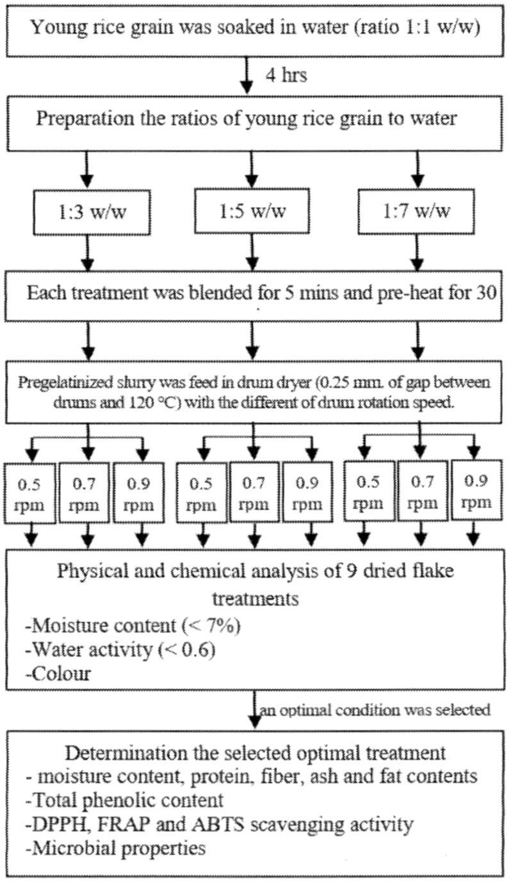

Fig. 2. Flow diagram of the experiment.

3 Results and discussion

3.1 Effect of young rice grain to water ratio and drum rotation speeds on physicals and chemical properties of rice flake

After conducting the drying experiments, the results showed that increasing of rotation speed gives significant effect ($p<0.05$) on the higher value of moisture content and water activity (Table 1). Application of lower water content with lower speed of rotation resulted in faster dehydration. Decreasing speed of rotation used in drum dryer allowed rice slurry to contact the surface of drum dryer in longer period of time and this, consequently, increased the dehydration rate of water in rice film. The water activity of all treatments was not exceeded the standard. The lowest water activity was 0.189±0.002 (treatment No. 4 at ratio 1 :5, 0.5 rpm). The lowest moisture content of the product was 3.28±0.13% (treatment No. 1 at ratio 1 :3, 0.5 rpm). According to Thai Industrial Standard No. 315-2548 and Thai Community product Standard No.1068/2558, the requirement moisture content of the young rice flake and water activity were less than 7% and 0.6, respectively. The water activity is a crucial characteristic for shelf life of the product. Therefore, the optimum condition of young rice flake from the study could be the treatments No. 1-5 and No. 7-8.

Results in Table 1 indicated that the colour of rice flake tends to be lighter when decreasing the speed of drum rotation and increasing amount of rice grain. All flake treatments were presented the lightness (L^*) range between 31.89±0.24 to 46.40±0.36, redness (a^*) between 1.27±0.03 to 1.85±0.03 and yellowness (b^*) between 7.77±0.13 to 14.82±0.07. The colour of rice flake treatments differs from the young rice grain with lightness, redness, and yellowness as 26.10±0.32, 1.05±0.06 and 7.61±0.35, respectively. Changes in colour of the treatments were possibly caused by the alteration of contacted surface areas in materials when exposed to high temperature during drum dryer process. In addition, chemical reaction such as Millard reaction induced by thermal treatment is one of the factors affecting the change in colour of the products. When temperature and heating time was increased, this browning reaction was also elevated accordingly. The colour quality and nutritional properties of rice is affected directly by thermal processing [11,16]. After drum drying, WAI of the treatments with high amount of rice grain were significantly ($p<0.05$) higher than those with the low grain content. In addition, the samples with high rice grain ratio were significantly decreased in water soluble index (WSI). The rice grain content effected the water absorption index (WAI) and WSI. While WAI and WSI value were not affected by rotation speeds. P. Wiriyawattana *et al.* [17] investigated the effects of drum drying on the physical and antioxidant properties of pregelatinized riceberry flour and their results demonstrated that the WSI and swelling power of the samples were significantly higher than the unheated rice flour. WAI was used as an indicator to measure water volume inside the granule or starch polymer after maximum absorption of water. WSI increased in parallel with moisture content that is mostly due to proper gelatinization and lateral expansion of the starch [25]. To determine the rate of penetration of water into solid particles and diffusion of soluble components outside the particles, the WSI is a common measurement to obtain these values. The common evaluation is to measure the degradation of molecular components generated by starch degradation which are easily dissolved. Therefore, the solubilization of power is indicated by an increase of WSI value [26].

The important parameter such as feed rate, thickness of film, drum speed, and drum surface temperature play the crucial role to maintain uniformity of the film on drum surface as well as to maximize the consistency of final moisture content in pregelatinized rice [27, 28]. From the results, the optimal condition of producing young rice flake product using drum dryer was retrieved from treatments No. 1, 2, 4 and 5. The treatment No. 4 yielding the lowest water activity was selected for further study including physical, chemical, and microbial properties.

Table 1. Moisture content, water activity, colour, water absorption index and water solubility index of rice flake in different treatments.

Sample No.	Ratio (Grain:Water)	Rotation speed (rpm)	% Moisture Content	a_w	Colour			WAI	WSI (%)
					L*	a*	b*		
1	1:3	0.5	3.28±0.13^{f}	0.238±0.003^{f}	46.40±0.36^{a}	1.85±0.03^{a}	14.82±0.07^{a}	2.56±0.04^{a}	4.99±0.57^{f}
2	1:3	0.7	3.66±0.17^{f}	0.413±0.001^{e}	40.73±0.16^{de}	1.35±0.07^{f}	10.29±0.34^{d}	2.55±0.06^{a}	5.43±0.06^{ef}
3	1:3	0.9	6.42±0.18^{b}	0.441±0.002^{d}	37.70±5.94^{e}	1.57±0.05^{c}	9.72±0.13^{e}	2.43±0.05^{b}	5.60±0.33^{e}
4	1:5	0.5	4.25±0.13^{e}	0.189±0.002^{g}	44.79±0.36^{abc}	1.46±0.01^{de}	12.17±0.16^{b}	2.37±0.02^{bc}	6.31±0.19^{d}
5	1:5	0.7	5.44±0.37^{c}	0.459±0.002^{c}	41.24±0.33^{cde}	1.58±0.01^{c}	11.15±0.16^{c}	2.37±0.08^{bc}	6.73±0.09^{cd}
6	1:5	0.9	7.50±0.38^{a}	0.519±0.003^{b}	44.58±0.47^{abc}	1.44±0.05^{e}	11.34±0.02^{c}	2.31±0.06^{c}	6.99±0.24^{c}
7	1:7	0.5	4.94±0.34^{d}	0.234±0.008^{f}	31.89±0.24^{f}	1.27±0.03^{g}	7.77±0.13^{f}	2.07±0.05^{d}	7.92±0.12^{b}
8	1:7	0.7	6.25±0.14^{b}	0.442±0.010^{d}	42.24±0.10^{bcd}	1.75±0.03^{b}	11.14±0.07^{c}	2.09±0.05^{d}	9.36±0.45^{a}
9	1:7	0.9	7.64±0.30^{a}	0.529±0.003^{a}	45.50±0.55^{ab}	1.52±0.03^{cd}	12.13±0.8^{b}	2.05±0.02^{d}	9.56±0.17^{a}

Data were presented as mean values. ^{a,b} values in the same column followed by different lowercase superscript letters are significantly different of each product treatment.

3.2 Physical, chemical and microbial properties of young rice flake

Immature kernel as show in Table 2, L^*, a^* and b^* values were 26.10±0.32, 1.05±0.06 and 7.61±0.35, respectively. Approximately, composition of immature kernel consisted of 1.43±0.12 % ash, 6.98±0.57 % protein, 0.107±0.002% fat and 81.53% carbohydrate. The grain had 8.75±0.19 % moisture content, under the Thai Hom Mali rice standard (moisture content less than 14%, Thai Agricultural Standard (TAS 4000-2017)) microbial properties of the grain were also evaluated as 1×10^3 CFU/g total plate count and 3×10^2 CFU/g yeast/mold.

In the present investigation, the results showed significant ($p<0.05$) changes in the color values of selected immature rice flake product compared to the immature kernel (Table 2). The flake product was lighter (higher L^* value), redder (higher a^* value) and yellower (higher b^* value) than the immature kernel. The possible changed colour of the young rice flake was because pre-heat of rice slurry exposed to the high temperature during the drum drying process. Moreover, using thermal treatment induces non-enzymatic browning that can lead to the formation of the brown colour. The browning colour reaction was enhancing by increased temperature degree and heating time. Rice colour also can be affected it and degrade during thermal processing which affects the nutritional properties of the rice [26, 29].

For the evaluation the total phenolic content (Table 2) and antioxidant free radical scavenging properties (Table 2 and Figure 3), the results presented immature kernel contain high concentration of phenolic compound (347.965±81.522 mg GAE/g) and rich in antioxidant activities (DPPH, FRAP and ABTS were 254.158, 123.172 and 372.298 mg TE/g, respectively). There are several studies reveal that Thai rice cultivars enriched with bioactive compound content and antioxidant capacities. This trend is similar the previous study of W. Daiponmark *et al.* [30] found high value of total phenolic compound and antioxidant activities of four white grain of Thai rice cultivar. White rice varieties contain total phenolic content range between 24.86-41.73 (mg GAE/g extract dw), ABTS range between 39.49-52.24 (mg Trolox/g extract dw) and DPPH range between 15.75-23.68 (mg Trolox/g extract dw). In addition, previous study has shown that young rice (red rice) at stages 2-3 has high antioxidant activity using FRAP assay [3]. After drum drying total phenolic content and antioxidant activity of the young rice flake significantly decreased ($p<0.05$). The flake product remains total phenolic content 280.930±3.995 mg GAE/g) and DPPH, FRAP and ABTS were 254.158±7.644, 123.172±3.137 and 372.298±3.334 mg TE/g, respectively. This was attributed to the degradation of phenolic compounds at high temperature [27]. A study by Z. Hou *et al.* [31] reported that anthocyanins degraded quickly by using high temperatures and increased the pH at 100 °C and pH 5.0.

For evaluation of microbial properties, the selected young rice flake had water activity 0.189±0.003, The product was not exceed the level from the microbial standard of Thai Industrial Standard No. 315-2548.

Table 2 Physical, chemical and microbial properties of young rice grain and rice flake

Properties	Young rice grain	Young rice flake
	Mean±SD	Mean±SD
Physical properties *		
L^*	26.10±0.32	44.79±0.36
a^*	1.05±0.06	1.46±0.01
b^*	7.61±0.35	12.17±0.16
Water activity	0.560±0.003	0.189±0.003
Chemical properties		
Moisture content (%)*	9.95±0.17	4.25±0.13
Protein (%)^{ns}	6.98±0.57	6.14±0.48
Fat (%)*	0.11±0.002	0.004±0.001
Ash (%)*	1.43±0.12	0.92±0.17
Carbohydrate (%)	81.53	88.686
Bioactive compound		
Total Phenolic)mg GAE/g)*	347.965±81.522	280.930±3.995
Antioxidant activity		
DPPH (mg TE/g)*	368.455±50.099	254.158±7.644
FRAP)mg TE/g)*	209.037±8.736	123.172±3.137
ABTS (mg TE/g)*	486.018±40.818	372.2983±3.334
Microbial properties		
Total plate counts (CFU/g)	1×10^3	$2x10^3$
Yeast-molds (CFU/g)	3×10^2	Not found

Data were presented as mean values. * Significant difference of young rice grain compared between young rice flake (1:5 ratio of grain to water, 0.5 rpm drum speed at 120°C) is using paired Student's t-test ($p<0.05$) for data analysis.

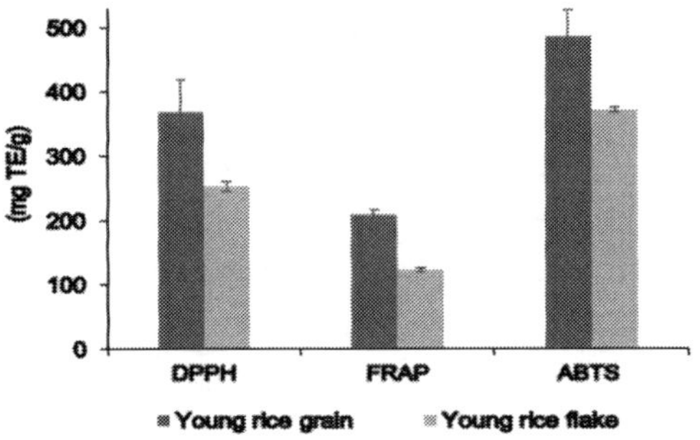

Fig. 3. Antioxidant activity against DPPH, FRAP and ABTS of young rice grain and rice flake.

4 Conclusion

Young rice grains contain the high nutritional value. Harvesting rice grains at the early stage might be an alternative process for the conservation of resources.

However, the high nutritional properties of immature rice grains may lead to shorter shelf life, colour changing, microbial contamination, and compromise the marketing value. Therefore, from this perspective, preservation and value-added agricultural products have major concerns. Drying has been known as one of the best methods for preserving raw materials and limiting food waste and loss of agricultural products. Dehydration process using drum dryer technology is suitable for slurry form, low economics costs and low energy consumption. The finding of this study demonstrated that there were four optimal conditions (the treatments No. 1, 2, 4 and 5) for producing the immature kernel flake. We selected one of the optimal treatments with ratios of young rice grain to water at 1:5 w/w with o.5 rpm drum rotation speed to be the representative of the flake product due to the lowest water activity. The selected treatment of young rice flake had 0.189±0.002 water activity and 4.250±0.130 % moisture content. The color values of L* , a* and b* were 44.790±0.360, 1.460±0.010 and 12.170±0.160, respectively. The proximate analysis was 0.004±0.001% fat, 6.140±0.480% protein, 0.920±0.170% ash, 88.686% carbohydrate. The total plate counts and yeast-molds in the milky rice flake were below the maximum standard. The total phenolic content and antioxidant capacity of rice flake were decreased after drying. Total phenolic content of the flake was 280.930±3.995 mg GAE/g) and antioxidant activities; DPPH, FRAP and ABTS were 254.158±7.644, 123.172±3.137 and 372.298±3.334 mg TE/g, respectively. These results indicate that young rice flakes were produced by drum dryer, have a great nutritional value retention which might has potential in the food industry and sustainable development of rice products.

Acknowledgments

This research was funded by King Mongkut's University of Technology North Bangkok. Contract no. KMUTNB-61-DRIVE-020.

References

1) O. Napasintuwong. *Rice Economy of Thailand*. Agricultural and Resource Economics Working Paper No. 2562/1 (January 2019). Department of Agricultural and Resource Economics, Faculty of Economics, Kasetsart University, Bangkok, Thailand (2019): 1-36
2) T. Pantoa, M. Kubota, P. Suwannaporn, M. Kadowaki. Characterization and bioactivities of young rice protein hydrolysates, Journal of Cereal Science, **95** (2020): 103049
3) S. Jiamyangyuen, N. Nuengchamnong, P. Ngamdee. Bioactivity and chemical components of Thai rice in five stages of grain development, Journal of Cereal Sciences, **74** (2017): 136-144
4) S. Ngamsuk, J.L Hsu, T-C. Huang, P. Suwannaporn. Ultrasonication of milky stage rice milk with bioactive peptides from rice bran: Its bioactivities and absorption food and bioprocess technology, **3** (2020): 462–474
5) S. Butsat, N. Weerapreeyakul, S. Siriamornpun. Changes in phenolic acids development, J. Agric. Food Chem, **57** (2009): 4566-4571
6) P.Y. Lin, H.M. Lai. Bioactive compounds in rice during grain development, Food Chemistry, **127**, 1 (2011): 86-93
7) T. Pantoa, I. Jones, P. Suwannaporn, M. Kadowaki, M. Kubota, S. Roytrakul, E.N. Clare Mills. Young rice protein as a new source of low allergenic plant-base protein, Journal of Cereal Science, 93 (2020): 102970
8) Future market insights. (2021, June 15). Instant cereals Market: Global Industry Analysis and Opportunity Assessment 2017-2027. https://www.futuremarketinsights.com/reports/instant-cereals-market
9) The insight partner. (2021, June 15). Market research reports. Porridge Market Forecast to 2028 - Covid-19 Impact and Global Analysis - by Type (Oat, Maize, Wheat, Rice, Millet); Distribution Channel (Supermarkets and Hypermarkets, Convenience Stores, Specialty Stores, Online) and Geography. https://www.theinsightpartners.com/reports/porridge-market
10) K. Srikaeo, P.A. Sopade. Functional properties and starch digestibility of instant Jasmine rice porridges, Carbohydrate Polymers, **82**, 3 (2010): 952-957
11) W. Jittanit, M. Chantara-In, T. Deying, W. Ratanavong. Production of tamarind powder by drum dryer using maltodextrin and Arabic gum as adjuncts. Songklanakarin Journal of Science and Technology. **33**, 1 (2011): 33-41
12) M. Majzoobi, M. Radi, A. Farahnaky, J. Jamalian, T. Tongtang, Gh. Mesbahi. Physicochemical Properties of Pre-gelatinized Wheat Starch Produced by a Twin Drum Drier, Journal of Analytical Science and Technology, **13**, 2 (2011): 193-202
13) C. Córdova, J. P. Vivanco, J. Quintero, A. Mahn. Effect of drum-drying conditions on the content of bioactive compounds of broccoli pulp, Foods, **9**, 9 (2020): 1224
14) A. S. Mujumdar (Ed). *Principles, classification, and selection of dryers*. In Handbook of industrial drying. (Taylor & Francis Group. LLC, New York, 2006)
15) F. H. Cosme-De Vera, A. N. Soriano, N. P. Dugos, R. V. C. Rubi. A comprehensive review on the drying kinetics of common tubers, J.asep, 14, 2 (2021): 146-155
16) K. Ruttarattanamongkol, S. Chittrakorn, M. Weerawatanakorn, N. Dangpium. Effect of drying conditions on properties, pigments and antioxidant activity retentions of pretreated orange and purple-fleshed sweet potato flours, Journal of Food Science and Technology. **53**, 4 (2016): 1811–1822
17) P. Wiriyawattana, S. Suwonsichon, T. Suwonsichon. Effects of drum drying on physical and antioxidant properties of riceberry flour, Agriculture and Natural Resources, **52** (2018): 445-450
18) P. Ngammuangtueng, N. Jakrawatana, P. Nilsalab, S. H. Gheewala. Water, energy and food nexus in rice production in Thailand, Sustainability, **11**, 5852 (2019): 1-21

19) N. V. Hung, M. V. Migo, R. Quilloy, P. Chivenge, M. Gummert. *Life cycle assessment applied in rice production and residue management.* In M. Gummert, N. V. Hung, P. Chivenge, B. Douthwaite (Eds). Sustainable Rice Straw Management. (Springer, Cham, 2020)
20) R. A. Anderson, H. F. Conway, A. J. Peplinski. Gelatinization of corn grits by roll cooking, extrusion cooking and steaming, Cereal Science Today, **14** (1969): 14-12
21) O. Folin, V. Ciocalteu. Tyrosine and tryptophane determinations in proteins. Journal of Biological Chemistry, **73** (1927): 627-650
22) W. Daiponmak, C. senakun, S. Siramornpun. Antiglycation capacity and antioxidant activities of different pigmented Thai rice, International Journal of Food Science and Technology, **49** (2014): 1805–1810
23) I. F. Benzie, J.J. Strain. Ferric reducing/antioxidant power assay: Direct measure of total antioxidant activity of biological fluids and modified version for simultaneous measurement of total antioxidant power and ascorbic acid concentration, Methods Enzymology, **18** (1999): 15-27
24) R. Re, N. Pellegrini, A. Proteggente, A. Pannala, M. Yang, C. Rice-Evans. Antioxidant activity applying an improved ABTS radical cation decolorization assy, Free Radical Biology and Medicine, **26** (1999): 1231-1237
25) N. Yousf, F. Nazir, Rehana Salim, Hafiza Ahsan and Adnan Sirwal . Water solubility index and water absorption index of extruded product from rice and carrot blend, Journal of Pharmacognosy and Phytochemistry, **6**, 6 (2017): 2165-2168
26) X. Qi , L. Chenga , X. Lia, D. Zhang, G. Wua , H. Zhanga, , L. Wanga, H. Qiana, Y. Wang. Effect of cooking methods on solubility and nutrition quality of brown rice powder. Food Chemistry, **274**, 15 (2019): 444-451
27) Y. Gat, L. Ananthanarayan. Effect of extrusion process parameters and pregelatinized rice flour on physicochemical properties of ready-to-eat expanded snacks, Journal of Food Science and Technology, **2**, 5 (2015): 2634–2645
28) W. Jittanit, C. Lalitmassakul, P. Charn-Utsar. Quality of instant congee and energy consumption in the drying process by Using Drum Dryer. The Journal of KMUTNB, 22, **2** (2012): 256-264
29) A. Patras, N.P. Brunton, C. O'Donnell, B.K. Tiwari. Effect of thermal processing on anthocyanin stability in foods; Mechanisms and kinetics of degradation, Trends in Food Science and Technology, **21** (2010): 3-11
30) W. Daiponmak, C. Senakun, S. Siriamornpun. Antiglycation capacity and antioxidant activities of different pigmented Thai rice, International Journal of Food Science and Technology, **49** (2014): 1805–1810
31) Z. Hou, P. Qin, Y. Zhang, S. Cui, G. Ren. Identification of anthocyanins isolated from black rice (*Oryza sativa* L.) and their degradation kinetics, Food Research International., **50** (2013): 691-697

Effect of hydrochloric acid extraction on yield and gel properties of gelatine from salted jellyfish by-products

Wiriya Charoenchokpanich[1], *Pratchaya* Muangrod[1], *Vilai* Rungsardthong[1], *Savitri* Vatanyoopaisarn[1], *Benjamaporn* Wonganu[2], *Sittiruk* Roytrakul[3] and *Benjawan* Thumthanaruk[1,*]

[1]Department of Agro-Industrial Food and Environmental Technology, Faculty of Applied Science, King Mongkut's University of Technology North Bangkok, Bangkok, Thailand
[2]Department of Biotechnology, Faculty of Applied Science, King Mongkut's University of Technology North Bangkok, Bangkok, Thailand
[3]National Center for Genetic Engineering and Biotechnology, National Science and Technology Development Agency (NSTDA), Pathum Thani, Thailand

Abstract. Salted jellyfish by-products have collagen protein that is mainly sold for animal feed at a low price. The change of jellyfish by-products into a food ingredient like gelatine could benefit food applications and reduce food waste. Indeed, jellyfish gelatine production is a time-consuming process that includes alkaline pre-treatment, acid pre-treatment, hot water extraction, and drying. Reduced times of acid pre-treatment and water extraction might deliver different gel properties. Therefore, this research aimed to investigate the effect of hydrochloric acid (HCl) pre-treatment on the gel quality of resultant gelatine. Desalted jellyfish by-products were immersed in 0.5 M sodium hydroxide at 4°C for 1 h and then were acid-treated by varying HCl concentrations (0.1, 0.2, and 0.3 M) at 25°C for 2 h. After that, samples were extracted at 60°C for 3 h and dried at 60°C for 3 days. Results showed that gelatine yield significantly increased with increasing HCl concentration. Gelatine yield were 2.97±0.97%, 5.60±1.01%, and 6.34±1.08%, after extraction with 0.1, 0.2, and 0.3 M HCl, respectively. Gel strength generally decreased as HCl concentration increased. Gel strength values were in the range of 118.89-223.60 g. The colour of jellyfish gelatine showed light to dark brown with no differences in Hue values. Thus, the short duration of HCl pre-treatment for 2 h and hot water extraction for 3 h was insufficient for the jellyfish gelatine process.

Keyword. Salted jellyfish by-products, Gelatine, Pre-treatment, Yield, Gel strength

1 Introduction

Gelatine is a protein-type hydrocolloid widely utilised in food industries, functioning as gelling, thickening stabilising, foaming, emulsifying, water binding, and crystal growth modifying [1-3]. Gelatine is derived from the denaturation of collagen [4]. Commercial gelatines are mainly produced from bovine bone and porcine skin [5]. However, bovine and porcine gelatine have faced some constraints related to religious issues and biological contaminants [6]. Therefore, alternative gelatine derived from marine animals, especially from fish, has been sold and substituted for bovine and porcine gelatines.

Many marine animals have been researched as alternative gelatine sources, especially from the by-products of a particular industry. Interestingly, the functional qualities of marine gelatine can be comparable to mammalian gelatine to some extent [7]. However, fish gelatine generally has lower gel strength than commercial bovine and porcine gelatines [8]. To date, several studies have reported gelatine extracted from marine by-products, including clown featherback [8], blacktip shark [9], brown-banded bamboo shark [9], salmon [10], tuna [11], and squid [12]. In general, gelatine production is a time-consuming process depending on the type of raw material, type of chemicals, duration time used for the alkaline and acid pre-treatments, water extraction, and drying. However, the previous studies took a long time for acid extraction ranging from 6 to 12 h. No study has been used to shorten the time for acid extraction, which could minimise the duration time of gelatine production.

Apart from marine species, jellyfish, which main protein is collagen and consists of little fat and cholesterol [13-15], have gained much attention for gelatine extraction. Jellyfish-based foodstuff is generally known as delicacies in Asian countries [16] which has recently gained attention in European countries [17] due to its unique jelly, crunchy texture, and its richness in collagen protein [16, 18]. As a consequence, the demand and preference for jellyfish as a portion of food has increased worldwide. The fresh jellyfish has been preserved with salt chemicals by professional fishermen and jellyfish companies. Salted jellyfish by-products are generated from various stages of the dehydration process [19]. In general, most salted jellyfish by-products have been sold for feed meals at a low price [19]. Research on

* Corresponding author: benjawan.t@sci.kmutnb.ac.th

jellyfish gelatine has been reported with a lower yield and gel strength than commercial gelatines (bovine and fish gelatine) [13-15, 20]. In addition, the drawback of jellyfish gelatine is not being able to retain the gel structure at room temperature [15, 20]. Upon jellyfish gelatine research (Fig. 1), chemicals used for pre-treatment and the extraction condition of temperature and time [12, 13, 20] affecting gelatine quality and yield have been reported elsewhere. The jellyfish gelatine extracted with 0.1 M HCl at 60°C for 12 h showed a maximum gel strength of 323.74 g [20]. According to the previous result, during the gelatine extraction, factors of the acid pre-treatment step and hot water extraction for a particular time could influence collagen extractability [21]. Therefore, this research aimed to investigate the reduced time process of gelatine extraction by the varying conditions of HCl concentrations (0.1, 0.2, and 0.3 M) for only 2 h and hot water extract at 60°C for only 3 h on yield and gel quality of the resultant gelatine.

2 Materials and methods

2.1 Preparation of desalted jellyfish by-products

Salted jellyfish by-products (*Lobonema smithii*) were obtained from Mahachai Food and Trading Co., Ltd., Samut Sakhon, Thailand. The by-products were stored in polyethylene bags, transported to the lab at King Mongkut's University of Technology North Bangkok, Thailand, and kept refrigerated until use. For the experiment, the jellyfish material was prepared with slight modifications [19, 22]. First, the material was mechanically washed with water by a washing machine at a ratio of 1:10 (w/v) at 2 cycles. Each cycle was set for 15 min. The sample was then drained for 10 min and ready for the next step of gelatine production.

2.2 Preparation of jellyfish gelatine

Fig. 1 displays the production of jellyfish gelatine. The gelatine production was modified according to the method of Lueyot et al. [20]. In this study, the gelatine preparation consists of multi-steps, including alkaline pre-treatment, washing, acid pre-treatment, washing, water extraction, and drying. Firstly, 1 kg of the desalted jellyfish by-products was continuously stirred with 0.05 M NaOH solution in the ratio of 1:3 (w/v) at 4°C for 1 h using a shaking incubator (WIS-20R, WiseCube, Korea) at 150 rpm. The sample was washed with tap water until neutral pH (6.5-7.0). The sample was agitated in HCl solution in a ratio of 1:2 (w/v) at 25°C for 2 h at a speed of 150 rpm. Three different HCl conditions of 0.1, 0.2, and 0.3 M were compared at this step. The HCl solution was changed every 60 min. Next, the swollen sample was thoroughly washed with tap water until the sample had neutral pH. After that, the sample was extracted with distilled water at 60°C for 3 h with a sample to water ratio of 1:2 (w/v) in a temperature-controlled water bath shaker (Memmert, Schwabach, Germany) at a speed of 100 rpm. Then, the sample was filtered using a Buchner funnel with Whatman No. 4 filter paper. Finally, the sample was dried with a tray dryer (ED 400, Binder, U.S.A.) at 60°C for 3 days. The obtained jellyfish gelatine samples were calculated for yield and subjected to further analysis. Each experiment was performed in triplicate.

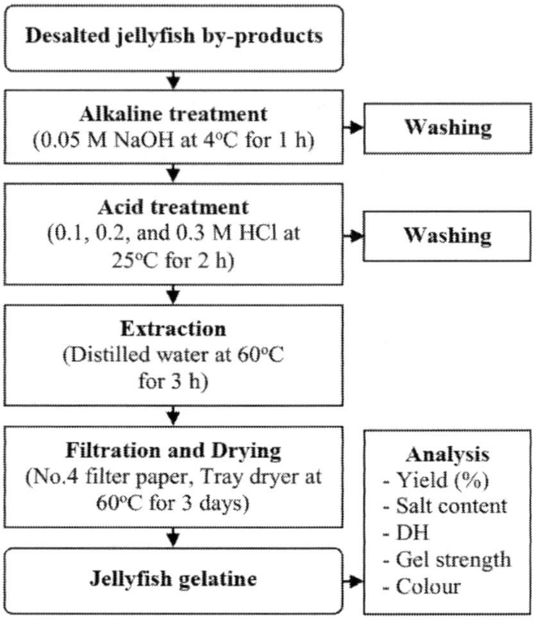

Fig. 1. Gelatine extraction from jellyfish.

2.3 Analysis

2.3.1 Chemical composition

The chemical composition of desalted jellyfish by-products was determined protein, ash, fat, and moisture content according to the AOAC standard methods [23]. The protein content was determined by the Kjeldahl method by using a factor of 5.55 to convert the nitrogen value to protein [23, 24]. The ash content, fat content, and moisture content were determined using the muffle furnace method, the Soxhlet extraction method, and the oven drying method, respectively [23].

2.3.2 Yield

The gelatine yield was calculated for all the treatments on a dry weight basis based using the following eq (1):

$$Yield\,(\%) = (M_2 / M_1) \times 100 \qquad (1)$$

where M_1 = Weight of the oven-dried salted jellyfish (g) and M_2 = Weight of the oven-dried gelatine (g)

2.3.3 Salt content

The salt (sodium chloride) content of jellyfish gelatine solution after filtration (200 mL) was measured with a conductivity meter (TDS Meter 308, Systronics, India). The calculation of the percentage of sodium chloride was extrapolated from the standard curve.

2.3.4 Degree of hydrolysis

The sample used for DH determination was the sample before drying, followed by the previous report [25]. The jellyfish gelatine solution after filtration (125 µL) was mixed with 1 mL of 0.01% TNBS solution and 2 mL of phosphate buffer (pH 8.2) and incubated at 50°C for 30 min in the dark using a temperature-controlled water bath (WNB 45, Memmert, Germany). The reaction was terminated by adding 2 mL of 0.1 M sodium sulphite and cooling at room temperature for 15 min. The absorbance was measured at 420 nm using a spectrophotometer (SP 830 plus, Metertech, Taiwan). The percentage of DH was calculated according to the following equation (2):

$$DH\,(\%) = [(L - L_b) / (L_{max} - L_b)] \times 100 \qquad (2)$$

where L = The α-amino groups content of hydrolysate sample, L_b = The α-amino groups content in the original substrate (blank), and L_{max} = The total α-amino groups the sample hydrolysed by 6 N HCl at 100°C for 24 h.

2.3.5 Determination of gel strength

The gelatine gel was prepared slightly modified according to the standard method [26]. The dried gelatine sample was dissolved in distilled water at 60°C and continuously stirred until gelatine was solubilised completely to obtain the final concentration of 6.67% (w/v). The gelatine sample (6.67%) was poured in a container with the size of 2.5 cm in height and 3 cm in diameter and then refrigerator at 4°C for 18 h before analysis. The gel strength was determined using a texture analyser (TA-XT2i, Stable Micro Systems, UK) equipped with a 1.27 cm diameter flat-faced cylindrical Teflon plunger (P/0.5R). The parameter of pre-test and post-test speeds were set at 1.5 and 1.0 mm/s, respectively. The maximum force was expressed to be defined as the gel strength (g). The gel strength of jellyfish gelatine was compared to commercial fish and bovine gelatine (McGarrett, JR F&B Co., Ltd, Thailand).

2.3.6 Colour measurement

The colour of jellyfish gelatine gel prepared at 6.67% (w/v) was evaluated by CIE L* a* b* system and measured using a colourimeter (ColorQUEST 45/0, HunterLab, USA). The results of the CIE L*a*b* colour parameters were displayed as L* (lightness 0-100), a* (redness (+)/greenness (-)), and b* (yellowness (+)/ blueness (-)). The colour of jellyfish gelatine was compared to commercial fish and bovine gelatine (McGarrett, JR F&B Co., Ltd, Thailand). The hue angle of the sample was calculated according to the following equation (3):

$$Hue\ angle = \tan^{-1}(b^* / a^*) \qquad (3)$$

2.4 Statistical analysis

All experiments were performed in triplicate, except for the gel strength and colour determination performed in quintuplicate. The data were subjected to analysis of variance (ANOVA) and Duncan's multiple range test for statistical significance ($p<0.05$) using the statistical package for social science (SPSS) 22.0.

3 Results and discussions

3.1 Quality of desalted jellyfish by-products

The desalted jellyfish by-products had protein, ash, fat, and moisture content of 8.68±0.19%, 1.04±0.23%, 0.98±0.08%, and 89.10±0.10%, respectively. These results were slightly different from other studies, which reported that the desalted jellyfish (*Lobonema smithii*) (washed 2 cycles) had protein, ash, fat, and moisture content of 4.79-5.55%, 1.19-1.39%, 0.91-0.95%, and 91.86-93.36%, respectively [19, 22]. Differences in proximate results might be due to differences in the seasonal catch or varying salting process [22]. Despite no determination of collagen in this study, previous reports showed that most of the jellyfish proteins were collagen protein [16, 27, 28].

3.2 Effect of pre-treatment with different HCl concentrations on yield, salt content, and DH

3.2.1 Yield

After pre-treatment with 0.1, 0.2, and 0.3 M HCl, the yields of jellyfish gelatine were 2.97%, 5.60%, and 6.34%, respectively, calculated on a dry weight basis (Table 1). The gelatine yield significantly increased with increased HCl concentration from 0.1 to 0.2 M but not with 0.3 M HCl. However, the jellyfish gelatine yields of this study showed lower values from the other works with different types and concentrations of acid used and extraction temperature and time. The highest yield of type A jellyfish gelatine was 39.47% (on a dry weight basis) when pre-treated with a sulfuric acid solution (pH 2) and extracted at 75°C for 12 h [15]. The tilapia skin gelatine pre-treated with 0.1 M NaOH and extracted at 55°C for 6 h showed the highest yield of 21.10% on a dry weight basis) [29]. The highest yield of blacktip shark and the brown-banded bamboo shark was 21.17-24.76% and 19.06–22.81% (based on wet weight), respectively when pre-treated with a 0.1 M NaOH, 1 M HCl, and 0.2 M acetic acid and extracted at 45, 60, and 75°C for 6 and 12 h [9]. Thus, compared to the yields from other studies, the shortened duration of HCl pre-treatment for only 2 h and hot water extraction for 3 h was insufficient for the jellyfish gelatine process due to the leftover jellyfish by-products greater than 90% remained.

Table 1. Yield, salt content, and degree of hydrolysis of jellyfish gelatine at different HCl concentrations pre-treatment.

Sample	Yield* (%)	Conductivity* (us/cm)	NaCl content* (%)	DH* (%)
0.1 M HCl	2.97±0.97^{b}	1,029.67±4.72^{a}	0.047±0.00^{a}	29.74±0.6^{c}
0.2 M HCl	5.60±1.01^{a}	996.67±5.50^{b}	0.046±0.00^{b}	33.63±0.45^{b}
0.3 M HCl	6.32±1.08^{a}	947.33±2.08^{c}	0.043±0.00^{c}	36.47±0.68^{a}

*Different superscript letters in the same column indicate a significant difference ($p<0.05$).

Table 2. Gel strength and colour of jellyfish gelatine at different HCl concentrations pre-treatment and commercial gelatines (bovine and fish gelatine).

Sample	Gel strength* (g)	Colour*			
		L*	a*	b*	h*
0.1 M HCl	223.61±3.72^{c}	4.43±0.61^{cd}	5.68±0.44^{a}	8.64±0.32^{a}	33.05±2.46^{a}
0.2 M HCl	214.47±4.12^{c}	5.70±0.93^{bd}	5.71±0.53^{a}	9.01±0.66^{a}	32.43±3.81^{a}
0.3 M HCl	118.89±0.48^{d}	7.24±1.18^{b}	5.94±0.02^{a}	9.19±0.85^{a}	32.98±2.31^{a}
Bovine	358.62±3.42^{a}	16.76±2.39^{a}	0.71±0.46^{c}	- 9.66±1.08^{b}	3.98±2.27^{b}
Fish	306.11±3.79^{b}	3.67±1.10^{d}	4.92±0.41^{b}	- 9.90±0.37^{b}	26.45±2.69^{c}

*Different superscript letters in the same column indicate a significant difference ($p<0.05$).

Fig. 2. Gelatine gel (A) jellyfish gelatine gel pre-treated at 0.1 M HCl, (B) jellyfish gelatine gel pre-treated at 0.2 M HCl, (C) jellyfish gelatine gel pre-treated at 0.3 M HCl, (D) commercial bovine gelatine gel, and (E) commercial fish gelatine gel.

3.2.2 Conductivity and salt content

The effect of electrolytes has directly influenced gelatine gel quality. For the jellyfish gelatine process, increased HCl concentration decreased conductivity and sodium chloride. The conductivity value and sodium chloride of jellyfish gelatine solutions were in the range of 947.33-1,029.67 μs/cm and 0.047-0.043%, after pre-treatment with 0.1, 0.2, and 0.3 M HCl (Table 1). Before washing, the sample used in this study was salted jellyfish that was cured by sodium chloride (NaCl), sodium bicarbonate ($NaHCO_3$), and potassium alum sulfate ($KAl(SO_4)_2^-$). Previous studies reported that the salt content, determined by the titration method, of the salted jellyfish before and after washing for two cycles was 14% and 1%, respectively, but elements of Na, Al, Si, P, S, Cl, K, Ca, Fe, and Br were also found in washed samples [19, 22]. The salt elements remained in the protein structure of desalted jellyfish via the binding of positively charged ions (e.g., Na^+, K^+, Al^{3+}, Ca^{2+}, Fe^{2+}) and negatively charged ions (e.g., Cl^-, Si^-, Br^-, PO_4^{3-}) to the negatively and positively charged amino acids in the jellyfish protein structure [22, 30]. Collagen in the salted jellyfish found 7 polar amino acids, including charged acidic amino acids (aspartic acid and glutamic acid), charged basic amino acids (arginine and lysine) and uncharged amino acids (serine, threonine, and tyrosine), and 10 nonpolar amino acids including alanine, cysteine, glycine, hydroxyproline, isoleucine, leucine, methionine, phenylalanine, proline, and valine [17]. All charged polar amino acids can be engaged in ionic bonds via electrostatic attraction, and all uncharged amino acids can be formed hydrogen bonds. In this study, during the gelatine process, it was noticed that the swollen jellyfish protein structure increased as the HCl concentration increased. As a result, in the system, the increases of H^+ ion and Cl^- ion might affect the charged and uncharged amino acids, causing the unfolded triple helix structure of collagen and increased binding water molecules to denatured jellyfish collagen [30]. Thus, the excess Cl^- ion might be reformed into NaCl, thereby reducing the salt content from the swollen collagen structure. In addition, the salt was diluted during the washing step. The effect of salt on fish gelatine showed that 1.5%

NaCl addition caused a loss in molecular order in secondary structure, affecting low gel strength [31]. Therefore, it is necessary to remove the excessive salt content before the drying step that might be interfering with the gel property.

3.2.3 DH

The DH of jellyfish gelatine solutions increased significantly with increasing HCl concentration. The results showed that DH of jellyfish gelatine was in the range of 29.74-36.47% (Table 1). The highest DH of jellyfish gelatine was pre-treated with 0.3 M HCl, which was lower than other reports that used 1 and 6 N HCl for 12 and 24 h and heated at 95°C, yielding high %DH of 73.70 and 75.40 [32]. The different chemicals affected DH. When pre-treated with a sulfuric acid solution (pH 2) and extracted at 60 and 75°C for 6 and 12 h, DH of jellyfish gelatine values were 65.30-77.13% [15]. During HCl treatment, the triple helix structure of jellyfish collagen was denatured with a higher HCl concentration by breaking down the hydrogen and covalent bonds [33], resulting in increased helix-to-coil transition and converted to soluble gelatine. Thus, the condition used in this study was relatively mild, compared to other studies [14, 15], and the shorten acid pre-treatment and water extraction. This condition might not be applicable for producing jellyfish gelatine when the desalted jellyfish was used as a raw material.

3.3 Gel strength of gelatine gel

Gelatine gel is a biopolymer network that is reversibly crosslinked and primarily stabilised by hydrogen-bonded junction zones [34]. Gel strength affects the texture quality in food products [14]. The gel strength of 6.67% jellyfish gelatine gel pre-treated at different HCl concentrations is shown in Table 2. The jellyfish gelatine decreased gel strength values from 223.61, 214.47, and 118.89 g when increased HCl concentration from 0.1, 0.2, and 0.3 M HCl. The gel strength of jellyfish gelatine was pre-treated with 0.1 M HCl, which was significantly different from those with 0.2 and 0.3 M HCl, while gel strength of jellyfish gelatine was pre-treated at 0.2 and 0.3 M HCl did not show significantly different. The gel strength values of jellyfish were lower than those of bovine gelatine (358.62 g) and fish gelatine (306.11 g). Previous studies reported that jellyfish gelatine's gel strength was inferior to bovine and fish gelatine due to differences in collagen type, amino acid composition, the alpha and beta chain ratio in the gelatine, and the gelatine peptides of the alpha chain [20]. However, the gel strength of jellyfish gelatine gel, in this study, was higher than the gel strength values of Cuttlefish (*Sepia Officinalis*) skins (198 g) [35] and squid (*Loligo formosana*) skins (156 g) [4]. Compared to the same jellyfish sample, the jellyfish gel strength reported in this study was higher than the previous work that showed the gel strength of 18-118 g [14, 15]. Recently, the highest gel strength of jellyfish (*Lobonema smithii*) gelatine was 323.74 g [20] which applied a higher extraction time of 12 h. As a result, the concentration of HCl pre-treatment, pre-treatment time, and water extraction time directly affected the gel strength. Therefore, in this study, applying a mild concentration of HCl (0.1 M) for 2 h and shortening hot water extraction time for only 3 h might not be suitable for the jellyfish gelatine process. Despite low gel strength, jellyfish gelatine might be used in frozen and refrigerated food products and used for crystal growth modification.

3.4 Colour of gelatine gel

The colour of jellyfish gelatine gel (6.67%) pre-treated at different HCl concentrations, commercial bovine gelatine gel, and commercial fish gelatine gel are presented in Table 2. The colour of gelatine gel was expressed as lightness (L*), redness (a*), yellowness (b*), and hue angle (h*). The lightness, redness, and yellowness value of jellyfish gelatine gel increased significantly with increasing HCl concentration. The change of redness and yellowness value related to hue angle (h*) of gelatine gel. The colour of jellyfish gelatine showed light to dark brown (tata. 2) with no differences in hue values that were 33.05±2.46, 32.43±3.81, and 32.98±2.31 after pre-treatment with 0.1, 0.2, and 0.3 M HCl, respectively. The colour of commercial bovine gelatine gel and commercial fish gelatine gel showed a transparent gel (Fig. 2) which the hue value was 3.98±2.27 and 26.45±2.69.

The intense brown of jellyfish gelatine gel could be due to a non-enzymatic browning reaction (Maillard reaction), in which reducing sugars, possibly ribose and deoxyribose that typically found in RNA and DNA [36], react with a free amino acid at the appropriate conditions of temperature during jellyfish gelatine drying [37]. Indeed, the drying condition for 3 days at 60°C might be too long, and no bleaching step was performed during jellyfish gelatine extraction. Then, the effect of the bleaching agent on jellyfish gelatine could be further studied.

4 Conclusion

Jellyfish by-products can be a source of gelatine production which reduced food waste and save time. The stepwise pre-treatment, including sodium hydroxide, hydrochloric acid, and hot water extraction, directly impacts gelatine qualities. The jellyfish treated with 0.1 M HCl showed the highest gel strength of 223.61 g but showed the lowest gelatine yield of 2.97%. The colour of jellyfish gelatine gel was light to intense brown colour compared to clear gelatine gel from fish and bovine gelatine. Therefore, improved jellyfish gelatine properties, including yield, colour, and gel strength, are still needed for further study.

Acknowledgement

This research has been financially supported by the KMUTNB-NSTDA High-Performance Post Graduate Scholarship in Science and Technology (Contract no. B.021-2563) and the

National Research Council of Thailand through the NRCT Senior Research Scholar Program (Contract no. 814-2020). The authors would like to thank Mahachai Food and Trading Co., Ltd., Samut Sakhon, Thailand, for their kind support of salted jellyfish.

References

1. S. Sultana, M.E. Ail, M.N.U. Ahamad, in *Preparation and Processing of Religious and Cultural Foods*, edited by M.E. Ail, N.N.A. Nizar (Woodhead Publishing, Massachusetts, 2018), pp. 215-239
2. L.M. Kasankala, Y. Xue, Y. Weilong, S.D. Hong, Q. He, Optimization of gelatin extraction from grass carp (*Catenopharyngodon idella*) fish skin by response surface methodology, Bioresource Technology, **98**, 17 (2007): 3338–3343
3. P. Kittiphattanabawon, S. Benjakul, W. Visessanguan, F. Shahidi, Cryoprotective effect of gelatin hydrolysate from blacktip shark skin on surimi subjected to different freeze-thaw cycles, Lebensmittel-Wissenschaft & Technologie,**4**, 2 (2012): 437-442
4. A. Hamzeh, S. Benjakul, T. Sae-Leaw, S. Sinthusamran, Effect of drying methods on gelatin from splendid squid (*Loligo formosana*) skins, Food Bioscience, **26** (2018): 96-103
5. J. Tkaczewska, M. Morawska, P. Kulawik, M. Zając, Characterization of carp (*Cyprinus carpio*) skin gelatin extracted using different pre-treatments method, Food Hydrocolloids, **81** (2018): 169-179
6. A. Aberoumand, Estimation of microbiological variations in minced lean fish products, World Journal of Fish and Marine Sciences, **2**, 3 (2010): 204-207
7. T. Huang, Z.C. Tu, H. Wang, W. Liu, L. Zhang, Y. Zhang, X.C. ShangGuan, Comparison of rheological behaviors and nanostructure of bighead carp scales gelatin modified by different modification methods, Journal of Food Science and Technology, **54**, 5 (2017): 1256–1265
8. P. Kittiphattanabawon, S. Benjakul, S. Sinthusamran, H. Kishimura, Gelatin from clown featherback skin: extraction conditions, Lebensmittel-Wissenschaft & Technologie, **66** (2016): 186-192
9. P. Kittiphattanabawon, S. Benjakul, W. Visessanguan, F. Shahidi, Comparative study on characteristics of gelatin from the skins of brownbanded bamboo shark and blacktip shark as affected by extraction conditions, Food Hydrocolloids, **24**, 2-3 (2010): 164-171
10. H. Fan, M.J. Dumont, B.K. Simpson, Extraction of gelatin from salmon (*Salmo salar*) fish skin using trypsin-aided process: optimization by Plackett–Burman and response surface methodological approaches, Journal of Food Science and Technology, **54**, 12 (2017): 4000-4008
11. E.T.A. Tümerkan, Ü. Cansu, G. Boran, J.M. Regenstein, F. Özoğul, Physiochemical and functional properties of gelatin obtained from tuna, frog and chicken skins, Food Chemistry, **287** (2019): 273-279
12. B.E. Abdelmalek, J. Gómez-Estaca, A. Sila, O. Martinez-Alvarez, M.C. Gómez-Guillén, S. Chaabouni-Ellouz, M.A. Ayadi, A. Bougatef, Characteristics and functional properties of gelatin extracted from squid (*Loligo vulgaris*) skin, Lebensmittel-Wissenschaft & Technologie, **65** (2016): 924-931
13. S. Cho, J.R. Ahn, J.S. Koo, S.B. Kim, Physicochemical properties of gelatin from jellyfish *Rhopilema hispidum*, Fisheries and Aquatic Sciences, **17**, 3 (2014): 299-304
14. U. Rodsuwan, B. Thumthanaruk, O. Kerdchoechuen, N. Laohakunjit, Functional properties of type A gelatin from jellyfish (*Lobonema smithii*), International Food Research Journal, **23**, 2 (2016): 507-514
15. P. Chancharern, N. Laohakunjit, O. Kerdchoechuen, B. Thumthanaruk, Extraction of type A and type B gelatin from jellyfish (*Lobonema smithii*), International Food Research Journal, **23**, 1 (2016): 419-424
16. Y.H.P. Hsieh, F.M. Leong, J. Rudloe, Jellyfish as food, Hydrobiologia, **451**, 1-3 (2001): 11-17
17. L. Torri, F. Tuccillo, S. Bonelli, S. Piraino, A. Leone, The attitudes of Italian consumers towards jellyfish as novel food, Food Quality and Preference, **79** (2020): 103782
18. M.T. Pedersen, T.A. Vilgis, Soft matter physics meets the culinary arts: From polymers to jellyfish, International Journal of Gastronomy and Food Science, **16** (2019): 100135
19. P. Muangrod, V. Rungsardthong, S. Vatanyoopaisarn, Y. Tamaki, E. Kuraya, B. Thumthanaruk, Effect of wash cycle on physical and chemical properties of rehydrated jellyfish by-products and jellyfish protein powder, Science, Engineering and Health Studies, **15** (2021): 21030004
20. A. Lueyot, V. Rungsardthong, S. Vatanyoopaisarn, P. Hutangura, B. Wonganu, P. Wongsa-Ngasri, S. Charoenlappani, S. Roytrakul, B. Thumthanaruk, Influence of collagen and some proteins on gel properties of jellyfish gelatin, PLOS One, **16**, 6 (2021): e0253254
21. M.C. Gómez-Guillén, J. Turnay, M.D. Fernández-Díaz, N. Ulmo, M.A. Lizarbe, P. Montero, Structural and physical properties of gelatin extracted from different marine species: a comparative study, Food Hydrocolloids, **16**, 1 (2002): 25-34
22. W. Charoenchokpanich, V. Rungsardthong, S. Vatanyoopaisarn, B. Thumthanaruk, Y. Tamaki, Salt reduction in salted jellyfish (*Lobonema smithii*) using a mechanical washing machine, Science,

Engineering and Health Studies, **14**, 3 (2020): 184-192

23. AOAC, *Official Methods of Analysis of AOAC International* (AOAC International, Maryland, USA, 2000)
24. C.S. James, *Analytical Chemistry of Foods* (Springer, London, 1995)
25. S. Benjakul, M.T. Morrissey, Protein hydrolysates from Pacific whiting solid wastes, Journal of Agricultural and Food Chemistry, **45**, 9 (1997): 3423-3430
26. GMIA, *Standard methods for the sampling and testing of gelatins* (Gelatin Manufacturers of America, Inc., New York, 1986)
27. T. Klaiwong, P. Hutangura, S. Rutatip, P. Wongsa-Ngasri, B. Thumthanaruk, Comparative properties of pepsin hydrolyzed jellyfish protein from salted jellyfish, Journal of Agricultural Science and Technology B, **4**, 7 (2014): 555-564
28. M. Omori, E. Nakano, Jellyfish fisheries in Southeast Asia, Hydrobiologia, **451**, 1-3 (2001): 19-26
29. T. Zhang, R. Sun, M. Ding, L. Tao, L. Liu, N. Tao, X. Wang, J. Zhong, Effect of extraction methods on the structural characteristics, functional properties, and emulsion stabilization ability of *Tilapia* skin gelatins, Food Chemistry, **328** (2020): 127114
30. H.D. Belitz, W. Grosch, P. Schieberle, Food Chemistry (Springer-Verlag Berlin Heidelberg, Berlin, 2009)
31. L.C. Sow, H. Yang, Effects of salt and sugar addition on the physicochemical properties and nanostructure of fish gelatin, Food Hydrocolloids, **45** (2015): 72-82
32. S. Silaprueng, B. Thumthanaruk, P. Wongsa-ngasri, Comparative functional properties of jellyfish (*Lobonema smithii*) protein hydrolysate as influenced by bromelain and hydrochloric acid, Journal of Food Science and Agricultural Technology, **1**, 1 (2015): 171-176
33. B. Wonganu, Application of gelatin derived from waste tilapia scales to an antibiotic hydrogel pad, E3S Web of Conferences, **141** (2020): 03004
34. S. Benjakul, P. Kittiphattanabawon, J.M. Regenstein, in *Food Biochemistry and Food Processing*, edited by B.K. Simpson, L.M. Nollet, F. Toldrá, S. Benjakul, G. Paliyath, Y.H. Hui (John Wiley and Sons, Inc, Iowa, 2012), pp. 388-405
35. M. Jridi, R. Nasri, I. Lassoued, N. Souissi, A. Mbarek, A. Barkia, M. Nasri, Chemical and biophysical properties of gelatins extracted from alkali-pretreated skin of cuttlefish (*Sepia officinalis*) using pepsin, Food Research International, 54, 2 (2013): 1680-1687
36. R.E. Farrell, *RNA* Methodologies: A Laboratory Guide for Isolation and Characterization (Academic Press, San Diego, 1993)
37. H. Kchaou, N. Benbettaied, M. Jridi, M. Nasri, F. Debeaufort, Influence of Maillard reaction and temperature on functional, structure and bioactive properties of fish gelatin films, Food Hydrocolloids, **97** (2019): 105196

Effect of pepsin hydrolysis on antioxidant activity of jellyfish protein hydrolysate

Pratchaya Muangrod[1], *Wiriya* Charoenchokpanich[1], *Vilai* Rungsardthong[1], *Savitri* Vatanyoopaisarn[1], *Benjamaporn* Wonganu[2], *Sittiruk* Roytrakul[3] and *Benjawan* Thumthanaruk[1,*]

[1]Department of Agro-Industrial Food and Environmental Technology, Faculty of Applied Science, King Mongkut's University of Technology North Bangkok, Bangkok, Thailand
[2]Department of Biotechnology, Faculty of Applied Science, King Mongkut's University of Technology North Bangkok, Bangkok, Thailand
[3]National Center for Genetic Engineering and Biotechnology, National Science and Technology Development Agency (NSTDA), Pathum Thani, Thailand

Abstract. Edible jellyfish have been consumed as food for more than a century with offering high protein and crunchy texture. The pepsin hydrolysis of jellyfish protein yields jellyfish protein hydrolysate (ep-JPH), reported for potential bioactivities such as antioxidant activity or antihypertensive activities. Due to the substantial number of by-products generated from jellyfish processing, the by-products were then selected as a raw material of JPH production. This research aimed to evaluate the effect of the hydrolysis time of pepsin on the antioxidant activity of ep-JPH. The dried desalted jellyfish by-products powder was enzymatically hydrolysed by 5% (w/w) pepsin, and the hydrolysis time was varied from 6, 12, 18, and 24 h at 37°C. Results showed that increased hydrolysis time increased the degree of hydrolysis (DH) and inhibition of DPPH radical. The 24 h ep-JPH possessed the highest DH and the highest inhibitory effect of DPPH radical. The results demonstrated that, in this experiment, all ep-JPHs were DPPH radical scavengers, exhibiting different inhibition activities depending on DH values.

Keyword. Salted jellyfish by-products, Protein hydrolysate, DPPH, Degree of hydrolysis

1 Introduction

Protein hydrolysates are products of protein degradation that yield various sizes of peptides and free amino acids [1, 2]. There are several methods of protein hydrolysate production, including acids or alkali hydrolysis and enzymatic hydrolysis [1-3]. Different methods of protein hydrolysate production result in different bioactivity of protein hydrolysate due to its amino acid composition and sequence [1, 4, 5].

Enzymatic hydrolysis is one of the practical protein hydrolysate productions via protease enzymes that cleave specific peptide bonds and provide short peptides with small molecular weights [1]. Most low molecular weight enzymatic hydrolysate peptides containing approximately 2-20 amino acid residues are considered bioactive peptides [5, 6]. Commercial proteases are practically used for protein hydrolyses, such as trypsin, pepsin, collagenase, alcalase, papain, and bromelain [7, 8]. Different types of protease enzymes also provided different bioactive peptides.

To date, the different sources of food proteins for producing bioactive peptides are meat [9, 10], milk [11-13], eggs [14, 15], nuts [16], cocoa bean [17], beef [18], pork [18], chicken [18], turkey [18], and fish [19]. Factors of type of protease enzymes, hydrolysis time, and raw material factors used for peptide production yield different amino acid compositions and amino acid sequences [20-23], which offer different bioactive activities such as antioxidant, anticancer, antimicrobial, and antihypertensive [3, 24, 25]. Antioxidant activity, the ability to reduce or scavenge free radicals, is one of the major biological activities of protein hydrolysates that have mostly been reported [3]. If the number of free radicals exceeds normal, free radicals can cause diseases such as arthritis, diabetes, atherosclerosis, ageing, and cancer [8, 26]. Then, the novel protein sources of bioactive peptides having antioxidant activities are currently of interest for reducing the risk of various diseases [26]. The marine source has become attractive with various organisms that have potentially been reported for biological activities. Edible salted jellyfish, the oldest marine plankton consumed for more than a century by Chinese people, is selected to produce jellyfish protein hydrolysate (JPH). Apart from water content in the body, jellyfish are considered high protein food, mainly collagen [27, 28]. Few jellyfish species used for producing JPH have been reported for antioxidant activities [6, 8, 24]. No antioxidant activity of protein hydrolysate from jellyfish (*Lobonema smithii*) hydrolysed with pepsin has not been studied. Therefore, the by-products of salted jellyfish from salted jellyfish processing were selected for producing hydrolysate by pepsin hydrolysis (ep-JPH). The research aimed to

* Corresponding author: benjawan.t@sci.kmutnb.ac.th

evaluate the effect of the hydrolysis time of pepsin on the antioxidant activity of JPH.

2 Materials and methods

2.1 Preparation of salted jellyfish by-products

The broken pieces or irregular shapes of jellyfish (*Lobonema smithii*) sold as by-products were kindly received from Chockdee Sea Products Co., Ltd., Samut Songkhram, Thailand. Due to the high salt in jellyfish by-products, washing and drying were used to prepare the raw material as previously described [27, 29]. First, the washing of salted jellyfish by-products with tap water at a ratio of 1:40 (w/v) was performed by a rotational jellyfish washing machine for 2 cycles, and 15 min per cycle and drained for 30 min. The desalted jellyfish by-products were dried in a tray dryer (ED 400, Binder, USA) at 60°C for 24 h and then ground into fine particles and filtered through a sieve of size 100 mesh. The jellyfish protein powder (JPP) was analyzed for the chemical compositions (moisture, protein, fat, and ash content) before use (Fig. 1). The powder was packed in sealed PE bags and kept at room temperature until use.

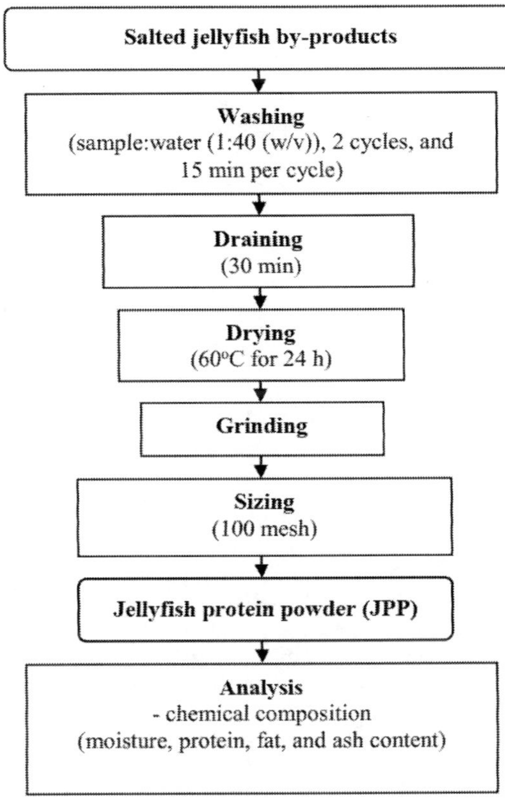

Fig. 1. Preparation of salted jellyfish by-products (JPP).

2.2 Preparation of jellyfish protein hydrolysate (JPH)

JPH was prepared as previously described [30, 31] with a slight modification (Fig. 2). Briefly, JPP was mixed in 0.05 M sodium acetate buffer at the ratio of 1: 25 w/v (4g/100mL) at 95°C for 10 min to inactivate the endogenous enzymes and then equilibrated to the optimum temperature conditions. Next, the sample was dissolved in 0.05 M sodium acetate buffer at the ratio of enzyme: substrate (1: 20 g: g; w/w). Then, the mixture was hydrolysed by pepsin at 37°C for 6, 12, 18, and 24 h using a shaking incubator (WIS-20R, WiseCube, Korea) at 150 rpm. After that, the enzyme was inactivated by heating at 95°C for 10 min in a temperature-controlled water bath shaker (Memmert, Schwabach, Germany) at 100 rpm. Next, the hydrolysate was centrifuged at 9,500 × g for 10 min. Finally, the supernatant of ep-JPH was filtered by Whatman filter paper No1. The samples were stored at −18°C until further analysis.

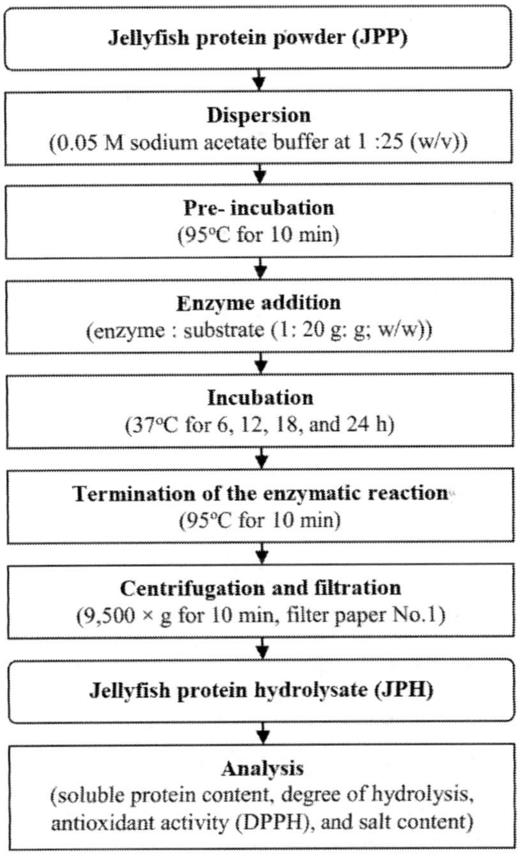

Fig. 2. Preparation of jellyfish protein hydrolysate (JPH).

2.3 Analysis

2.3.1 Chemical composition

The proximate composition of JPP including, moisture, protein, fat, and ash content, was measured according to the AOAC standard methods [32]. The protein content was measured using the Kjeldahl method. A factor of 5.55 was used to convert the nitrogen value to protein [33].

2.3.2 Soluble protein content

The soluble protein content of JPH was determined according to the Lowry method [34]. Bovine serum albumin was used as a standard.

2.3.3 Degree of hydrolysis

The analysis of DH was slightly modified by the method of trinitrobenzyl sulfonic acid (TNBS) [35]. Briefly, a 125 μL of jellyfish protein hydrolysate was mixed with 2 mL of phosphate buffer (pH 8.2) and 1 mL of 0.01% TNBS reagent. Then, the mixture solution was incubated at 50°C for 30 min in the dark using a temperature-controlled water bath (WNB 45, Memmert, Germany). The reaction was stopped by adding 2 mL of 0.1 M sodium sulfite and incubated at room temperature for 15 min. The absorbance was measured at 420 nm using a spectrophotometer (SP 830 plus, Metertech, Taiwan). The DH was calculated according to the following equation (1):

$$DH\,(\%) = [(L - L_0) / (L_{max} - L_0)] \times 100 \qquad (1)$$

where L = The amount of α-amino groups of hydrolysate sample, L_0 = The amount of α-amino groups in the original substrate (blank), and L_{max} = The total α-amino groups in the sample hydrolysed by 6 N HCl at 100°C for 24h.

2.3.4 Antioxidant activity (DPPH radical scavenging assay)

DPPH radical scavenging activity was analyzed as previously described [26, 36] with slight modifications. A 0.1 mL of JPH was mixed with 1.9 mL of 0.15 mmol/L DPPH in 70% ethanol. The mixture was then incubated in the dark for 30 min at room temperature. The absorbance was measured at 517 nm using a spectrophotometer (SP 830 plus, Metertech, Taiwan). In this measurement, ascorbic acid was a positive control. The inhibition of DPPH radical scavenging activity was calculated by the following equation (2):

$$\%inhibition = [(A_{control} - A_{sample}) / A_{control}] \times 100 \qquad (2)$$

where $A_{control}$ is the absorbance of the control that contained all reagents except the test samples. A_{sample} is the absorbance of the JPH with reagents added.

2.3.5 Salt content

The salt content of jellyfish protein hydrolysate referred to the percentage of sodium chloride (NaCl) in JPH was determined using a conductivity meter (TDS Meter 308, Systronics, India). The percentage of sodium chloride in JPH was predicted from the standard curve.

2.4 Statistical analysis

All experiments were performed in triplicate. Results were shown as mean±standard deviations. The data were subjected to analysis of variance (ANOVA), and Duncan's multiple range test was used to measure the significant difference ($p<0.05$) using the statistical package for social science (SPSS) 22.0.

3 Results and discussions

3.1 Quality of jellyfish protein powder

The jellyfish protein powder (JPP) had moisture, protein, fat, and ash content of 7.42±0.20%, 76.59±0.11%, 1.32±0.05%, and 15.54±0.36%, respectively. Results of proximate composition of JPP were similar to the previous study that JPP had 7.69%, 76.41%, 1.35%, and 15.76%, respectively [27], but slightly different from the work of Emrerk et al. (2021) [37], who reported that JPP had 5.27%, 68.8%, 3.18%, and 9.11%, respectively. Therefore, according to the high protein content, the JPP was selected for producing jellyfish protein hydrolysate.

3.2 Effect of hydrolysis time of pepsin on soluble protein content, degree of hydrolysis, antioxidant activity (DPPH), and salt content of jellyfish protein hydrolysate (JPH)

Pepsin is an aspartic protease that has an optimum pH of 1.8. Pepsin hydrolyzes peptide bonds between large hydrophobic amino acid residues and yields peptides. Table 1 shows the soluble protein content of enzymatic pepsin hydrolysed JPH (ep-JPH) for different hydrolysis times (6, 12, 18, and 24 h) at 37°C. Increased hydrolysis time of pepsin increased soluble protein content of ep-JPH. The soluble protein contents of ep-JPH derived from 6, 12, 18, and 24 h were 1,428.47, 1,461.25, 1,491.80, and 1,589.30 mg/L, respectively. During the hydrolysis process, jellyfish collagen protein was denatured by pepsin and acidic conditions. As a result, the jellyfish protein was hydrolysed, thus generating the soluble proteins.

DH is a parameter used for determining how well the pepsin hydrolysed jellyfish protein. In this experiment, increased hydrolysis time of pepsin increased the values of DH. The hydrolysis condition of ep-JPH for 6, 12, 18, and 24 h showed the DH values of 52.20%, 54.01%, 54.50%, and 55.41% (Table 1). During the pepsin hydrolysis, the peptide bonds of a jellyfish protein were hydrolysed and provided a short-chain peptide and free amino acid [38], depending on hydrolysis factors, including pepsin concentration, hydrolysis time, and temperature. In this study, the pepsin hydrolysed jellyfish sample at the hydrolysis time of 12, 18, 24 h showed similar DH values, and these values were not significantly different. The DH results were similar to the previous studies that hydrolysed jellyfish protein with bromelain at 50°C for 18 h and trypsin at 50°C for 9 h [8, 37]. Thus, the soluble protein and DH values are affected by pepsin hydrolysis. In this study, the pepsin hydrolysis for 24 h at 37°C was sufficient for producing ep-JPH.

Table 1. Soluble protein content and degree of hydrolysis of enzymatic pepsin-jellyfish protein hydrolysate (ep-JPH) at different hydrolysis times.

Sample	Soluble protein content* (mg/mL)	Degree of hydrolysis* (%)
ep-JPH 6 h	1,428.47±21.75^{c}	52.20±0.53^{d}
ep-JPH 12 h	1,461.25±12.27^{bc}	54.01±0.07^{c}
ep-JPH 18 h	1,491.80±23.48^{b}	54.50±0.10^{b}
ep-JPH 24 h	1,589.30±51.14^{a}	55.41±0.10^{a}

*Different superscripts (a, b, c, d) in the same column mean significant difference in value (p<0.05)

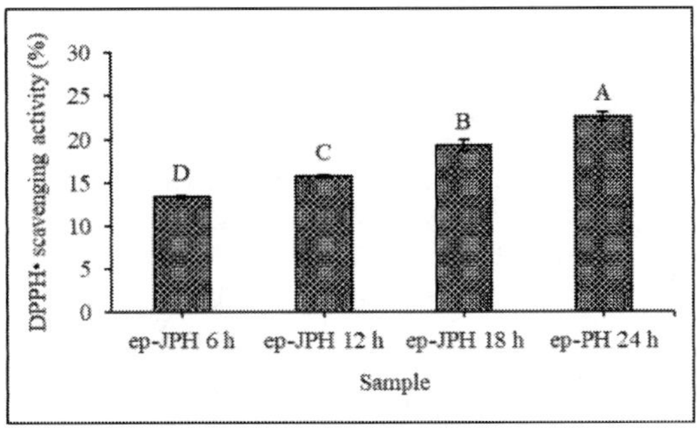

Fig. 3. DPPH radical scavenging activity (%) of jellyfish protein hydrolysate (JPH) at different duration times of hydrolysis (6, 12, 18, and 24 h); bars with the different superscripts were significantly different (p<0.05).

DPPH is an organic nitrogen radical that shows maximal absorbance at 517 nm in ethanol when DPPH encounters a protein donating substance that is an indicator of the antioxidant capacity of the sample [6, 39]. Results showed increased pepsin hydrolysis time of JPH increased % inhibition of DPPH radical scavenging activity. The inhibition values of DPPH radical scavenging activity by ep-JPHs obtained from 6, 12, 18, and 24 h of hydrolysis were 13.38±0.07%, 15.73±0.13%, 19.31±0.66%, and 22.56±0.51%, respectively (Fig. 3). Results of DPPH radical scavenging activity of ep-JPH were slightly different from that value of trypsin hydrolysed ribbon jellyfish (*Chrysaora* sp.) for 7 h, approximately 25% [8]. The factors that affected the antioxidant activity of protein hydrolysate are highly dependent on hydrolysis condition (temperature and duration time), sequence and amino acid composition, and protease enzyme (types and concentration) [6]. Thus, in this study, all ep-JPHs were able to scavenge DPPH radicals, and the 24 h ep-JPH showed the highest inhibition effect on DPPH radicals.

Results of DPPH radical scavenging activity were correlated with the results of DH. The small peptides generated by pepsin hydrolysis may donate electrons to DPPH radical, thereby reducing free radical action [6]. Thus, the highest DH value showed the highest inhibition of the DPPH radical. Previous research has reported that short-chain peptides containing 2 to 20 amino acids may positively impact health [17]. Therefore, the amino acid sequence of ep-JPHs is needed for further analysis.

The conductivity values measuring the salt ion content of ep-JPH increased as pepsin hydrolysis time increased, as shown in Table 2. The conductivity values of ep-JPH at different hydrolysis times were 1,853.33, 1,894.67, 1,960.67, and 1,997.00 µs/cm, respectively. All JPH samples presented a salt content of approximately 0.09%. Results also related to the previous work that the salt content of enzymatic bromelain-JPH at 50°C for 6-18 h were 1.50-2.22% [37]. Despite desalted jellyfish used, the salt content that remained in the solution must be removed to prevent the interferences of electrophoresis or LC-MS/MS analysis in the future.

Table 2. Salt content of enzymatic pepsin-jellyfish protein hydrolysate (ep-JPH) at different hydrolysis times.

Sample	Conductivity* (µs/cm)	Salt content* (%)
ep-JPH 6 h	1,853.33±1.52^{d}	0.090±0.00^{d}
ep-JPH 12 h	1,894.66±1.52^{c}	0.092±0.00^{c}
ep-JPH 18 h	1,960.66±7.02^{b}	0.096±0.00^{b}
ep-JPH 24 h	1,997.00±2.64^{a}	0.098±0.00^{a}

*Different superscripts (a, b, c, d) in the same column mean significant difference in value (p<0.05)

4 Conclusion

The 24 h ep-JPH at 37°C showed the highest soluble protein content, %DH, and %inhibition of DPPH radical. All ep-JPHs were able to scavenge DPPH radicals. Thus, jellyfish protein hydrolysate may be used in functional food applications due to its antioxidant ability. However, %inhibition of DPPH radical of ep-JPH was lower when compared with other studies. Therefore, further studies on ep-JPH conditions, such as an increase enzyme concentration, different hydrolysis times, and temperature are needed to offer the higher bioactivity of ep-JPH. The results of this study lead to an added value of salted jellyfish by-products and reduce waste from the salted jellyfish industry.

Acknowledgement

This research was financially supported by the KMUTNB-NSTDA High Performance Post Graduate Scholarship in Science and Technology (Contract no. B.020-2563) and the National Research Council of Thailand through the NRCT Senior Research Scholar Program (Contract no. 814-2020). In addition, the authors would like to thank Ms.Wanlee Muensawat, managing director of Chockdee Sea Products Co.,

Ltd., Samut Songkhram, Thailand, for supporting salted jellyfish and the jellyfish washing machine used in this study.

References

1. S. Severin, W.S. Xia, Enzymatic hydrolysis of whey proteins by two different proteases and their effect on the functional properties of resulting protein hydrolysates, Journal of Food Biochemistry, **30**, 1 (2006): 77-97
2. T. Julmanlik, S. Kongruang, Functional properties and applications of egg white protein hydrolysates, Journal of Food Technology, Siam University, **14**, 1 (2019): 69-87
3. K.M.I. Bashir, J.H. Sohn, J.S. Kim, J.S. Choi, Identification and characterization of novel antioxidant peptides from mackerel (*Scomber japonicus*) muscle protein hydrolysates, Food Chemistry, **323** (2020): 126809
4. P. Ambigaipalan, F. Shahidi, Bioactive peptides from shrimp shell processing discards: Antioxidant and biological activities, Journal of Functional Foods, **34** (2017): 7-17
5. N. Ktari, R.B.S.B. Salem, I. Bkhairia, S.B. Slima, R. Nasri, R.B. Salah, M. Nasri, Functional properties and biological activities of peptides from zebra blenny protein hydrolysates fractionated using ultrafiltration, Food Bioscience, **34** (2020): 100539
6. S. Silaprueng, B. Thumthanaruk, P. Wongsa-ngasri, Comparative functional properties of jellyfish (*Lobonema smithii*) protein hydrolysate as influenced by bromelain and hydrochloric acid, Journal of Food Science and Agricultural Technology, **1** (2015): 171-176
7. G. Guillén, M.E. López Caballero, A. Alemán, A.L.D. Lacey, B. Giménez, P. Montero García, in *Sea by-Products as Real Material: New Ways of Application*, edited by E.L. Bihan (Transworld Research Network, Trivandrum, 2010) pp. 89-115
8. Z. Barzideh, A.A. Latiff, C.Y. Gan, M. Abedin, A.K. Alias, ACE inhibitory and antioxidant activities of collagen hydrolysates from the ribbon jellyfish (*Chrysaora* sp.). Food Technology and Biotechnology, **52**, 4 (2014): 495-504
9. L. Mora, M. Reig, F. Toldrá, Bioactive peptides generated from meat industry by-products, Food Research International, **65** (2014): 344-349
10. T. Sayd, C. Dufour, C. Chambon, C. Buffière, D. Remond, V. Sante-Lhoutellier, Combined in vivo and in silico approaches for predicting the release of bioactive peptides from meat digestion, Food Chemistry, **249** (2018): 111-118
11. F. Tonolo, A. Folda, L. Cesaro, V. Scalcon, O. Marin, S. Ferro, A. Bindoli, M.P. Rigobello, Milk-derived bioactive peptides exhibit antioxidant activity through the Keap1-Nrf2 signaling pathway, Journal of Functional Foods, **64** (2020): 103696
12. M. Amorim, J.O. Pereira, L.B. Silva, R.C.S.C. Ormenese, M.T.B. Pacheco, M. Pintado, M., Use of whey peptide fraction in coated cashew nut as functional ingredient and salt replacer, Lebensmittel-Wissenschaft & Technologie, **92** (2018): 204-211
13. C. Zhao, T.J. Ashaolu, Bioactivity and safety of whey peptides, Lebensmittel-Wissenschaft & Technologie, **134** (2020): 109935
14. J. Zheng, T. Bu, L. Liu, G. He, S. Li, J. Wu, Naturally occurring low molecular peptides identified in egg white show antioxidant activity, Food Research International, **138** (2020): 109766
15. S. Moreno-Fernández, M. Garcés-Rimón, M. Miguel, Egg-derived peptides and hydrolysates: a new bioactive treasure for cardiometabolic diseases, Trends in Food Science & Technology, **104** (2020): 208-218
16. J. Sealey-Voyksner, J. Zweigenbaum, R. Voyksner, Discovery of highly conserved unique peanut and tree nut peptides by LC–MS/MS for multi-allergen detection, Food Chemistry, **194** (2016): 201-211
17. L.A. Domínguez-Pérez, L.M. Beltrán-Barrientos, A.F. González-Córdova, A. Hernández-Mendoza, B. Vallejo-Cordoba, Artisanal cocoa bean fermentation: From cocoa bean proteins to bioactive peptides with potential health benefits, Journal of Functional Foods, **73** (2020): 104134
18. S. Martini, A. Conte, D. Tagliazucchi, Comparative profile and bioactivities of cooked beef, pork, chicken and turkey meat after In vitro gastro-intestinal digestion, Journal of Proteomics, **208** (2019): 1-10
19. A. Cerrato, A.L. Capriotti, F. Capuano, C. Cavaliere, A.M.I. Montone, C.M. Montone, S. Piovesana, Z. Chiozzi, A. Laganà, Identification and antimicrobial activity of medium-sized and short peptides from yellowfin tuna (*Thunnus albacares*) simulated gastrointestinal digestion, Foods, **9** (2020): 1-12
20. H. Korhonen, A. Pihlanto, Bioactive peptides: production and functionality, International Dairy Journal, **16**, 9 (2006): 945-960
21. K. Fields, T.J. Falla, K. Rodan, L. Bush, Bioactive peptides: signaling the future, Journal of Cosmetic Dermatology, **8**, 1 (2009): 8-13
22. E. Maestri, M. Marmiroli, N. Marmiroli, Bioactive peptides in plant-derived foodstuffs, Journal of Proteomics, **147** (2016): 140-155
23. H.M. Rawel, G. Huschek, S.T. Sagu, T. Homann, Cocoa bean proteins-characterization, changes and modifications due to ripening and post-harvest processing, Nutrients, **11**, 2 (2019): 428
24. Y.L. Zhuang, X. Zhao, B.F. Li, Optimization of antioxidant activity by response surface methodology in hydrolysates of jellyfish (*Rhopilema esculentum*) umbrella collagen, Journal of Zhejiang University Science B, **10**, 8 (2009): 572-579
25. B. Li, F. Chen, X. Wang, B. Ji, Y. Wu, Isolation and identification of antioxidative peptides from porcine

collagen hydrolysate by consecutive chromatography and electrospray ionization–mass spectrometry, Food Chemistry, **102**, 4 (2007): 1135-1143

26. P. Khositanon, N. Panya, S. Roytrakul, S. Krobthong, S. Chanroj, W. Choksawangkarn, Effects of fermentation periods on antioxidant and angiotensin I-converting enzyme inhibitory activities of peptides from fish sauce by-products, Lebensmittel-Wissenschaft & Technologie, **135** (2021): 110122
27. P. Muangrod, V. Rungsardthong, S. Vatanyoopaisarn, Y. Tamaki, E. Kuraya, B. Thumthanaruk, Effect of wash cycle on physical and chemical properties of rehydrated jellyfish by-products and jellyfish protein powder, Science, Engineering and Health Studies, **15** (2021): 21030004-21030004
28. Y.H.P. Hsieh, F.M. Leong, J. Rudloe, Jellyfish as food, Hydrobiologia, **451** (2001): 11-17
29. W. Charoenchokpanich, V. Rungsardthong, S. Vatanyoopaisarn, B. Thumthanaruk, Y. Tamaki, Salt reduction in salted jellyfish (*Lobonema smithii*) using a mechanical washing machine, Science, Engineering and Health Studies, **14** (2020): 184-192
30. C.F. Chi, F.Y. Hu, B. Wang, X.J. Ren, S.G. Deng, C.W. Wu, Purification and characterization of three antioxidant peptides from protein hydrolyzate of croceine croaker (*Pseudosciaena crocea*) muscle, Food Chemistry, **168** (2015): 662-667
31. K.O. Lima, C.D.C.D. Quadros, M.D. Rocha, J.T.J.G.D. Lacerda, M.A. Juliano, M. Dias, M.A. Mendes, C. Prentice, Bioactivity and bioaccessibility of protein hydrolyzates from industrial byproducts of Stripped weakfish (*Cynoscion guatucupa*), Lebensmittel-Wissenschaft & Technologie, **111** (2019): 408-413
32. AOAC, *Official Methods of Analysis of AOAC International* (AOAC International, Maryland, USA, 2000)
33. C.S. James, *Analytical Chemistry of Foods* (Springer, London, 1995)
34. O.H. Lowry, N.J. Rosebrough, A.L. Farr, R.J. Randall, Protein measurement with the Folin phenol reagent, Journal of Biological Chemistry, **193** (1951): 265-275
35. S. Benjakul, M.T. Morrissey, Protein hydrolysates from Pacific whiting solid wastes, Journal of Agricultural and Food Chemistry, **45**, 9 (1997): 3423-3430
36. P. Molyneux, The use of stable free radical diphenylpicrylhydrazyl (DPPH) for estimating antioxidant activity, Songklanakarin Journal of Science and Technology, **26** (2004): 211–219
37. T. Emrerk, V. Rungsardthong, S. Vatanyoopaisarn, B. Thumthanaruk, Y. Tamaki, E. Kuraya, Processed flavors derived from combined bromelain hydrolyzed jellyfish protein hydrolysate, reducing sugars and arginine, Science, Engineering and Health Studies, **15** (2021): 21030002-21030002
38. I. Batish, D. Brits, P. Valencia, C. Miyai, S. Rafeeq, Y. Xu, M. Galanopoulos, E. Sismour, R. Ovissipour, Effects of enzymatic hydrolysis on the functional properties, antioxidant activity and protein structure of black soldier fly (*Hermetia illucens*) protein, Insects, **11**, 876 (2020): 1-12

Acute oral toxicity of zein nanoparticles with encapsulated gamma oryzanol in Sprague Dawley rats

Ubonphan Rodsuwan[1], *Krittiya* Thisayakorn[2], *Benjawan* Thumthanaruk[1], *Savitri* Vatanyoopaisarn[1], *Dudsadee* Uttapap[3], *Qixin* Zhong[4], and *Vilai* Rungsardthong[1,*]

[1]Department of Agro-Industrial, Food and Environmental Technology, Faculty of Applied Science, Food and Agro-Industrial Research Center, King Mongkut's University of Technology North Bangkok, 1518, Pracharat 1 Road, Bangsue, Bangkok 10800, Thailand
[2]Expert Center of Innovative Herbal Products (InnoHerb), Thailand Institute of Scientific and Technological Research (TISTR), Khlong Luang, Pathum Thani 12120, Thailand
[3]Division of Biochemical Technology, School of Bioresources and Technology, King Mongkut's University of Technology Thonburi, Bangkhuntian Bangkok 10150, Thailand
[4]Department of Food Science, The University of Tennessee, 2510 River Drive, Knoxville, TN 37996, United States

Abstract. Gamma oryzanol (GO) is well known for its antioxidant activity and health-promoting benefits. The gamma oryzanol-loaded zein nanoparticles (GOZNs) were successfully prepared in our previous study. In the present work, the acute oral toxicity of GOZNs was evaluated based on OECD guideline 420. GOZNs were fabricated by the liquid-liquid dispersion and lyophilized. The samples displayed a mean diameter of 311.20±3.01 nm and high loading capacity of 311.22±7.97 to 322.69±5.67 mg-GO/g-powder after 42-day storage at -18°C. Healthy female Sprague Dawley rats (8 weeks of age) were used for the experiments. Five female rats were administered a single dose of 2,000 mg GO/kg body weight via the oral route. Observations of toxicity signs were recorded for the first 24 hours, and the changes in the general physical conditions were monitored for 14 days before the gross necropsy on day 15. The results show that Sprague Dawley rats exhibited normal growth, and neither mortality nor acute toxicity signs were observed throughout the study period. The findings revealed that GOZNs did not have acute toxicity and were safe when administered orally in Sprague Dawley rats for short periods with LD50 > 2,000 mg/kg.

Keyword. Gamma oryzanol, Zein nanocapsules, Encapsulation, Acute oral toxicity, Sprague Dawley rats

1 Introduction

Rice is the most widely consumed staple food in Asia, especially in Thailand, where the rice production in 2020 was 18.6 million tonnes [1]. During industrial rice milling process, rice bran was regarded as a by-product (8-11% w/w) and used as low-grade animal feed. Gamma oryzanol (GO) is a mixture of ferulic acid esters with phytosterol and can be extracted from rice bran [2]. The potent antioxidant capacity, inhibition effects on lipid peroxidation, and therapeutically health-promoting functions have been reported for GO. Gamma oryzanol was proved as safe for consumption with no significant side effects [3,4]. However, the applications of GO as a functional ingredient or in food systems have been limited because of its sensitivity to degradation by heat and oxidation, lower bioavailability, and poor water-solubility [5]. Encapsulation of GO within a biopolymeric material could overcome these limitations.

The utilization of nanoscale materials has been increasing exponentially throughout the world in various fields such as pharmaceuticals, cosmetics, and food. Nanoencapsulation is defined in the pharmaceutical area as forming nanometric carriers with a diameter from a few nm to just below 1000 nm [6]. Encapsulation is beneficial for targeted delivery, controlled release, and protection of the bioactive components against destructive environments during processing or delivery [7].

Nowadays, many researchers are interested in amphiphilic molecules-based nano-sized drug delivery systems due to their unique core-shell structures and their ability to transport through cell membranes [8]. Zein has been approved for oral use by the FDA and is a safe, non-toxic, biodegradable, and generally recognized as safe (GRAS) biopolymer [6]. More than 50% amino acid residues of zein consist of proline, glutamine, and asparagine [9]. The tasteless, odourless, distinctive thermal resistance and excellent oxygen barrier characteristics of zein may be the advantages as carrier materials for encapsulating food antioxidants. Moreover, zein is a core biomaterial for delivery systems of hydrophobic nutrients. Zein can self-assemble into different structures, including spheres, sponges, and films [10]. It has been used to encapsule compounds significant to the food, pharmaceutical and textile industries such as tocopherol [11], vitamin D3 [12], flax oil [13], essential oil [14], puerarin [15], sacha inchi oil [16], and fabric softener [17].

* Corresponding author: vilai.r@sci.kmutnb.ac.th

GO loaded-zein nanoparticles (GOZNs) were prepared using a liquid-liquid dispersion technology in our previous study [18]. GOZNs can be prepared as a colloidal suspension or dried into powders using several drying technologies such as freeze drying and spray drying, and also provide sustained release in gastrointestinal conditions. However, nanoparticles might cause significant concerns for adverse effects on deliberation to the human body or environment. The toxicity of nanoparticles is closely related to their dose, size, surface area, concentration, chemical properties, and crystalline structure [4]. Nanoparticles could activate biological responses in terms of cellular toxicity and genotoxicity higher than microparticles [19].

The stability, internalization efficiency, and low toxicity are essential requirements for their successful applications. The evaluation of in vivo toxicity of GOZNs is considered essential for their application in food products. Humans have exceedingly complicated metabolic systems. The interactions between each compound in the consumed products and the biological systems could lead to adverse effects, requiring the assessments of the immune response, metabolism patterns, biodistribution, and obliteration of nanoparticles [20]. Pan and Zhong [21] suggested that applications of nanoparticles require in vitro and in vivo studies to guide the development of safe food products that utilize the unique functionalities of nanoparticles. Toxicological studies in animals were raised to identify the possibility of adverse effects from the accumulation of nanoparticles in the body on the immune system, oxidative stress-related disorders, and other diseases [22,23]. Subacute toxicity studies of GO have already been performed by Moon and co-workers [24]. Sprague Dawley rats received oral gavage administration of GO at doses of 0, 1000, and 2000 mg/kg body weight/day survived throughout the study period with no toxicological and histopathological changes. Andersen [25] reported that GO showed no chronic oral toxicity, reproductive toxicity, genotoxicity, and carcinogenicity in vivo. However, to the best of our knowledge, the in vivo toxicology including the acute toxicity of GOZNs has not been reported.

The present study aimed to evaluate the acute oral toxicity of GOZNs in Sprague Dawley rats according to OECD No. 420 [26]. According to literature surveys of conventional LD50 tests in both male and female rats [27], female rats are slightly more sensitive than males. Thus, the female rats-based method in the OECD No. 420 was adopted in this study [26]. The GOZNs were prepared and determined for properties such as particle size, polydispersity index (PDI), and GO content before being fed to the rats. Toxicity signs of each rat were recorded for the first day, and changes in general physical conditions were investigated for 14 days before examining internal organs by gross necropsy on day 15.

2 Materials and methods

2.1. Materials

Zein was purchased from Sigma Aldrich Corp., (St. Louis, MO, USA). GO was a product from Tsuno Rice Fine Chemicals Co., Ltd. (Wakayama, Japan). Tween 80 and lecithin were supplies from Union Chemical 1986 Co., Ltd. (Thailand). All other chemicals used were of analytical grade.

2.2 Preparation of GOZNs and measurement of their properties

The GOZNs were prepared as described by Rodsuwan and co-worker [18]. Briefly, zein (0.4 g) was dissolved in 20 ml of 85% (w/v) aqueous ethanol with a magnetic stirrer. GO (0.2 g; 50% loading of zein) was dissolved at 1% w/v in isopropanol that was dispersed dropwise to the zein solution using a magnetic stirrer. The dispersion was added to a citrate solution containing Tween 80 and lecithin with pH 8 to 9 and mixed using a homogenizer (Ultra-Turrax T25, IKA, Germany). The final concentration of GOZNs in dispersion was set at 0.3% zein and 50% GO loaded. The resulting mixture was concentrated to 100 ml using a rotary evaporator. The resulting GOZNs dispersions were stored at 4 °C for further uses. The particle size and PDI of freshly prepared colloidal dispersions were determined by dynamic light scattering using a DelsaTM nanoparticle analyzer (Beckman Coulter, Fullerton, CA). The ζ-potential measurements were performed using a Laser Doppler Velocimeter (Zetasizer Nano ZS90, Malvern, UK). The dispersion was also freeze-dried (Alpha 1–4 LSCplus, Christ, Germany).

2.3 Morphological observation

The morphology of GOZNs was investigated using scanning electron microscopy SEM (JSM-6610LV, JEOL, Japan) and FESEM (JSM-7610F, JEOL, Japan). Prior to the observation, the lyophilized GOZNs powder was placed on the stubs and subsequently coated with a thin layer of gold using a sputter coater to avoid charging under the electron beam.

2.4 Determination of GO content in GOZNs

High performance liquid chromatography (HPLC) was used to quantify the amount of GO content in GOZNs following a protocol previously described by Rodsuwan and co-workers [18]. Briefly, GOZNs were dissolved in 4 ml of a mixture of methanol and dichloromethane (1:1 v/v) and filtered through a 0.45 μm filter. The GO content was determined using HPLC (Agilent Technology 1,200 series, Germany) with an Eclipse Zorbax XDB-C 18 (Agilent, 250 × 4.6 mm, 5 μm) analytical column and a Zorbax XDB-C18 (Agilent, 12.5 × 4.6 mm, 5 μm) guard column. Sample was then eluted at a flow rate of 1.4 ml/min with a mobile phase composed of methanol, acetonitrile, dichloromethane, and acetic acid (55:35:9.5:0.5). The freeze-dried GOZNs were kept in an aluminum foil under -18oC and taken out every 1, 7, 14, and 42 days to determine the GO content (mg/g sample) remaining in GOZNs.

https://doi.org/10.1051/e3sconf/202130202011

2.5 Procurement of Sprague Dawley rat

Healthy female Sprague Dawley rats (8 weeks of age) were obtained from the Nomura Siam International Co., Ltd., Thailand. There were five rats per cage under standard laboratory conditions, with a period of 12 h light/12 h dark, at 23±1°C under 40-60% relative humidity. The animals were provided with commercial food and had tap water ad libitum. According to the National Act on the Use of Experimental Animals (Thailand) requirement, the protocols for all animals studies were reviewed and approved by the Animal Ethics Committee of Thailand Institute of Scientific and Technological Research (No. T-62021).

2.6 Administration and observations

The acute oral toxicity study was conducted following OECD 420 Guideline for Testing of Chemicals, acute oral toxicity-fixed dose procedure [26]. After one week of acclimatization period, the rats were 9 weeks old at the start of the study (Day 1) and had an average body weight of 222.60±3.03 g. The animals were fasted for 16 hours with ad libitum water prior to dosing. They were administered with a single oral gavage dose of GOZNs at 2000 mg GO/kg body weight. The GO content in the GOZNs for the feeding was calculated based on the HPLC assay as described above. The GOZNs powders were dissolved in distilled water and mixed well before the oral administration. Toxicity signs were recorded systematically starting the first 30 min and periodically during the first 24 hours, with particular attention during the first 4 hours. The surviving rats were monitored for 14 days for any changes in the general physical conditions such as appearance, fur condition, behavior, and mortality. The body weight of the animals was measured before the dosing and then weekly after (Day 1, 8, and 15). Surviving animals were sacrificed using a carbon dioxide method of euthanasia on day 15 and undergone gross necropsy.

2.7 Statistical analysis

All data were expressed as mean ± SD unless otherwise stated. Statistical significance was determined by one-way analysis of variance (ANOVA) and post hoc least significant difference (LSD) test using the SPSS program (IBM SPSS Statistics 22). Statistical significance was determined at p-values of less than 0.05.

3 Equations and mathematics

3.1 GOZNs and their properties

GOZNs displayed an average diameter of 311.20±3.01 nm. Zein is known to self-assemble as nanospheres during anti-solvent precipitation. The low PDI value, 0.10±0.02 (lower than 0.05), displayed a highly monodispersed distribution of GOZNs in the dispersion [28]. The mean ζ-potential of GOZNs was -54.91±0.60 mV. This value exhibited that the freshly prepared dispersion had good colloidal stability [29]. Moreover, at above pH 8.0, the carboxylic groups of glutamate in zein are deprotonated to result in numerous negative charges [30]. The properties of the GOZNs prepared for the acute toxicity test in animals were in the same range as our previous study [18].

3.2 Morphological observation

The SEM images of the zein, GO and GOZNs are presented in Fig. 1. Zein and GO had irregular structures, while GOZNs had a spherical shape with a smooth surface. These observations were consistent with our previous study [18]. The mean diameter of particles estimated from SEM images (218.86±0.04 nm) was smaller than that estimation from DSL. Similar observations were reported by Zhong and Jin [31], due to the effect of freeze drying.

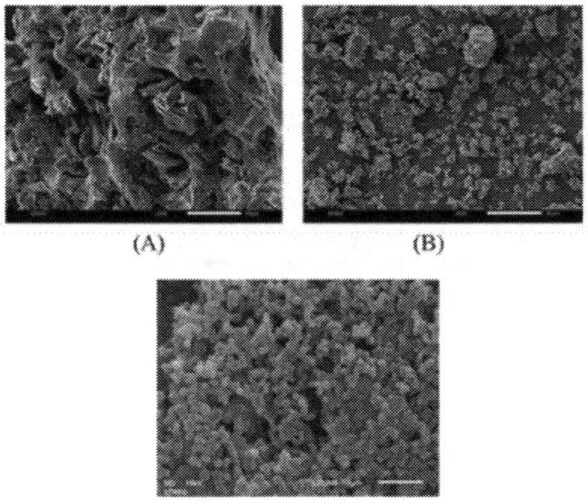

Fig. 1. Scanning electron micrographs (SEM) of zein (A), gamma oryzanol (B), and GOZNs (C). Scale bars represent 10 μm in (A) and (B), and 1 μm in (C).

3.3 GO content in GOZNs

The HPLC chromatogram used to determine GO content in the freeze-dried GOZNs powder exhibited five peaks at retention times of 15.96, 17.87, 19.20, 20.67, and 21.95 min. GO consisted of five major compounds which were cycloartenyl ferulate (32.65%), 24-methylenecycloartanyl ferulate (38.62%), Δ7campestenyl ferulate (16.63%), campesteryl ferulate (0.83%), and sitosteryl ferulate (9.65%) (Fig. 2). During the storage at -18°C, the GO content (mg/g sample) in GOZNs was not significantly different on day 1, 7, 14, and 42 (Fig. 3). GO content decreased from 322.69±5.67 to 311.22±7.97 mg-GO/g-powder, which were equivalent to GO retention of 100% to 96.44% compared to the first day of preservation. The previously reported GO retention in GOZNs, decreasing to 90.13% when stored in a clear glass vial at -18₀C for 45 days [18]. However, free GO obtained lower GO contents (87.53%) than both GOZNs samples after storage for 40 days (data do not shown). Zein could improve the stability of the encapsulated bioactive by delaying deteriorative reactions and thus enhancing chemical stability during storage [32].

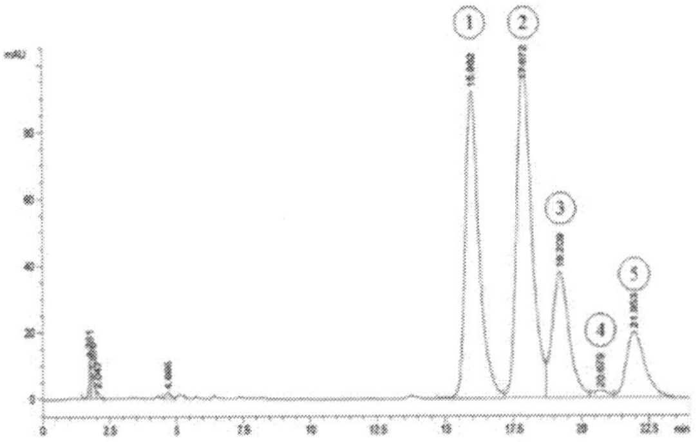

Fig. 2. HPLC chromatogram of gamma oryzanol. Peaks 1 to 5 indicate the following ferulate esters of triterpenoids: (1) cycloartenol, (2) 24-methylenecycloartanol, (3) campesterol, (4) cycloblanol, and (5) sitosteryl ferulate.

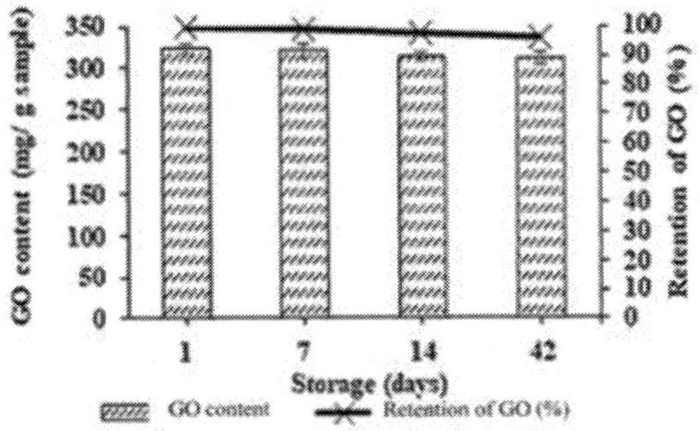

Fig. 3. Changes of GO content (mg/g sample) and % retention of GO during storage at -18°C.

3. 4 Animal observation and body weight

Behavior toxicity signs in Sprague Dawley rats were observed after a single oral dose at 2,000 mg GO/kg body weight. No changes of skin, fur, eyes, mucous membranes, respiration, circulatory system, autonomic and central nervous systems were found after the feeding. In addition, daily examination of all rats for their physical conditions showed none of the above-mentioned toxicity signs, and no mortality was recorded throughout the 14-day observation period (Table 1).

Moon and co-worker [24] investigated the effects of GO in Sprague Dawley rats with the dose of 2,000 mg/kg body weight in which no adverse effects and no histopathological changes in various organs were noticed. The same dose for GO in the GOZNs in the present study, also revealed no negative effects as acute toxicity in the after receiving GOZNs, the body weight of all tested animals increased throughout the study and demonstrated to be significantly different between day 1 and 8, and day 8 and 15 ($P < 0.05$) (Fig. 4) with 15.61% and 7.42% body weight change, respectively. From the gross pathological examination, no significant abnormality of all internal organs including lungs, heart, thymus, esophagus, trachea, stomach, small intestine, large intestine, liver, spleen, pancreas, adrenals, kidneys, ovary, and uterus was observed. rats, implying that GOZNs are non-toxic.

Table 1. Toxicological effect of the acute exposure of GOZNs in Sprague Dawley rats at 2,000 mg/kg body weight.

Time	Clinical signs*
30 min	None
4 hours	None
24 hours	None
Up to 8 days	None
Up to 15 days	None

In addition, the oral administration of GO in GOZNs at a dose limit of 2,000 mg/kg produced neither mortality nor acute toxicity signs throughout 14 days of the observation. The slight increase in body weight of the treated animals can still be accepted as normal, as it can arise due to differences in food consumption. However, the study on long-term consumption, which might increase the cytotoxic risk related to subchronic toxicity of the GOZNs in animals, would be beneficial for food applications in the future.

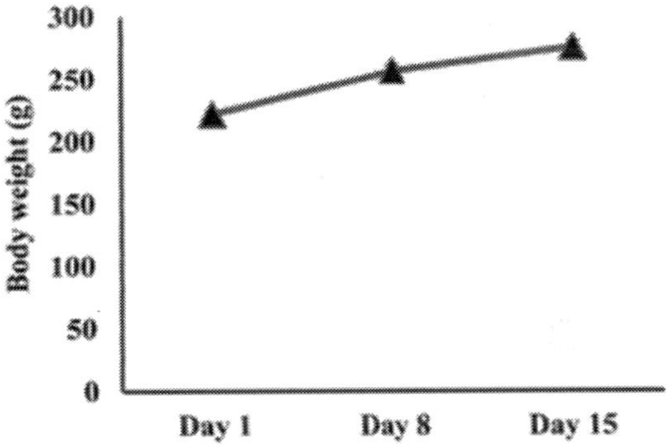

Fig. 4. Body weight (g) of female rats (age= 9 weeks, n = 5) after receiving a single dose of GOZNs in an acute toxicity (at day 1).

Note: * Significantly different from the Day 1 group ($P<0.05$), ** Significantly different from the Day 8 group ($P<0.05$)

4 Conclusion

The GOZNs prepared in this study exhibited a nanoscale diameter, low polydispersity index, and highly negative zeta-potential, as well as the good retention of GO during 42-day storage at -18°C. The acute toxicity test of GOZNs in Sprague Dawley rats, using OECD 420 via single oral administration at 2,000 mg/kg, showed no mortality and toxicity signs during the 14-day experiments, and the average body weight of rats increased during observations. The results provide preliminary safety information for GOZNs, which should be investigated for subchronic toxicity in the future. The present work is significant to the development of GOZNs as value-adding products using zein (corn) and GO (rice bran) from agricultural raw materials for use in nutraceuticals or food and beverage industries to promote human health.

Acknowledgement

The authors would like to thank the Thailand Institute of Scientific and Technological Research (grant number 886109) for their facilities and Thailand Research Fund, under the Research and Researchers for Industries program (grant number PHD59I0053) for their financial support.

References

1. H. Demaree-Saddler https://www.world-grain.com/articles/14832-thailand-sees-uptick-in-rice-production (2021)
2. A. Moongngarm, J. Amornpan, A. Homduang, C. Haoteng, A. Jelekovac, Rice Starch as Coating Materials for Microencapsulation of Gamma Oryzanol to Improve its Stability, American Journal of Applied Sciences, **13**, 8 (2016): 900-906
3. K. A. Szcześniak, P. Ostaszewski, A. Ciecierska, T Sadkowski, Investigation of Nutriactie Phytochemical Gamma Oryzanol in Experimental Animal Models, Journal of Animal Physiology and Animal Nutrition, **100**, 4 (2016): 601-617
4. A. Sukhanova, S. Bozrova, P. Sokolov, M. Berestovoy, A. Karaulov, I. Nabiev, Dependence of Nanoparticle Toxicity on their Physical and Chemical Properties, Nanoscale Research Letters, 13, 1 (2018): 1-21
5. J.S. Lee, J.S. Kim, H.G. Lee, Dependence of Nanoparticle Toxicity on their Physical and Chemical Propertiees, Colloids and Surfaces B: Biointerfaces, **70**, 2 (2009): 213-217
6. M.R. Kasaai, Trends in Food Science and Technology, **79** (2018): 184-197
7. G. Spigno, F Donsì, D. Amendola, M. Sessa, G. Ferrari, D. M. De Faveri, Nanoencpsulattion Systems to Improve Solobility and Antioxidant Efficiency of a Grape Marc ExtracT into Hazelnut Paste, Journal of Food Engineering, **114**, 2 (2013): 207-214
8. A. Jain, G. Sharma, V. Kushwah, G. Ghoshal, A. Jain, B. Singh, U.S. Shivhare, S. Jain, O.P. Katare, Beta Carotene-Loaded Zein Nanoparticles to Improve the Biopharmaceutical Attributes and to Abolish the Toxicity of Methotrexate: a Preclinical Study for Breast Cancer, Artificial Cells, Nanomedicine, and Biotechnology, **46**, sup 1 (2018): 402-412
9. A.O. Elzoghby, M.M. Elgohary, N.M. Kamel, Implications of Protein and Peptide based Nanoparticles as Potential Vehicles for Anticancer Drug, In Advances in Protein Chemistry and Structural Biology, (2015):169-221
10. J. He, L. Wang, Z. Wei, Y. Yang, C. Wang, X. Han, Z. Nie, Vesicular Self-Assembly of Colloidal Amphiphiles in Microfluidics, ACS Applied Materials and Interfaces, **5**, 19 (2013): 9746-9751
11. L.Y. Luo, B. Zhang, M. Whent, L.L. Yu, Q. Wang, Preparation and Characterization of Zein/Chitosan Complex for Encapsulation of Alpha-Tocopherol, and Its in Vitri Controlled Release Study, Colloids and Surfaces B: Biointerfaces, **85**, 2 (2011): 145-152
12. X.Y., Luo, M.H. Yang, F.X. Wu, L.J. Wu, L. Chen, Z. Tang, N.T. Liu, X.F. Zeng, J.L. Guan, G.H. Yuan, Vitamin D Receptor gene *Bsml* Polymorphism B allele, but Not BB genotype, is associated with Systemic Impus Erythematosus in a Han Chinese Population, Lupus, **21**, 1 (2012): 53-59
13. S. Quispe-Condori, M.D.A. Saldaña, F. Temelli, Microencapsulation of Flax Oil with zein Using Spray and Freeze Drying, LWT - Food Science and Technology, **44**, 9 (2011): 1880-1887
14. N. Parris, P.H. Cooke, K.B. Hicks, Encapsulation of Essential Oils in Zein Nanospherical Particles, Journal of Agricultural and Food Chemistry, **53**, 12 (2005): 4788-4792
15. V. Rungsardthong, U. Pithanthanakul, C. Puttanlek, D. Uttapap, K. Boonpisuttinant, Preparation of Puerarin Loaded Zein Nanoparticles: Characterization and Stability Study, Journal of Current Science and Technology, **11**, 1 (2021): 60-70
16. S. Suwannasang, B. Thumthanaruk, Q. Zhong, D. Uttapap, C. Puttanlek, S. Vatanyoopaisarn, V. Rungsardthong, Food and Bioprocess Technology, (2021): under review
17. U. Pithanthanakul, S. Vatanyoopaisarn, B. Thumthanaruk, C. Puttanlek, D Uttapap, B. Kietthanakorn, V. Rungsardthong, Encapsulation of Fragranced in Zein Nanoparticles and Use as Fabric Softener Textile Application, Flavour and Fragrance Journal, **36**, 3 (2021): 365-373
18. U. Rodsuwan, U. Pithanthanakul, K. Thisayakorn, D. Uttapap, K. Boonpisuttinant, S. Vatanyoopaisarn, B. Thumthanaruk, V. Rungsardthong, Preparation and Charecterization of Gamma Oryzanol Loaded Zein Nannoparticles and Its Improved Stability, Food Science and Nutrition, **9**, 2 (2021): 616-624
19. S.S.P. Singh, M.F. Rahman, U.S.N, Murty, M. Mahboob, P. Grover, Comparative Study of Genotoxicity and Tissue Distribution of Nano and Miron Size Iron Oxide in Rats After Acute Oral Treatmente, Toxicology and Applied Pharmacology, **266**, 1 (2013): 56-6

20. B. Mangalampalli, N. Dumala, P. Grover, Genotoxicity, Biochemical, and Biodistribution Studies of Magnesium Oxide Nano and Microparticles in Albino Wistar Rats After 29 Day Repeated Oral Exposure , Toxicology and Pharmacology, 90 (2017): 170-184
21. K. Pan, Q. Zhong, Annual Review of Food Science and Technology, **7** (2016): 245-266
22. C.D. Walkey, J.B. Olsen, H. Guo, A. Emili, W.C. Chan, Nanoparticle Size and Surface Chemistry Determine Serum Protein Adsorption and Macrophage Uptake, Journal of the American Chemical Society, **134**, 4 (2012): 2139-2147
23. H.K. Biesalski, L.O Dragsted, I. Elmadfa, R. Grossklaus, M. Müller, D. Schrenk, P. Walter, P. Weber, Bioactive Compounds: Definition and Assessment of Activity, Nutrition, **25**, 11-12 (2009): 1202-1205
24. S.H. Moon, D. Kim, N. Shimizu, T. Okada, S. Hitoe, H. Shimoda, Ninety-Day Oral Toxicity Study of Rice Derived Gamma Oryzanok in Sprague Dawley Rats, Toxicology Reports, **4** (2017): 9-18
25. F. Andersen, International Journal of Toxicology, **25**, Sup 2 (2006): 91-120
26. OECD Guideline for Testing of Chemicals, Section 4, Health Effect. Test No 420; Acute Oral Toxicity-Fixed Dose Procedure, Paris: Organization for Economic Corporation and Development (2001).
27. R.L. Lipnick, J.A. Cotruvo, R.N. Hill, R.D. Bruce, K.A. Stitzel, A.P. Walker, I. Chu, M. Goddard, L. Segal, J.A. Springer, R.C. Myers, Food and Chemical Toxicology. **33**, 3 (1995): 223-23
28. M. Danaei, M. Dehghankhold, S. Ataei, F. Hasanzadeh Davarani, R. Javanmard, A. Dokhani, S. Khorasani, M.R. Mozafari, Impact of Particle Size and Polydispersity Index on the Clinical Application of Lipidic Nanocarrier Systems, Pharmaceutics, **10**, 2 (2018): 57
29. E. Joseph, G. Singhvi, Nanomaterials for Drug Delivery and Therapy, (2019): 91-116
30. B. Zhang, Y. Luo, Q. Wang, Effect of Acid and Base Treatments on Structural, Rheological, and Antioxidant Properties of Alpha-Zein, Food Chemistry, **124**, 1 (2011): 210-220
31. Q. Zhong, M. Jin, Zein nanoparticles Produced by Liquid-Liquid Dispersion, Food Hydrocolloids, **23**, 8 (2009): 2380-2387
32. S. Torres-Giner, A. Martinez-Abad, M.J. Ocio, J.M. Lagaron, Stabilization of a Nutraceutical Omega-3 Fatty Acid by Encapsulation in Electrosprayed Zein Prolamine, Journal of Food Science, **75**, 6 (2010): N69-N79

Optimization of Enzyme Assisted Extraction of Chondroitin Sulfate from *Bohadschia argus* by Response Surface Methodology

Thanagorn Ruensodsai[1], *Elizabeth Jayex* Panakkal[2], *Prawpan* Teerapornnarong[2], *Wawat* Rodiahwati[3], *Malinee* Sriariyanun[2,4]*, *Kittipong* Rattanaporn[1]*

[1] Department of Biotechnology, Faculty of Agro-industry, Kasetsart University, Bangkok, Thailand

[2] Department of Mechanical and Process Engineering, The Sirindhorn International Thai-German Graduate School of Engineering (TGGS), King Mongkut's University of Technology North Bangkok (KMUTNB), Bangkok, Thailand

[3] Department of Chemistry, University of New England: Armidale, New South Wales, Australia

[4] Biorefinery and Process Automation Engineering Center (BPAEC), King Mongkut's University of Technology North Bangkok, Bangkok, Thailand

Abstract. Chondroitin sulfate is currently one of the bioactive compounds obtained from different natural materials with high benefit in human medical treatment and pharmaceutical. However, the natural concentration of chondroitin sulfate in high cost raw material is low and this situation makes the commercialization become difficult. This study focused on production of chondroitin sulfate isolated from *Bohadschia argus* using enzyme-assisted extraction methodology. Optimization experiments were conducted based on Response Surface Methodology (RSM) using Box–Behnken design (BBD). Three important extraction parameters, including enzyme concentration (X_1: 0.05-0.5 %), Time (X_2: 0.25–3 h), and temperature (X_3: 55–65 °C) were varied to obtain maximum chondroitin sulfate yields. The RSM optimized model obtained from statistical analysis presented the high correlation coefficiency (R2) at 0.7508, advocating the significance of the model. The optimum extraction conditions were selected as enzyme concentration of 0.48% with extraction time of 1.01 h. at an extraction temperature of 56.53°C. Under these conditions, the experimental yield of chondroitin sulfate was 415.59 mg/100g dry, which is in good agreement with the value predicted by the model.

Keyword. Sea cucumber, Chondroitin sulfate, Response surface methodology, Papain, Extraction

1 INTRODUCTION

Osteoarthritis clinical studies confirmed the therapeutic effects of orally taken chondroitin sulfate on osteoarthritis patients with improvement of joint function and pain reduction [1-2]. Chondroitin sulfate is present abundantly in the extracellular matrix of cartilaginous tissues and connective tissue, which are composed of collagen networks and amorphous ground substance including glycoproteins, hyaluronan, proteoglycans [3]. Chondroitin sulfate is an acid mucopolysaccharide, a macrobiomolecule, that is comprehensively used in medicine, biological materials, cosmetics, and other fields for its unique pharmacological activities [4], such as anti-inflammatory, anticancer [5]. In general, chondroitin sulfate can be added in cosmetic products as a moisturizing agent. Due to the large amount of negative charge within the chondroitin sulfate molecule, the repulsions create a gap causing water and nutrients to retain within the molecule's empty space. Therefore, chondroitin sulfate is very effective in maintaining moisture [6]. Being a water-soluble biopolymer, it has the potential to retain the moisture in skin and replace the conventionally used glycerin from moisturizers.

Currently, the commercial product of chondroitin sulfate is derived from shark, cow, pig and chicken cartilage, etc. The main benefits for utilizations of these biomaterials are to increase the value of bio-wastes and to reduce the disposal of wastes to environment and to reduce the workloads of waste management. It is widely being used in cosmetic industry and as a dietary supplement. However, the current problem is due to the chance of infectious disease outbreaks such as Mad cow disease and H1N1 virus caused by poor preparing raw materials. Furthermore, the religious or social restrictions on animal sources such as pork, cow or shark are the reasons to seek for alternative source for chondroitin sulfate production.

Marine sources have been demonstrated to be the raw materials of bioactive compounds with numerous health benefits and high level of biodiversity. In these recent years, extensive studies have been conducted to investigate biological activities, nutritional value, and potential health benefits of marine based bioactive

* Corresponding author: author@e-mail.org, kittipong.r@ku.th

https://doi.org/10.1051/e3sconf/202130202012

substances [7-8]. Marine living organisms produced biological compounds, especially sulfated polysaccharides with anticoagulant, antithrombotic activities [9]. Several researches are conducted to harvest bioactive compounds from sea cucumbers because this natural meat source is one of staple food in Asian countries due to its high good fatty acid and collagen and it is included in Ayuravedic medicine [10-11].

Sea cucumbers are a good alternative to the future production of chondroitin sulfate. Among large variation of chondroitin sulfate, there are two main types of sulfated polysaccharide in the body wall sea cucumber namely fucans (SF) often named "fucoidans" and fucosylated chondroitin sulfates (FuCS) (Figure 1). FuCS obtained from sea cucumbers are characterized to be polymeric molecules of D-glucuronic acid, N-acetyl-D-galactosamine, L-fucose, and sulfate residues [12]. Interestingly, FuCS isolated from the body wall of sea cucumbers are structurally different from sulfated polysaccharides in other invertebrates [13]. The molecular structures and biological activities of FuCS can vary depending on the species. However, the extraction process of chondroitin sulfate has been demonstrated mostly in laboratory scale.

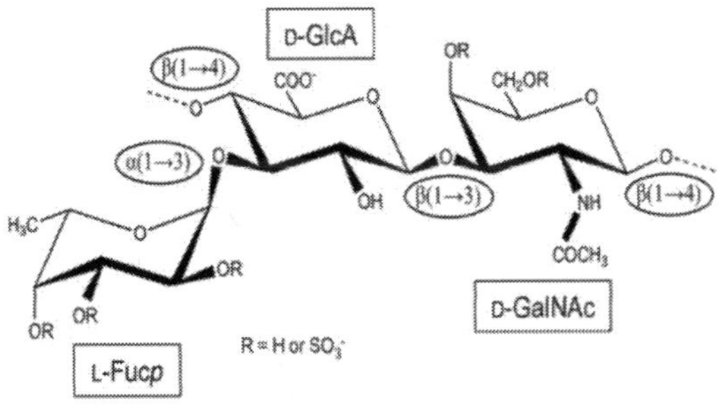

Fig. 1. Fucosylated chondroitin sulfates (FucCS)

From the past reports of chondroitin sulfate extraction, there are different extraction methods depending on the type of raw material, such as extraction of chondroitin sulfate using acids, alkaline [14] or extraction under high pressure and enzymatic extraction. Generally, proteolysis methods are used to release peptides of glycosaminoglycans (GAGs) from their tissues [3]. For this purpose, many researchers use an overly expensive proteinase namely, papain. An alternative and more economical enzyme is required to substitute this costly enzyme to reduce the cost of extraction. Moreover, cleavage of core protein of proteoglycan depends on the types of proteolytic enzymes. The size of peptides greatly depends on the proteolytic enzyme used [3].

The major drawbacks of conventional methods, acid or alkali extraction, include breaking of bonds between xylose and serine resulting in the release of sulfate and chondroitin from their extracellular matrix. This makes it difficult to detect CS after digestion. However, these CS molecules which are released free, can be easily absorbed by human digestive system. Enzymatic extraction was demonstrated to be able to overcome this problem because enzymes digest only in the specific regions of the substrates. Hence, using different enzymes provide different end products of CS and CS biological activities.

Therapeutic capability of the orally ingested CS has been proved by the clinical studies conducted on osteoarthritis patients. These patients had pain reduction and improvement in joint function by intake of CS [15-17]. Therapeutic effects of bovine derived CS (C4S and C6S) as a drug has also been proved by its oral intake at a dose of 800 mg/day in regular intervals, twice a year [18]. The cartilaginous rings of bovine trachea, pork ears and snout and shark cartilage are used as sources of CS in nutritional supplements. Other available sources of CS have been investigated by many researchers [19–22], also different methods for extraction and analysis of CS have been reported [23–27]

This study was aimed to investigate the Papain-assisted extraction method of chondroitin sulfate from sea cucumber, Bohadschia argus, based on Response Surface Methodology (RSM) with Box–Behnken design (BBD), a widely used statistical technique [28][29-30]. This study could provide new insights in the production of CS from sea cucumber under optimized conditions using papain. The extracted CS can be further applied in the nutraceutical and functional food industry [31].

2 MATERIAL AND METHODS

2.1. Sample prepareation

Samples of dried Tigerfish (*Bohadschia argus*) were delivered from Papua New Guinea to the laboratory (Courtesy provided by Wonnapob Co., Ltd., Bangkok, Thailand) (Figure 2). The dried weight and the wet weight were measured and recorded. For the wet weight, the sea cucumbers were rehydrated, by putting sea cucumber into deionized water for 96 hours at 4 °C. Deionized water was changed every 24 hours. Weight was recorded and the moisture content was analyzed. The sea cucumbers were sliced into small pieces and stored at -20 °C. Food grade Papain (EC 3.4.22.2), chondroitin-4-sulfate from bovine trachea (Fluka) was used as standards. 1,9-Dimethylmethylene blue was purchased from Aldrich. All the other reagents used in the experiment were analytical grade.

Fig. 2. Tiger fish sea cucumber (*Bohadschia argus*)

2.2 Enzyme extraction of chondroitin sulfate

Extraction of proteins depends on two vital elements, namely extraction condition and the specificity of the enzyme. A previous study on defatted Lemna minor has demonstrated that protein recovery, degree of hydrolysis values and antioxidant properties of hydrolysed proteins could be enhanced by increasing enzyme concentration and hydrolysis time [32]. During enzyme assisted extraction, protein and glycan sulfate present in the connective tissue of sea are broken-down to release chondroitin sulfate free. Chondroitin sulfate was extracted by a method modified from Garnjanagoonchorn et al. [30].

Initially, the sea cucumber was cut into about 1 cm-sized slices. 2.0 g of sliced sea cucumber was blended in a homogenizer with 0.1 M sodium phosphate buffer containing 0.005 M Ethylenediaminetetraacetic acid, 0.005 M cysteine hydrochloride, and 0.02% sodium azide for preservation of raw material in extraction process. In addition to this, sodium phosphate buffer can maintain the ionic environment required for extraction process. pH of the buffer was adjusted to 7.0 and Papain enzyme was added into mixture. Enzyme hydrolysis treatment was carried out according to Box-Behnken Design (BBD). The sample was taken after stopping the enzyme reaction with boiling water at 95 °C.

2.3 Experimental design

Optimization of parameters for the extraction of chondroitin sulfate from *Bohadschia argus* was carried out based on RSM with Box-Behnken design (BBD) by using the Design-Expert software (V.7.0.0, Stat- Ease, Inc, Minneapolis, USA). This model has been widely used to evaluate the relationships between response values and factors [5] Furthermore, this method has the benefit of reducing the number of experiments. The selected extraction parameters in this work included the concentration of enzyme, extraction time, and extraction temperature. The range and center point values of the three independent variables were presented in (Table 1.) The three factors were assigned as enzyme concentration (X_1), extraction times (X_2), and extraction temperature (X_3) and varied to three levels, coded +1, 0, and 1 for high, intermediate, and low values, respectively. A total of 17 experimental runs were carried out according to this design in (Table 2) for optimization of the extraction process. The extraction yield of chondroitin sulfate was selected as the response (Y).

Table 1. Factor and level of Box-Behnken design

Factor	Coded symbols	Levels		
		-1	0	+1
Concentration (%)	X_1	0.05	0.253	0.5
Time (min)	X_2	0.25	1.625	3
Temperature (°C)	X_3	55	60	65

The ranges of each factor were selected based on our preliminary studies. The temperature range was set between 55 to 65°C. Even though the enzyme is active in room temperature (+/-20°C), the optimal papain activity was observed between 55 to 65°C. Higher temperature above 65 °C may lead to inactivation of enzyme which led to setting temperature range between 55 and 65 °C. Extraction time can depend on raw material used for study. Extraction of chondroitin sulfate from bovine cartilage require an extraction time was 48 hours due to differences in the raw material used in the extraction process. However, extraction from sea cucumber does not require too long incubation. Therefore, the extraction time was limited between 0.25 to 3 minutes. Enzyme concentration was chosen according to the previous study from Nguyen et al. within range of 0.2 to 1%. [33]

Table 2. Experimental design and responses of the dependent variables to extraction conditions

Run	Factor 1 X_1 : Concentration (%)	Factor 2 X_2: Time (min)	Factor 3 X_3 : Temperature (°C)	Yields (mg/100g dry)
1	0.50	1.63	55	426.98
2	0.01	0.25	60	252.28
3	0.50	3.00	60	343.69
4	0.25	0.25	55	395.28
5	0.25	1.63	60	392.24
6	0.25	3.00	65	349.58
7	0.25	1.63	60	410.03
8	0.01	1.63	55	272.92
9	0.25	1.63	60	393.07
10	0.01	1.63	65	407.82
11	0.25	1.63	60	429.93
12	0.25	1.63	60	443.93
13	0.25	3.00	55	432.88
14	0.01	3.00	60	362.85
15	0.25	0.25	65	409.29
16	0.50	1.63	65	437.30
17	0.50	0.25	60	404.13

2.4 Determination of chondroitin sulfate

The content of chondroitin sulfate was determined by dimethylmethylene blue assay (DMMB) according to the procedure defined by Thomas-Coulson et al. [34]. Different concentrations 0, 2, 4, 6, 8 µg/ml of chondroitin-4-sulphate (C4S) was used as standard curve for quantitative calculation (Figure 3). Stock solutions were prepared along with DMMB dye and was pipetted into 96 well microplates. The volume was adjusted to 20 µl with deionized water. A volume of 20 µl from each extracted chondroitin sulfate sample was pipetted out in subsequent wells containing 200 µl of DMMB dye in the microplate followed by shaking of the microplate for 5 seconds using a microplate shaker. The absorbance was recorded using UV–Vis spectrophotometer at 525 nm immediately and plotted against the concentration on a graph and the CS content was determined by referring to the standard graph. The standard curve selected the chondroitin sulfate as a standard, and the regression equation was $Y = 0.0382X+0.0116$ with $R^2 = 0.9901$

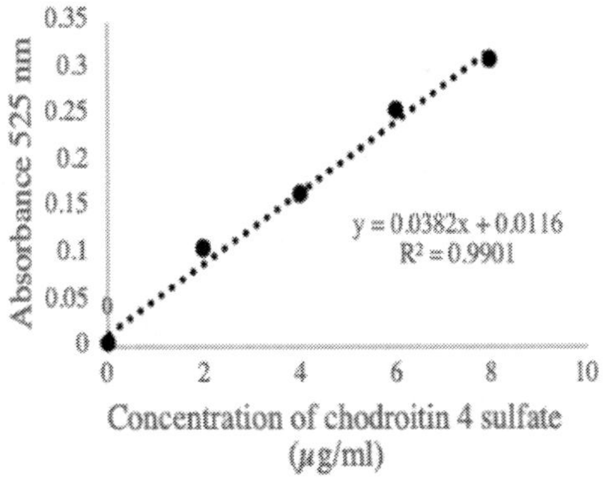

Fig. 3. Standard curve for quantitation of chondroitin sulfate

3 RESULT AND DISCUSSION

Optimization of extraction conditions of chondroitin sulfate from *Bohadschia argus* was conducted based on RSM. Before conducting the RSM, the levels of RSM independent variables for chondroitin sulfate extraction were determined based on the preliminary experiments (data not shown). After preliminary experiments, 17 experimental runs with three independent variables, and three levels were conducted according to the Box–Behnken design (BBD) as shown in (Table 2). The independent variables were enzyme concentration (X_1), extraction times (X_2), and extraction temperature (X_3) while the response variable (Y) was the amount of chondroitin sulfate from the Bohadschia argus.

The application of the RSM offered a model representing an empirical relationship between the response variable (extraction rate of CS) and the test variables under targeted investigation. Table 2 represented the yields of CS under different 17 extraction conditions. It was found that the highest yield of CS from RSM experiment was 443.93 mg/100 g from run number 12 and the lowest yield was 252.28 mg/100 g from run number 2.

The significance of the model was determined by ANOVA analysis (Table 3) suggesting the quadratic model to fit with experimental data. The P-value of the model was less than 0.0001, which suggested the significance of the used model. The high correlation coefficiency (R^2) at 0.7508 of the predicted model was obtained, advocating the significance of the model. Furthermore, Lack of Fit test showed insignificant p-value at 0.2619, which was in good agreement to model p-value and R^2. In case of term models, enzyme concentration (X_1) and Time x Temp were statistically significant (P-value < 0.1). By applying regression analysis on the experimental data, the response variable (CS yield) and the test variables (extraction parameters) were related by the following second order polynomial equation $Y = - 2120.22564 + 1535.51868 * X_2 + 6872.55469 * X_1 + 36.47787 * X_3 - 94.02424 * X_1 * X_3 - 463.95372 * Time^2 - 1637.79206 * X_1$

Table 3. ANOVA for response surface quadratic model

Source	Sum of Squares	df	Mean Squares	F Value	p-value Prob > F
Model	3.521E+0.006	6	5.868E+005	50.73	< 0.0001
X_2	11578.38	1	115793.38	1.00	0.3407
X_1	79990.00	1	79990.00	6.91	0.0252
X_3	32444.96	1	32444.96	2.80	0.1249
X_2X_3	54153.94	1	54153.94	4.68	0.0558
X_1^2	3.249E+006	1	3.249E+006	280.82	< 0.0001
X_2^2	42497.11	1	42497.11	3.67	0.0843
Residual	1.157E+005	10	11568.36		
Lack of Fit	86751.68	6	14458.61	2.00	0.2619
Pure Error	28931.88	4	7232.97		
Cor Total	3.637E+006	16	5.868E+005		

The normal distribution plot was conducted to examine the residuals of the data and analyzed the linearity of response distribution. The distribution of the residual values in Figure 4 showed that there was a linear distribution. It confirmed that the residues from the experimental results of the chondroitin sulfate yield extracted from the *Bohadschia argus* were normally distributed.

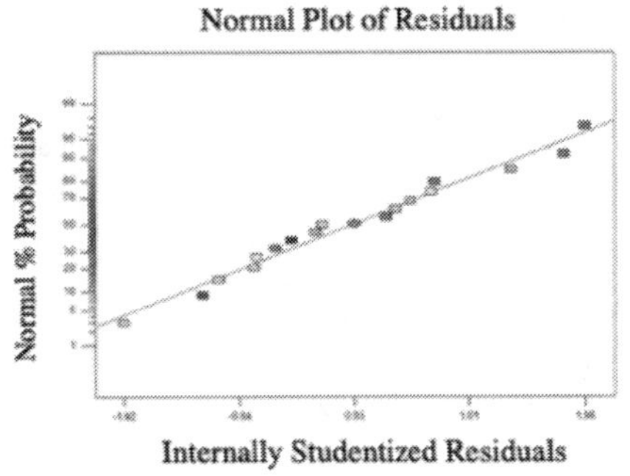

Fig. 4. Normal plot of residuals

The response surface plots between different independent variables and response were plotted to monitor the interactions of the variables and their effects on target response, CS yield. This response surface plot also determined the optimum level of each variable for the maximum response. The optimum values of the variables were obtained by solving the regression equation using the Design-Expert software. The 3D response surface plots and contour plots between different variables and yield of CS were represented in (Figure 5) and (Figure 6). Figure 5 showed the effect of enzyme concentration and extraction time on the yield of CS. From figure 5A, at the center point, where enzyme concentration was 0.25% and time 1.63 hours, a slight increase in the yield of CS was visible. It was also observed that the CS yield was increased with the increasing enzyme concentration.

This could be explained that the higher enzyme concentration caused a greater number of site-specific digestions leading to the release of more polysaccharides in a short extraction time. However, when the extraction time was extended, the trend of CS yield was different. With increasing extraction time, the yield of CS decreased, which could be due to the partial decomposition of the polysaccharide. This indicated that the shorter extraction time of enzyme concentration was suggested in this study for improving the extraction yield.

(A)

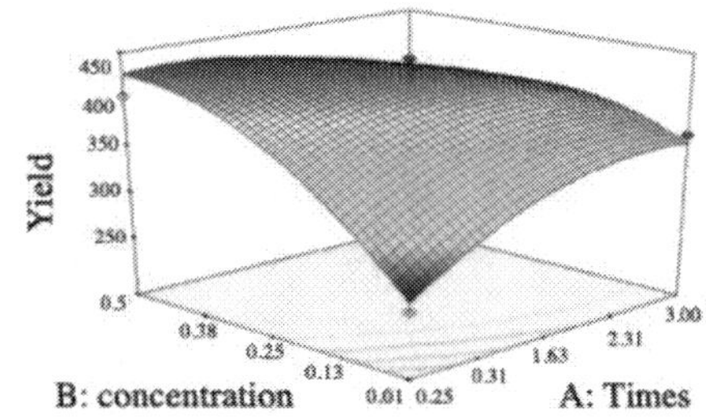

(B)

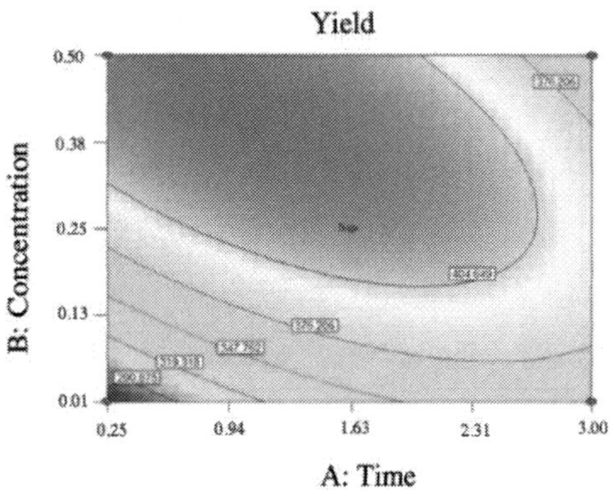

Fig. 5. (A) 3D response surface and (B) contour picture plots between extraction parameters (concentration, times) and yield.

On the other hand, when enzyme concentration was fixed at the center point (0.25%), CS yield was increased by increasing the extraction temperature from 55 °C to 60 °C (Figure 6). Figure 6 showed the correlation between enzyme concentration and extraction temperature on the CS yield at a fixed extraction temperature at 60 °C. CS yield increased steadily when enzyme concentration increased from 0.035 to 0.25%. However, the CS yield decreased when temperature exceeded 62 °C, which may be due to decomposition of CS.

(A)

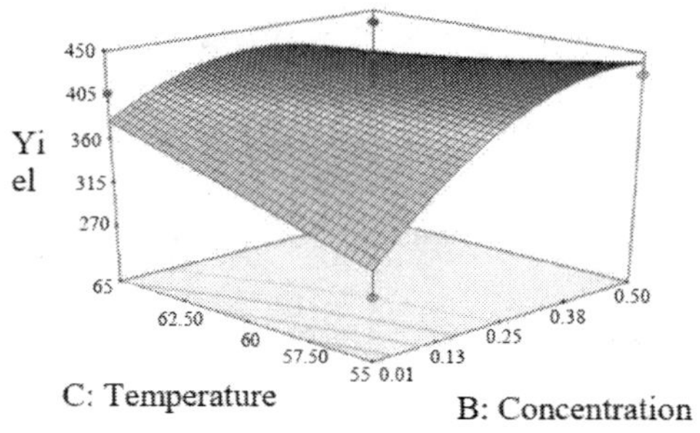

(B)

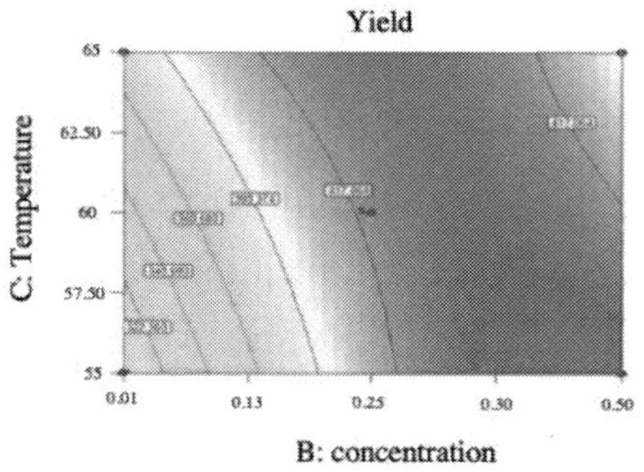

Fig. 6. (A) 3D response surface and (B) contour picture plots between extraction parameters (concentration, temperature) and yield.

Table 4. Factor and level of Box-Behnken design

Parameter	Optimum value	
	Predicted values[a]	Experimental values[b]
Concentration (%)	0.48	0.48
Time (min)	1.01	1.01
Temperature (°C)	56.53	57
Chondroitin sulfate yield (mg/100g dry)	444.566	415.59

Additionally, model validation experiments were performed using the predicted optimum extraction conditions from RSM (Table 4). The predicted optimum extraction conditions for the maximum yields of chondroitin sulfate contents were determined at an enzyme concentration 0.48%, extraction temperature of 56.53 °C and extraction times 0.96 hours. This predicted optimum extraction conditions will provide an extraction yield of 444.566 mg/100g dried sample. From this validation experiment, the chondroitin sulfate yield at 415.59 mg/100g dried was obtained, which showed model error of about 6.51%. This result showed that the actual experimental value was in good agreement with the predicted values obtained from the quadratic model [35-38]. Hence, the response surface modeling can be applied effectively to the prediction of extraction of chondroitin sulfate from *Bohadschia argus*.

4 Conclusion

RSM was employed to understand the relationship between the extraction parameters and yield of chondroitin sulfate. According to ANOVA, the effects of enzyme concentration on extraction yield were significant. And the best conditions to obtain a high yield of chondroitin sulfate content in the extracts were determined as enzyme concentration 0.48%, extraction time of 1.01 h., extraction temperature of 56.53 °C, with the maximum yield of chondroitin sulfate 415.59 mg/100g dry. Further studies are required to deduce the structure and the physiological activity of the chondroitin sulfate from *Bohadschia argus*.

5 Acknowledgment

Authors would like to thank to Wonnapob Co., Ltd. for financial support and raw materials used in this study, also King Mongkut's University of Technology North Bangkok (KMUTNB-PHD-64-02, KMUTNB-BasicR-64-37) and National Research Council of Thailand (NRCT) for financial supports.

References

1. V. Lobo, A. Patil, A. Phatak, N. Chandra, Free radicals, antioxidants and functional foods: Impact on human health, Pharmacognosy Reviews, **4**,8 (2010): 118-126
2. D. Uebelhart, M. Malaise, R. Marcolongo, F. Vathaire, M. Piperno, E. Mailleux, Intermittent treament of knee osteoarthritis with oral chondroitin sulfate: a one-year, randomized double-blind, multicenter study versus placebo, Osteoarthritis Cartilage, **12**,4 (2004): 269-76
3. T. Nakano, M. Betti, Z. Pietrasik, Extraction, Isolation and Analysis of Chondroitin Sulfate Glycosaminoglycans, Recent Patents on Food, Nutrition and Agriculture, **2**,1 (2010): 61-74
4. S. Bordbar, F. Anwar, N. Saari, High-value components and bioactives from sea cucumbers for functional foods-a review, Marine Drugs, **9**,10 (2011): 1761-805
5. C.C. yu, W.Y. yan, D.Z. Hua, C. Hong S. Jun, Optimization of microwave-assisted extraction of chondroitin sulfate from tilapia byproduct by r esponse surface methodology, (International Conference on Consumer Electronics, Communications and Networks ,2011)
6. M. Vigan, Allergic contact dermatitis caused by sodium chondroitin sulfate contained in a cosmetic cream. Contact Dermatitis, **70**,6 (2014): 383-384
7. T. Qiu, D. Wu, L. Yang, H. Ye, Q. Wang, Z. Cao, Exploring the Mechanism of Flavonoids Through systematic Bioinformatics Analysis, Frontiers in Pharmacol, **9** (2018): 918
8. K. Krebuansang, A. Swatdipong, S. Sammipak, Genetic diversity of *Saccostrea forkali* rock oyster in the gulf of Thailand, Applied Science and Engineering Progress, **13**,2 (2020): 158-165
9. L. Luo, M. Wu, L. Xu, J. Xiang, F. Lu, Comparison of physicochemical characteristics and anticoagulant activities of polysaccharides from three sea cucumbers, Marine Drugs, **11**,2 (2013): 399-417
10. E. Siahaan, R. Panggestuti, H. Munandar, S.K. Kim, Cosmeceuticals Properties of Sea Cucumber : Prospects and trends, Cosmetics, **4**,3 (2017): 26
11. P. Budzinski, M. Maimeun, P. Mutrakilcharoen, B. Wonganu, M. Sriariyanun, Profiling analysis of fatty acid and collagens obtained from sea cucumber, E3S Web of conference **141**,03006 (2020)
12. J. Vazquez, J. Fraguas, R. Novoa-Carvallal,R. Reis, L. Antelo,R. Perez-Martin, Isolation and Chemical Characterization of Chondroitin Sulfate from Cartilage By-Products of Blackmouth Catshark (*Galeus melastomus*), Marine Drugs. **16**,10 (2018): 344
13. G. Yoo, I. Lee, S. Park, N. Kim, J. Park, S. Kim, Optimization of Extraction Conditions for Phenolic Acids from the Leaves of Melissa officinalis L. Using Response Surface Methodology, Pharmacognosy Magazine, **14**,54 (2018):155-61
14. S. Akkaravathasinp, P. Narataruksa, C. Prapainainar, Optimization of semi-batch reactive distillation using response surface method: case study of esterification of acetic acid eith methanol in a process simulation, Applied Science and Engineering Progress, **12**,3 (2019): 209-215
15. P. Morreale, R. Manopulo, M. Galati, L. Boccanera, G. Saponati, L. Bocchi, Comparison of the anti-inflammatory efficacy of chondroitin sulfate and diclofenac sodium in patients with knee osteoarthrist osteoarthritis, J. Rheumatol. Journal Rheumatol, **23**,3 (1996): 1385-1391
16. L. Bucsi, G. Poor, Efficacy and tolerability of oral chondroitin sulfate as a symptomatic slow-acting drug for osteoarthritis (SYSADOA) in the treatment of knee osteoarthritis, Osteoarthritis Cartilage 6 (Suppl. A), **31**,36 (1998)
17. F. Ronca, L. Palmieri, Anti-inflammatory activity of chondroitin sulfate, Osteoarthritis Cartilage **14**,21 (1998)
18. D. Uebelhart, M. Malaises, R. Marcolongo, F. DeVathairell, M. Piperno, E. Mailleux, A. Fioravanti, L. Matoso, E. Vignon, Intermittent treatment of knee osteoarthritis with oral chondroitin sulfate: a one year, randomized, double blind, multicenter study versus placebo, Osteoarthr, Cartilage, **12**,4 (2004) 269-276 DOI: 10.1016/j.joca.2004.01.004.
19. X.M. Luo, G.J. Fosmire, R.M. Leach, Chicken keel cartilage as a source of chondroitin sulfate, Poultry Science, **81**,7 (2002): 1086–1089.
20. S.K. Sikder, A. Das, Isolation and characterization of glycosaminoglycans (mucopolysaccharides)

from the skin of the fish *Labeo rohita*, Carbohydrate Research, **71**,1 (1979): 273–285.
21. A. Hjerpe, B. Engfeldt, T. Tsegenidis, Analysis of the acid polysaccharides from squid cranial cartilage and examination of a novel polysaccharide, Biochimica et Biophysica Acta, **754**,1 (1983): 85-91.
22. D.H. Vynios, A. Aletras, C.P. Tsiganos, T. Tsegenidis, C.A. Antonopoulos, A. Hjerpe, B. Engfeld, Proteoglycans from squid cranial cartilage extraction and characterization, Comparative Biochemistry and Physiology, **80**,4 (1985): 761–766.
23. G.S. Harper, P.G. Allingham, X. Qiu, Cartilage Co-Products Commercial Development from Alternative Production Species, (RIRDC Publication, Germany, 2000)
24. T. Sumi, H. Ohba, T. Ikegami, M. Shibata, T. Sakaki, I. Sallay, S.S. Park. Method for the preparation of chondroitin sulfate compounds. (US Patent 6,342,367, 2002)
25. P. Hoffman, T.A. Mashburn, Protein–polysaccharide of bovine cartilage. I: Extraction and electrophoretic studies, Journal of Biological Chemistry, **242**,17 (1967) 3799-3804.
26. S.W. Sajdera, V.C. Hascall, Protein polysaccharide complex from bovine nasal cartilage, Journal of Biological Chemistry, **244**,1 (1969): 77-87.
27. N.K. Karamanos, A. Syrokou, P. Vanky, M. Nurminen, A. Hjerpe, Determination of 24 variously sulfated galactosaminoglycan and hyaluronan derived disaccharides by high performance liquid chromatography, Analytical Biochemistry, **221**,1 (1994) 189-199
28. Y. Lye, N. Salih, J. Salimon, Optimization of partial epoxidation on *Jatropa curcas* oil base methyl linoleate using urea hydrogen peroxide and methyltrioxorhenium catalyst. Applied Science Engineering Progress, **14**,1 (2021): 89-99
29. R. Pangestuti, Z. Arifin, Medicinal and health benefit effects of functional sea cucumbers. Journal Tradit Complement Medicine, **8**,3 (2018): 31-51
30. W. Garnjanagoonchorn, L. Wongekalak, A. Engkagul, Determination of chondroitin sulfate from different sources of cartilage. Chemical Engineering and Processing: Process Intensification, **46**,5 (2007): 465-71
31. A. Ahmad, M.U. Rehman, A.F. Wali, H.A. ElSerehy, F.A. Misned, S.N. Maodaa, H.M. Aljawdah, T.M. Mir, Box- Behnken Response Surface Design of Polysaccharide Extraction from *Rhododendron arboreum* and the Evaluation of Its Antioxidant Potential Molecules, **25**,17 (2020): 3835
32. H. C. Tran, H. A. T. Le, T. T. Le, and V. M. Phan, Effects of enzyme types and extraction conditions on protein recovery and antioxidant properties of hydrolysed proteins derived from Defatted Lemna minor, Applied Science and Engineering Progrss, (2021).
33. N.T.L. Vien, P.B. Nguyen, L.D.C. Cuong, T.T.T. An, D.T.A. Dao, Optimization of papain hydrolysis condition for release of glycosaminoglycans from the chicken keel cartilage. Chemical engineering food and biotechnology, **1878**,1 (2017): 020009
34. V. Coulson-Thomas, T. Gesteira, Dimethylmethylene Blue Assay (DMMB). (Bio Protocol, 4,18 2020)
35. P. Pangsri, T. Wuttipornpun, W. Songserm, Mannanase and Cellulase Enzyme Production from the Agricultural Wastes by the Bacillus subtilis P2-5 Strain, Applied Science Engineering Progress, **14**,3 (2021): 425-434
36. E.J. Panakkal, M. Sriariyanun, J.Ratanapoompinyo, P. Yasurin, K. Cheenkachorn, W. Rodiahwati, P. Tantayotai, Influence of Sulfuric Acid Pretreatment and Inhibitor of Sugarcane Bagasse on the Production of Fermentable Sugar and Ethanol. Applied Science Engineering Progress, **15**,1 (2022)
37. L.K. Akula, R.K. Oruganti, D. Bhattacharyya, K.K. Kurilla. Treatment of Marigold Flower Processing Wastewater Using a Sequential Biological-Electrochemical Process, Applied Science Engineering Progress, **14**,3 (2021): 525-542
38. M. Bahadi, N. Salih, J. Salimon. D-Optimal Design Optimization for the Separation of Oleic Acid from Malaysian High Free Fatty Acid Crude Palm Oil Fatty Acids Mixture Using Urea Complex Fractionation, Applied Science Engineering Progress, **14**,2 (2021): 175-186

Improvement of potassium permanganate pretreatment by enzymatic saccharification of rice straw for production of biofuels

Parita Mutrakulcharoen[1], *Peerapong* Pornwongthong[1], *Anne Sahithi S.T* [2], *Theerawut* Phusantisampan[3], *Atthasit Tawai*[4], *Malinee* Sriariyanun[4,5*]

[1]Department of Agro-Industrial, Food, and Environmental Technology, Faculty of Applied Science, King Mongkut's University of Technology North Bangkok (KMUTNB), Bangkok, Thailand
[2]School of Biosciences and Technology (SBST), Vellore Institute of Technology (VIT) Vellore, Tamilnadu, India
[3]Department of Biotechnology, Faculty of Applied Science, King Mongkut's University of Technology North Bangkok (KMUTNB), Bangkok, Thailand
[4]Department of Chemical and Process Engineering, The Sirindhorn International Thai-German Graduate School of Engineering (TGGS), King Mongkut's University of Technology North Bangkok (KMUTNB), Bangkok, Thailand
[5]Biorefinery and Process Automation Engineering Center (BPAEC), King Mongkut's University of Technology North Bangkok (KMUTNB), Bangkok, Thailand

Abstract. Commonly, the agricultural waste, i.e. lignocellulosic biomass is disposed through combustion causing air pollution with production of PM2.5 and PM10 particles. However, it has been found that these biomasses can be used as source for the production of biofuels and other valuable biochemicals. Though deconstruction of lignocellulosic biomass is challenging due to its complex structure. In this study, rice straw (RS) was pretreated using potassium permanganate ($KMnO_4$) to enhance the enzymatic saccharification efficiency. The study was carried out by varying the operational factors in pretreatment, including temperature (30-90°C), time (30-360 min) and concentration of $KMnO_4$ (0.5-3.0, % w/v), respectively, based on Box-Behnken design (BBD). Through multi-regression analysis of the experimental data obtained after pretreatment, the optimum conditions were determined. The optimum conditions for temperature, time and potassium permanganate concentration were 48.09°C, 360 min, and 1.36% w/v, respectively. The saccharifications of pretreatment and untreated rice straw were carried out using Cellic Ctec2. The reducing sugar was determined by using DNS method and the yields of the untreated and pretreated RS were 32.38 and 49.011 mg/mL, respectively. The results showed that the sugar for pretreated RS were 1.51 fold times higher compared to untreated RS. Therefore, this work illustrates the pretreatment efficiency for KMnO4 to enhance the reducing sugar yield during saccharification, which can be used for biofuel and value-added product productions.

Keywords. Potassium permanganate, Pretreatment, Enzymatic saccharification, Biorefinery, Biofuels

1 Introduction

Lignocellulosic biomass is an inedible organic material source having the potential to be utilized as an alternative source for fossil fuels. Common lignocellulosic biomass includes forest residues, agricultural residues and dedicated crops. It has been reported in 2013 that 62 million tons of waste biomass are generated from the agricultural activities of Thailand. The biomass waste accounts for 13.3 million tons and is likely to increase further in coming years. Increase in the waste generation is related to the continuous increase in the food demand due to the rapid growth of population. The industrial growth is also proportional to the population growth but it greatly affects the environments [1]. It was reported that the air was highly polluted in Thailand (2007-2016) due to the presence of particulate matters (PM10 an PM2.5) [2]. One of the causal agents of these polluted particles was due to the incineration of waste biomass generated after the harvesting seasons. This tradition is being followed worldwide to prepare the soil for next harvesting season. Incinerations of agricultural residues also increase in the emission of carbon dioxide (CO_2), an important of greenhouse gas.

To lessen combustion of agricultural waste, lignocellulose biomass has been proposed to utilized as raw materials for productions of biofuels and biochemical [3]. Lignocellulose is composed of cellulose, hemicellulose, and lignin. Each of this component varies in different ratio from biomass to biomass. In general lignocellulose composes of cellulose, hemicellulose and lignin in the range of 40-50%, 20-30%, and 10-25%, respectively [4-5]. Cellulose is a polysaccharide chain of glucose units linked via β-

*Corresponding author: macintous@gmail.com

1,4 glycosidic bonds. On the other hand, hemicellulose is a polysaccharide with C_5 sugars (xylose and arabinose) and C_6 sugars (glucose, galactose, and mannose). Lignin is an organic polymer composed of phenylpropane units. [6-8]. However due to the recalcitrance nature of the lignocellulosic biomass, it has been a challenge for the researchers for converting biomass to biofuel and other by-products. Conversion of lignocellulose to the valuable products, therefore requires three major steps such as pretreatment, saccharification, and fermentation. [9-11]

In recent years, several pretreatment methods have been developed for the fractionation of lignocellulosic biomass. Pretreatment methods include physical pretreatment (mechanical size reduction, pyrolysis), chemical pretreatment (ozonolysis, acid pretreatment, alkali pretreatment, oxidative delignification), physicochemical pretreatment (stream explosion, ammonia fiber explosion, microwave pretreatment, carbon dioxide explosion, wet oxidation) and biological pretreatment (fungus and bacteria) [4], [12], [13-17]. Pretreatment is an essential step prior to saccharification process. Pretreatment leads to fractionation of biomass through solubilization and structure modification of lignin and hemicellulose, respectively. After pretreatment, cellulose is converted to glucose by the cellulase enzymes due the ease in accessibility. Though pretreatment methods have advantages in the fractionation of biomass, each pretreatment method has its own disadvantage, for instance, some pretreatment methods require high energy or use toxic substances or have high operational cost [4].

Among the different chemical pretreatment methods, acid and alkali pretreatments are studied worldwide. However strong acid and alkaline are considered to be harmful to both the operator and the environment. In addition, acid pretreatment degrades sugars to inhibitory compounds (furfural and 5-hydrozymethylfurfural [18]) to the functions of enzymes, and therefore decreases the conversion efficiency from cellulose to glucose monomers. On the other hand, lignin and hemicellulose is solubilized completely or partially during alkaline pretreatment, respectively [17-20]. Lee et. al., reported delignification and saccharification efficiency of 89.4% and 75.9% in corn stover after diluted alkaline pretreatment [21]. On the other hand, Ravindran et. al., reported delignification and cellulose recovery of 46% and 98%, from $KMnO_4$ pretreated spent coffee waste, respectively [22]. Ma et. al., reported 46.79%, and 81.47% removal of lignin and hemicellulose from pretreated corncob using $KMnO_4$ [23].

In this study, $KMnO_4$ pretreatment of the rice straw was carried out to improve saccharification efficiency. $KMnO_4$ is inexpensive, readily available, and less toxic compared to acids such as sulphuric acid and it is one of the common chemicals in kitchen application. $KMnO_4$ pretreatment of rice straw was carried out by varying the reaction time, temperature and concentration of $KMnO_4$. The $KMnO_4$ pretreatment conditions were optimized based on Response Surface Methodology (RSM) with Box Behnken Design (BBD). The operational parameters were optimized through multiple regression analysis that generated the mathematical model. The optimal condition of pretreatment obtained from the model was validated by experiments to confirm the reliability of the predicted pretreatment conditions.

2 Material and methods

2.1. Raw materials

Rice straw was obtained from Phra Nakhon Si Ayutthaya province, Thailand. The collected rice straw was subjected to particle reduction using food processor after drying in a hot air oven at 60°C to remove moisture. The dried sample was screened through 20 mesh sized aluminium sieve for the homogeneous nature of particle size.

Cellic Ctec2, a commercial cellulase enzyme, was obtained from Sigma-Aldrich (aqueous solution, density 1.15 g/mol). This enzyme is a cocktail mixture of cellulases, ß-glucosidases and hemicellulase. Potassium permanganate ($KMnO_4$), and other analytical reagents were obtained from Alex Finechem (Univar).

2.2 Pretreatment of rice straw

Pretreatment of rice straw using $KMnO_4$ was carried out under different reaction time, temperature and concentration of $KMnO_4$. A solid loading of 10% (1:10 w/w) was maintained throughout the study. Firstly, the testing ranges of pretreatment factors, such as reaction time, temperature, and concentration of $KMnO_4$ were varied to 2-8 h, 30 – 75°C, and 1 – 4% w/v, respectively to preliminarily evaluate the boundary of testing ranges for RSM experiments. Secondly, after the determination of the suitable testing ranges for RSM, the experimental designed for optimization was carried out based on BBD. Three independent variables, composing of time (30-360 min), temperature (30 - 90°C), and $KMnO_4$ concentration (0.5 - 3% w/v), were set with three testing levels (-1, 0, and +1) as summarized in Table 1. The pretreatments of rice straw were carried out by following the individual runs as summarized in Table 2. The pretreated solid biomass was harvested by filtration with No.1. Whatman filter paper to remove the liquid hydrolysate. After collection of the hydrolysate, the solids were washed using deionized water until pH was neutral. The pretreated solids were dried in a hot air oven at 60°C until the weight was constant. The dried pretreated and untreated samples were subjected to enzymatic saccharification, and reducing sugars released from hydrolysis reactions were analysed by DNS assay to determine the pretreatment efficiency.

The amounts of reducing sugars obtained from each experimental runs were analyzed by ANOVA analysis to observe the effect of pretreatment parameters on yields of reducing sugars with the significant level of P-value <0.05 for the model and term model. The mathematical model derived from ANOVA analysis was then generated to predict the optimal pretreament condition. The correlation coefficiency (R^2) was calculated to confirm the reliability of the predicted model. All statistical analysis of the experimental results obtained for the BBD design were analyzed using Design-Expert software (version 7.0)

Table 1. Independent variables in pretreatment condition of rice straw based on RSM.

Independent variable	Coded symbols	Levels		
		-1	0	+1
Time (min)	X_1	30	195	360
Temperature (°C)	X_2	30	60	90
Concentration (%)	X_3	0.50	1.75	3.00

Table 2. RSM experiment design for $KMnO_4$ pretreatment of rice straw

Run	Pretreatment condition			Reducing sugar (mg/mL)
	X_1: Time (min)	X_2: Temperature (°C)	X_3: Concentration (%)	
1	195	60	1.75	47.86
2	195	30	0.50	43.36
3	360	60	3.00	30.53
4	30	60	0.50	42.60
5	360	90	1.75	52.70
6	360	60	0.50	45.25
7	30	30	1.75	43.99
8	195	60	1.75	47.86
9	195	30	3.00	31.25
10	195	60	1.75	45.59
11	30	60	3.00	31.61
12	195	60	1.75	51.02
13	195	90	0.50	45.72
14	30	90	1.75	46.68
15	360	30	1.75	49/04
16	195	60	1.75	49.75
17	195	90	3.00	30.36

2.3 Enzymatic hydrolysis

Enzymatic saccharification process was carried out for untreated and pretreated rice straw to determine the efficiency of $KMnO_4$ pretreatment. For each reaction, 0.2 g of the biomass (untreated and pretreated) was added to 8 mL of citrate buffer (50 mM, pH 4.8) in a 50 mL centrifuge tube. In addition, 80 μL of 2 M sodium azide were added to avoid the contamination of the samples due to growth of microbes during saccharification process. To the mixture, 1.4 μL Cellic Ctec2 enzyme mix was added. The mixed samples were placed in a shaking incubator with the temperature and mixing speed at 50°C and 150 rpm, respectively. Enzymatic saccharifications of the untreated and treated rice straw were carried out for 72 h. After 72 h, the liquid hydrolysate was taken from the mixture by centrifugation at 10,000×g for 5 min. The supernatant was collected for the determination of the reducing sugar by following the modified dinitrosalicylic acid (DNS) method.

2.4 Analysis of reducing sugar

The reducing sugar concentration in the hydrolysate was measured by following the dinitrosalicylic acid (DNS) method. In a 1.5 mL centrifuge tube, 50 μL of hydrolysate and 3,5-dinitrosalicylic acid (DNS) reagent was added. The tubes were incubated at 95°C using a water bath for 5 min. The tubes were allowed to cooled down in an ice bath for 5 min. After cooling of the sample, 1 mL of distilled water was added and mixed with a vortex shaker. After the samples were prepared, the reducing sugar was analyzed using UV/Vis spectrophotometer at 540 nm. The concentration of the reducing sugar (mg/mL) was calculated by using a standard glucose curve.

3 Result and discussion

3.1 Determine the testing levels of pretreatment parameters

The effects of reaction, temperature and $KMnO_4$ concentration on pretreatment of rice straw were preliminary evaluated to set the boundary of testing level for RSM. Three sets of experiments to pretreat rice straw were conducted by varying each pretreatment factor at a time (Figure 1). Firstly, the pretreatment time was varied between 2 to 8 hr, and the pretreatment temperature was fixed at 40°C using 2% (w/v) $KMnO_4$ (Figure 1a). Secondly, the pretreatment temperature was varied between 30 to 75°C, and the pretreatment time was fixed at 4 h, using 2% (w/v) $KMnO_4$ (Figure 1b). Thirdly, the $KMnO_4$ concentration was varied between 1-4%, and the pretreatment time and temperature were set at 40°C for 4 h (Figure 1c). It was observed from Fig. 1a that the reducing sugar yield varied for pretreated samples at different reaction time (2 – 8h). The reducing sugar yields increased 1.15 fold times (18.36 to 21.13 mg/mL), when the reaction time was varied between (2-6 h) and this increasing yield was saturated at between 6-8 h. Therefore, the upper boundary level of pretreatment time of 6 h was selected for RSM experiments. Next, the testing boundary level of pretreatment temperature (30–75°C) on saccharification efficiency was evaluated (Fig. 1b). Maximum reducing sugar yield of 23.68 mg/mL was obtained when rice straw was pretreated at 60°C. As the pretreatment temperature was increased to 75°C, it was found that the reducing sugar yield decreased 1.01 fold times. Therefore, it was implied as similar to the effect of reaction time that, sugars were degraded at higher temperature. Due to the lower effect of temperature on reducing sugar from rice straw pretreated at 75°C. The temperature for BBD design was considered to be 30 – 90°C. Meanwhile, the independent effect of KMNO4 concentration on reducing sugar yield during saccharification of pretreated rice straw was also investigated (Fig.1c). The concentration of $KMNO_4$ was varied from 1-4% w/v. It was noted that the reducing sugar yield was higher initially (27.04 mg/mL), when the $KMnO_4$ concentration was maintained at 1% w/v

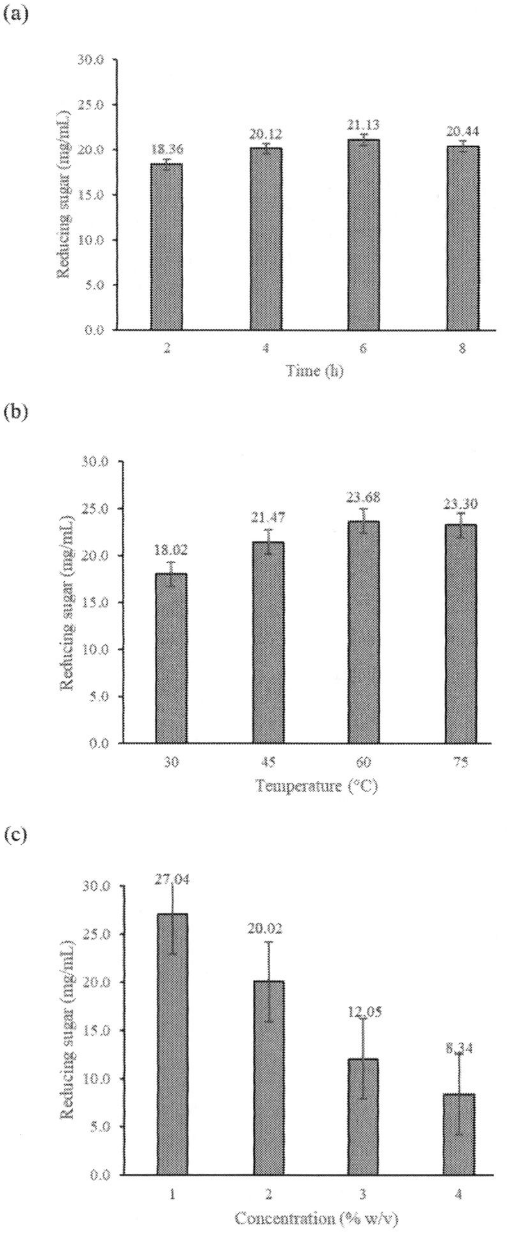

Fig. 1. Effect of pretreatment time, temperature and $KMnO_4$ concentration on the enzymatic hydrolysis of rice straw. (A) The pretreatment was conducted using 2% (w/v) $KMnO_4$ at 40°C, (B) The pretreatment was conducted using 2% (w/v) for 4 hours and (C) The pretreatment were conducted at 40°C for 4 hours.

However, as the $KMNO_4$ concentration increased from 1 – 4% w/v, the reducing sugar yield decreased 3.24 fold times (27.04-8.34 mg/mL). Thus, the saccharification efficiency was negatively impacted by the increase in concentration of $KMnO_4$ during pretreatment. Therefore, the $KMnO_4$ concentration in the BBD design was varied from 0.5-3.0% w/v. The results or released reducing sugars showed that the parameters with the preliminary tested ranges were not statistically significant. Therefore, the independent variables levels were modified and used in the BBD design of RSM study (Table 1).

Overall, it was observed that the testing levels of independent variables (pretreatment time, pretreatment temperature and $KMnO_4$ concentration) had effects on enzymatic saccharifications of rice straw. This finding was in agreement with different studies carried out by several published works. Romero et al performed pretreatment of rapeseed straw using $FeCl_3$ under various conditions of time, temperature and $FeCl_3$ concentration. A direct correlation between reducing sugar content and operational parameters were reported [24]. In addition, Dahunsi et al performed alkali pretreatment of pineapple peel and reported similar correlation between reducing sugar contents and operational parameters [25].

3.2 RSM design and experimental testing

The RSM experiment has been proposed as an effective method of designing and conducting experiments to determine the optimal condition for production process. The RSM study assists evaluation of effects of individual independent variables and their interactions on the dependent variables. RSM reduces the numbers of experimental runs, in turn, reducing the experimental cost. Furthermore, RSM is often used for generation of the empirical model and prediction of the desired response under different conditions of independent variables [15]. In recent years, several studies have illustrated the significance of using RSM in reducing the experimental runs and improving the process efficiency by varying numbers of parameters conditions simultaneously [26-27].

In the present study, the RSM study was carried out by generating a BBD matrix (Table 3) with three testing levels (-1, 0, +1) for three pretreatment parameters (time, temperature and $KMnO_4$ concentration) (Table 1) with a total 17 experimental runs (Table 2). The statistical analysis was carried out using Design-Expert software (ver 7.0.0) throughout the whole study. The experimental data were used in determining the regression coefficient (β) using least square method. In addition, the second order model was generated from the experimental data. The optimum pretreatment conditions were determined for achieving maximum reducing sugar yield from the multi-regression analysis of the second-order model. The second order model significance were evaluated by (a) determining the coefficient of determination (R^2), (b) significance of model terms in ANOVA, and (c) non-significance of the lack-of-fit. The second-order model is generally considered to be significant when the R^2 value is greater than 0.9. In this study, the R^2 value of the second-order model was 0.9646 implying lower variance between the experimental data and the representative model. In fact, the R^2 value of higher than 0.9 does not always indicate the significance of the model. Therefore, ANOVA was carried out to determine the significance of the second-order model (Table 3).

Table 3. ANOVA analysis of the second-order model obtained from RSM experiments.

Source	Sum of Squares	df	Mean Squares	F Value	P-value Prob > F
Model	855.98	3	285.33	80.13	< 0.0001
A-Time	19.77	1	19.77	5.55	0.0348
C-Conc	352.76	1	352.76	99.07	< 0.0001
C^2	483.46	1	483.46	135.76	< 0.0001
Residual	46.29	13	3.56		
Lack of Fit	29.14	9	3.24	0.76	0.6677
Pure Error	17.15	4	4.29		
Cor Total	902.27	16			

The ANOVA results indicated that the model with F-value of 80.13 was statistically significant ($P<0.05$). There was only a 0.01% chance that a "Model F-value" could occur due to noise (Table 3). Furthermore, the model terms of time, and $KMnO_4$ concentration were also significant (since $P<0.05$). Based on the same criteria, it was noted that neither the interaction terms for time, temperature and $KMnO_4$ concentration was significant. However, significance for the quadratic term of $KMnO_4$ concentration was seen. Therefore, the accuracy of the second-order model was improved by including the model terms that were significant. Additionally, the model significance was also evaluated through determination of the lack-of-fit test. The F-value of 0.76 implied that the lack of fit is not significant relative to the pure error. There is a 66.77% chance that a "Lack of Fit F-value" this large could occur due to noise. Therefore, the second order model was considered significant since the lack of fit was not significant and the errors were normally distributed. Therefore, through evaluation of the second-order model, time and $KMnO_4$ concentration were considered to have more effect on the reducing sugar yield released during saccharification after pretreatment of rice straw.

3.3 Effects of pretreatment parameters on enzymatic saccahrification and validation of mathematical models for pretreatment

The effect of each significant model terms (time, and $KMnO_4$ concentration) on response (reducing sugar concentration) was monitored (Fig. 2). In addition, the interaction effects of the independent variables on response were visualized through three-dimensional contour plots (Fig. 3). It can be seen that the reducing sugar yield obtained from saccharification was increased when rice straw was pretreated with longer pretreatment time (Fig 2a). Under variations of $KMnO_4$ concentration, the reducing sugar yield was increased when the $KMnO_4$ concentration increased. However, when $KMnO_4$ concentration was higher than 1.36%, the more $KMnO_4$ concentration, the less reducing sugar yields (Fig 2b). Meanwhile, interaction effects of time, temperature, and $KMnO_4$ resulted in similar observation (Fig 3).

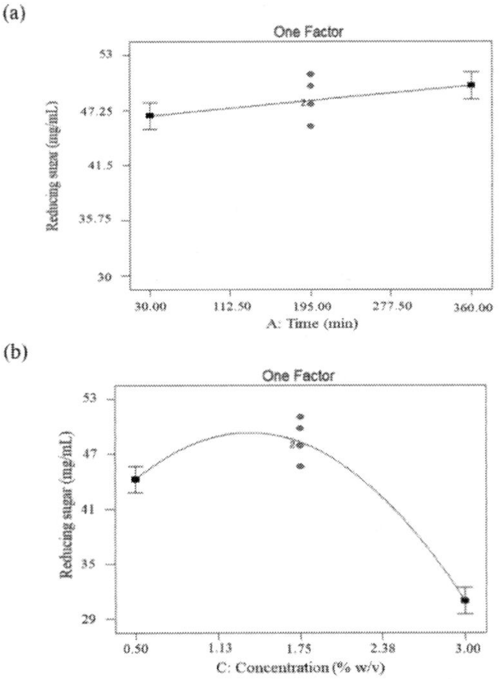

Fig. 2. The relationship between each pretreatment factor, including (A) pretreatment time (min) and (B) $KMnO_4$ concentration (%) and reducing sugar concentration (mg/mL) obtained from pretreated rice straw.

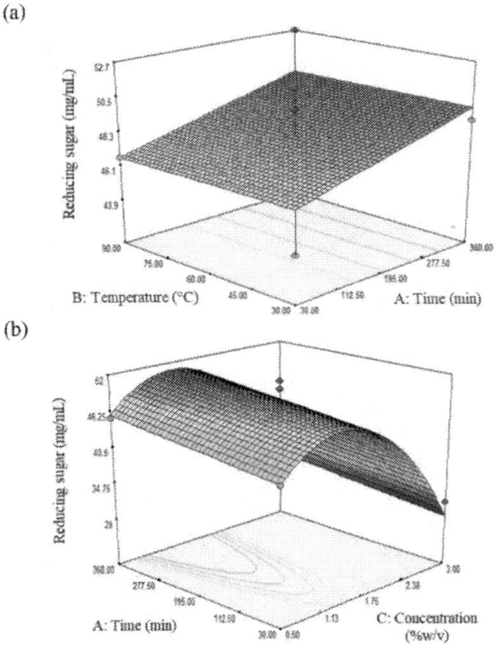

Fig. 3. Response surface plot represents the effects of pretreatment factors on reducing sugar yield (mg/mL). (A) pretreatment time vs. pretreatment temperature and (B) $KMnO_4$ concentration vs pretreatment time.

Through ANOVA analysis of the second order model, the relationships of pretreatment conditions on reducing sugar concentrations were represented in the form of mathematical model (Table 4). The optimum pretreatment conditions for obtaining maximum reducing sugar yield at 50.88 mg/mL were predicted and summarized in Table 4. To confirm the reliability of this model, the rice straw was pretreated again using this predicted optimum pretreatment conditions. A reducing sugar yield of 49.01 mg/mL was obtained from the enzymatic saccharification of the pretreated rice straw under optimum pretreatment conditions. Thus, there was only 3.49% error obtained from experimental result compared to predicted yield from the model, indicating high validity of RSM model. The influence of time and $KMnO_4$ concentration on reducing sugar yield were in agreement with studies carried out by Romero at al. [24] and Zhao et al. [28]. Therefore, it was found that increase in reaction time with optimum $KMnO_4$ concentration can improve enzymatic saccharification efficiency.

Table 4. Mathematical models and optimal pretreatment condition obtained from RSM experiment.

Optimal pretreatment parameter			Reducing sugar yield (mg/mL)	
Time (min)	Temperature (°C)	$KMnO_4$ Concentration (% w/v)	Predicted	Experimental
360	84.09	1.36	50.88	49.01
Mathematical Model Reducing Sugar = +34.77502 + 9.52745 × 10-3 × Time + 18.62003 × Conc – 6.83781 × Conc²				

4 Conclusion

The present study was focused on the objective to obtain maximum reducing sugar from rice straw during saccharification after pretreatment using $KMnO_4$. These reducing sugars have the potential to be converted to biofuels and other valuable products. The individual effect of the operational parameters (time, temperature and $KMnO_4$ concentration) was studied based on RSM with BBD. The optimum pretreatment conditions for time, temperature, and $KMnO_4$ concentration were 84.09°C, 360 min, and 1.36% (concentration of $KMnO_4$). The maximum reducing sugar yield from pretreated rice straw under optimum conditions were 49.01 mg/mL. In conclusion, $KMnO_4$ pretreatment had significant effect on rice straw to improve the saccharification efficiency and had potential to be applied in further uses due to its less toxicity and cost.

Acknowledgement

The authors would like to thank the King Mongkut's University of Technology North Bangkok (Grant Contract No. KMUTNB-BasicR-64-37) for financial support of this work.

References

1. A. Sukkaew, P. Boonsong, S. Thongpradistha, M. Intan, Physical and Chemical Pretreatment of Lignocellulosics in Pineapple (Ananus comosus) Peels Dried for Investment, AIP Conference Proceedings, **1868**, 090001 (2017)
2. R. Janta, K. Sekiguchi, R. Yamaguchi, K. Sopajaree, B. Plubin, T. Chetiyanukornkul, Spatial and Temporal Variations of Atmospheric PM10 and Air Pollutants Concentration in Upper Northern Thailand During 2006–2016, Applied Science and Engineering Progress, **13**,3 (2020): 256-267
3. EJ. Panakkal, M Sriariyanun, J. Ratanapoompinyo P. Yasurin, K. Cheenkachorn, W. Rodiahwati, P. Tantayotai, Influence of Sulfuric Acid Pretreatment and Inhibitor of Sugarcane Bagasse on the Production of Fermentable Sugar and Ethanol, Applied Science and Engineering Progress, **15**,1(2022)
4. S. Cheng, P. Mutrakulcharoen, S. Chuetor, K. Cheenkachorn, T. Tantayotai, E. J. Panakkal, M. Sriariyanun, Recent Situation and Progress in Biorefining Process of Lignocellulosic Biomass: Toward Green Economy, Applied Science and Engineering Progress, **13**,1 (2020): 3-10
5. R. Akkharasinphonrat, T. Douzou, M. Sriariyanun, Development of ionic liquid utilization in biorefinery process of lignocellulosic biomass, King Mongkut's University of Technology North Bangkok International Journal of Applied Science and Technology, **10**,2 (2017): 89-96
6. H. Chen, J. Liu, X. Chang, D. Chen, Y. Xue, P. Liu, H. Lin and S. Han, A review on the pretreatment of lignocellulose for high-value chemicals, Fuel Processing Technology, **160** (2017): 196-206
7. A.K. Obeng, D. Premjet, and S. Premjet, Fermentable Sugar Production from the Peels of Two Durian (Durio zibethinus Murr.) Cultivars by Phosphoric Acid Pretreatment, Resources, **7**,4 (2018): 60
8. P. Moyer, K. Kim, N. Abdoulmoumine, S.C. Chmely, B.K. Long, D.J. Carrier, and N. Labbé, Structural changes in lignocellulosic biomass during activation with ionic liquids comprising 3-methylimidazolium cations and carboxylate anions, Biotechnology for Biofuels. **11** (2018): 265
9. MP. Gundupalli, YS. Cheng, S. Chuetor, D. Bhattacharyya, M. Sriariyanun. Effect of Dewaxing on saccharification and ethanol production from different lignocellulosic biomass. Bioresource Technology. **125596.**
10. M. Sriariyanun and K. Kitsubthawee. 2020. Trends in Lignocellulosic Biorefinery for Production of Value-added Biochemicals. Applied Science and Engineering Progress, **13**, 2 (2020)

11. P.L. Show and M. Sriariyanun. 2020. Prospect of Liquid Biphasic System in Microalgae Research. Applied Science and Engineering Progress, **14**, 2 (2020): 1–2
12. L. Shao, H. Chen, Y. Li, J. Li, G. Chen, G. Wang, Pretreatment of corn stover via sodium hydroxide–urea solutions to improve the glucose yield, Bioresource Technology, **307** (2020)
13. Y.S. Cheng, Z.Y. Wu, and M. Sriariyanun, Evaluation of Macaranga tanarius as a biomass feedstock for fermentable sugars production, Bioresource Technology, **294** (2019): 122195
14. A. Boontum, J. Phetsom, W. Rodiahwati, K. Kitsubthawee, and T. Kuntothom. Characterization of Diluted-acid Pretreatment of Water Hyacinth, Applied Science and Engineering Progress, **12**, 4 (2019): 253–263
15. K. Rattanaporn, S. Roddecha, M. Sriariyanun, and K. Cheenkachorn. Improving Saccharification of Oil Palm Shell by Acetic Acid Pretreatment for Biofuel Production, Energy Procedia, **141** (2017): 146–149
16. K. Rattanaporn, P. Tantayotai, T. Phusantisampan, P. Pornwongthong, and M. Sriariyanun, Organic acid pretreatment of oil palm trunk: effect on enzymatic saccharification and ethanol production, Bioprocess and Biosystems Engineering, **41**, 4 (2018), 467–477
17. K. Cheenkachorn, T. Douzou, S. Roddecha, P. Tantayotai, and M. Sriariyanun, Enzymatic Saccharification of Rice Straw under Influence of Recycled Ionic Liquid Pretreatments, Energy Procedia, **100** (2016): 160–165
18. P. Rachmontree, T. Douzou, K. Cheenkachorn, M. Sriariyanun and K. Rattanaporn, Furfural: A sustainable platform chemical and fuel Applied Science and Engineering Progress, **13**, 1 (2020): 3-10
19. S. Sahu and K. Pramanik, Evaluation and Optimization of Organic Acid Pretreatment of Cotton Gin Waste for Enzymatic Hydrolysis and Bioethanol Production., Applied Biochemistry and Biotechnology, **186**, 4 (2018): 1047–1060
20. S.M.G. Saerens, F.R. Delvaux, K.J. Verstrepen and J.M. Thevelein, Production and biological function of volatile esters in Saccharomyces cerevisiae, Microb Biotechnology, **3**, 2 (2010): 165–177
21. J.W. Lee, J.Y. Kim, H.M. Jang, M.W. Lee and J.M. Park, Sequential dilute acid and alkali pretreatment of corn stover: Sugar recovery efficiency and structural characterization, Bioresource Technology, **188** (2015): 296-301
22. R. Ravindran, S. Jaiswal, N.A. Ghannam and A.K. Jaiswal, Evaluation of ultrasound-assisted potassium permanganate pretreatment of spent coffee waste, Bioresource Technology, **224** (2017): 680-687
23. I. Romero, J.C. López-Linares, M. Moya and E. Castro, Optimization of sugar recovery from rapeseed straw pretreated with $FeCl_3$, Bioresource Technology, **268** (2018): 204-211
24. L. Ma, Y. Cui, R. Cai, X. Liu, C. Zhang and D. Xiao, Optimization and evaluation of alkaline potassium permanganate pretreatment of corncob, Bioresource Technology, **180** (2015): 1-6
25. S.O.Dahunsi, Liquefaction of pineapple peel: Pretreatment and process optimization, Energy, **185** (2019): 1017-1031
26. LK. Akula, RK Oruganti, D. Bhattacharyya and KK. Kurilla. Treatment of Marigold Flower Processing Wastewater Using a Sequential Biological-Electrochemical Process. Applied Science Engineering Progress, **14**,3 (2021): 525-542
27. P. Amnuaycheewa, W. Rodiahwati, P. Sanvarinda, K. Cheenkachorn, A. Tawai and M. Sriariyanun, Effect of Organic Acid Pretreatment on Napier Grass (Pennisetum purpureum) Straw Biomass Conversion, Applied Science and Engineering Progress, **10**,2 (2017): 107-117
28. X. Zhao, L. Wang, X. Lu and S. Zhang, Pretreatment of corn stover with diluted acetic acid for enhancement of acidogenic fermentation, Bioresource Technology, **158** (2014): 12-8

Comparative evaluation of DNA extraction from rice's root-associated bacterial consortium for population structure study

Nichapat Kittiborwornkul[1], *Anne Sahithi* S.T [2], *Jirapa* Phetsom[3], *Theerawut* Phusantisampan[4], *Wasinee* Pongprayoon[5], *Malinee* Sriariyanun[1,6*]

[1]Department of Chemical and Process Engineering, The Sirindhorn International Thai-German Graduate School of Engineering (TGGS), King Mongkut's University of Technology North Bangkok (KMUTNB), Bangkok, Thailand
[2]Department of Biomedical Sciences, School of Bio-Sciences and Technology (SBST), Vellore Institute of Technology (VIT) Vellore, Tamilnadu, India
[3]Department of Biology, Faculty of Science, Mahasarakam University, Mahasarakam, Thailand
[4]Department of Biotechnology, Faculty of Applied Science, King Mongkut's University of Technology North Bangkok (KMUTNB), Bangkok, Thailand
[5]Department of Biology, Faculty of Science, Burapha University, Chonburi, Thailand
[6]Biorefinery and Process Automation Engineering Center (BPAEC), King Mongkut's University of Technology North Bangkok (KMUTNB), Bangkok, Thailand

Abstract. Understanding in population structure of a plant's root-associated microbiome is applied in good practices in agricultural activities to improve production yields and enhance plant immune responses. The molecular analysis of bacterial populations inhabited in soil faces difficulties in obtaining high yield and high purity of DNA, and different commercial DNA extraction kits have been developed for this purpose. This study focuses on the comparison of DNA extraction of six different rice root-associated bacterial consortium using three commercial kits with two key technologies, spin-column adsorption and magnetic bead adsorption. The quality and quantity of genomic DNA obtained from these extraction methods were analyzed and compared based on DNA concentration, DNA purity and efficiency to be used as a template for 16sRNA amplification. The results showed that the extraction kit with magnetic bead adsorption technology showed the highest concentration (101.32 ng/μl) compared to other DNA extraction kits (32.67 and 1.89 ng/μl). The purity values (A260/A280) were assessed by using Nano-drop spectrophotometer and resulted in purities of nucleic acids in the range of 1.4-1.7. Thus, it was concluded that the extracted DNA obtained from the extraction kit with magnetic bead adsorption technology can be valuable for molecular analysis of microbial communities present in the soil.

Keyword. DNA extraction method, Bacterial consortium, Rice, Root-associated microbiome, Population analysis

1 Introduction

Cereals play an essential role in the human daily diet, which is rich in nutrients and calorie composition [1]. Rice is one of the most economically and nutritionally important cereals, with about 60% of the world's population consuming rice as a basic diet. In addition, rice is the most cultivated crop in the Asia-Pacific region and is the staple food of some developing countries [2] and its by-products could be converted to various value-added products [3]. Some rice cultivars were characterized to contain beneficial nutrition to improve human health [4]. In rice cultivation, several microorganisms are involved for the growth and can also cause disease resulting in reduce yields [5], [6-7]. The study of microorganisms present in the soil helps to improve their productivity. Some species of these microbes provide the benefit to rice by activating the defense mechanism to protect plants from pathogens [8]. However, culturing microorganisms from soil samples in laboratory is difficult, and understanding the diversity and ecology of microorganisms is a fundamental obstacle [9]. Several factors are demonstrated to involved in soil pollution, which reduce the viability of soil microbes [10]. In addition, microorganisms are important to produce carbon, nitrogen and many other organic matters for plants. Due to the high contamination of the soil with organic substances, it can lead to rupture of cells and the removal of humic acids that may interfere with the polymerase activity during DNA amplification reaction in laboratory for studying soil microbe's population [11]. Therefore, many strategies and techniques have been developed to help researchers to be able to handle the samples for further research.

DNA-based techniques can overcome this obstacle by understanding the genes involved in microorganisms [12]. Extraction of large amounts of high-quality DNA from the rice plant can be used for further genetic

*Corresponding author: macintous@gmail.com

analysis. Methods for extracting DNA from soil are quite complex because the soil is contaminated to the level that is not suitable to be manipulated at the molecular level [13]. DNA is primarily characterized by high molecular weight (MW) fragments with an A260 / 280 ratio of 1.8-2.0, and the presence of contaminants such as polysaccharides and phenols can reduce DNA quality [14]. Polysaccharides released from plant's exudates can be contaminated in DNA extraction resulting in DNA precipitation and reduce the extraction yields of DNA. Anionic impurities inhibit restriction enzymes and interfere with enzymatic DNA analysis [15] and RFLP and PCR-based molecular analysis of dried seeds of different cereals [16]. Rapid and pure extraction of DNA is a required for assessing advanced techniques such as fingerprinting, marker-assisted breeding, gene mapping, and the authenticity of exported variety of cereals [17]. There are several methods for extracting DNA from plant tissue, but these methods produce small amounts or varying qualities of DNA. Several methods raised for the DNA extraction are altered versions of cetyltrimethyl ammonium bromide (CTAB) extraction, with some limitations, which vary in time and cost. CTAB is a cation surfactant added to a DNA extraction buffer that dissociates DNA from histone proteins and selectively precipitates it [18]. Even though DNA extraction protocols have become much more advanced in recent years, they still have their advantages and disadvantages. In addition, commercially available genome isolation kits have significant limitations when extracting genomic DNA of higher concentrations (especially from soil samples) [19].

This study focuses on the comparison of DNA extraction using different commercial kits, which refer to different extraction technologies for DNA from contaminated soil samples. Six different rice root associated bacterial consortium samples were taken to extract DNA using three commercial kits with two key technologies, spin-column adsorption and magnetic bead adsorption. These kits are (1) Bioneer MagListo™ 5M Genomic DNA Extraction Kit, (2) PureDireX Genomic DNA Isolation Dual Kit and (3) QIAamp® DNA Mini Kit. The quality and quantity of genomic DNA/g of contaminated soil were analyzed and compared using these kits. This comparative study was performed to determine the effectiveness of the kit in extraction DNA from soil to be preliminary data to select suitable method for further study in soil microbial population.

2 Material and methods

2.1 Raw materials

Rice berry roots with soil samples were obtained from research rice field in Eastern part of Thailand. The soil samples were separated from roots to three different compartments, root associated soil, root proximity soil and bulky soil, as shown in (Figure 1). Two biological replicates were collected from the same rice field, resulting in a total of 6 different samples Table 1. Firstly, to obtain root proximity sample, soil particles attached with rice root (not thicker than 10 mm from root's perimeters) were taken out by hands. Then these rice roots were dipped in phosphate buffer (pH 7.0) and vortex for 5 min to detach the root-associated soils from root. Then the root-associated soils were collected by centrifugation at 8,000 rpm for 10 min. After that 1 g of each soil sample was taken and 20 ml of nutrient broth were added to enrich bacterial growth. The cultures were incubated in rotary shakers 30 °C for 72 hours. The growths of soil's bacterial cultures obtained from 6 soil samples were measured by using spectrophotometer at the wavelength of 600 nm to monitor culture's turbidity. Then, the cultures were diluted with phosphate buffer to prepare the same amounts of bacterial cultures for testing with 3 different commercial kits. 15 ml of diluted sample was taken and centrifuged at 6000 rpm for 10 mins, pellet was stored for further analysis of DNA extraction.

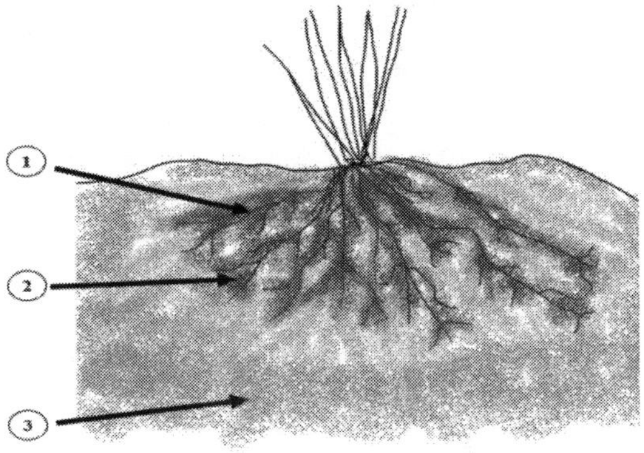

Fig. 1. A schematic diagram of the rice berry root. 1. Root associate soil (within 2 mm. from root's perimeters) 2. Root proximity soil (within 10 mm. from root's perimeters) 3. Bulky soil (more than 10 mm.).

Table 1. Six soil samples used for the extraction of bacterial consortium's DNA.

No	Sample	Source
1	Soil from Rice berry rep. 1	root proximity soil
2	Soil from Rice berry rep. 1	root associate soil
3	Soil from Rice berry rep. 2	root proximity soil
4	Soil from Rice berry rep. 2	root associate soil
5	Soil from Rice berry rep. 1	bulky soil
6	Soil from Rice berry rep. 2	bulky soil

2.2 Extraction DNA

The DNA was extracted using three commercial kits with two key technologies, spin-column adsorption (PureDireX Genomic DNA Isolation Dual Kit, QIAamp® DNA Mini Kit) and magnetic bead adsorption (Bioneer MagListo™ 5M Genomic DNA Extraction

Kit). The major steps for these DNA extraction kits were composed of cell harvesting, cell lysis, DNA binding, Wash, and elution (Figure 2.). The experimental details of each step in each kit were followed by vendor's instructions as follows.

2.2.1 Bioneer MagListo™ 5M Genomic DNA Extraction Kit

The genomic DNA was extracted from cultured cells ($1x10^6$) by centrifugation for 10 minutes at 3000 x g. The pellet was suspended in 200µl of 1X PBS. 20µl of Proteinase K and 10 µl of RNase A was added and incubated for 2 min at room temperature. Then 200 µl of GB buffer added and mixed by vortexing, incubated at 60°C for 10 min. After incubation 400 µl of absolute ethanol and 100 µl of Magnetic Nano Bead Solution was added and mixed thoroughly by vortexing. The samples were placed in MagListo™-2 Magnetic Separation Rack and gently inverted for 3-4 times. The supernatant was collected without removing the tube from MagListo™ rack. Then the magnet plate was detached from the stand and 700 µl of WM1 buffer was added and vortexed. Magnet plate was attached again and inverted the tube gently for 3-4 times the beads started binding tightly to the magnet. The supernatant was collected without removing tube from the rack. This was repeated for 2 times. Supernatant was washed by adding W2 buffer of 700 µl. The 3rd wash was done by adding 700 µl WE Buffer to the opposite side of the bead pellet and gently inverted twice after closing the cap. Later, EA buffer of 100 µl was added and mixed vortexing at 60°C on elution stand for 1 min.

2.2.2 PureDireX Genomic DNA Isolation Dual Kit

Cultured bacterial cells (up to 109) were transferred to 1.5 ml micro centrifuge tube. The tube was centrifuged at 12000 x g for 1 min. Supernatant was removed and the pellet was suspended in 50 µl of buffer CR by pipetting. Then 300 µl of CC buffer was added and incubated at 60°C for 10 mins by inverting the tube for every 3 min. for the removal of protein 400 µl of buffer CB was added mixed thoroughly and centrifuged at 12000 x g for 1 min. The sample was placed in a 2 ml collection tube and supernatant was transferred to the previous step of CC column. Centrifuged at 14000 x g for 30 secs and discarded the flow collected in the tube. The sample was washed using 400 µl of buffer W1 into the column x g for 2 mins. The liquid collected in the tube column CC was discarded. DNA was eluted by transferring the dried CC to a new 1.5 ml of micro centrifuge tube. TE buffer of 200 µl was added to the center of the column and incubated at 60 °C for 3 min. Column was centrifuged for 2min at 14000 x g to elute the purified DNA.

2.2.3 QIAamp® DNA Mini Kit

Extraction of DNA was done using 220 mg of sample in a 2 ml centrifuge tube place on ice. 1 ml of EX buffer was added to the sample and vortexed thoroughly until homogenized. The sample was centrifuged for 1 min. 15 µl of proteinase K was taken in a 1.5 ml tube and the supernatant of 200 µl was collected from centrifuged tube. 200 µl of buffer AL was added to the sample and

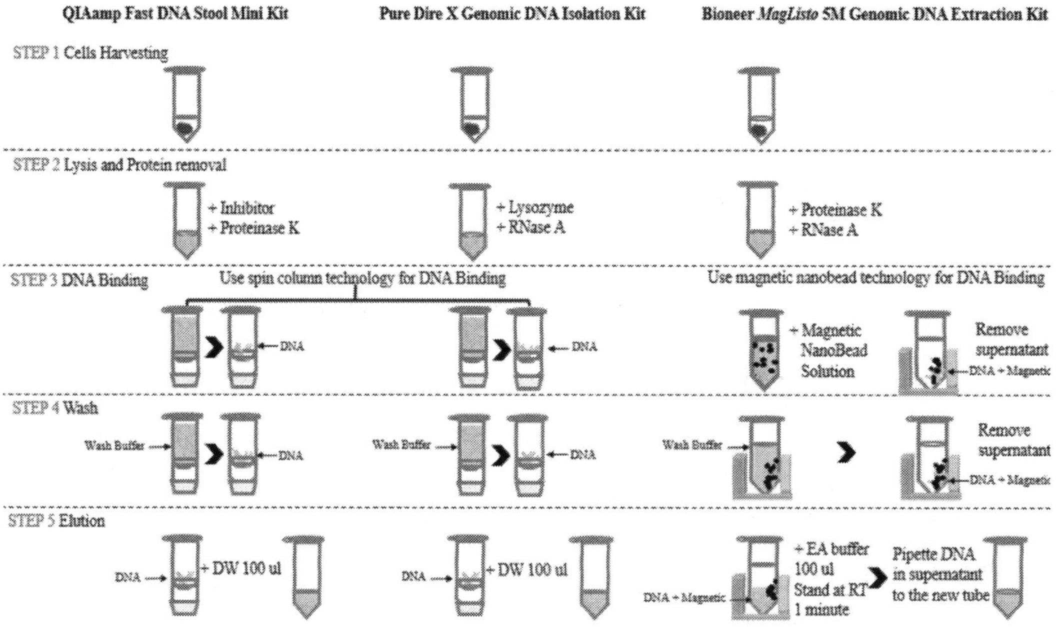

Fig. 2. Summarization of three DNA extraction kits with two key technologies; spin column and magnetic nanobead.

vortexed for 15 secs, incubated at 70°C for 10 min. 600 µl of ethanol was added and mixed thoroughly. Then the lysate of 600 µl was added to QIAamp spin column in a 2 ml collection tube. The filtrate was discarded by repeating 2 times. 500 µl of buffer AW1 was added to the spin column and centrifuged for 1min. QIAamp spin column was placed in a new 2ml collection tube and the filtrate was discarded. QIAamp spin column was opened carefully, 500 µl of buffer AW2 was added and centrifuged foe 3 min, filtrate was discarded. Then spin column was placed in a new collection tube and the old collection tube filtrate was discarded. Spin column was centrifuged for 3 min. QIAamp spin column was transferred in to a new 1.5 ml centrifuge tube and 200 µl of ATE buffer was added on to the QIAamp membrane, incubated at RT for 1 min. the tube was centrifuged for 1 min to elute the DNA.

2.3 DNA quantitation and analysis

To evaluate the quantity and quality of the DNA samples (3 soil samples x 2 biological replicates x 3 kits) Nano-400A Micro-Spectrophotometer was used (ALLSHENG). 2 µl of DNA extracted from the sample was used to determine the DNA purity based on the values of A260/A280 ratio and A260/A230 ratio. The concentration of each DNA sample was determined automatically from the Nano-400A Micro-Spectrophotometer. To determine the integrity of genomic DNA and analysis of PCR products, the sample was analyzed by gel electrophoresis with 2% agarose (for genomic and amplified DNA). Electrophoresis was performed using 1x TBE buffer along with SyBR® Green staining. The setup was performed by constant voltage of 120 V for 50 mins.

2.4 Polymerase Chain Reaction (PCR)

The amplification of the 16S rRNA fragment of extracted DNA sample was conducted using 16S rRNA primers. For PCR analysis, the sample was amplified with forward primer P1: 5-AGAGTTTGATCCTGGCTCAG-3 and reverse primer P2: 5-GGTTACCTTGTTACGACTT-3 [8]. Each PCR reaction contains 25 µl containing mixture of (10x buffer, dNTP, primers, Taq DNA Polymerase, MgCL2, H2O, and DNA template). The PCR reaction setup per reaction and conditions are shown in the (Table 2 and Table 3) was analyzed by gel electrophoresis with 2% agarose (for genomic and amplified DNA). Electrophoresis was performed using 1x TBE buffer along with SyBR® Green staining. The setup was performed by constant voltage of 120 V for 50 mins.

3 Results

To prepare the same number of bacterial cells for DNA extraction with different extraction kits, the bacterial cell density of each cultured sample (after 72 hr) was determined by taking 1 ml from each sample with spectrophotometer at the wavelength of 600 nm. The absorbance values of cell density were shown in the Table 4. It was observed that soil sample in the region of root proximity and bulky soil had higher absorbance value when compared to root associated samples. These variations of cell turbidities of each soil samples could imply to variations in starting numbers of cells as well as the species in original samples, when all of them were cultured under the same nutrients and conditions.

Table 2. PCR Master mix components per reaction

PCR Mixer Components	Volume (µl)
10x Buffer	2.5
dNTP	0.5
Forward Primer	0.5
Reverse Primer	0.5
Taq DNA Polymerase	0.25
$MgCl_2$	0.75
H_2O	19
DNA Template	1
Total volume	**25**

Table 3. Condition for PCR amplification

Step	Temperature (°C)	Time	Cycles
Hot Start	95	5 min	
Denature	95	50 sec	
Annealing	57	50 sec	40 cycles
Extension	72	90 sec	
	72	5 min	

Then, all bacterial cultures were subjected to DNA extractions with 3 commercials kits. The quality and quantity of final product DNA from each sample (3 soil samples x 2 biological replicates x 3 kits) was determined by using Nano-400A Micro-Spectrophotometer (Table 5-7). It was found that the highest concentration of DNA was found in Bioneer MagListo ™ 5M Genomic DNA extraction kit (magnetic bead technology) of about 101.32 ng/µl compared to other DNA extraction kits PureDireX Genomic DNA Isolation Dual Kit and QIAamp® DNA Mini Kit (spin-column technology) 32.67 ng/µl and 1.89 ng/µl. The purity values (A260/A280) indicate significant levels for contaminations of nucleic acids in the range of 1.4-1.7. The concentration of DNA extracted using Bioneer MagListo ™ 5M Genomic DNA extraction kit for all six samples was shown in the Table 5. Among them, DNA extracted from soil sample in the region of root proximity was observed (101.32 ng/µl) comparatively

 https://doi.org/10.1051/e3sconf/202130202014

higher than the region of bulky soil (54.33 ng/µl). The yield of DNA extracted using PureDireX Genomic DNA Isolation Dual Kit was determined for all six samples was shown in the Table 6. Among them, DNA extracted from soil sample 5 has the highest yield with 35.56 ng/µl). DNA concentration of all six samples extracted using QIAamp® DNA Mini Kit was determined and shown in the Table 7. Among them, it was observed that DNA extracted from root proximity showed lowest value of 1.89 ng/µl when compared to bulk soil (6.53 ng/µl). The purity values of DNA extracted using kits for all six samples were also observed. It was found that the purity of the DNA was not significant. This implies the samples are contaminated by protein.

Table 4. Absorbance at 600 nm of cultured bacterial samples

Sample	Source	Absorbance 600 nm
1	Root proximity soil-rep1	1.174
2	Root associate soil-rep1	0.892
3	Root proximity soil-rep2	1.176
4	Root associate soil-rep2	0.945
5	Bulky soil-rep1	1.262
6	Bulky soil-rep2	0.905

Table 5. Concentration of DNA obtained from Bioneer MagListo ™ 5M Genomic DNA extraction kit.

Sample	Absorbance			DNA concentration (ng/µl)
	A260	A280	A260/A280	
1	2.026	1.296	1.56	101.32
2	1.703	1.195	1.43	85.19
3	1.109	0.719	1.54	55.48
4	1.014	0.618	1.64	50.72
5	1.086	0.671	1.62	54.33
6	0.368	0.239	1.53	18.41

Table 6. Concentration of DNA obtained from PureDireX Genomic DNA Isolation Kit.

Sample	Absorbance			DNA concentration (ng/µl)
	A260	A280	A260/A280	
1	0.653	0.402	1.62	32.67
2	0.578	0.400	1.44	28.95
3	0.419	0.254	1.64	20.95
4	0.581	0.381	1.52	29.07
5	0.711	0.391	1.82	35.56
6	0.326	0.229	1.42	16.31

Table 7. Concentration of DNA obtained from QIAamp Fast DNA Stool Mini Kit.

Sample	Absorbance			DNA concentration (ng/µl)
	A260	A280	A260/A280	
1	0.140	0.090	1.55	1.89
2	0.067	0.039	1.71	3.38
3	0.084	0.060	1.41	4.23
4	0.075	0.050	1.50	3.79
5	0.089	0.052	1.70	4.49
6	0.130	0.092	1.41	6.53

The evaluations of intact genomic DNA were analyzed by gel electrophoresis. 1 kb ladder was used to determine the molecular size of the DNA samples (Figure 3). It was found that the DNA extracted using PureDireX Genomic DNA Isolation Dual Kit and QIAamp® DNA Mini Kit were observed while no band was shown in Bioneer MagListo ™ 5M Genomic DNA extraction kit. This implies that the soil samples were highly contaminated with the organic compounds or chaotropic salts present in the purified DNA. In addition, it can cause the extraction process of DNA or during the removal of protein after the lysis step. Therefore, the sample extracted using Bioneer MagListo ™ 5M Genomic DNA extraction kit is not efficient based on this criterion. DNA purity is required for PCR and sequencing, as well as a high molecular weight with less shear, no contamination of proteins, RNA or polysaccharides, and an absorbance of 260/280 nm of approximately 1.8–2.0. To overcome this problem, PCR process was done to further evaluate DNA quality by using 16S rRNA primers.

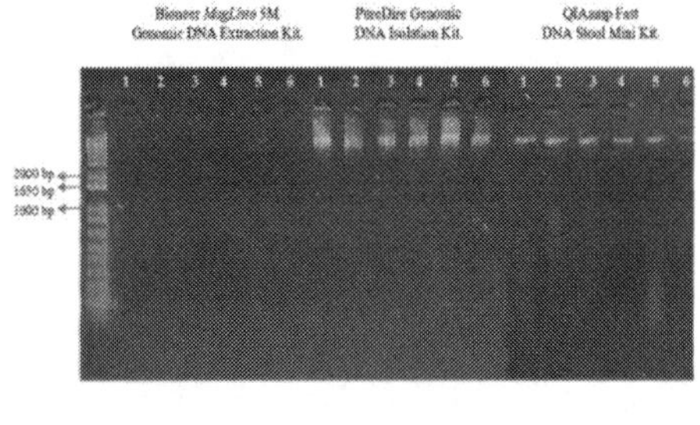

Fig. 3. Genomic DNA analysis in agarose gel electrophoresis (Lane 1-6 represents sample No. 1 - No. 6)

PCR reactions were set up to amplify the DNA obtained by extracting all three methods. The amplified product was separated on the Gel Electrophoresis (Figure 4). It was observed that 2 soil samples of (No.1 and No.4) extracted using the QIAamp® DNA Mini Kit band were shown and bp size was observed at about 1.4 kbp, which is the targeted size of PCR reaction. Despite, this QIAamp® DNA Mini Kit spin column technology showed lowest measured concentration using Nano-400A Micro- Spectrophotometer. Meanwhile, there were no PCR bands at all from samples extracted using Bioneer MagListo ™ 5M Genomic DNA extraction kit that had the highest concentration observed in the Nanodrop.

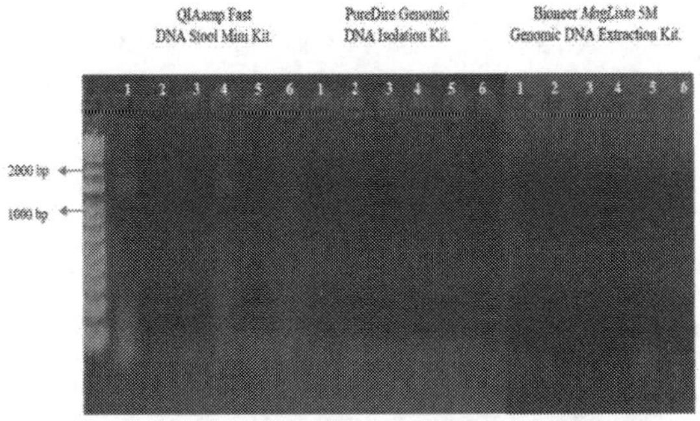

Fig. 4. Analysis of PCR products in agarose gel electrophoresis (Lane 1-6 represents sample No. 1 - No. 6)

Altogether, based on analysis of Nano-400A Micro-Spectrophotometer, gel electrophoresis of genomic DNA and PCR amplification, the efficiency of different commercial DNA extraction kit was demonstrated. To avoid the interference of the magnetic beads and to reduce the loss of DNA samples. Some additional steps are recommended to remove the magnetic beads carefully to achieve DNA concentration of higher purity. However, methods are still yet to be developed to eradicate the interference of trace Nano-beads during the analysis of DNA. Therefore, it is concluded that the magnetic Nano-bead method must further be improvised to achieve the higher quantity of DNA for further molecular analysis. On the other hand, purity of the DNA samples is further necessary for the existing method. In conclusion, a more effective extraction technique must be developed for extraction of DNA in higher proportion and purity from soil samples. Further application of this work could incorporated to development of bioeconomy of using agricultural waste to produce value-added products [20].

4 Conclusion

In this study, the DNA was extracted from 3 different soil samples (Root associate soil, Root proximity soil and Bulky soil). The extraction of DNA was carried using two key technologies namely, spin column and magnetic Nano bead methods. This study was concluded with higher concentration of DNA extracted using Bioneer MagListo ™ 5M Genomic DNA extraction kit from different soil samples. The concentration of DNA varied for soils collected from different regions. Though the DNA concentration was higher for magnetic Nano bead method. It was observed that the DNA bands were not identified or not visible when tested at a molecular level using PCR technique. It was concluded that DNA bands were not shown due to contaminated soil (organic compounds or chaotropic salts present in the purified DNA).

Acknowledgment

The authors would like to thank the King Mongkut's University of Technology North Bangkok (Grant Contract No. KMUTNB-BasicR-64-37) and Thailand Science Research and Innovation (Basic Research Fund Contract No. 4.3/2564) for financial support of this work.

References

[1] A. Abdel-Latif and G. Osman, Comparison of three genomic DNA extraction methods to obtain high DNA quality from maize, Plant Methods. **13**,1(2017): 1.
[2] A. Cachada, T. Rocha-Santos, and A.C. Duarte, Soil and Pollution, Soil Pollution. (2018): 1-28.
[3] Y.S. Cheng, P. Mutrakulcharoen, S. Chuetor, K. Cheenkachorn, P. Tantayotai, E.J. Panakkal, M. Sriariyanun, Recent Situation and Progress in Biorefining Process of Lignocellulosic Biomass: Toward Green Economy, Applied Science and Engineering Progress. **13**, 4 (2020): 299-311.
[4] N. Chairerk, P. Pongyeela, J. Chungsiriporn, and N. Rakmak, Ethanol Extraction of Active Ingredients and Antioxidants from Germinated Sangyod Rice, Applied Science and Engineering Progress. **14**(1), (2021).
[5] H. Liu, L.C. Carvalhais, M. Crawford, et al., Inner Plant Values: Diversity, Colonization and Benefits from Endophytic Bacteria., Frontiers in Microbiology. **8**, (2017).
[6] P. Songkumarn, P. Chaijuckam, V. Tongsri, and J.J.G. Guerrero, Expression Analysis of Defense Related Genes in Rice Response to Bipolaris oryzae, the Causal Agent of Rice Brown Spot, Applied Science and Engineering Progress. **12**(2), (2018).
[7] P. Songkumarn, P. Chaijuckam, and J.J.G. Guerrero, Genetic Diversity and Aggressiveness of Bipolaris oryzae in North-Central Thailand, Applied Science and Engineering Progress. **12**(2), (2018).
[8] W. Pongprayoon, T. Siringam, A. Panya, and S. Roytraku, Application of Chitosan in Plant Defense Responses to Biotic and Abiotic Stresses, Applied Science and Engineering Progress. **15**(1), (2022)
[9] J.M. Awika, Major Cereal Grains Production and Use around the World, Presented at the January (2011).
[10] M.A. Mohamad Roslan, M.A. Naim Mohamad, and S. Mohd Omar, High-Quality DNA From Peat Soil For Metagenomics Studies: A Minireview On DNA Extraction Methods, Science Heritage Journal. **1**, 2(2017): 01-06.
[11] M.D. Braid, L.M. Daniels, and C.L. Kitts, Removal of PCR inhibitors from soil DNA by chemical flocculation, Journal of Microbiological Methods. **52**,3(2003): 389-393.
[12] M. Majaneva, O.H. Diserud, S.H.C. Eagle, M. Hajibabaei, and T. Ekrem, Choice of DNA extraction method affects DNA metabarcoding of

unsorted invertebrate bulk samples, Metabarcoding and Metagenomics. **2**, (2018).

[13] M. Sriariyanun, P. Yasurin, P. Pornwongthong and, K. Cheenkachorn, Production, purification and characterization of an ionic liquid tolerant cellulase from Bacillus sp. Isolated from rice paddy field soil, Electronic Journal of Biotechnology. **19**, (2015).

[14] N. Atheeqah-Hamzah, C.S.Y. Yong, and U.K. Yusuf, Extraction of High-quality RNA from Metabolite and Pectin Rich Recalcitrant Calyx Tissue of Hibiscus sabdariffa L, Pertanika Journal of Tropical Agricultural Science. **43**,4(2020).

[15] R. Backer, J.S. Rokem, G. Ilangumaran, et al., Plant Growth-Promoting Rhizobacteria: Context, Mechanisms of Action, and Roadmap to Commercialization of Biostimulants for Sustainable Agriculture, Frontiers in Plant Science. **9**, (2018).

[16] R.S. Pickard and B.J. McKevith, The Role Of Cereals In The Diet, Using Cereal Science and Technology for the Benefit of Consumers. 89 (2005).

[17] R. Von Post, L. Von Post, C. Dayteg, M. Nilsson, B.P. Forster, and S. Tuvesson, A high-throughput DNA extraction method for barley seed, Euphytica. **130**,2(2003): 255–260.

[18] S. Khalid, M. Shahid, N.K. Niazi, B. Murtaza, I. Bibi, and C. Dumat, A comparison of technologies for remediation of heavy metal contaminated soils, Journal of Geochemical Exploration. **182**, (2017): 247–268.

[19] X. Ren, B. Kang, and Z. Zhang, Understanding tumor ecosystems by single-cell sequencing: promises and limitations, Genome Biology. **19**,1(2018): 211.

[20] R. Akkharasinphonrat, T. Douzou, M. Sriariyanun, Development of ionic liquid utilization in biorefinery process of lignocellulosic biomass, King Mongkut's University of Technology North Bangkok International Journal of Applied Science and Technology, 10,2 (2017): 89-96.

https://doi.org/10.1051/e3sconf/202130202015

Effect of sodium hydroxide pretreatment on released sugar yields from pomelo peels for biofuel production

Chaichana Chatkaew[1], *Elizabeth Jayex* Panakkal[2], *Wawat* Rodiahwati[3], *Suchata* Kirdponpattara[1,4], *Santi* Chuetor[1,4], *Malinee Sriariyanun*[2,4]*, *Kraipat Cheenkachorn*[1,4]*

[1]Department of Chemical Engineering (ChE) Faculty of Engineering, King Mongkut's University of Technology North Bangkok (KMUTNB), Bangkok, Thailand
[2]Department of Chemical and Process Engineering, The Sirindhorn International Thai-German Graduate School of Engineering (TGGS), King Mongkut's University of Technology North Bangkok (KMUTNB), Bangkok, Thailand
[3]Department of Chemistry, University of New England: Armidale, New South Wales, Australia
[4]Biorefinery and Process Automation Engineering Center (BPAEC), King Mongkut's University of Technology North Bangkok (KMUTNB), Bangkok, Thailand

Abstract. Most of the agricultural wastes in developing countries are disposed of by on-site combustion leading to unmanaged environmental pollutions. Conversion of agricultural wastes to value-added products, such as bioethanol and biogas, is a promising method to reduce agro-waste after harvesting seasons. In this study, *Citrus maxima* peels (Pomelo peels) was selected to be converted to reducing sugars, which could be a raw material to produce other value-added products. To promote enzymatic hydrolysis reactions, pomelo peels were pretreated with sodium hydroxide by variations of three pretreatment parameters, including temperature (50-100 °C), time (0.5-6 h), and concentration of NaOH (0.5-3.0 M). Box-Behnken design (BBD) was applied in Response Surface Methodology (RSM) to determine the optimized pretreatment conditions and to find the relationship between pretreatment factors and reducing sugar yields. The predicted optimal pretreatment condition was determined to be at 94.28 °C, 4.5h, 2.17M with reducing sugar yield of 98.9 mg/g of dried pomelo peels. The results clearly showed that reducing sugar yields obtained from pretreated pomelo peels were 1.87 folds higher than untreated biomass (52.81 mg/g of pomelo peels). Therefore, this study demonstrated the potential of pomelo peels to be used as an alternative raw material for value-added products rather than being a landfill or causal agent of pollution.

Keywords: Sodium hydroxide, Pretreatment, Enzymatic saccharification, Biorefinery, Biofuels

1 INTRODUCTION

Energy consumption has increased in the past decade across the globe. An increase in population rate along with industrial development has led to more energy consumption than in the past years. Energy production using non-renewable sources is insufficient to meet the rising energy demand. This has led to an increase in the research on renewable energy produced from hydropower, solar energy, wind power, and biomass. Many researchers have shifted their focus to biomass for energy production since it is widely available [1].

Thailand is a country rich in agricultural resources and produces a large amount of agricultural waste, such as straw, rice husk, coconut shell, sawdust, leaves, bark, fruit husk, corncob, and bagasse, etc. According to the data in 2013, Thailand produced 62 million tons of biomass per year, which had the potential to generate 13,348 thousand tons of crude oil [1]. Statistic data show the possibility of utilizing agricultural waste biomass for energy production. For instance, fruit peels from canned fruit factories could be utilized to produce biogas or can be processed into fertilizers or animal feeds. Currently, most of the residual biomass after processing in the industry is either landfilled or incinerated, which causes fine particle release with serious health and environmental concerns. Instead of this, residual biomass could be used to generate renewable energy for local use. This not only provides health benefits for people but also reduces environmental pollution [2].

Biomass materials, such as agricultural processed fruit peels, are mainly composed of lignocellulose materials. Lignocellulose is a complex material with a rigid structure. Lignocellulose is composed of 3 mains components, namely cellulose, hemicellulose, and lignin [3,4]. Cellulose is the main component, which is a homopolymer of glucose connected with β-1, 4 linkages to form a long polymeric chain. Hemicellulose is a heteropolymer composed of repeated units of pentoses and hexoses. Lignin is an aromatic polymer that provides the rigidity for biomass. Each plant material has a varying amount of these components depending on its species, growth stage, and climate [5-7]. These components are arranged in a specific manner in the biomass to make it rigid and recalcitrant. The rigid

* Corresponding author: macintous@gmail.com, kraipat@gmail.com

nature of biomass greatly hinders the activity of enzymes or microorganisms to degrade biomass to produce energy. Therfore, a preliminary process to facilitate the disintegration of biomass, before its utilization is necessary.

In general, there are four important steps in bioenergy production, namely the pretreatment process, followed by hydrolysis, fermentation, and bioenergy extraction [8,9]. The pretreatment process is conducted to make biomass become vulnerable to enzyme activity. Pretreatment can be done in several ways, such as physical pretreatment)milling or grinding([10-11], chemical pretreatment)alkali, dilute acid, oxidizing agents and organic solvents, ionic liquid([12-20], biological pretreatment [21-24] and, physicochemical pretreatment [10,11]. Each method has its advantages and disadvantages. However, these pretreatment methods can increase porosity and surface area to the biomass structure [25]. This provides more surface area for the enzyme to react and increases the hydrolysis efficiency by increasing sugar yield at a reduced cost of enzyme for cellulose digestion [26]. Chemical pretreatment using acid helps in the digestion of hemicellulose and lignin to produce high sugar yield and generate inhibitors or furfural [7]. Whereas, alkaline pretreatment, using solutions such as sodium hydroxide, calcium hydroxide, and ammonia helps in the removal of lignin and some amount of hemicellulose. Pretreatment using these chemicals can cause degradation of ester and glycosidic side chains causing the cellulose to swell. This also helps in decreasing crystallinity and polymerization thereby increasing internal surface area by destroying the lignin structures [27]. Pretreatment is followed by intensive washing to remove any trapped impurities in the solids after pretreatment [28]. This can enhance the efficiency of the enzyme in the hydrolysis process [29] and thereby increases the yield of required products, such as bioethanol [26].

In this research, pomelo peels was targeted in our investigation in the production of fermentable sugars, which could subsequently be converted to value-added products or biofuels. To improve the efficiency of enzymatic saccharification of pomelo peels, the pretreatment process with sodium hydroxide was optimized based on Response Surface Methodology (RSM) with Box-Behnken design (BBD). The maximum amount of sugars produced from this work could be used as critical data for process design in the biorefining process of lignocellulosic biomass.

2 Material and methods.

2.1. Raw materials

Pomelo peels were obtained from a local market in Nonthaburi province in Thailand. Pomelo peels were dried at 60 °C for 12 h. In hot air oven to remove moisture from them. After drying, it was cut by using house blender thoroughly and particle sizes were screened through a 20 mesh-sized aluminum sieve to obtain sample consistency. The biomass was stored in a sealed plastic bag at room temperature until further use.

CelluClast 1.5 L from *Trichoderma reesei* was purchased from Sigma-Aldrich (aqueous solution 700 units / g), β-glucosidase was obtained from Megazyme, Sodium hydroxide (NaOH), and other analytical reagents were purchased from Alex Fincohem (Univar).

2.2 Alkaline pretreatment

A pretreatment was carried out by immersing 5 g of pomelo peels in 50 ml of NaOH solution at a solids loading of 10%. A treatment was performed in hot air-oven using varied NaOH concentration (0.5 to 3 M), temperature (70 to 100°C) and time (1 to 24 hours) to preliminarily find the boundary of testing factors. After pretreatment, the solid residues were removed from the liquid fraction by filtration with Whatman No. 1 filter paper. The solid residues were further washed with distilled water until neutral pH. The solids were dried in a hot-air oven at 60 °C for 8 hours. Pretreated samples were stored in a desiccator before proceeding to enzymatic saccharification.

2.3 RSM Experiment design

To determine the optimal pretreatment condition for producing maximum sugar yield (Y), BBD of RSM was selected. RSM experiments were designed with 3 pretreatment factors, namely pretreatment temperature (X_1), Pretreatment time (X_2), and NaOH Concentration (X_3). Pretreatment was carried out at varied NaOH concentrations of 0.5-3 M, the temperature of 50-100 °C, and time of 0.5-6 h. Each of these factors was set at three different levels -1, 0, and +1 (Table 1). A total of 17 runs were performed by varying conditions of pretreatment factors as in Table 2. The pretreated biomass was further hydrolyzed by the enzyme to determine the reducing sugar yield. The RSM experiments were design and experimental data were analyzed by using Design Expert version 7.0.0 software.

Table 1. Independent variable factors in pretreatment condition of pomelo peels on RSM.

Independent variable factors	Coded symbols	Levels		
		-1	0	+1
Temperature (°C)	X_1	50	75	100
Time (h)	X_2	0.50	3.25	6
Concentration (%)	X_3	0.50	1.75	3

2.4 Enzymatic hydrolysis

Untreated and pretreated pomelo peels were enzymatic hydrolyzed by commercial cellulase. The hydrolysis reaction was carried out using 0.1 g biomass in 4 ml of hydrolysis buffer (containing 50 mM citrate buffer, 40 µl of 2M sodium azide, 35µl of CelluClast 1.5 L, and 10µl of β-glucosidase enzyme). The reaction mixture was incubated at 50 °C, 150 rpm for 72 hours. The reaction was stopped by incubating the reaction mixture at 100 °C for 10 minutes. Biomass hydrolysates were harvested by centrifugation at 10,000 g for 5 minutes. The supernatant was used to analyze for reducing sugar concentration after enzymatic hydrolysis.

Table 2. RSM experimental design for NaOH pretreatment of pomelo peels

Run	Pretreatment condition			Reducing sugar (g/100g biomass)
	X_1: Temperature (°C)	X_2: Time (h)	X_3: Concentration (M)	
1	100	3.25	0.5	56.80
2	50	6	1.75	43.44
3	75	0.5	0.5	37.85
4	100	3.25	3	97.76
5	50	3.25	3	47.07
6	75	0.5	3	41.69
7	75	3.25	1.75	75.04
8	50	0.5	1.75	40.44
9	100	0.5	1.75	45.05
10	75	3.25	1.75	79.68
11	75	6	3	98.15
12	100	6	1.75	91.31
13	75	6	0.5	57.96
14	75	3.25	1.75	78.73
15	50	3.25	0.5	29.94
16	75	3.25	1.75	80.27
17	75	3.25	1.75	80.84

2.5 Analysis of reducing sugar

The reducing sugar concentration was analyzed by the dinitrosalicylic acid (DNS) method [30]. Briefly, 50 mL of supernatant from enzymatic hydrolysis reaction was mixed with 150 mL of DNS solution. The mixture was incubated at 95°C in a water bath for 5 minutes. The mixture was incubated on ice for 5 minutes. After incubation, 1 mL of distilled water was added to the mixture and it was mixed well with a vortex shaker. The reducing sugar concentration was measured using a UV / VIS spectrophotometer at 540 nm. The concentration of reducing sugar was calculated based on a glucose standard curve (Figure 1). The effects of various pretreatment conditions on reducing sugar concentration were statistically analyzed by ANOVA.

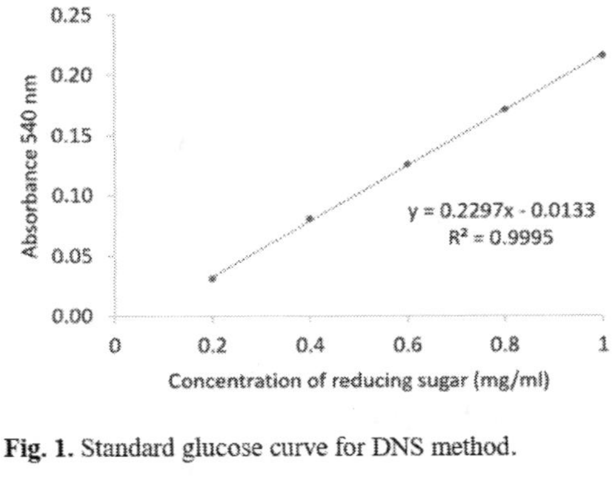

Fig. 1. Standard glucose curve for DNS method.

3 Result and discussion

3.1. Alkaline pretreatment of Pomelo peels

An alkaline pretreatment of pomelo peels was carried out to understand the range of pretreatment factors to be used in the RSM study for optimization of pretreatment conditions based on the sugar yield. Initially, pomelo peels was pretreated varying pretreatment temperature from 70 to 100 °C while keeping NaOH concentration and pretreatment time constant at 2 M and 1 hour respectively. The results clearly show that the yield of reducing sugar has increased from 52.81 g to 91.01 g/ 100 g pomelo peels when the pretreatment temperature was increased (Figure 2a).

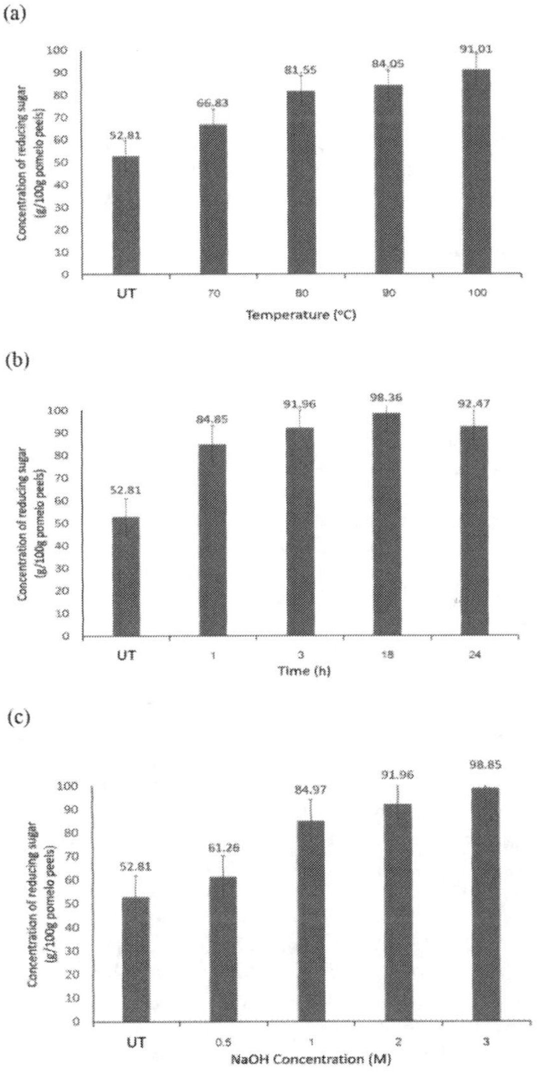

Fig. 2. Effect of pretreatment temperature, time and NaOH concentration on the enzymatic hydrolysis of Pomelo peels. (A) The pretreatment was using 2M of NaOH for 1 hour, (B) The pretreatment was using 2M of NaOH at 80°C and (C) The pretreatment were conducted at 80°C for 3 hours.

In comparison with the untreated biomass, there was a 1.72-fold increase in the sugar yield when pretreated at 100°C. Similarly, when the pretreatment was carried out by varying the NaOH concentration from 0.5 to 3 M, at 80 °C for 3 hours, the result showed that the reducing sugar yield increased with the increasing NaOH concentration (Figure 2b). A maximum sugar yield of 99.52 g/ 100 g of biomass was obtained at 3 M NaOH concentration, which is 1.88 times more than untreated pomelo peels. Unlike this, when the pretreatment was performed by varying time from 1-24 hours, with 2 M NaOH at 80 °C, the reducing sugar yield increased up to a certain pretreatment time, after that the yield was reduced (Figure 2c). The reducing sugar yield increased from 52.81 g/100 g biomass to 98.36 g/ 100 g pomelo peels by 18 hours of pretreatment. However, when the pretreatment time was further extended to 24 hours, the reducing sugar yield decreased to 92.47g / 100 g pomelo peels. This could be possibly due to the destruction of the internal structure of the biomass caused by the long pretreatment time. Also, the sugar yield did not show much variation when the pretreatment time was varied from 3 hours (91.96 g/ 100 g biomass) to 18 hours (98.36 g/ 100 g biomass). Hence, to reduce the pretreatment cost and time, the time range for pretreatment at the RSM study was chosen between 0.5 and 6 hours.

3.2. RSM design and experimental testing

The RSM is considered an efficient method to design and conduct experiments to achieve the optimal state in the production process. RSM evaluates the effects of individual independent variables and their interactions with RSM-dependent variables and also helps to reduce the number of experimental trials. It is widely used to reduce the time and cost of the experiment. In addition to this, RSM can also predict results and can create surface models under different conditions of the independent variables. Several studies have applied RSM to improve the efficacy of their studies [31-34].

In this study, RSM created the BBD matrix with 3 levels of testing: low (-1), mid (0), and high (+1) for three factors used for the pretreatment (pretreatment time, pretreatment temperature, and NaOH concentration). A total of 17 runs were conducted. The statistical analysis was performed using Design-Expert software (ver. 7.0.0) throughout this research. The experimental data were used to find the experimental coefficient (β) using the least-squares method. In addition to this, the second-order model was generated from the experimental data. The optimum pretreatment conditions were determined based on the highest yield of reducing sugar from the multiple regression analysis of the second-order model. The correlation of the second-order model was assessed by determination of the coefficient (R^2), the significance of the test condition in ANOVA, and the lack of fit, which is generally considered significant when the R^2 value is greater than 0.9 [31-34]. In this paper, R^2 value of model was 0.9396, which indicated the lower variance during the experiment and reliability of the model. ANOVA was further conducted to determine the importance of the second-order model (Table 3).

Table 3. ANOVA analysis of the second-order model obtained from RSM experiments.

Source	Sum of Squares	df	Mean Squares	F Value	P-value Prob > F
Model	8018.51	9	890.95	28.65	< 0.0001
A-Time	2113.48	1	2113.48	67.96	< 0.0001
B-Time	1979.15	1	1979.15	63.64	< 0.0001
C-Conc	1303.56	1	1303.56	41.92	0.0003
A^2	651.18	1	651.18	20.94	0.0026
B^2	548.74	1	548.74	17.65	0.0040
C^2	310.22	1	310.22	9.98	0.0160
Residual	217.68	7	31.10		
Lack of Fit	196.5	3	65.50	12.37	0.0172
Pure Error	21.18	4	5.29		
Cor Total	8236.19	16			

The ANOVA analysis of the model was found to have an F value of 28.65 with statistical significance ($p < 0.05$) and a 0.01% chance of the model F-value noise. Moreover, pretreatment temperature, pretreatment time, NaOH concentration were also significant pretreatment factors according to ANOVA analysis. The lack of fit is 12.37 which was also considered significant and has a 1.72% chance of noise. However, according to the data we have found that there was a small chance that the model will not fit and distributes anomalies. According to the ANOVA analysis above, it was estimated that all three parameters : pretreatment temperature, pretreatment time, NaOH Concentration affected the reducing sugar production after pretreatment of pomelo peels. Other studies have also used such parameters as L. Kim et al [35] performed alkaline pretreatment of Rice straw by using NaOH under various conditions of time, temperature and NaOH concentration to find a correlation between reducing sugar content and pretreatment parameters, In addition, P. Amnuaycheewa *et. al.* [36] performed acid pretreatment of Napier grass by using HCL under various conditions of time, temperature and HCL concentration to reported a similar correlation between reducing sugar contents and pretreatment parameters, so it was found that these parameters were related to reducing sugar content.

3.3 Effect of pretreatment parameters on the enzymatic saccharification on mathematical models

The impact of each factor, pretreatment temperature, pretreatment time, NaOH Concentration on the response of the curve was analyzed based on the concentration of reducing sugar (Figure 3). The model also analyzed the effect of the interaction between independent variables (pretreatment factors) in the form of a three-dimensional contour plot. The data showed that the yield of reducing sugar increased as the pretreatment temperature (Figure 4a), pretreatment time (Figure 4b), and NaOH concentration increased (Figure 4c). This implied that all three parameters affected the sugar yield. The more severe pretreatment conditions could lead to the swelling in the cellulose structure, leading to an increase in the surface area. This provided more accessibility for the enzyme to cellulose leading to enhanced sugar production.

(a)

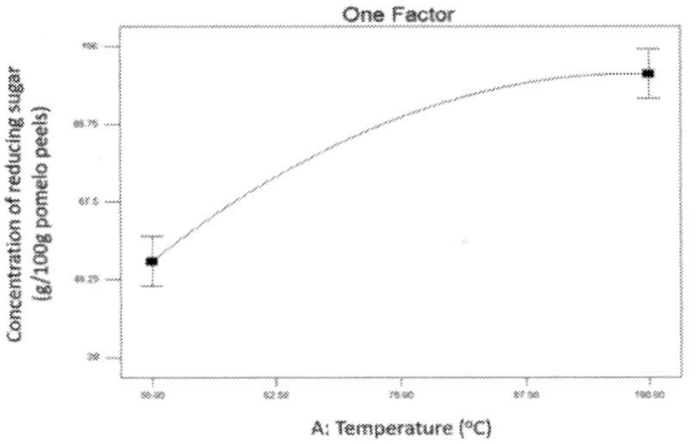

(b)

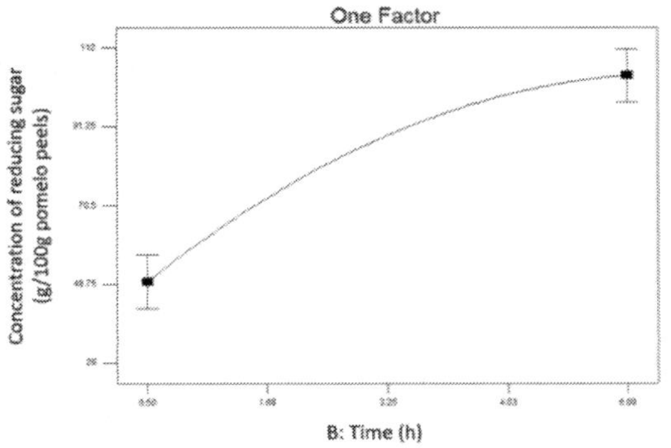

(c)

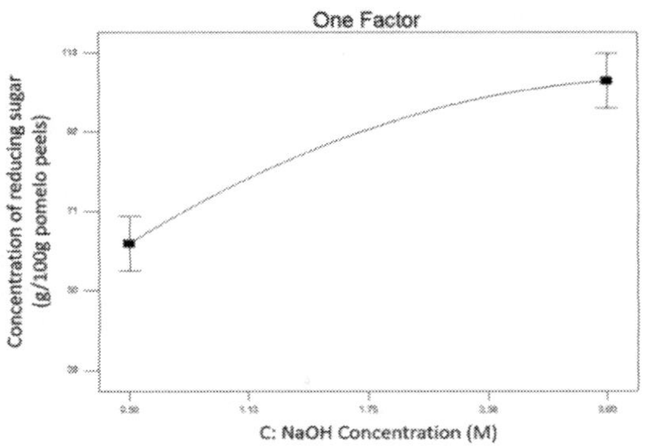

Fig. 3. Relationship between each pretreatment factor, (A) pretreatment temperature (°C), (B) pretreatment time (h), (C) NaOH concentration (M) and Y-axis is concentration of reducing sugar (g/100g pomelo peels)

(a)

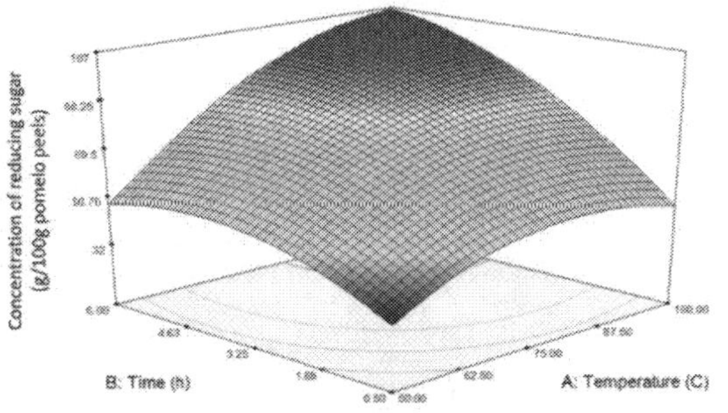

(b)

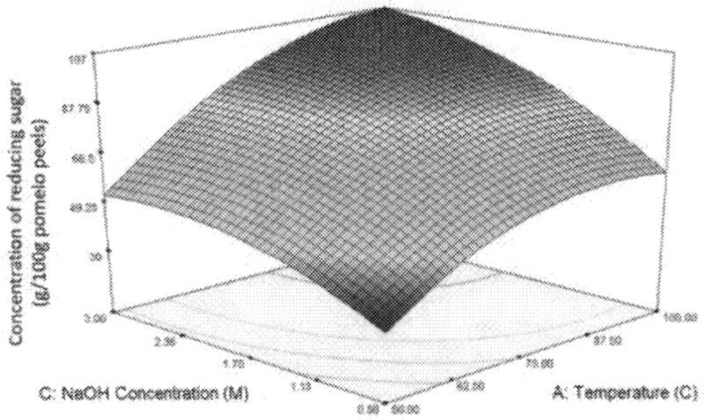

(c)

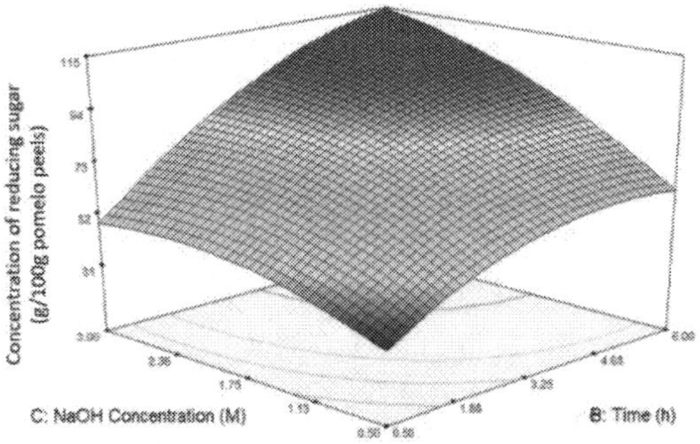

Fig. 4. Response surface plot represents the effects of pretreatment factors on reducing sugar yield (g/100g pomelo peels). (A) pretreatment temperature vs. pretreatment time, (B) pretreatment temperature vs NaOH concentration and (C) pretreatment time vs NaOH concentration

ANOVA analysis could simulate a model and predict optimal parameters for the pretreatment. The mathematical model suggested pretreatment using 2.17 M NaOH at 94.28 °C, for 4.5 hours as the optimal pretreatment conditions for pretreating Pomelo peels to yield maximum sugar yield. (Table 4). The model also predicted that the optimal pretreatment condition could yield 98.9 mg reducing sugar per gram of pomelo peels. To confirm the reliability of this model, pretreatment was conducted again at the predicted optimal pretreatment conditions. It could yield 99.1 mg reducing sugar per gram pomelo peels with only 0.2% error from the predicted value. When comparing the yield of reducing sugar, it could be deduced that the model from ANOVA prediction was highly accurate

Table 4. Mathematical models and optimal pretreatment condition obtained from RSM experiment.

Mathematical Model		
Reducing sugar = -72.59982 + 2.78992 x Temp - 0.89290 x Time + 6.54922 x Conc + 0.15731 * Temp * Time + 0.19064 * Temp * Conc + 2.64364 x Time x Conc - 0.019898 x Temp ^{2} - 1.50955 * Time^2 - 5.49344 x Conc^2		
Optimal pretreatment parameter		
Temperature(°C)	**Time (h)**	**NaOH Concentration(M)**
94.28	4.5	2.17
Predicted Concentration of sugar (g/100g pomelo peels)	Experimental Concentration of sugar (g/100g pomelo peels)	Difference (%)
98.90	99.10	0.2

4 Conclusion

The study was conducted to achieve maximum yield of reducing sugar from optimally pretreated pomelo peels and these sugars could be further converted to products such as biogas, bioethanol, etc. In this research, the optimization of pretreatment conditions was carried out using the RSM model. The preliminary alkaline pretreatment studies could help us in determining the range of pretreatment factors to be used in the RSM model. BBD of RSM was applied to design the experiments with various pretreatment conditions by varying NaOH concentration, pretreatment temperature, and pretreatment time. The BBD design along with the ANOVA analysis predicted NaOH concentration of 2.17M, pretreatment temperature of 94.28 °C, and pretreatment time of 4.5 hours as optimal pretreatment condition, which could yield of reducing sugars of 98.9 mg / g of pomelo peels. This implied a 1.87 fold increase in sugar yield obtained from pretreated biomass compared to untreated biomass. This study demonstrated the potential of pomelo peels to be used as feedstock in the biorefining process. Also, it provided an alternative way to reduce agriculture waste and air pollution.

Acknowledgement

The authors would like to thank King Mongkut's University of Technology, North Bangkok (Research Grant No. KMUTNB-BasicR-64-37, KMUTNB-PHD-64-02) for financial support of this work.

References

1. Ministry of Energy Annual Report, Biomass production.http://biomass.dede.go.th/biomass_web/index.html.
2. YS. Cheng, P. Mutrakulcharoen, S. Chuetor, K. Cheenkachorn, P. Tantayotai, EJ. Panakkal, M. Sriariyanun, Recent situation and progress in biorefining process of lignocellulosic biomass: toward green economy, Applied Science and Engineering Progress, **13**,4(2020): 299-311.
3. EJ. Panakkal, M Sriariyanun, J. Ratanapoompinyo P. Yasurin, K. Cheenkachorn, W. Rodiahwati, P. Tantayotai, Influence of Sulfuric Acid Pretreatment and Inhibitor of Sugarcane Bagasse on the Production of Fermentable Sugar and Ethanol, Applied Science and Engineering Progress, **15**,1(2022)
4. M. Sriariyanun, K. Kitsubthawee, Trends in Lignocellulosic Biorefinery for Production of Value-added Biochemicals, Applied Science and Engineering Progress. **13(4)**, (2020): 283-284.
5. H. Chen, J. Liu, X. Chang, D. Chen, Y. Xue, P. Liu, H. Lin and S. Han, A review on the pretreatment of lignocellulose for high-value chemicals., Fuel Processing Technology. **160**, (2017): 196-206.
6. M. Sriariyanun, JH. Heitz, P. Yasurin, S. Asavasanti, P. Tantayotai, Itaconic acid: A promising and sustainable platform chemical, Applied Science and Engineering Progress. **12(2)**, (2019) : 75-82.
7. P. Rachmontree, T. Douzou, K. Cheenkachorn, M. Sriariyanun, Furfural : A sustainable platform chemical and fuel, Applied Science and Engineering Progress. **13(1)**, (2020): : 3-10.
8. MP. Gundupalli, YS. Cheng, S. Chuetor, D. Bhattacharyya, M. Sriariyanun. Effect of Dewaxing on saccharification and ethanol production from different lignocellulosic biomass. Bioresource Technology. **125596.**
9. R. Akkharasinphonrat, T. Douzou, M. Sriariyanun, Development of ionic liquid utilization in biorefinery process of lignocellulosic biomass, King Mongkut's University of Technology North Bangkok International Journal of Applied Science and Technology, 10,2 (2017): 89-96
10. W. Rodiahwati and M. Sriariyanun, Lignocellulosic Biomass to Biofuel Production: Integration of Chemical and Extrusion)Screw Press(Pretreatment, KMUTNB International Journal of Applied Science and Technology., **9)4(**, (2016): 289-298.
11. S. Chuetor, T. Ruiz, A. Barakat, N. Laosiripojana, V. Champreda, M. Sriariyanun, Evaluation of rice straw biopowder from alkaline-mechanical pretreatment by hydro-textural approach, Bioresource Technology. **323**, (2021)
12. M. Sriariyanun, Q. Yan, I. Nowik, K. Cheenkachorn, T. Phusantisampan, M. Modigell, Efficient pretreatment of rice straw by combination of screw press and ionic liquid to enhance enzymatic hydrolysis, Kasetsart Journal (Natural Science(, **49**, (2015): 146-154.
13. K. Cheenkachorn, T. Douzou, S. Roddecha, P. Tantayotai, and M. Sriariyanun, Enzymatic saccharification of rice straw under influence of

recycled ionic liquid pretreatments, Energy Procedia. **100**,)2016(: 160-165.

14. K. Rattanaporn, S. Roddecha, M. Sriariyanun, K. Cheenkachorn, Improving saccharification of oil palm shell by acetic acid pretreatment for biofuel production, Energy Procedia, **141C**,)2017(: 146-149.
15. K. Rattanaporn, P. Tantayotai, T. Phusantisampan, P. Pornwongthong, M. Sriariyanun, Organic acid pretreatment of oil palm trunk: effect on enzymatic saccharification and ethanol production, Bioprocess and Biosystem Engineering, **41**, (2018): 467-477.
16. MP. Gundupalli, A. Sahithi , YS. Cheng, P. Tantayotai, M. Sriariyanun. 2021. Differential effects of inorganic salts on cellulase kinetics in enzymatic saccharification of cellulose and lignocellulosic biomass. Bioprocess and Biosystems Engineering. Published online. (2021).
17. A. Boontum, J. Phetsom, W. Rodiahwati, K. Kitsubthawee and T. Kuntothom, Characterization of Diluted-acid Pretreatment of Water Hyacinth, Applied Science and Engineering Progress. **12(4)**, (2019): 253-263.
18. YS. Cheng, ZY. Wu, M. Sriariyanun, Evaluation of Macaranga tanarius as a biomass feedstock for fermentable sugars production, Bioresource Technology. **294**, (2019).
19. P. Tantayotai, K. Rattanaporn, S. Tepaamorndech, K. Cheenkachorn, M. Sriariyanun "Analysis of an ionic liquid and salt tolerant microbial consortium which is useful for enhancement of enzymatic hydrolysis and biogas production, Waste and Biomass Valorization, **10)6(**, (2019): 1481-1491.
20. P. Tantayotai, P. Pornwongthong, C. Muenmuang, T. Phusantisampan and M. Sriariyanun, Effect of cellulase-producing microbial consortium on biogas production from lignocellulosic biomass, Energy Procedia. **141C**,)2017(: 180-183.
21. M. Sriariyanun, P. Tantayotai, P. Yasurin, P. Pornwongthong, and K. Cheenkachorn, Production, purification and characterization of an ionic liquid tolerant cellulase from Bacillus sp. isolated from rice paddy field soil, Electronic Journal of Biotechnology. **19**, (2016): 23-28.
22. P. Tantayotai, P. Rachmontree, W. Rodiahwati, K. Rattanaporn, M. Sriariyanun, Production of ionic liquid-tolerant cellulase produced by microbial consortium and its application in biofuel production, Energy Procedia. **100**,)2016(: 155-159.
23. P. Chandranupap, P. Chandranupap, Effect of Hemicellulase Enzyme in Flotation Deinking of Laser-printed Paper, Applied Science and Engineering Progress. **14(3)**, (2021): 370-377.
24. R. Runajak, S. Chuetor, W. Rodiahwati, M. Sriariyanun, P. Tantayotai, and S. Phornphisutthimas, "Analysis of Microbial Consortia with High Cellulolytic Activities for Cassava Pulp Degradation, E3S Web of Conferences. **141**, (2020)
25. Y. Zheng, J. Zhao, F. Xu and Y. Li, Pretreatment of lignocellulosic biomass for enhanced biogas production, Progress in Energy and Combustion Science **42**, (2014):35-53
26. W. Siri-anusornsak, N. Sinbuathong, Chemical pretreatment of rice straw for a raw material in the production of renewable energy, Proceedings of 51st Kasetsart University Annual Conference: Science, Natural Resources and Environment, (2014).
27. R. Pawongrat, Pretreatment processes for enhancing the efficiency of ethanol production from lignocellulosic agricultural wastes, Veridian E-Journal Science and Technology Silpakorn University, (2015).
28. C. Martin, B. Alriksson, A. Sjode, N. Nilverbrant and L.J. Jonsson, Dilute Sulfuric Acid Pretreatment of Agricultural and Agro-Industrial Residues for Ethanol Production, Applied Biochemistry and Biotechnology, (2011).
29. T. Li, Production of bioethanol from unwashed-pretreated rapeseed straw at high solid loading, Chengdu Institute of Biology. Bioresource
30. G.L. Miller, Use of dinitrosalicylic acid reagent for determination of reducing sugar, Analytical Chemistry **31**, (1959): 426-428.
31. S. Akkaravathasinp, P. Narataruksa, C. Prapainainar, Optimization of Semi-batch Reactive Distillation Using Response Surface Method: Case Study of Esterification of Acetic acid with Methanol in a Process Simution, Applied Science and Engineering Progress. **12(3)**, (2019): 209-215.
32. M. Bahadi, N. Salih, J. Salimon, D-Optimal Design Optimization for the Separation of Oleic Acid from Malaysian High Free Fatty Acid Crude Palm Oil Fatty Acids Mixture Using Urea Complex Fractionation, Applied Science and Engineering Progress. **14(2)**, (2021): 175-186.
33. L.K. Akula, R.K. Orauganti, D. Bhattacharyya, Treatment of Marigold Flower Processing Wastewater Using a Sequential Biological-Electrochemical Process, Applied Science and Engineering Progress. **14(3)**, (2021).
34. P. Pangsri, T. Wuttipornpun, W. Songserm, Mannanase and Cellulase Enzyme Production from the Agricultural Wastes by the Bacillus subtilis P2-5 Strain, Applied Science and Engineering Progress. **14(3)**, (2021).
35. L. Kim and J. Han, Optimization of alkaline pretreatment conditions for enhancing glucose yield of rice straw by response surface methodology, Department of Civil and Environmental Engineering. KAIST, **373-1**, (2012).
36. P. Amnuaycheewa, W. Rodiahwati, P. Sanvarinda, K. Cheenkachorn, A. Tawai and M. Sriariyanun, Effect of Organic Acid Pretreatment on Napier Grass (Pennisetum purpureum) Straw Biomass Conversion, KMUTNB International Journal of Applied Science and Technology. **10(2)**, (2017).

Use of waste mushroom beds for the production of value-added biodegradable fiber sheet

Walaiporn Suksai[1], *Chaniga* Chuensangjun[2,*], *Jittimon* Wongsa[3], *Vilai* Rungsardthong[1], *Savitri* Vatanyoopaisarn[1], *Benjawan* Thumthanaruk[1], *Rungsima* Yeetsorn[4], and *Buddhi* P. Lamsal[5]

[1]Department of Agro-Industrial, Food and Environmental Technology, Faculty of Applied Science, Food and Agro-Industry Research Center, King Mongkut's University of Technology North Bangkok, Bangkok 10800, Thailand.
[2]Science and Technology Research Institute, King Mongkut's University of Technology North Bangkok, Bangkok 10800, Thailand.
[3]Department of Agricultural Engineering for Industry, Faculty of Industrial Technology and Management, King Mongkut's University of Technology North Bangkok, Prachinburi 25230, Thailand.
[4]The Sirindhorn International Thai-German Graduate School of Engineering, King Mongkut's University of Technology North Bangkok, Bangkok 10800, Thailand.
[5]Department of Food Science and Human Nutrition, Iowa State University, 536 Farm House Lane, Ames, IA 50011-1054, USA.

Abstract. This research focused on the utilization of waste mushroom beds (WMB) after the harvesting of oyster (WMB-O) and lingzhi mushrooms (WMB-L) for the preparation of eco-friendly materials, fiber sheets. The WMB were sterilized and determined for their chemical compositions. The dry fiber of the sterilized WMB were pretreated by a steam explosion, comparing with alkaline pretreatment before the fiber sheet forming process. The results showed that α-cellulose contents of the WMB were in the range of 27-35% by dry weight basis. The fiber from WMB-L treated by alkaline at 13.5% w/w of NaOH for 120 min showed better fiber sheet appearance, compared to the steam explosion. Afterward, tapioca starch was added as a natural binder during the fiber sheet forming and their physical properties were determined. The analytical results indicated that an increase of NaOH concentration in the pretreatment led to an increase in the toughness and water absorption of the fiber sheet. The additional tapioca starch promoted the interaction between cellulose fiber networks, corresponding to the decrease of water absorption and a compressed appearance after water immersion. These finding results disclosed a potential use of the WMB as eco-friendly materials, e.g. biodegradable packaging, packing materials, cultivation vase in the future.

Keyword. Bio-material, Eco-friendly material, Lingzhi mushroom, Oyster mushroom, Pretreatment, Waste mushroom bed

1 Introduction

Mushroom are popularly consumed in many countries because of their texture, aroma, taste as well as their high nutritional composition. Mushrooms indicate high fiber content, low energy, high nutrition, and low calories. Several varieties of mushrooms contain bioactive compounds and present medicinal properties and health benefits. Generally, mushrooms are grown on timber rotting decay, manure, or compost. Some species need specific nutrients for their growth. Moreover, most species of mushrooms require high humidity up to 70-90%, especially during their caps opening. Water holding capacity of planting containers and humidity control are important factors for mushroom cultivation [1, 2]. Nowadays, mushroom farmings are widely expanded. There are three types of mushroom cultivations in Thailand including the cultivation in the wooden pickup, in the plastic bag, and on timber. The plastic bags for mushroom cultivation are commonly used because they have heat resistance, require a small cultivation area with convenient preparation and easy removability [2]. Lignocellulosic materials such as rice bran and sawdust are generally used as raw materials for mushroom cultivation in the plastic bag because they are abundant agricultural waste resources and degradable under the natural environment [3]. After the first mushroom cultivation, the nutrients in the culture medium were reduced and some types of mushroom beds could be reused with the addition of necessary nutrients, while some species of edible mushroom bed could be reused up to 5 times [4]. The mushroom beds after cultivation still contain some components or nutrition needed for another cultivation, especially fiber cellulose, hemicellulose, and lignin [5]. The waste mushroom beds (WMB) could be utilized as raw materials for the production of reducing sugars for biodiesel production [6], planting crops [7], and hydrogen production for the growth of *Clostridium thermocellum* [8]. According to the information obtained from the Thai Intellectual Learning Center Samutprakarn (Samutprakarn Province, Thailand), one of the large mushroom farms in Thailand, the oyster and lingzhi mushrooms are the two most popularly cultivated as edible, and medicinal mushrooms in the market, respectively. Generally, after-harvesting in mushroom

* Corresponding authors: chaniga.c@stri.kmutnb.ac.th

cultivation, the WMB are discarded rather than reused due to the insufficient nutrients for cultivation to the growth of the mushrooms. Consequently, there are a lot of WMB and the discarded plastic bags in many mushroom farms presently. The wastes are underuse and require waste management or utilization into more valuable products. Substitution of plastic planting containers using degradable materials is one expected alternative to reduce the waste plastics in mushroom farms.

Cellulose fibers from the lignocellulosic material and agricultural wastes, such as mulberry inner bark, rubberwood sawdust, spent coffee grounds, straw, etc. are reported for the preparation of high value-added eco-friendly materials, such as natural paper, the substrate for biorefining process, natural fiber-reinforced composites or degradable cultivation vase [9-12]. Eco-friendly fiber sheets prepared from cellulose fibers or agricultural wastes were reported, such as the use of Phyllostachys (bamboo) to produce carbonaceous fiber sheets for preparing the gas diffusion layers of proton exchange membrane fuel cells [13] and production of the bio-composite films from cassava starch by using cellulose extracted from durian rind as reinforcement agent [14]. However, there were no the published reports about the preparation of fiber sheets from WMB. Therefore, the preparation of water-absorbent and biodegradable fiber sheets from WMB would be interesting for further application in the future. The high water holding capacity and biodegradability of obtained fiber sheets are expected for forming the mushroom cultivation vase, lead to reduced frequency for watering, and promote zero waste in the mushroom farms.

Cellulose ($C_6H_{10}O_5$) is a polysaccharide that contains D-glucose linked together by ß-1,4-glycosidic bonds [15-16]. Each chain of cellulose is interacted by hydrogen bonds, resulting in high strength structure of the cellulose network [17]. Natural cellulose can be extracted from the plant (wood and non-wood), animal, and some microorganisms [15], therefore, cellulose is the most abundant lignocellulosic materials in the world [18]. Moreover, there are many reports on the properties of natural cellulose fiber including its advantages, potential use for renewable resources, strong and tough structure, and water absorbability [19-21]. Cellulose has been used in several industries including food, pharmaceutical, and textile industries [22]. However, the pretreatment of raw materials is an important process to remove lignin, hemicellulose, and other extractives from the raw materials to obtain high purity of cellulose [23]. For the preparation of fiber sheets from agricultural waste, high cellulose fiber contents and remained fibrous morphology are main characteristics of the pretreated raw materials [13], especially for the production of fiber sheets with high water absorption. Alkaline pretreatment and steam explosion could break the strong interaction between hemicellulose, lignin, and cellulose, resulting in an increase of porous structure in the cellulose fiber [24].

In this study, the potential use of WMB from oyster (WMB-O) and lingzhi mushrooms (WMB-L) cultivation as fiber sheets, the biodegradable material was investigated. Both fibers were pretreated by a steam explosion and compared to the alkaline treatment. The fiber sheets were formed and their physical properties such as toughness, water absorption, swelling, and porosity were measured according to the Thai Industrial Standard (876-2547) [10, 25] to evaluate their potentials for further applications in the future.

2 Materials and methods

2.1 Materials

Both WMB after harvesting for 7 days, WMB-O from oyster (*Pleurotus ostreatus*) and WMB-L from lingzhi (*Ganoderma lingzhi*) cultivation (Fig. 1.) were supplied from Thai Intellectual Learning Center (Samutprakarn Province, Thailand).

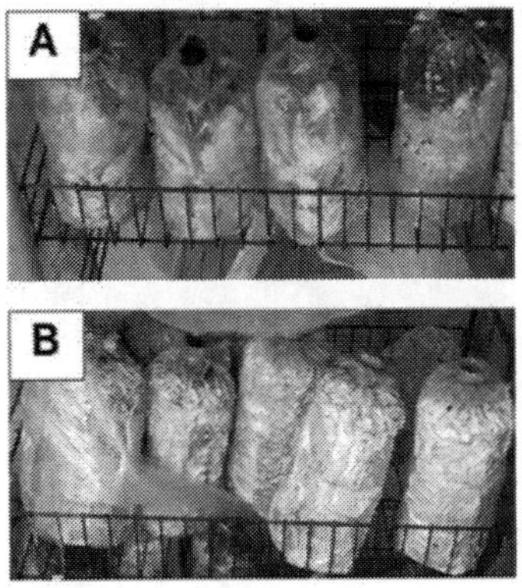

Fig. 1. Waste oyster (A) and lingzhi (B) mushroom beds after harvesting for 7 days.

Tapioca starch to be used as a natural binder for preparing the fiber sheet was derived from Tongchan Co., Ltd. (Bangkok, Thailand). Sodium hydroxide (NaOH) to be used for the alkali pretreatment was the commercial-grade product obtained from Chemiall Co., Ltd. (Bangkok, Thailand).

2.2 Methods

2.2.1 Preparation of natural fibers from WMB-O and WMB-L

The sterilization of WMB-O and WMB-L was carried out in an autoclave at 121°C for 1.5 h. Thereafter, the sterile WMB-O and WMB-L were dried in a hot air oven at 60°C for 24 h. The dried WMB samples were ground and kept in closed plastic bags at room temperature until the composition analysis and preparation of the fiber sheets.

2.2.2 Chemical composition analysis

The chemical composition analysis including moisture content, ash, solvent extractives, lignin, holocellulose (alpha-cellulose and hemicellulose) of the dried WMB were determined according to the Technical Association of the Pulp and Paper Industry (TAPPI) [26], as followed.

2.2.2.1 Moisture content

Moisture cans were weighed, then 1 g of the WMB-O and WMB-L sample was added into each moisture can. The samples were dried in a hot air oven at 105°C for 24 h after they were cooled down in a desiccator. The moisture content (%) of the samples was calculated according to equation (1).

$$\text{Moisture content (\%)} = [(A-B)/A] \times 100 \quad (1)$$

when A is the weight of the sample before drying (g) and B is the weight of the sample after drying (g).

2.2.2.2 Ash

A crucible was burnt at 550°C for 15 min, cooled down in the desiccator, and weighed. A 1 g (dry matter basis) of the WMB-O, and WMB-L was added into the crucible. The samples were burnt at 550°C until the white ashes were obtained. The samples were cooled down and weighed. The ash content of the samples was calculated according to equation (2).

$$\text{Ash (\%)} = (A/B) \times 100 \quad (2)$$

when A is the weight of ash (g) and B is the weight of the sample (g).

2.2.2.3 Solvent extractives

Extractives which are the plant resin, wax, alkaloid, and flavonoid that could be dissolved in an organic solvent were extracted from both WMB-O and WMB-L (3 g dry matter basis) using a soxhlet extractor for 4 h. A mixture of ethanol and benzene (1:2) was used as solvents for the extraction. The solvents were then vaporized from the extracted samples overnight followed by the drying at 105°C in a hot air oven and cooled down in a desiccator before their weight measurement. The solvent extractives were calculated according to equation (3).

$$\text{Solvent extractives (\%)} = [(A-B)/C] \times 100 \quad (3)$$

when A is weight of extracted sample (g), B is weight of blank (solvent) (g) and C is weight of the WMB (g).

2.2.2.4 Lignin

The extracted samples (3 g) were secondly extracted for 4 h using 200 mL of the 95% ethanol. Samples were left in a fume hood at room temperature overnight and then filtered with a Buchner funnel. The precipitates were washed several times with distilled water and left in the fume hood at room temperature (25 ± 2°C) for 2 h before drying in a hot air oven at 60°C for 24 h. The extractive-free samples of 0.35 g were added to the beaker in an ice bath. The concentrated H_2SO_4 (72% w/w) of 5.21 mL was gently added into the beaker. The mixture was stirred every 15 min for 2 h, at room temperature. Subsequently, the sample was added into the distilled water of 139.13 mL, and the total volume was adjusted to 200 mL by distilled water. Sample solutions were condensed for 4 h by a reflux technique, after that the solutions were poured into a beaker and left for 16 h at room temperature. The sample was filtered and dried at 105°C for 3 h. The beaker containing the sample was cooled down in the desiccator and weighed. The lignin was calculated according to equation (4).

$$\text{Lignin (\%)} = (A/B) \times 100 \quad (4)$$

when A is the weight of lignin (g) and B is the weight of the sample (g).

2.2.2.5 Holocellulose

The extractive-free samples of 3 g were suspended in a mixture of water (350 mL), acetic acid (1.1 mL), and NaCl (3.3 g). The sample suspension was incubated in a water bath shaking at 70°C for 3 h. The acetic acid of 1.1 mL and NaCl of 3.3 g were added into the suspension every hour (3 times). At the end of the reaction, the suspension was cooled down to 10°C in the ice bath before the filtration. The precipitates were dried at 105°C for 16 h and then cooled down in a desiccator before the weight measurement. The holocellulose was calculated according to equation (5).

$$\text{Holocellulose (\%)} = (A/B) \times 100 \quad (5)$$

when A is the weight of holocellulose (g) and B is the weight of the sample (g).

2.2.2.6 Alpha-cellulose and hemicellulose

The holocellulose of 1.5 g was mixed with 75 mL of NaOH (17.5% w/v). The mixture was cooled down to 2.5 ± 0.2°C in the ice bath and stirred with a magnetic bar for 30 min. The magnetic bar was washed with 2.5 mL of NaOH (17.5% w/v) and then adjusted the total volume of the mixture to 100 mL by distilled water. The mixture was continuously stirred for 30 min at 2.5 ± 0.2°C. The sample mixture was filtered and washed with distilled water until the neutral pH was obtained. The acetic acid (10% v/v) of 40 mL was then added to the sample precipitates and filtered. The precipitates were dried at 105°C for 16 h and cooled down in the desiccator before being weighed. The alpha-cellulose was calculated according to equation (6).

$$\text{Alpha-cellulose (\%)} = (A/B) \times 100 \quad (6)$$

when A is the weight of alpha-cellulose (g) and B is the weight of the holocellulose (g).

Hemicellulose was calculated according to equation (7).

$$\text{Hemicellulose (\%)} = A-B \quad (7)$$

when A is the percentage of holocellulose and B is the percentage of the alpha-cellulose.

2.2.3 Optimization of the fiber pretreatment

The optimal conditions for an alkaline pretreatment of the WMB fibers were investigated using the sterilized WMB-O and WMB-L fibers as the raw materials. The concentration of NaOH was studied at 1.5, 5.5, 9.5, and 13.5% w/w (g NaOH / g dry fiber) in this study. The fibers (30 g dry basis) were boiled in the NaOH solution for variable times, namely 5, 60, 90, and 120 min. In addition, a steam explosion technique was used to compare with the alkaline pretreatment. Briefly, the fibers (30 g dry basis) were steamed in an autoclave at 121°C for 5, 60, 90, and 120 min. The pretreated fibers from both methods were used for the preparation of the natural fiber sheet.

2.2.4 Preparation of fiber sheet

The pretreated pulps from WMB-O and WMB-L were soaked in water (1 g dry basis:30 mL) overnight, and then the pulp suspension was agitated homogeneously. The binder, tapioca starch was dissolved in water to obtain 10% w/v. The pulp was mixed with the tapioca starch solution at the weight ratio of 60:40 [10]. The mixture was thoroughly spread on a mold (21.59 cm × 27.94 cm), and the natural fiber sheet was then dried under the sunlight for 1 day. Thereafter the fiber sheets obtained from each WMB were peeled carefully from the mold. The significant properties of the fiber sheet, such as water absorption and swelling, were measured with the standard method of Thai Industrial Standards Institute (876-2547) [25]. The mulberry paper and printing paper (80 grammages) with the same thickness were used for the comparison of their properties.

2.2.5 Characterization of fiber sheet

2.2.5.1 Toughness

The fiber sheet was cut into pieces (50 × 50 mm). The top side of the fiber sheet was held on a shoulder pole whereas the bottom side of the fiber sheet was bound vertically to a plastic bag. The sand was slowly added into the plastic bag until the fiber sheet started to tear. The toughness of the fiber sheet was measured as the weight of the sand was put in the plastic bag. All measured data were given as mean values standard deviation (SD) of triplicate measurements.

2.2.5.2 Water absorption and swelling

The three pieces of fiber sheet (50 × 50 mm) were weighed and immersed in 120 mL of water and kept at room temperature. The fiber sheet piece was removed from the water every 30 min, dabbed with absorbent tissue paper to remove the surface moisture, and weighed until a constant weight was observed. The absorption was calculated according to equation (8).

$$\text{Absorption (g/g)} = (W_a - W_b)/W_b \tag{8}$$

when W_a is the weight of the fiber sheet after immersion (g) and W_b is the weight of the fiber sheet before immersion (g) [27]. For the determination of swelling, the thickness of the fiber sheet (50 × 50 mm) was measured using a hand-held manual micrometer to prevent compression, after that the fiber sheet was immersed in water for 180 min at room temperature. The fiber sheet piece was removed from the water, determined for its thickness and swelling ability according to equation (9)

$$\text{Swelling (\%)} = [(T_a - T_b)/T_b] \times 100 \tag{9}$$

when T_a is the thickness of the fiber sheet after immersion (cm) and T_b is the thickness of the fiber sheet before immersion (cm) [28].

2.2.5.3 Porosity

The volume of the fiber sheet (V_s) was measured using a micrometer and its density (ρ_b) was calculated according to [28]. The fiber sheet was ground, added into a volumetric flask (250 mL), and weighed. Then 120 mL of water was added into a volumetric flask and the suspension of the sample was mixed thoroughly. The sample was incubated in a water bath at 80°C until the steam was observed and then the sample was cooled down in an ice bath. Water was added into the volumetric flask to adjust the volume, and the flask was weighed again [25]. The porosity (ε) was calculated according to equation (10)-(13), respectively.

$$V_w = (m_2 - m_1)/\rho_w \tag{10}$$

$$V_s = V_F - V_w \tag{11}$$

$$\rho_s = m_s/V_s \tag{12}$$

$$\varepsilon = [1 - (\rho_b/\rho_s)] \times 100 \tag{13}$$

when m_1 is weight of the volumetric flask contained sample (g); m_2 is weight of the volumetric flask contained suspension after incubation (g); m_s is mass of sample (g); V_w, V_s, V_F are volume of water suspension, sample and volumetric flask (mL); ρ_w, ρ_s, ρ_b are density of water (at 80°C), sample and fiber sheet (g/cm^3), respectively.

3. Results and discussion

3.1 Chemical compositions

The chemical compositions of WMB-O and WMB-L i.e. ash, extractives, lignin, holocellulose, alpha-cellulose, and hemicellulose, are presented in Table 1. The chemical compositions of WMB-O were significantly different from those of WMB-L, except extractives, as shown in Table 1. The WMB-L contained a lower amount of ash, extractives, and lignin than the WMB-O. This implied that the WMB-L comprises fewer residue nutrients, compared to the WMB-O. According to the information from the mushroom farm (Thai Intellectual Learning Center, Samutprakarn Province, Thailand), the oyster mushroom bed could be reused as many times up to 5 times while the lingzhi mushroom bed is a disposable bed due to some insufficient substances for re-cultivation. In terms of holocellulose which is comprised of alpha-cellulose and hemicellulose, the WMB-O showed a higher amount of alpha-cellulose, but contained a lower amount of hemicellulose compared to the WMB-L. However, this result might due to the growth of bacteria and fungi in the lingzhi mushroom beds which could produce a cellulosic enzyme to degrade the cellulose [29]. In addition, the xylanase and laccase enzyme activity were also found in the oyster mushroom bed [30], resulting in possible degradation of hemicellulose and lignin in the mushroom beds too [31].

Table 1. Chemical compositions of waste mushroom beds.

Chemical compositions	Waste oyster mushroom bed (WMB-O) (% dry basis)	Waste lingzhi mushroom bed (WMB-L) (% dry basis)
Ash	16.01±2.43^{a}	9.38±0.74^{b}
Extractives	2.94±0.99^{ns}	1.46±0.48^{ns}
Lignin	15.13±0.64^{a}	7.29±0.87^{b}
Holocellulose	42.44±0.69^{a}	46.36±1.42^{b}
α-Cellulose	34.82±1.85^{a}	27.33±2.30^{b}
Hemicellulose	7.61±1.43^{a}	19.03±2.3^{b}

Values within the same row followed by the same letter are not significantly different ($p>0.05$)
ns: not significantly different ($p>0.05$)

3.2 Optimization of the fiber pretreatment

Fiber pretreatment aimed to remove undesirable elements such as lignin, hemicellulose, and extractives from the WMB before the forming process for the fiber sheet. Comparison of the methods for WMB pretreatment (steam explosion and alkaline solution) and type of WMB (WMB-O and WMB-L), demonstrated a different appearance of the pretreated fiber sheets as presented in Fig. 2. The pretreated fiber from the WMB-O showed a very rough surface, dark color, brittle, and unable to prepare as the fiber sheets by both pretreatment methods. This might be due to residual lignin after the pretreatment [32]. The results of steam explosion pretreatment of WMB-L at 121°C indicated that an increase in the pretreatment time could promote the fiber sheet forming (Fig. 2. L1-L4).

However, a slightly brittle fiber sheet was observed under the steam explosion at 121°C for 120 min. Comparing with alkaline pretreatment, the fiber sheets from WMB-L after boiling with NaOH at the concentration of 1.5% w/w for 120 min showed a better appearance including a smooth surface, light brown color, and increased toughness (Fig. 2. L8). Therefore, the effects of NaOH concentration were further investigated to optimize the fiber pretreatment of WMB-L. From Fig. 2. (L9-L11), the increase of NaOH concentration at 5.5, 9.5, and 13.5% w/w promoted better fiber sheet forming, especially with NaOH at 13.5% w/w for 120 min (Fig. 2. L11). Using a high concentration of NaOH at a high temperature and long time could completely remove lignin and hemicellulose from the raw fiber. These results corresponded to previous reports, for example, increasing alkali concentration gave the proper properties of the physic nut (*Jatropha curcas* Linn.) pulps (15% alkalinity) [33], and cornstalk paper (20% NaOH) [34]. Accordingly, the fiber sheets prepared by alkaline pretreatment of WMB-L at 9.5 and 13.5% w/w of NaOH for 120 min were selected and analyzed for their physical properties including toughness, water absorption, swelling, and porosity, respectively.

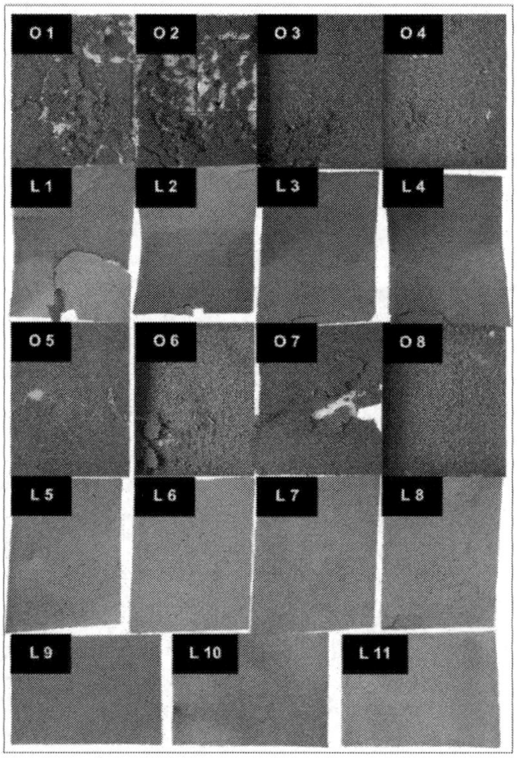

Fig. 2. Photographs of natural fiber sheets prepared from the waste mushroom bed of oyster mushroom (WMB-O) and lingzhi mushroom (WMB-L) with different pretreatment. O1-O4: WMB-O with the steam explosion at 121°C for 5, 60, 90, and 120 min; L1-L4: WMB-L with the steam explosion at 121°C for 5, 60, 90, and 120 min; O5-O8: WMB-O with NaOH at 1.5% w/w for 5, 60, 90, and 120 min; L5-L8: WMB-L with NaOH at 1.5% w/w for 5, 60, 90, and 120 min; L9-L11: WMB-L with NaOH at 5.5, 9.5, and 13.5% w/w for 120 min.

3.3 Toughness

The toughness of the fiber sheet prepared by alkaline pretreatment of WMB-L at the concentration of 9.5 and 13.5% w/w NaOH for 120 min was compared with the sheet of mulberry paper and printing paper (Fig. 3.). The results revealed that the fiber sheet from WMB-L pretreated with NaOH at 9.5 and 13.5% w/w, could support the sand weight of 50.75±5.22 and 76.47±18.35 g, respectively whereas, the printing paper and mulberry paper could support the sand weight of 1522.58±152.00 and 1585.91±376.90 g, respectively. These results presented that the fiber sheets from the pretreated WMB showed less toughness as compared with the commercial paper (the mulberry and printing paper). Due to the multi-step for commercial papermaking (i.e. pulp extraction, digestion, bleaching, paper formation), almost lignin and hemicellulose were removed from the raw materials [35]. The cellulose content of the mulberry paper and the printing paper from *Eucalyptus* has been reported as over 62-90% by oven dry weight [9, 36], which was higher than the WMB-L. This demonstrated that the higher cellulose content could promote the higher toughness of the paper. From this result, the tapioca starch was therefore applied as a natural binder to further improve the production of fiber sheets from the WMB.

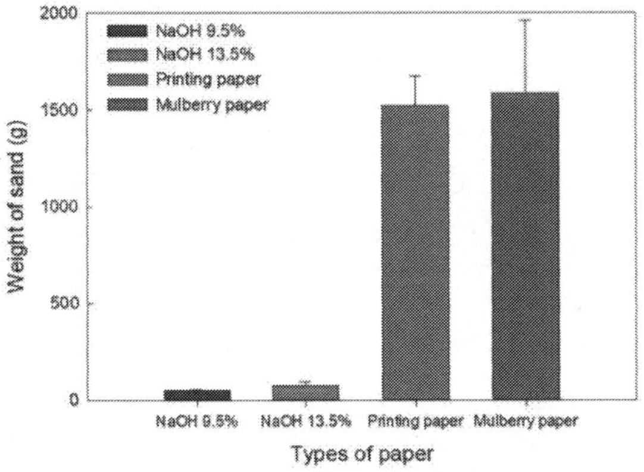

Fig. 3. Toughness of fiber sheet prepared from waste lingzhi mushroom bed after pretreatment with 9.5 and 13.5% w/w NaOH for 120 min, compared with the sheet of printing paper and mulberry paper.

3.4 Water absorption and swelling

The fiber sheet prepared from WMB-L pretreated with NaOH at 9.5 and 13.5% w/w showed the water absorption after water immersion for 30 min at 4.61±0.13 and 5.14±0.14 g/g, respectively (Fig. 4.). Their water absorption was higher than that of the printing paper (1.93±0.05 g/g), but lower than the mulberry paper (7.99±0.15 g/g). The addition of 40 wt% of natural binder (i.e. tapioca starch) into the fiber sheet prepared from WMB-L treated with NaOH at 13.5% w/w could reduce its water absorption to 3.47±0.03 g/g. The water absorption of all samples was quite constant, even though the water immersion time increased. The results were corresponding with other reports that an increased amount of cellulose in the composite materials could improve water absorption due to the binding of the hydroxyl group in cellulose to the hydrogen bonds of water [37-40]. However, into the commercial process for printing paper production, surface strengthening, surface sizes or hydrophobic agents (e.g. starch) are usually added [41] to improve resistance level to the water absorption of water. Meanwhile, the printing paper needs a liquid holding capacity to quickly absorb and retain water or inks by micro capillarity [36].

The results of the swelling analysis are shown in Fig. 5. The printing paper showed the highest swelling of 42.76%. Whereas the natural fiber sheet from the WMB-L with/without adding the natural binder and the mulberry paper showed contrary results, namely the compressed appearance and traces of decay due to water holding capacity after water immersion were observed (Fig. 6.). This phenomenon could be described that different raw materials and processes of pretreatment affected the structure of the fiber. The swellability depends upon the cell wall of the fiber, in other words, the higher microfibrils at the secondary wall of the fiber decrease swellability after water absorption but promote increasing strength [33-34, 14]. Moreover, the residual lignin after the pretreatment process, the addition of binder (such as starch), or introducing additives during the papermaking caused the different structure of the fiber, water absorption, and swelling properties [36-38].

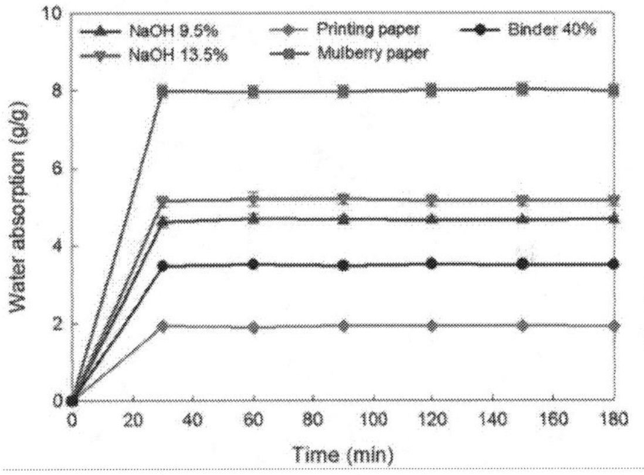

Fig. 4. Water absorption of fiber sheet prepared from waste lingzhi mushroom bed after pretreated with 9.5 and 13.5% w/w NaOH for 120 min, and fiber sheet with 40wt% binder compared with the sheet of printing paper and mulberry paper.

Although, good properties of the degradable cultivation vase from spent coffee grounds, shell lime, and rubberwood sawdust have ever been reported, their high water absorption and swelling, resulted in faster degradation (16 days) [10]. The strength of the cultivation vase is also one important property for production in the commercial. Moreover, the lifetime of the cultivation vase should be consistent with the growth time of cultivated plants, for example, the lingzhi and oyster mushroom require 75-90 days for their growth. Therefore, the optimal condition for fiber sheet forming to obtain the suitable properties for mushroom cultivation vase production is required for further investigation.

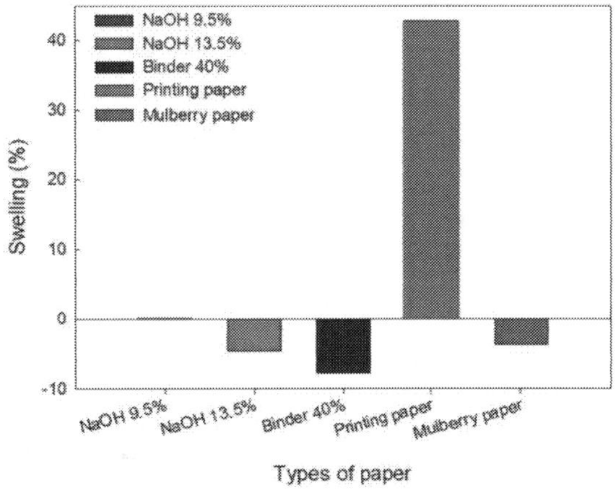

Fig. 5. Swelling of fiber sheet prepared from waste lingzhi mushroom bed after pretreated with 9.5 and 13.5% w/w NaOH for 120 min, and fiber sheet with 40wt% binder compared with the sheet of printing paper and mulberry paper.

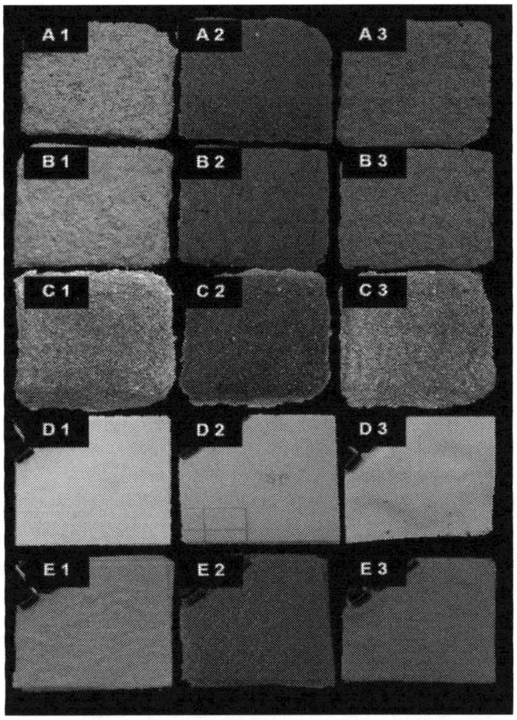

Fig. 6. Photographs of the fiber sheets prepared from waste lingzhi mushroom bed after pretreated with 9.5% w/w NaOH for 120 min (A), and 13.5% w/w NaOH (B); C) fiber sheet with 40wt% binder; D) printing paper and; E) mulberry paper. The number of 1-3 mean the sample before soaking in water, after soaking in water and after soaking in water and left at room temperature for 16 h, respectively.

3.5 Porosity

The porosity of the fiber sheet prepared by alkaline pretreatment of WMB-L with/without adding the natural binder, printing paper, and the mulberry paper are shown in Fig. 7. The porosity of the fiber sheet from the WMB-L without binder was approximately 168-172% and slightly decreased in the fiber sheet with the binder, due to the increase of interaction between cellulose and binder (starch) [42-43]. Corresponding to the water absorption results in Fig. 4, the decrease of capacity of water uptake was found in the fiber sheet with the binder. This implied that the high porosity of fiber sheets could promote the increase of capacity to absorb and retain water in their structure [36]. While the printing paper and mulberry paper showed higher porosity than the fiber sheet with/without binder due to their higher cellulose contents [9, 36].

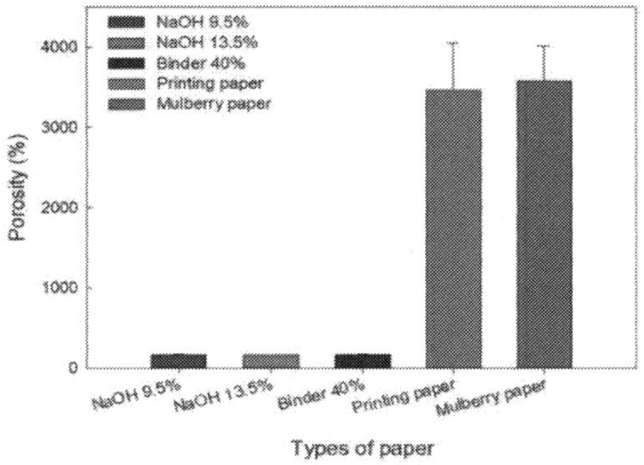

Fig. 7. The porosity of fiber sheet prepared from waste lingzhi mushroom bed after pretreated with 9.5 and 13.5% w/w NaOH for 120 min, and fiber sheet with 40wt% binder compared with the sheet of printing paper and mulberry paper.

From the results, the fiber sheet prepared from WMB-L demonstrated appropriate water absorption property and degradable behavior. This would be new alternative material for the possible production of value-added materials, such as a water-absorbent and biodegradable fiber sheet for mushroom cultivation vase production. In addition, the use of WMB for preparing as alternative materials could promote zero waste in the mushroom farms. The summary of finding results and expected applications from this study are also compared with the other lignocellulosic materials, as shown in Table 2.

4. Conclusions

The WMB-O and WMB-L samples after harvesting for 7 days showed different chemical compositions due to the different use-life of each mushroom bed. The amount of α-cellulose of the WMB were in the range of 27-35% dry basis, whereas, the lignin and hemicellulose were still high and might be the limitation of fiber sheet forming. The pretreatment processes by steam explosion and alkaline solution were optimized for the removal of lignin, hemicellulose, and other extractives. The fiber sheets derived from the alkaline pretreatment of the WMB-L at 9.5 and 13.5% w/w of NaOH for 120 min disclosed an opportunity for the production of value-added materials, such as a water-absorbent and biodegradable fiber sheet for mushroom cultivation vase production. Natural binder, i.e. tapioca starch was added to improve the properties of the fiber sheet from the WMB-L. Although the finding results of fiber quantity/quality of the WMB-L have not yet qualified for papermaking as compared to commercial paper, such as mulberry paper and printing paper. Some physical properties results, such as water absorption, swelling, and porosity demonstrated the possibility for utilization in agricultural application. The high water absorption and traces of decay after water immersion of the fiber sheet from the WMB-L revealed a suitable water holding capacity for producing mushroom cultivation vase, and lead to zero waste in the mushroom farms. However, further study on the optimal condition for natural binder addition and the degradation behavior of the obtained fiber sheet from the WMB are required.

Table 2. Summary of the expected applications from several lignocellulosic materials.

Material	Application	Characteristic	Reference
Bamboo	Carbonaceous fiber sheets for use as the gas diffusion layers	- Fibrous morphology remained after carbonization. - Carbonaceous fiber sheet was able to electric conductivity.	Matsumura et al. [13]
Banana fiber	Cellulose nanofiber for use as reinforcing elements	- α-cellulose increased from 64% to 95% after pretreatment of banana fiber. - High aspect ratio, crystallinity, and thermal stability	Deepa et al. [44]
WMB of shiitake (*Lentinus edodes*)	Cellulose nanofiber preparation	- Cellulose content of WMB was 25.4%. - Dry films of cellulose nanofiber prepared from WMB were transparent.	Konno et al. [45]
WMB of lingzhi (*Ganoderma lingzhi*)	Fiber sheets preparation for further application in mushroom cultivation	- α-cellulose of WMB was 27.33% dry basis. - Fiber sheet from WMB showed high water absorption and biodegradability.	This study

Acknowledgement

Special thanks go to Thai Intellectual Learning Center (Samutprakarn province, Thailand) for their kind support and information, the National Research Council of Thailand through the NRCT Senior Research Scholar Program (Contract No. 814-2020) for the financial support.

References

1. P. Kalac, *Mineral composition and radioactivity of edible mushrooms* (Elsevier Inc., London, 2019)
2. M.C. Tseng, J.H. Luong, Mushroom cultivation – technology for chemical production, Fermentation Processes, **7** (1984): 45–79
3. P. Karpilanondh, *Mushroom technology* (Textbook Publishing Center KMUTNB, 2018)
4. S. Ritnuch, *Straw mushroom cultivation in a plastic bag* (Offset Creation, 2009)
5. Z. Lou, Y. Sun, X. Zhou, S.A. Baig, B. Hu, X. Xu, Composition variability of spent mushroom substrates during continuous cultivation, composting process and their effects on mineral nitrogen transformation in soil, Geoderma Regional, **307**, 19 (2017): 30–37
6. S. Wu, Y. Lan, Z. Wu, Y. Peng, S. Chen, Z. Huang, L. Xu, I. Gelbic, X. Guan, L. Zhang, S. Zou, Pretreatment of spent mushroom substrate for enhancing the conversion of fermentable sugar, Bioresource Technology, **148** (2013): 596–600
7. Z. Run-Hua, D. Zeng-Qiang, L. Zhi-Guo, Use of spent mushroom substrate as growing media for tomato and cucumber seedlings, Pedosphere, **22**, 3 (2012): 333–342
8. H.N. lin, Y.T. Wang, M.J. Zhu, Evaluation of spent mushroom compost as a lignocellulosic substrate for hydrogen production by *Clostridium thermocellum*, International Journal of Hydrogen Energy, **42** (2017): 26687–26694
9. W. Anapanurak, B. Puangsin, Paper mulberry pulp properties by various alkaline processes, The research project for higher utilization of forestry and agricultural plant materials in Thailand (HUFA), (n.d.): 117–125
10. W. ChaiChan, W. ChaiChan, A. Sawain, *Plant pots from co-production of spent coffee grounds, shell lime and rubber wood sawdust* (Rajamangala University of Technology Srivijaya, 2017)
11. Y.S. Cheng, P. Mutrakulcharoen, S. Chuetor, K. Cheenkachorn, P. Tantayotai, E.J. Panakkal, M. Sriariyanun, Recent situation and progress in biorefining process of lignocellulosic biomass: toward green economy, Applied Science and Engineering Progress, **13**, 4 (2020): 299–311
12. B.D.S. Deeraj, K. Joseph, J.S. Jayan, A. Saritha, *Dynamic mechanical performance of natural fiber reinforced composites: a brief review* (Applied Science and Engineering Progress, 2021)
13. T. Matsumura, T. Kinumoto, M. Matsuoka, T. Tsumura, M. Toyoda, *Preparation of carbonaceous*

fiber sheets derived from bamboo and application to gas diffusion layer of proton exchange membrane fuel cells (Oita University, 2015)

14. P. Kuntadong, M. Peekoh, O. Sindanjark, Bio-composite films based on cassava starch reinforced with durian rind cellulose fibers, RMUTP Research Journal, **13**, 1 (2019): 39–50
15. H. Kargarzadeh, M. Loelovich, I. Ahmad, S. Thomas, A. Dufresne, *Nanocellulose and cellulose nanocomposites* (Wiley-VCH Verlag GmbH & Co. KGaA - Weinheim, Germany, 2017)
16. C. Doree, *The methods of cellulose chemistry including methods for the investigation of substances associated with cellulose in plant tissues* (Chapman and Hall, London, 1947)
17. P. Teerapronchaisit, *Polymers and biomolecules in organic compounds*, (Department of Chemistry, Faculty of Applied Science, Khon Kaen University, 2010)
18. Y. Habibi, Key advances in the chemical modification of nanocelluloses, Chemical Society Reviews, **43** (2014): 151–1542
19. R.A. SHANKS, *Natural fiber, composites materials, processes and applications*, Woodhead Publishing Limited, (2014): 66-83
20. B. Boontima, A. Noomhorm, C. Puttanlek, D. Uttapap, V. Rungsardthong, Mechanical properties of sugarcane bagasse fiber-reinforced soy based biocomposites, Polymers and the Environment, **23**, 1 (2014): 97–106
21. P. Saenghirunwattana, A. Noomhorm, V. Rungsardthong, Mechanical properties of soy protein based "green" composites reinforced with surface modified cornhusk fiber, Industrial Crops and Products, 60 (2014): 144–150
22. M.P. Arrieta, E. Fortunati, N. Burgos, M.A. Peltzer, J. López, L. Peponi, *Nanocellulose-based polymeric blends for food packaging applications, multifunctional polymeric nanocomposites based on cellulosic reinforcements* (Elsevier Inc., London, 2016)
23. R. Pawongrat, Pretreatment processes for enhancing the efficiency of ethanol production from lignocellulosic-agricultural wastes, Veridian E-Journal, **2**, 1 (2015): 143-157
24. S. Sriwongchai, *Biodiesel production from agricultural wastes by using biotechnological process* (Burapha University, 2017)
25. T. Piyang, W. Chaichan, K. Sagulsawasdipan, Environment-friendly plant pot production from palm oil sludge and mushroom cultured waste, RMUTSV Research Journal, 10, 3 (2018): 497-511
26. TAPPI's (Technical Association of the Pulp and Paper Industry) International Nanotechnology Division, Roadmap for the Development of International Standards for Nanocellulose, (2011)
27. N. Kongklom, C. Chuensangjun, Y. Chisti, S. Sirisansaneeyakul, Improved keeping quality of *Dendrobium* "Bom" orchids using nutrients entrapped in a biodegradable hydrogel, Scientia Horticulturae, **234** (2018): 184–192
28. A. Atiqah, M. Jawaid, M.R. Ishak, S.M. Sapuan, Moisture absorption and thickness swelling behaviour of sugar palm fiber reinforced thermoplastic polyurethane, Procedia Engineering, **184** (2017): 581–586
29. P. Seephueak, C. Preecha, W. Seephueak, The study diversity of fungi and bacteria occurring on spent mushroom compost and utilization, Agricultural Technology, **13**, 5 (2017)
30. S.H. Lim, Y.H. Lee, H.W. Kang, Efficient recovery of lignocellulolytic enzymes of spent mushroom compost from oyster mushroom, *Pleurotus spp.*, and potential use in dye decolorization, Mycobiology, **41**, 4 (2013): 214–220
31. J. Kumla, N. Suwannarach, K. Sujarit, W. Penkhrue, P. Kakumyan, K. Jatuwong, S. Vadthanarat, S. Lumyong, Cultivation of mushroom and their lignocellulolytic enzyme production through the utilization of agro-industrial waste, Molecules, **25**, 12 (2020)
32. K.O. Reddy, C.U. Maheswari, M. Shukla, E. Muzenda, Preparation, chemical composition, characterization, and properties of Napier grass paper sheets, Separation Science and Technology, **49** (2014): 1527–1534
33. S. Butrat, *Steam explosion for chemi-thermo mechanical pulping of physic nut (Jatropha curcas Linn.) for paper packaging applications, (Master of Science (Packaging Technology)*, (Department of Packaging and Materials Technology, Kasetsart University (n.d.))
34. Y. Fahmy, T.A. Fahmy, F. Mobarak, M.E. Sakhawy, M. Fadl, Agricultural residues (wastes) for manufacture of paper, board, and miscellaneous products: background overview and future prospects, ChemTech Research, Sphinx Knowledge House, **10**, 2 (2017): 424–448
35. A. Guadie, A. Mulat, A. Kokeb, N. Gabbiye, Synthesis and characterization of Eucalyptus pulp by kraft process for paper sheet formation, Environmental Science and Technology, **1**, 2 (2015): 20–26
36. V. Rana, G. Joshi, S.P. Singh, P.K. Gupta, *Eucalypts in pulp and paper industry* (Eucalypts, India, 2016)
37. M. Chaisupakitsin, W. Wongsirojanakul, A. Ruengkul, O. Siriwan, Acoustical panel from coconut coir / EPS hybrid composites, Science and Technology, **15**, 2 (2007): 54–60
38. C. Homkhiew, S. Rawangwong, W. Boonchouytan, W. Thongruang, Mechanical and physical properties of thermoplastic natural rubber composites reinforced with rubberwood sawdust, SWU Engineering Journal, **13**, 1 (2018): 107–122
39. C. Ngaowthong, V. Rungsardthong, S. Siengchin, Polypropylene/hemp woody core fiber composites:

Morphology, mechanical, thermal properties, and water absorption behaviors, Mechanical Engineering, **8**, 3 (2016): 1–10

40. V. Dogra, C. Kishore, A. Verma, A.K. Rana, A. Gaur, *Fabrication and experimental testing of hybrid composite material having biodegradable bagasse fiber in a modified epoxy resin: evaluation of mechanical and morphological behavior* (Applied Science and Engineering Progress, 2021)
41. J. Brander, *Surface Application of paper chemicals* (Chapman & Hall, London, 1997)
42. I. Spiridon, C.A. Teaca, R. Bodirlau, M. Bercea, Behavior of cellulose reinforced cross-linked starch composite films made with tartaric acid modified starch microparticles, Polymers and the Environment, **21** (2013): 431–440
43. B. Nasri-Nasrabadi, M. Mehrasa, M. Rafienia, S. Bonakdar, T. Behzad, S. Gavanji, Porous starch/cellulose nanofibers composite prepared by salt leaching technique for tissue engineering, Carbohydrate Polymers, **108** (2014): 232–238
44. B. Deepa, E. Abraham, B.M. Cherian, A. Bismarck, J.J. Blaker, L.A. Pothan, A.L. Leao, S.F. Souza, M. Kottaisamy, Structure, morphology and thermal characteristics of banana nano fibers obtained by steam explosion, Bioresource Technology, **102** (2011): 1988–1997
45. N. Konno, M. Kimura, R. Okuzawa, Y. Nakamura, M. Ike, N. Hayashi, A. Obara, Y. Sakamoto, N. Habu, Preparation of cellulose nanofibers from waste mushroom bed of shiitak (Lentinus edodes) by TEMPO-mediated oxidation, Mokuzal Hozon (Wood Protection), **42** (2016): 157–164

Preparation and properties of biodegradable cat litter produced from cassava (*Manihot esculenta* L. Crantz) trunk

Nistha Saikeaw[1], *Vilai* Rungsardthong[1,*], *Peerapong* Pornwongthong[1], *Savitri* Vatanyoopaisarn[1], *Benjawan* Thumthanaruk[1], *Napisa* Pattharaprachayakul[1], *Jittimon* Wongsa[2], *Solange I.* Mussatto[3] and *Dudsadee* Uttapap[4]

[1]Department of Agro-Industrial, Food and Environmental Technology, Faculty of Applied Science, Food and Agro-Industry Research, King Mongkut's University of Technology North Bangkok, Bangkok 10800, Thailand
[2]Department of Agricultural Engineering for Industry, Faculty of Industrial Technology and Management, King Mongkut's University of Technology North Bangkok, Prachinburi 25230, Thailand
[3]Department of Biotechnology and Biomedicine, Technical University of Denmark, Søltofts Plads, Building 223, 2800, Kongens Lyngby, Denmark
[4]Division of Biochemical Technology, School of Bioresources and Technology, King Mongkut's University of Technology Thonburi, Bangkok 10150, Thailand

Abstract. Organic based cat litter becomes more and more popular in many countries including Thailand because of the concern on environmental and health impacts in cats and cat owners. Most of the cat litter sold in Thailand is manufactured and imported from abroad. Cassava trunk pith, the stele of cassava trunk, is the biomass obtained from agricultural waste which presents excellent water absorption. Thus, the purpose of this article was to investigate the preparation of the biodegradable cat litter from cassava trunk pith. The composition of cassava trunk pith was analyzed by standard methods. The preparation of the cat litter was performed by mixing cassava trunk pith with the binder (glycerol, and palm oil) at the ratio 49.7:40.0 by weight. The effects of two clumping agents, guar gum and xanthan gum at 5-15%, on the properties of cat litter were determined. The cassava trunk pith consisted of cellulose (47.12%), lignin (30.18%), hemicellulose (14.07%), extractives (5.69%) and ash (2.94%) by wet weight basic. The litter from all formulations obtained a high rating scale for clump strength with the use of palm oil. The addition of guar gum, and xanthan gum as a clumping agent resulted to the products with desirable physical properties. These results demonstrated that cassava trunk pith has a high potential for product development as biodegradable cat litter.

Keyword. Cat litter, Cassava trunk pith, Lignocellulose, Absorbent, Clump strength, Clumping agents

1 Introduction

Society and lifestyle have been changed presently. Household is getting smaller while the elderly and singles are increasing. Pets are friends and members of the family, treated like celebrities in this pet trends. As a result, pet business has grown by 10-15% from 2014 to 2018 in Thailand [1].

Cats are one of the most popular pets. According to statistics of cat breeders around the world, Thailand has the 8th highest number of cat breeders in the world [2], which accounts for 33 percent of the country's population. Cat urine and feces have a naturally strong smell inside the house. Therefore, the cat litter is commonly used for cats to excrete their waste nowadays [3]. In general, cats can generate 40 grams of fecal waste per day [4], which can be estimated that 281,952 tons of fecal waste are produced per year in Thailand.

The cat litter is an absorbent material that can absorb liquid, odors and help to eliminate cat waste more easily and comfortable to clean. Inorganic based cat litter such as bentonite cat litter and pumice cat litter are the products obtained from the mining industry and volcano [5]. Both products are able to retain odors and clump well, but still make small dust and could negatively affect to the health of cats and owner. In previous studies, it was found that using the litter for a long time, cat got sick and then died from inhalation and eating of the litter [6]. Cat's health concerns [7] and more landfill space requirement for the non-biodegradable cat litter [8] led to the needs for biodegradable cat litter. Consequently, cat owners currently tend to choose products that are safe and environmental-friendly, resulting in less sale amount for the inorganic cat litter [9]. Hence, organic based cat litter becomes more popular [10] and has been sold in a larger number nowadays [11]. Moreover, most of the cat litters sold in Thailand are manufactured and imported from abroad [12]. The products from local production still require more improvement for the quality.

Organic based cat litter could be made from starch, modified starch [13] and plant fiber such as rice straw, water hyacinth, coffee grounds, coconut husk, palm oil mesocarp, olive oil waste [14], and cassava pulp waste [15] which is lignocellulosic material. The lignocellulose can be found as the main composite of plant cell walls that consist of cellulose, hemicellulose and lignin. The

* Corresponding author: vilai.r@sci.kmutnb.ac.th

structure of lignocellulose also contains many hydrophilic functional groups, which obtain high water holding capacity [16]. Additionally, cellulose has a structure and ion exchange with organic substance as the resin, which is a desired property of absorbent. Therefore, from these aspects, lignocellulose had been studied in cat litter as an absorbent [17].

Cassava is an important economic and energy crop in Thailand [18]. Interestingly, almost all parts of the cassava can be used. For instances, the seed extracts are applied in the pharmaceutical industry, and root and leaf are generally used to make cassava starch and animal feed, respectively. Moreover, cassava pulp can be supplied for making dietary fiber noodles [19], the super-absorbent [20] and cat litter [15]. In terms of cassava trunk, it is estimated that around 116 million tons of fresh cassava trunk are produced worldwide each year approximately 10–20% [21] of the harvested cassava roots. Cassava trunks are used for propagation, while most of them have to be removed by landfilling or open-burning which leads to the air pollution. The trunk mainly consists of lignocellulose [22], glucan-rich feedstock which makes them especially suited for bioethanol production [23] or convert into biochar [24]. Moreover, the cassava trunk pith is a stele of cassava trunks that are abundant agricultural wastes and obtain an excellent water absorption [25], high porous and good adsorbent. In addition, there is no report on the use of cassava trunk pith as adsorbent in cat litter yet.

Thus, this study aimed to develop the biodegradable cat litter from the cassava trunk pith. The determination of the chemical compositions of the cassava trunk pith have been studied. The effects of binders and clumping agents on the properties of the cat litter such a water absorption, clumping percentage and clump strength were also determined.

2 Materials and methods

2.1 Raw materials and composition

Fresh cassava trunks collected from Nakhon Nayok province were cut into 8-15 cm long (Fig. 1) with a diameter of 5-21 mm (Fig. 2). Two types of binders used in this study, glycerol was supplied from Krungthepchemi Co., Ltd. (Bangkok, Thailand) while palm oil was bought from Sime Darby Oils Morakot Public Co., Ltd. (Bangkok, Thailand). Sodium propionate, guar gum and xanthan gum were supplied by Chemipan Co., Ltd. (Bangkok, Thailand).

2.2 Preparation of cassava trunk piths and chemical analysis

Fresh cassava trunk piths (Fig. 3) were pushed out of the cassava trunk by a round stick and dried at 40 °C for 24 hours. Then, cassava trunk piths were ground into fine particles by cutting mills (Retsch cross beater mill, Germany) and screened through a 2 mm sieve. The ground samples were separated into various sizes using a sieve shaker (Minor M200 Endecotts Ltd., London, United Kingdom) fitted with 12, 14, 35 and 80 mesh sieves (USA Std Test Sieve, Endecotts Ltd., London, United Kingdom). The samples retained on the top of the 14 mesh sieves were then used in cat litter experiments. The finer particles retained on 35 mesh sieves were used for further analysis. The chemical compositions of the samples the methods of Technical Association of the Pulp and Paper Industry [26] with the preparation of wood for chemical analysis (TAPPI-T-264 om-88), determination of ash (TAPPI-T-211 om-93), extractives (TAPPI-T-203), and lignin (TAPPI-T-222 om-98). Holocellulose (cellulose and hemicellulose) was analyzed by Browing method, while hemicellulose was measured followed TAPPI-T-203 om-88. The chemical composition was calculated by dry weight basic of raw materials. The functional groups and basic structure of the samples were analyzed using Fourier transform infrared spectroscopy (FTIR).

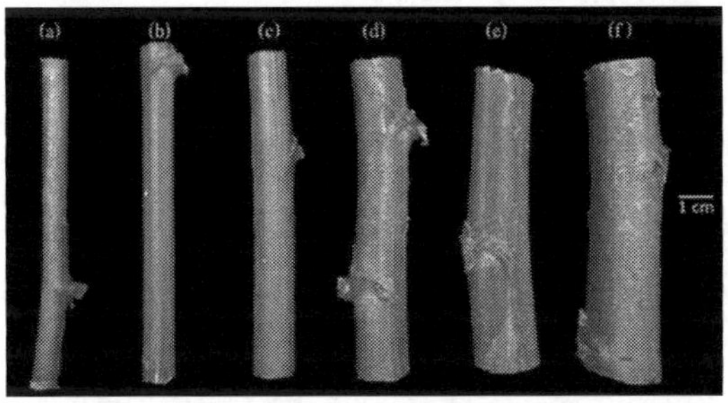

Fig. 1. Cassava trunks with a length of 8-15 cm: (a) 5.85 mm, (b) 7.88 mm, (c) 10.47 mm, (d) 13.07 mm, (e) 16.08 mm and (f) 18.47 mm.

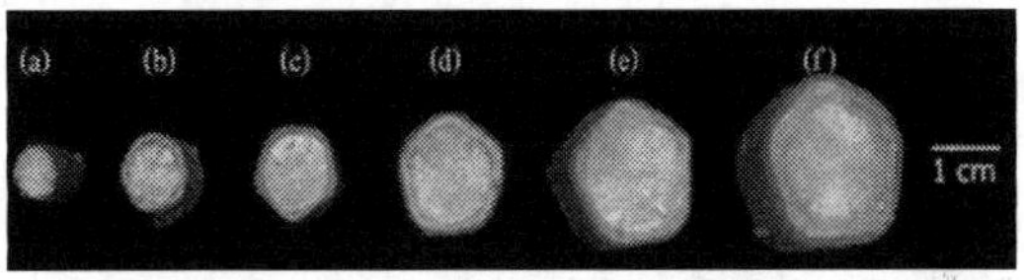

Fig. 2. Cross-section of cassava trunk with the diameter 5.00-21.00 mm: (a) 5.85 mm, (b) 7.88 mm, (c) 10.47 mm, (d) 13.07 mm, (e) 16.08 mm and (f) 18.47 mm.

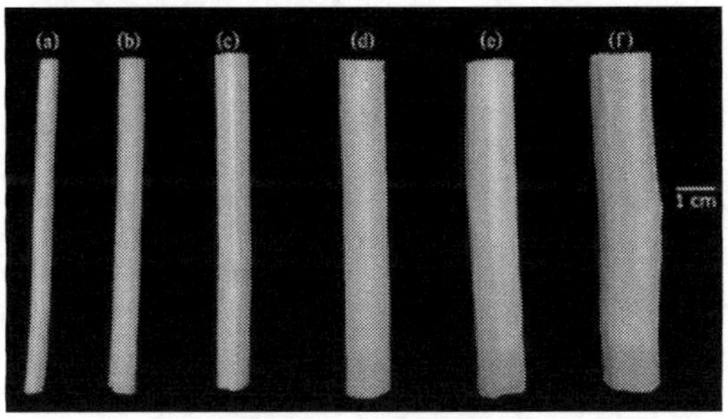

Fig. 3. Cassava trunk pith with the diameter of 5.00-21.00 mm: (a) 4.22 mm (b) 6.14 mm, (c) 7.68 mm, (d) 10.00 mm, (e) 11.50 mm and (f) 13.14 mm.

2.3 Preparation of cat litter

To investigate the effects of binder on clump strength, the cat litters were prepared from 49.7% (w/w) of the ground cassava trunk pith 14 mesh, 40% (w/w) of binder, 10 % (w/w) of guar gum and 0.3% (w/w) of sodium propionate.

To observe the effects of clumping agent on bulk densities, hydration capacities and clumping activity, the cat litters were prepared with the formulation listed above with the guar gum, and xanthan gum at 0.0, 5.0, 7.5, 10.0, 12.5 and 15.0% (w/w). The palm oil was applied as binder in the range 35-50% (w/w).

The ingredients for the cat litter preparation were mixed with the ground cassava trunk pith 14 mesh and each binder by a Food Mixer (Kenwood Electronics, Kenwood Ltd., LDN, UK) for 3 min. Thereupon, the clumping agent was then added to the mixture and mixed for 5 min and packed in a plastic zip lock until further analysis.

2.4 Physical Characteristics

2.4.1 Clump strength

Clump strength of the cat litters was determined by Standard Drop Method Test [27] with slight modification. Two grams of cat litters were placed in a plastic cup (diameter of 47 mm and a height of 29 mm) and then 15 mL of distilled water were added onto the samples, the clumped cat litters were then dried in a hot air oven at 45 °C for 24 hours. The clump strength test of the clumped samples was carried out at a height of 12 inches by dropping the samples onto the tray, followed by examination for the breakage by assigning a rating on a scale from 1 to 3, where 1: clump intact (no breakage), 2: clump was slightly broken and there were small fragments, and 3: clump was broken into more than two pieces with scattered small fragments, as shown in Fig. 4.

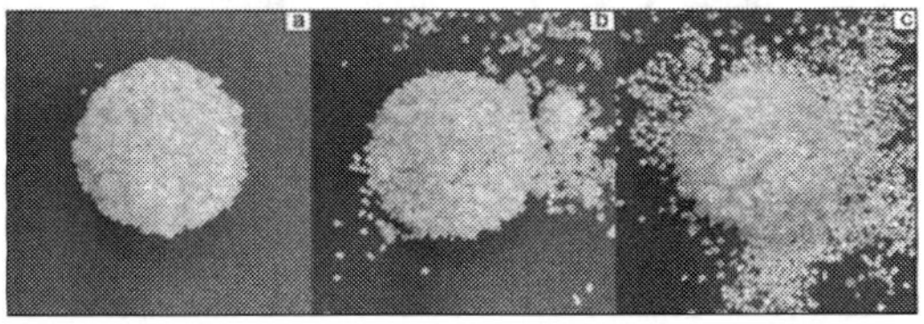

Fig. 4. Rating for the clump breakage: (a) clump intact (no breakage), (b) slightly broken and (c) moderately broken.

2.4.2 Bulk densities and hydration capacities

Ten grams of the sample (M) were placed in a dry 250 mL graduated cylinder. The apparent volume (V) was measured to the nearest graduated unit. The bulk densities ($g \cdot cm^{-3}$) of each sample were determined from the samples weight divided by the apparent volume.

To determine hydration capacity, 0.5 g of the samples were placed in 27 mL test tubes, then 15 mL of deionized distilled water were added and suspended by a shaking incubator (Daihan Labtech Co., Ltd.) at 150 rpm for 15 min and leave it for 15 min to precipitate the cat litter. The supernatants were carefully removed from the test tubes using a pipette dropper and the sample tubes were weighed. The hydration capacity ($g_{water} \cdot g_{litter}^{-1}$) were determined from the hydrated samples weight divided by 0.5 g.

2.4.3 Clumping percentage

Clumping activity was determined by adding 2.0 g of each formulation into a plastic cup with the same dimension as described above. Pipette 15 mL of distilled water onto the cat litters, then placed the clumps of cat litter in a hot air oven at 45 °C for 24 hours. The clumps were then put onto a 4 mesh (4.75 mm) sieve and placed on a shaking incubator (Daihan Labtech Co., Ltd.) at 250 rpm for 1 min. Clumping percentage [28] was finally calculated as followed (1)

$$Clumping\ \% = (Weight\ of\ clumps\ on\ sieve/2.0) \times 100 \quad (1)$$

2.5 Statistical design and analyses

Analyses of Variance (one-way ANOVA) were used to analyze differences among the treatments and Duncan test for hydration capacities, bulk densities and clumping percentages. Comparison between different treatments was made using differences of least squares means, when significant F-test values from the ANOVA were obtained at $p \leq 0.05$. All statistical analyses were performed using IBM SPSS Statistics Version 19.

Table 1. Chemical composition of cassava trunk pith.

Chemical component	% wet weight basis
Cellulose	47.12
Lignin	30.18
Hemicelluloses	14.07
Extractives	5.69
Ash	2.94

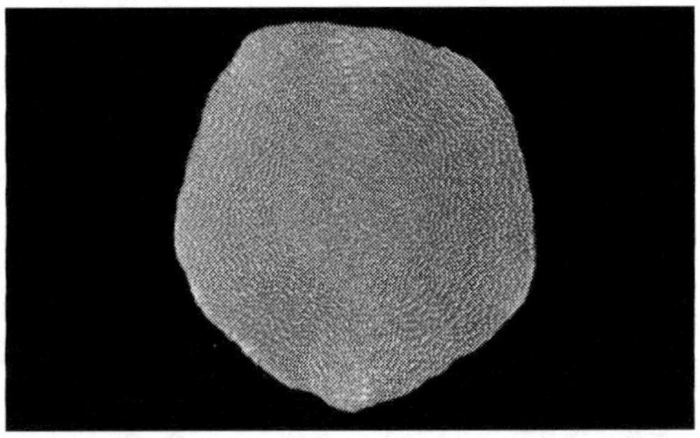

Fig. 5. Cassava trunk pith cross-section.

3 Results and discussion

3.1 Chemical composition of cassava trunk pith

The chemical composition of the cassava trunk piths is showed in Table 1. The cassava trunk piths consisted of cellulose 47.12%, lignin 30.18%, hemicellulose 14.07%, extractives 5.69% and ash 2.94%. The lignocellulose contents of the cassava trunk piths shown here are rather similar to the cassava stem [29], which contain 4.45% starch, 37.61% cellulose, 29.84% hemicelluloses, 19.11% lignin and 3.34% ash. In addition, lignocellulosic profile of cassava pulp residues contains 32% starch, 26% cellulose, 17% hemicelluloses and 16% lignin [30].

The spongy core of the cassava trunk pith tissue (Fig. 5) is formed by the parenchyma cell (pith), xylem and phloem [31]. These plant cell walls are majorly composed of lignocellulose, a natural polymer with many hydrophobic functional groups that are able to absorb moisture and exchange ions in the solution [32]. For instance, the functional groups are the acetyl and ester group of uronic ester in the hemicellulose structure, the carboxyl group and the aryl group in the lignin structure, the hydroxyl groups (-OH) in the cellulose structure, thus lignocellulose or cellulose is used as absorbent because of their capacity for water absorption [20].

The analysis of cassava trunk pith by FT-IR spectroscopy is presented in Figure 6. For cellulose structure, the wavelength between 3,350-3,500 cm^{-1} represents vibration of -OH group, 2,900-2,890 cm^{-1} indicating the C-H group [33] and 890-900 cm^{-1} representing peaks of β-1,4 glycosidic bond between glucose of cellulose. For hemicellulose spectra, the wavelength 1,160-1,110 cm^{-1} were derived from C-O-C of glycosidic bonds, 1,045 cm^{-1} assigned to C-O and C-O-C stretching and 1,414-1,467 cm^{-1} attributed to the CH_2, and 1,600-1,638 cm^{-1} showing as OH bending [34].

The wavelength of 1,250 cm^{-1} indicates the C-O-C groups representing for the Arty-alkyl ether of lignin, while that of 1,450-1,520 cm^{-1} showing as C-H in the aromatic band of phenylpropane [35], which is the main chain of lignin.

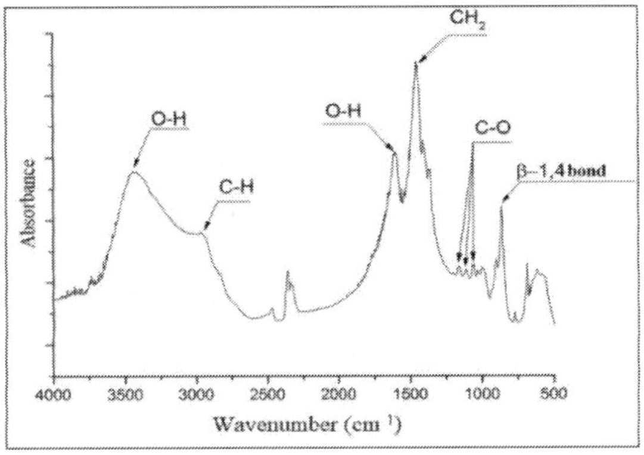

Fig. 6. FTIR spectra for lignocellulose from cassava trunk pith.

3.2 Physical characteristic of litter formulations

3.2.1 Effect of binder on clump strength

The clump strengths the cat litters with different binder by average rating, are presented in Table 2. the clump strength of the samples using glycerol, palm oil, and non-binder was rated as 1 (clump intact), 2 (slightly broken), and 3 (moderately broken into many small pieces) respectively. Apparently, the types of binder significantly affected to the strength of the clumped cat litters.

There are several types of binders such as oil, water, modified starch, etc. The cat litter without the binder addition contained dust particles from the mixture (Fig. 7 (b)). The binder could help the mixture to blend well and result in higher strength of the clump. However, the use of the binder depends on the composition of the cat litter and the manufacturing process [36, 37].

In this research, glycerol exhibited the highest potential clumping agent. However, it was found the litter grains absorbed more moisture and swollen with the increasing storage time (Fig. 7 (d)) due to a hydrophilic function group in glycerol causing the moisture absorption from the environment. This occurrence would affect the shelf life and ability to absorb water of the cat litters during the long storages. At room temperature, the cat litters immediately absorbed the moisture, clumped and scooped out. The cat litters prepared with glycerol were cracked more easily than the samples with palm oil. Therefore, palm oil is selected to be used as the binder for the preparation of biodegradable cat litters from cassava trunk pith.

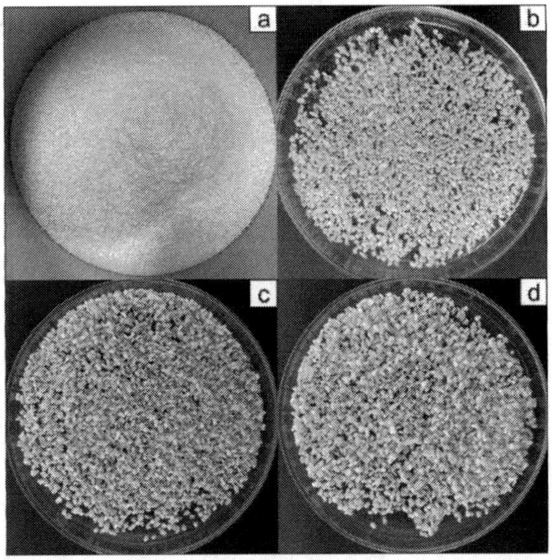

Fig. 7. (a) Ground cassava trunk pith, (b) the cat litter prepared without binder, (c) with palm oil, and (d) glycerol as a binder, respectively.

Table 2. Bulk densities and hydration capacities of cat litter prepared from each formulation.

Guar gum (%w/w)	Xanthan gum (%w/w)	Bulk densities ($g \cdot cm^{-3}$)	Hydration capacities ($g_{water} \cdot g_{litter}^{-1}$)
-	-	0.067	21.90
5.0	-	0.076	21.00
7.5	-	0.078	21.55
10.0	-	0.065	23.35
12.5	-	0.064	25.80
15.0	-	0.067	26.98
-	5.0	0.065	27.15
-	7.5	0.061	30.00
-	10.0	0.069	29.50
-	12.5	0.064	28.60
-	15.0	0.068	28.25

3.2.2 Bulk densities & hydration capacities

Cat litters prepared with guar gum, and xanthan gum revealed their bulk densities ranged 0.064-0.078 and 0.061-0.069 $g \cdot cm^{-3}$ and hydration capacities between 21.00-26.98, and 27.15-30.00 $g_{water} \cdot g_{litter}^{-1}$, respectively as presented in Table 2.

The hydration capacities of the cat litters with xanthan gum were higher than cat litters with guar gum. While the bulk density of cat litters with xanthan gum was significantly a lower, the addition of clumping agent could increase the hydration capacities. These results indicate that the types and quantity of clumping agent had significantly affected to the density and water absorption capacity of the cat litter.

The density of the cat litters is related to their porosity and the rate of moisture absorption [37]. Thus, the samples with higher density could absorb less moisture than lower density due to the smaller surface area and lower porosity [37]. While the product with low density has a larger surface area and high porosity, thus it can absorb more moisture with faster rate. Additionally, the additives, adsorption carriers and absorbents could also increase the water absorption capacity in cat litter [38]. The comparison with other studies was shown in Table 3, which different raw materials such as eastern red cedar [28], flour pellet [36], corn dried distiller grains [39], and bentonite [40] were used. Different raw materials led to the cat litters with different ranges of bulk density and hydration capacity since each material composted of different compositions and characteristics.

The cat litters prepared from the cassava trunk pith in this study with 10.0-15.0% of xanthan gum and palm oil (as binder) had presented in extremely high hydration capacities as 28.25-29.50 $g_{water} \cdot g_{litter}^{-1}$, which were more preferable than the cat litters that prepared from other raw materials.

Table 3. Bulk densities and hydration capacities of cat litter made from different materials.

Cat litter	Bulk densities ($g \cdot cm^{-3}$)	Hydration capacities ($g_{water} \cdot g_{litter}^{-1}$)
Cassava trunk pith	0.065-0.068	21.00-30.00
Eastern red cedar [28]	0.163	8.42
Corn dried distiller grains [39]	0.430	2.17
Flour pellet [36]	0.50-0.90	0.53-0.78
Bentonite [40]	0.52-0.74	0.52-1.17

Table 4. Clumping percentage of cat litter from each formulation.

Guar gum (%w/w)	Xanthan gum (%w/w)	Clumping percentage
-	-	27.17
5.0	-	27.98
7.5	-	26.89
10.0	-	36.42
12.5	-	34.17
15.0	-	32.49
-	5.0	43.70
-	7.5	59.39
-	10.0	67.79
-	12.5	77.87

3.2.3 Clumping percentage

The clumping percentage of the cat litters is shown in Table 4. the increased content of guar gum and xanthan gum led to an increase of clumping percentage of the samples. Using xanthan gum resulted to a significantly higher clumping percentage than the use of guar gum.

Clumping percentage is the ability of agglomeration with litter particles that contact with liquid and resulted to the adhesion force of clumping agents and other additives that stick on the particle surface. Thus, cat litter with a high percentage of coagulation can be easily scooped out of uncontaminated cat litter. Additionally, the minimum clumping of cat litters required by consumers is at least 60 percent [28]. In this study, it was found that 10.0-15.0% of xanthan gum can provide a sufficient clumping capacity for the cat litters.

4 Conclusion

In this study, the development of the biodegradable cat litter has been successfully derived using cassava trunk pith as main material. The cassava trunk is majorly composed with lignocellulose, which demonstrated as a good adsorbent for cat litter. Moreover, the sample with 10.0-15.0% of xanthan gum and palm oil (as binder) has the hydration capacities as 28.25-29.50 $g_{water} \cdot g_{litter}^{-1}$, which are higher than other cat litter products. Additionally, the cat litters also provided a sufficient clumping capacity much as 88.80% and suitable to be used for the preparation of biodegradable cat litters as commercial products. Some characteristics have been still in process such as analysis of odor absorption and dustiness.

Acknowledgement

The authors sincerely thank to the National Research Council of Thailand through the NRCT Senior Research Scholar Program (Contract No.814-2020) for the financial support.

References

1. Posttoday, Pet business for aging society, https://www.posttoday.com/economy/news/540479 (2018)
2. Niall McCarthy, Which Countries Have The Most Cat Owners?, https://www.statista.com/chart/10267/ which-countries-have-the-most-cat-owners (2017)
3. J.C. Neilson, AAFP-AAHA Feline life stage guidelines, Journal of Veterinary Medicine. **104** (2009)
4. H.A. Dabritz, P.A. Conrad, Serological prevalence of Toxoplasma gondii infection in cats (Belém, Pará, Brazil), Zoonoses and Public Health **57** (2010)
5. R.L. Virta, Mineral Commodity Summaries 2005, *Clays: U.S. Geological Survey Mineral Commodity Summaries* (United States, 2005)
6. C.S. Hornfeldt, M.L. Westfall, Suspected bentonite toxicosis in a cat from ingestion of clay cat litter, Veterinary and Human Toxicology. **38** (1996)
7. E.D. Mizer, Why natural cat litters are gaining market share, *Pet Product News* (2019)
8. R.L. Virta, Mineral Commodity Summaries 2001, *Geological Survey Minerals Yearbook 2001* (United States, 2001)
9. G. Godley, Cat litter, *Grocery Headquarters*, **70** (2003)
10. Market Watch, https://www.marketwatch.com/press release/cat-litter-industry-2019-global-market-growth-size-share-demand-trendsand-forecasts-to-2025-2019-03-19 (2019)
11. H. Miller, Looking at litter, *Pet Business Magazine* (2018)
12. Mybest, Top 10 Cat litters in 2020, https://my-best.in.th/48738 (2020)
13. J. M. Lipscomb, Services Reform and Manufacturing Performance: Evidence from India, US Patent No. 2016/0007561 A9 (2016)
14. T. A. Wendland, A thesis: Development of a novel cat litter from olive oil waste products, Master of Animal Science, Graduate College, Massy University (2011)
15. B. Phusae, Cat litter from cassava pulp, https://www.bangkokbiznews.com/news/detail/897171 (2020)
16. O. Wisetrat, R. Ngamsombat, P. Saueprasearsit, J. Prasara, Adsorption of suspended oil using bagasse and modified bagasse, Journal of Science and Technology Mahasarakham University. **31**, 4 (2012)
17. P. Saueprasearsit, M. Nuanjarae, M. Chinlapa, Biosorption of lead (Pb2+) by Luffa cylindrical fiber, Environ. Res. J. **4**, 1 (2010)
18. F.A., The State of Food and Agriculture, Organization, Production Crop Data (2017)
19. Y. Wandee, D. Uttapap, S. Puncha-arnon, C. Puttanlek, V. Rungsardthong, N. Wetprasit, Enrichment of rice noodles with fibre-rich fractions derived from cassava pulp and pomelo peel, Int. Journal of Food Science and Technology. **49** (2014)
20. Z.A. Mas'ud, M. Khotib, M. Farid, A. Nur, M. Amroni, Superabsorbent derived from cassava waste pulp, International Journal of Recycling of Organic Waste in Agriculture. **2**, 8 (2013)
21. C. Martína, M. Wei, S. Xiong, L.J. Jönsson, Enhancing saccharification of cassava stems by starch hydrolysis prior to pretreatment, Industrial Crops and Products, **97** (2017)
22. P. Dewi, R. Millati, R. Indrati, S. Sardjono, Effect of Lime Pretreatment on Microstructure of Cassava Stalk Fibers and Growth of Aspergillus niger, Biosaintifika, **10**, 1 (2018)
23. K. Tanakaa, M. Koyamaa, P.T. Phamb, A.P. Rollonc, H. Habakia, R. Egashiraa, K. Nakasakia, Production of high-concentration bioethanol from cassava stem by repeated hydrolysis and intermittent yeast inoculation, International Biodeterioration and Biodegradation, **138** (2019)
24. S. Y. Foong, N.S. Abdul Latiff, R.K. Liew, P. N. Yuh Yek, S. Shiung Lam, Microwave wet torrefaction: A catalytic process to convert waste palm shell into porous biochar, Materials Science for Energy Technologies. **3** (2020)

25. C. Ngaowthong, V. Rungsardthong, S. Siengchin, Polypropylene/hemp woody core fiber composites: Morphology, mechanical, thermal properties, and water absorption behaviors, Advances in Mechanical Engineering. **8**, 3 (2016)
26. Technical association of the pulp and paper industry (TAPPI) international nanotechnology division, *Roadmap for the Development of International Standards for Nanocellulose* (2011)
27. M. Kuras, B. Goss, D. Jaffee, B. Soral, E. Grace, R. Hunt, D. Yancey, K. Jones, M. Donatello, Cat litter product, WO Patent No. 2012/174356 A2 (2012)
28. S.F. Vaughn, M.A. Berhow, J.K. Winkler-Moser, Byarsa J.A., S.X. Liua, M.A. Jacksonb, An odor-reducing, low dust-forming, clumping cat litter produced from Eastern red cedar (Juniperus virginiana L.) wood fibers and biochar, S.C. Petersonc, F.J. Ellera, Industrial Crops and Products. **147** (2020)
29. S. Sivamani, R. Baskar, Bioconversion of cassava stem to ethanol: oxalic acid pretreatment and co-culture fermentation, Biofuels **9** (2018)
30. X.Q. Xu, X.B. Wu, Y. Cui, Y.X. Cai, R.W. Liu, M.N. Long, Q.X. Chen, Enzymatic Saccharification of Cassava Residues and Glucose Inhibitory Kinetics on β-Glucosidase from Hypocrea orientalis, Journal of Agricultural and Food Chemistry. **62** (2014)
31. Y. Chaweewan, N. Taylor, Anatomical assessment of root formation and tuberization in cassava (Manihot esculenta Crantz), Tropical Plant Biology. **8**, 2 (2015)
32. O. Wisetrat, R. Ngamsombat, P. Saueprasearsit, J. Prasara, Adsorption of suspended oil using bagasse and modified bagasse, Journal of Science and Technology Mahasarakham University. **31**, 4 (2012)
33. C. Areeprasert, P. Zhao, D. Ma, Y. Shen, K. Yoshikawa, Alternative solid fuel production from paper sludge employing hydrothermal treatment, Energy Fuels **28** (2014)
34. A. Salam, R.A. Venditti, J.J. Pawlak, K. El-tahlawy, A. Ayoub, Development of an acetylation reaction of switchgrass hemicellulose in ionic liquid without catalyst, *4th International Conference on Pulping, Papermaking and Biotechnology* (Nanjing Forestry University, Nanjing, China, 2012)
35. P. Manara, A. Zabaniotou, C. Vanderghem, A. Richel, Lignin extraction from Mediterranean agro-wastes: Impact of pretreatment conditions on lignin chemical structure and thermal degradation behavior, Catal. Today **223** (2014)
36. R. F. House, US Patent No. 5,230,305 (1993)
37. D.E. Ellis, D. Seekins, B. Gunter, S. McDonald, US Patent No. 2013/0112150 A1 (2013)
38. V. Y. Wong, US Patent No. 7,059.273 B2 (2006)
39. S. F. Vaughn, M.A. Berhow, J. K. Winkler-Moser, E. Lee, Formulation of a biodegradable, odor-reducing cat litter from solvent-extracted corn dried distillers grains, Industrial Crops and Products. **34** (2011)
40. M. Cliff, H. Heymann, DESCRIPTIVE ANALYSIS OF ORAL PUNGENCY, Journal of Sensory Studies. **6**, 4 (1991)

Study on the effects of green-based plant extracts and water-proofers as anti-corrosion agents for steel-reinforced concrete slabs

Jagadeesh Bhattarai[1*], *Madan* Somai[1], *Nirmal* Acharya[1], *Ajaya* Giri[1], *Akash* Roka[1], and *Nav Raj* Phulara[1]

[1]Central Department of Chemistry, Tribhuvan University, Kirtpur, Kathmandu, Nepal

Abstract. Widespread applications of reinforced concrete structures have been practiced since the 20th century because of their excellent properties despite their early corrosion degradation. For the control of such a problem, a design strategy of corrosion-resistant environments of the reinforced concrete structures is highly desirable for extending of a lifetime. The present research work was focused to investigate the effects of the green plant extract-based inhibitors from *Vitex negundo* and *Catharanthus roseus* leaves, and one waterproofing chemical (PtS) for controlling the corrosion susceptibility of concrete rebar using a half-cell potential method following the ASTM C876-91 standard. Both plant extracts have good anti-corrosion properties, and hence could be applied as green concrete additives to increase the corrosion resistance of the steel reinforcing bars. The anti-corrosion performance of the steel rebars in concrete is remarkably higher with the additions of 1000 and 2000 ppm plant extracts than the additions of waterproofing chemicals used, based on the shifting of corrosion potential ($\phi_{corr.}$) values to a more positive direction than −126 mV (SCE). The results agreed that both the plant extracts could be promising for the formulation of effective, eco-friendly anti-corrosion additives to delay the corrosion susceptibility of the concrete infrastructures.

Keyword. *Catharanthus roseus*, Green-based inhibitor, Rebar corrosion, *Vitex negundo*, Waterproofing agent

1 Introduction

The use of concrete placed 1st position among other manufactured materials, and it is one of the most widely used construction materials in our world after water [1]. The amount of concrete consumption is estimated to be between 25 and 33 × 10^9 tons annually [2, 3], and will continue to be used as a construction material in the future [4]. In 2016, around 4.7 tons of concrete produced each year per person in the world [3]. The concrete matrixes have high compressive strength but very poor tensile and elasticity strengths [5], and for the reasons of being such concrete properties, it is usually practiced to reinforce different carbon and stainless steel rods in the concrete matrixes to retain enough tensile, elastic including compressive strengths combined with durability. The reaction between concrete matrix comprised of cement, fine sand, and coarse gravel with impurities-free water results in the setting and hardening of the cement paste so that it binds both (coarse gravel and fine sand) aggregates [6].

Chloride-free concrete generally provides excellent corrosion protection properties to the reinforced steel or stainless steel rods, due to the high alkalinity nature of cement paste having 12.5- 13.5 pH [7]. As a result, the formation of a thin, protective diffusion barrier, stable, and tightly adhering thin oxide film (i.e., passive film) starts to appear spontaneously on the surface of the reinforced rebars. Details about the mechanism of the process of being formed such passive films on the surface of reinforced steel in a highly alkaline concrete matrix are discussed elsewhere [8]. However, shortcomings associated with reinforcement concrete infrastructures becoming matters of safety and economic point of view for human beings in recent years by depassivation of such diffusion barrier passive layers, mostly due to the loss of alkalinity of the concrete matrix [9].

Most of the reinforcement concrete infrastructures are porous, and different environmental gases and ions including water molecules (moisture) can penetrate the concrete matrix causing damages and failures, which reduced the life of the materials. The most crucial causes of loss of alkalinity in the concrete are atmospheric carbon dioxide (CO_2), chloride, or sulfate which shifts the pH of the concrete pore solution to less alkaline

* Corresponding author: bhattarai_05@yahoo.com, jagadeesh.bhattarai@cdc.tu.edu.np

around the reinforcing steel, making the concrete infrastructures susceptible to corrosion damages [10, 11]. Besides, there are several types of physico-chemical, electrochemical, and biological processes occurring on the reinforced steel (rebar) surface in concrete structures in presence of aggressive environmental factors. They accompanied by reduction of pH values of the concrete structures below 10 [12, 13], which starts depassivation and initiation of corrosion on the rebar surface leading to the loss of functionality, durability, structural integrity and finally resultant safety risks, and economic losses to human beings [14]. The corrosion products formed on the rebar surfaces reduce the gripping force with the concrete matrix and affect the quality of the whole reinforcing concrete infrastructures [15].

As a further matter, there is a growing appeal for developing different green-based corrosion inhibitors acquired from plant extracts in combating the corrosion of reinforced steel in concrete structures because of the less toxic and biodegradable nature of the extracts [11, 16-20]. Studies reported that the suitability of the N-, S-, and O-containing and π electron-rich leaves extracts of *Phyllanthus muellerianus* [18], *Chamaerops humilis L.* [20], and *Olea europaea* [21] acted as an effective anti-corrosion agent for reinforcing steel in concrete. The leaf extracts of *Davidia involucrate* [22], *A. djalonensis* [23], and *Pseudomonas aeruginosa* [24] acted as a mixed-type corrosion inhibitor from the electrochemical analysis. They enhance the performance of anti-corrosion action on the carbon steel plate in concrete pore solution with the formation of an organic layer to restrain both anodic and cathodic reactions. Corrosion protection efficiencies of leaves extracts of *Rosa damascena* [25], *Vernonia amygdalina* [26], and *Morinda citrifolia* [27] to the rebar in aggressive chloride solution reported about 82%, 90%, and 58%, respectively. Mostly due to the presence of high amounts of phytochemical constituents containing heteroatoms, aromatic rings, and π-electrons, which interact with iron ions resulting in the shielding of the steel bar from corrosion in concrete matrix. Water absorptivity of steel-reinforced concrete in the presence of *Bambusa arundinacea* leaves extract reported < 0.07 mL/m^2 s after 2 h, as required by ISAT (initial surface absorption test) standard for low permeability concrete [11].

In this condition, the potency of leaves of Nepal origin two plants (i.e., *Catharanthus roseus* (synonyms: *Vinca rosea)* and *Vitex negundo*) explored the first time for minimizing the corrosion level of the reinforced steel rebars in concrete infrastructures. Although previous works reported the effects of these two plant extracts as good anti-corrosion agents to different metals or alloys in acidic [28-30], neutral [31, 32], and alkaline [33, 34] electrolytes. Likewise, the *C. roseus* extracts acted as a mixed type of inhibitor for mild steel corrosion in an acidic medium due to the formation of an adsorptive film of electrostatic character [29, 33]. Methanol extract of *C. roseus* leaves acted as an efficient and mixed corrosion inhibitor to the mild steel in NaCl solutions [31, 32], usually due to the presence of several functional entities like hydroxyl, carbonyl groups, and fused heterocyclic compounds in the *C. roseus* extracts which enhanced the corrosion inhibition action. The corrosion inhibition of carbon steel in 1 M HCl containing aqueous extract of *V. negundo* leaves has been studied using chemical and electrochemical tests, and the results showed that the best concentration of 800 ppm of plant extract showed about 79 % of the inhibition efficiency [30]. Likewise, the corrosion inhibiting the action of *V. negundo* leaves extracts to Cu metal was investigated in nitric acid [33] and Al metal in 1 M NaOH [34], and the studies concluded that the extract acted as efficient corrosion inhibitors and predominantly cathodic (to Cu in HNO_3) and mixed (to Al in NaOH) inhibitor from potentiodynamic polarization tests.

Besides, the methanol extract of *V. negundo* plant leaves proved as an efficient and environmentally friendly corrosion inhibitor to improve the anti-corrosion properties of Cu and Al metals with about 83-96% efficiency in pure biodiesel and its 10% blend with petro-diesel at room temperature in airtight condition [35]. Moreover, the effects of Nepal origin plant species of *Vitex negundo* and *Catharanthus roseus* extracts on mild steel corrosion in different bioethanol fuels studied using corrosion, adsorption isotherm, and potentiodynamic polarization tests. The results revealed that the maximum inhibition efficiency reported about 89-86 and 71-75 % in *V. negundo* and *C. roseus* leaves, respectively, at 2000 ppm [36]. Many researchers have studied the reasons, control strategies, and techniques for such concrete corrosion calamities recently, though very little research works on the assessment of corrosion degree of some reinforcement concrete infrastructures used in the Kathmandu Valley [37], and Pokhara Valley [38] of Nepal. Notwithstanding, some research works carried out on corrosion phenomena and prevention techniques in Nepal in the last decade [39-47].

In this context, the present work aimed to substantiate the efficacy of methanol extracts of Nepal origin two plants of *Vitex negundo*, and *Catharanthus roseus* as a green-based corrosion inhibitor in preventing the steel reinforcement corrosion in concrete slabs using a simple and non-destructive half-cell corrosion potential (HCCP) method following the ASTM standard [48]. In addition, a comparative study investigated to identify the effects of commercially used two waterproofing agents on the corrosion degree of the reinforced concrete slabs in ambient environments. The findings of the present works would render a categorization of solution methods for corrosion protection, maintenance, and repair in Nepal first time.

2 Materials and methods

2.1. Preparation of reinforcement concrete slab

Ordinary Portland cement 53 grade was used for the preparation of rectangular steel-reinforced concrete slabs having the dimensions of 40 cm × 24 cm × 4 cm in the laboratory following the ASTM standard [49]. For the purposes, locally available fine river sand, coarse aggregate (2 cm ϕ) and rust-free steel rod (Premier-500 with 1.2 cm ϕ) were used after removing the rust using 200, 400, 600, 1000 and 1500 grit numbers sandpapers followed by ultrasonic washing with ethanol to degrease and acetone to dry.

The cement, sand, and gravel were mixed in the ratio of 1:2:3, and then added the distilled water (0.45 w/c ratio) without and with 500, 1000, 2000, 4000 ppm of each plant extracts or waterproofing agents were separately used to prepare different sets of the concrete mixture. Two steel rods having 12 mm diameter (ϕ) with 8 cm spacing between the rods were embedded in the concrete matrix to prepare a reinforcement concrete slab, and one end of each embedded steel rod was left exposed for open circuit corrosion potential (OCCP) measurements, as shown in Fig. 1.

Fig. 1. Set up for the measurement of OCCP.

The OCCP values of each steel bar were recorded on a digital multi-meter (UNI-T Business, Hong Kong) using a saturated calomel electrode (SCE) as a reference, and the exposed steel rod of the concrete slab as working electrodes, as shown in Fig. 1. Eight points, marked as 1, 2, 3, 4, 5, 6, 7, and 8 at equidistant from the sides were fixed on the surface of each steel-reinforced concrete slab. The SCE electrode and steel rod were connected to the negative and positive terminals, respectively, of the voltmeter for the record of the OCCP values of all these eight points of each slab after 7 days of the slab preparation up to about 5 months.

2.2 Preparation of plant extracts

The leaves of *Vitex negundo* and *Catharanthus roseus* plants, as shown in Figs. 2a and 2b, respectively, were collected from Pokhara-21 (28.1503°N, 83.9903°E) of Kaski district, Nepal. They thoroughly washed with distilled water to remove dirt substances attached to the leaves surface, and then shade dried for few weeks until it completely dried to make dry powder (Fig. 2) with the help of an electric grinder. About 200 g of each pulverized leaves powder was transferred into a corked 500 mL conical flask containing 400 mL of methanol (Fig. 2c). The mixture of each separate plant was stirred well and then was kept in a dry dark place for about two weeks. Then the methanol extract of each *V. negundo* and *C. roseus* leaves powder was obtained by the soxhlet extraction using rotavapor (Fig. 2d), dried to obtain solid plant extract and stored at 4° C, which was added in the concrete matrix as a green-based corrosion inhibitor to the steel-reinforced concrete slabs. For the purpose, the required amount of each solid plant extract mixed with distilled water that was used to prepare the reinforced concrete slabs.

2.3 Premise of corrosion condition of rebars

Different corrosion conditions of the reinforcing steel rod (rebar) in concrete slab were established with their recorded OCCP values according to the ASTM standards [48] in the HCCP technique. In this standard method, the probability of the corrosion level of the rebars ascertains with the recorded OCCP values, which is feasible to make a diagnosis of the corrosion level of the rebar in concrete infrastructures [51].

Fig. 2. Preparation processes of the methanol extracts of *V. negundo* and *C. roseus* plant leaves.

The corrosion level of the concrete rebars consider to be low (Lcr) or less than 10 % at the time of measurement if the recorded OCCP values are more positive than −126 mV vs SCE. Besides, the corrosion level of the reinforced steel in the analyzed concrete structures at the measured areas assumes be uncertain or mild corrosion risk (Mcr) if its OCCP value recorded between −276 and −126 mV vs SCE. Similarly, if the recorded OCCP is more negative than −276 mV vs SCE then the degree of the corrosion risk should be high (Hcr), i.e., more than 90 % corroded the reinforcing concrete slabs at the time of measurement according to the ASTM standard [48], as summarized in Table 1.

2.4 Phytochemical screening and FTIR analysis

To know the constituents of secondary metabolites present in the prepared methanol extracts of both plants leaves quantitatively, the phytochemical screening of leaves extracts of both *V. negundo* and *C. roseus* plants were carried out using the test methods as described elsewhere [50]. In addition, functional groups present in methanol extracts of *V. negundo* and *C. roseus* leaves, including the cement powder used in the study were analyzed using the recorded Fourier transform infrared (FTIR) spectra recorded in the range of 4000-400 cm^{-1} using IRPrestige-21 spectrometer (Shimadzu, Japan).

Table 1. OCCP value and corrosion level of reinforced steel.

OCCP (mV vs SCE)	Corrosion level of rebar steel in concrete slab
< −276	- High corrosion risk (Hcr); i.e., more than 90 % probability of corrosion risk
−276 to −126	- Mild corrosion risk (Mcr); i.e., corrosion risk level uncertain
> −126	- Low corrosion risk (Lcr); i.e., more than 90 % probability of corrosion risk

3 Results and discussion

3.1 Characterization of extracts and cement

From the phytochemical screening test, major chemical compounds present in the methanolic extracts of the *V. negundo* and *C. roseus* leaves contain phenols, alkaloids, saponin, flavonoids, glycosides, and tannins (Table 2), which may help to form a corrosion-resistant passive film on the surface of rebars steel in concrete. The results of the phytochemical screening tests of the present work are in agreement with the previously reported data in the methanol extract of *V. negundo* [52, 53], and *C. roseus* [54] leaves.

Table 2. Phytoconstituents in *V. negundo* and *C. roseus* leaves extracts.

Constituent	Test name/agent	*V. negundo*	*C. roseus*
Phenols	Lead acetate	+	+
Alkaloids	Dragendroffs	+	+
Saponins	Foam	+	+
Flavonoids	Alkaline	+	+
Glycosides	Legal's	+	+
Tannins	$FeCl_3$	+	+
Terpenoids	Salkowski	−	−

+ = presence, and − = absence of phytoconstituents

FTIR analysis makes it possible to find out the functional groups present in the materials like plant extracts and cement powder. Figures 3(a) and 3(b) show the IR spectra recorded, respectively, for methanol extracts of *V. negundo* and *C. roseus* plants, and cement powder, as summarized in Table 3. The symmetric and broad absorption peak at 3320 cm^{-1} (*V. negundo*), and 3288 cm^{-1} (*C. roseus*) attributes for O-H (stretch) of hydroxyl compounds. The asymmetric and symmetric stretching of C-H vibrations of the methanolic extracts of both plants ascribe by the peaks around 2926 cm^{-1} and 2850 cm^{-1}, respectively, as shown in Fig. 3(a). Similarly, C≡N (stretch) or P-H bonding in the vicinity 2376-2327 cm^{-1} noted in the methanolic extracts of both plants.

Table 3. FTIR band of the *Vitex negundo* and *Catharanthus roseus* extract.

Wavenumber (cm^{-1})		Assignements
V. negundo	*C. roseus*	
3320	3288	O-H stretching due to hydrogen bonding
2926	2926	Methylene (C-H) asymmetrical stretching
2854	2850	Methyl (C-H) symmetrical stretching
2376- 2327	2376-2327	C≡N stretching or P-H bonding
1692	–	C=O stretching (lowe value may be conjugated with other groups)
1608, 1515	–	C-C ring stretching
–	1572	C-C ring stretching
1448	1448	C-O-H in-plane bending
1370	1383	C-H bending
1268	1256	C-O bending
1165	–	P=O stretching
1028	1028	O-H bending
925	925	O-H bending
852	852	C-H (para-benzene)
816	816	C-H bending
780	780	C-H (meta-benzene)

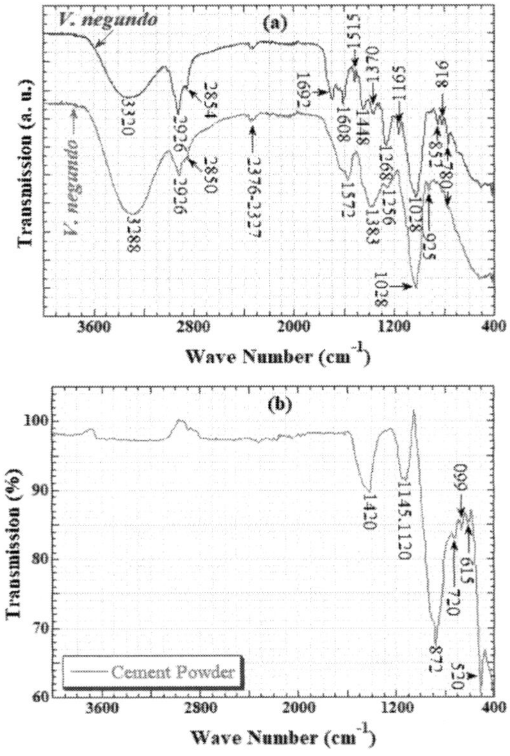

Fig. 3. FTIR spectra of the (a) methanol extracts of *V. negundo* and *C. roseus* leaves, and (b) OPC cement powder.

Besides, the C=O (carboxylic acid) group for *V. negundo* extract indicated by the presence of IR peaks at 1692 cm^{-1}, 1608 cm^{-1}, 1448 cm^{-1}, and of the C-O group at 1268 cm^{-1}, which are less pronounced in spectra of *C.*

roseus extract. The results of FTIR analysis of both plant extracts are in full agreement with the phytochemical screening tests as explained aforementioned, and also confirmed the presence of phenol, alkane, alkene, carboxylic acid, aromatic, nitro compounds including alcohol, and benzene, which is also in agreement with previously reported data [55, 56]. On the other hand, Fig. 3(b) shows the IR spectra of the cement powder used for preparing the reinforcement concrete slabs to accomplish this work.

It notice that IR peaks around 3200-3500 cm^{-1} and 1620-1640 cm^{-1} do not appear, which disclosed that the cement used for the study was free of (bulk) water or nearly dehydrated powder [57]. The appearance of absorption peak around 1145-1120 cm^{-1} wave number regions revealed that the cement sample contained remarkable amounts of carbonates and silicates. The FTIR absorption peaks at 1420 cm^{-1} and 872 cm^{-1} assigned to the C-O stretching and C-O bending, respectively, of the CO_3^{2-} ions present in the cement powder [58]. Especially, a strong absorption band exhibited between 1000-800 cm^{-1} with a peak value at about 872 cm^{-1} revealed that the cement powder contains the C_2S phase calcium silicate. The peak at 520 cm^{-1} assigned for bending vibration of O-Si-O bonds within the tetrahedral SiO_4 groups present in the cement sample [59].

3.2 Plant extracts as rebar corrosion inhibitor

The efficacy of different concentrations (i.e., 500, 1000, 2000, and 4000 ppm) of methanol extracts of *Vitex negundo*, and *Catharanthus roseus* plants leaves were studied for viewing their corrosion inhibiting actions on the reinforcement steel rebars in concrete slabs at ambient laboratory conditions using half-cell corrosion potential (HCCP) method following the ASTM standard [48]. For the present study completion, changes of open-circuit corrosion potential (OCCP) of a total of nine steel-reinforced concrete slabs without and with four different concentrations of two plant extracts (separately) were recorded at the different time interval between 7 to 150 days. Figures 4 (a) and 4(b) show the changes of the recorded OCCP values of the reinforced rebars from eight points marked on the surface of the concrete slab (see Fig. 1), which was prepared without the addition of the plant extract. All the recorded OCCPs are also given in Table 4.

The OCCP values of all eight points are more noble regions than -125 mV vs SCE up to the first seven weeks, as depicted in Fig. 4(a), which suggests that there was a low corrosion risk (less than 10 % probability of corrosion risk) during this period based on the ASTM C876-91 standard [48]. Nevertheless, with increasing the exposure time between 49-150 days, all the OCCP values are in the range of -126 mV to -275 mV vs SCE, suggesting the high probability of the corrosion risk of the reinforced steel rebars in the concrete slab without plant extract addition. Moreover, Fig. 4(b) represents the mean OCCP values with error bar as a function of exposure times of the concrete slabs, which inferred that the mean OCCP values shifted to corrosive regions with exposure times. The standard deviation from the mean OCCP values for each interval time found between 10 to 36 mV, as summarized in Table 4, and depicted in Fig. 4(b). The results indicate that the localized types of corrosion could expect to the steel rebar in the concrete slab without the addition of plant extract. However, the OCCP of the steel rebars of all the concrete slabs is remarkably shifted to a more noble direction than the value of −125 mV (SCE) with the additions of 500-4000 ppm of each of both plant extracts, as summarized in Table 4.

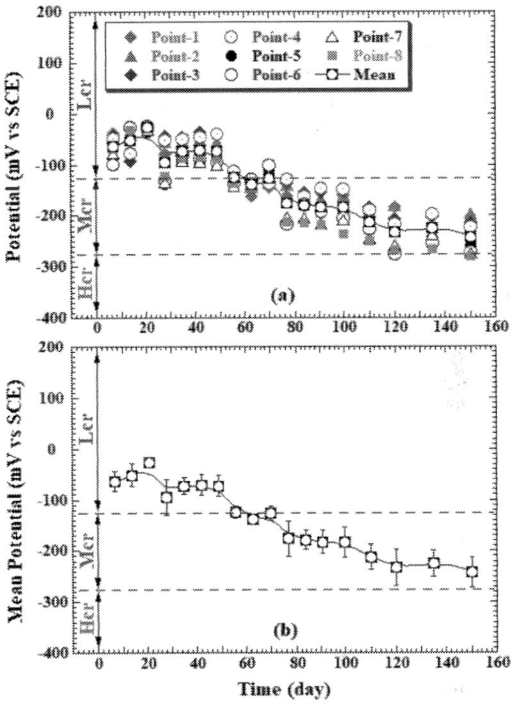

Fig. 4. Changes in the (a) OCCP values of all eight points, and (b) mean OCCP with the error bar of the steel rebar in concrete slab without plant extract.

Figures 5(a) and 5(b), respectively, display the potency of different concentrations of *Vitex negundo* and *Catharanthus roseus* extracts for the corrosion inhibition actions on the steel rebars in concrete. Most of all the mean OCCP values of all concrete slabs with plant extracts fluctuated inconsistently up to 4-6 weeks, and after then it attained a steady state, as shown in Fig. 5. Such behavior of the initial variations of the OCCP value is mainly due to the formation of unstable films on the rebar surface at initial concrete setting periods, and the time required for the occurrence of a stable and diffusion barrier passive film. Besides, the mean OCCP values of all the steel rebars in concrete slabs with plant extract additions significantly shifted to the Lcr region with more positive potential values than -126 mV (SCE), especially after 6-7 weeks exposure times.

 https://doi.org/10.1051/e3sconf/202130202018

Table 4. Mean OCCP and standard deviation values of different steel reinforced concrete slabs with and without plant extracts.

	OCCP (mV vs SCE)													
Time (day)	Without extract		Conc. of *V. negundo* extract in concrete (ppm)						Conc. of *C. roseus* extract in concrete (ppm)					
			1000		2000		4000		1000		2000		4000	
	$\bar{x}$	SD*	$\bar{x}$	SD*	$\bar{x}$	SD*	$\bar{x}$	SD*	$\bar{x}$	SD*	$\bar{x}$	SD*	$\bar{x}$	SD*
7	-63	±20.1	-61	±26.0	-110	±6.8	-113	±15.1	-116	±13.4	-109	±9.3	-59	±25.1
14	-52	±22.6	-120	±8.7	-30	±4.1	-45	±12.9	-27	±5.0	-32	±11.9	-37	±6.9
21	-27	±4.9	-116	±9.4	-51	±8.7	-63	±18.7	-55	±8.3	-28	±15.4	-34	±9.4
28	-95	±36.2	-121	±19.7	-113	±4.8	-151	±17.5	-149	±16.0	-75	±6.5	-129	±17.4
35	-73	±17.1	-108	±15.5	-98	±5.1	-148	±17.6	-122	±13.4	-64	±8.2	-89	±11.3
42	-71	±20.9	-82	±16.3	-96	±5.5	-125	±18.9	-90	±7.8	-55	±5.2	-96	±12.3
49	-73	±21.4	-68	±9.4	-92	±13.3	-115	±17.4	-87	±10.1	-65	±6.9	-86	±15.3
56	-124	±11.6	-54	±7.2	-31	±3.9	-60	±18.3	-50	±12.8	-21	±6.1	-73	±28.5
63	-138	±10.1	-50	±7.6	-71	±4.1	-87	±9.0	-95	±10.2	-75	±15.6	-94	±16.1
70	-127	±13.5	-51	±9.7	-72	±20.9	-84	±8.2	-102	±12.4	-74	±17.6	-78	±16.3
77	-176	±34.0	-48	±6.5	-63	±9.2	-63	±11.8	-77	±9.5	-56	±13.6	-63	±9.5
84	-179	±18.6	-51	±9.2	-50	±10.8	-64	±7.7	-84	±6.8	-102	±14.3	-75	±12.3
91	-183	±23.3	-52	±9.1	-63	±10.2	-71	±12.7	-87	±8.3	-76	±9.8	-89	±7.7
100	-183	±28.7	-51	±10.2	-63	±10.9	-72	±16.8	-89	±9.6	-83	±9.0	-87	±13.6
110	-212	±23.7	-37	±9.0	-73	±12.2	-55	±14.2	-95	±4.3	-74	±8.1	-79	±10.5
120	-233	±35.5	-36	±6.8	-81	±13.1	-78	±10.8	-92	±12.8	-89	±9.4	-93	±9.8
135	-225	±25.4	-53	±6.6	-93	±8.3	-91	±9.1	-88	±9.9	-79	±5.5	-84	±6.4
150	-242	±29.4	-49	±13.5	-87	±11.3	-87	±11.2	-103	±18.8	-116	±8.7	-99	±12.4

**SD= standard deviation (n= 8), $\bar{x}$ = mean value*

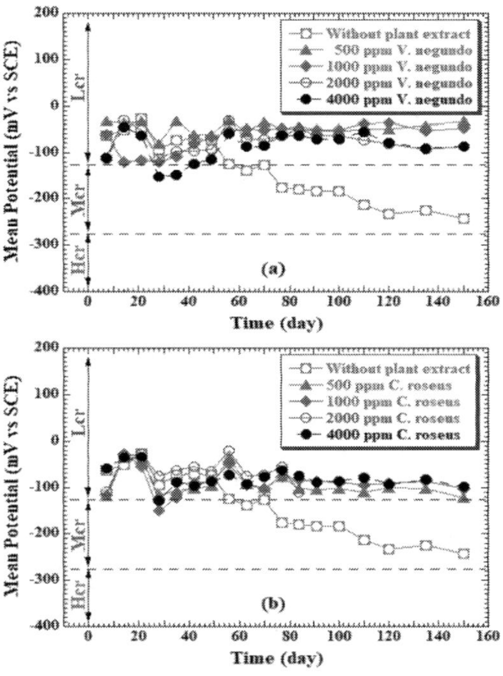

Fig. 5. Changes in the mean OCCP values of the steel rebar in concrete slab with different concentrations of (a) *Vitex negundo*, and (b) *Catharanthus roseus* extracts.

Correspondingly, Fig. 5(a) exhibits that the additions of 500 ppm and 1000 ppm *Vitex negundo* plant extract in the reinforced concrete seem to more effective in mitigating and controlling the corrosion of the steel rebars in concrete structures as compared with the concrete slabs with 2000 and 4000 ppm. However, the effectiveness of the anti-corrosion behavior of the *Catharanthus roseus* plant extract increased with increasing its concentration up to 4000 ppm. Mainly, the addition of 2000 ppm and 4000 ppm *Catharanthus roseus* extracts proved to be more effective in shifting the mean OCCP to the Lcr region where it assumes for occurring less than 10 % possibility of the steel rebars corrosion in concrete slabs based on the ASTM C876-15 standard [48].

A comparative study of the efficacy of the different concentrations of the green-based two plants extracts and one locally used synthetic waterproofing agent was checked over for controlling or minimizing the corrosion degree of steel rebars in concrete slabs, and the results are displayed in Fig. 6. The concrete slabs with the additions of 500 and 1000 ppm *Vitex negundo* extract exhibited a higher corrosion inhibition action than of *Catharanthus roseus* extract in the corresponding concentrations by shifting the OCCP values towards a more positive direction, i.e., in the areas of less than 10 % probability of corrosion rate, as shown in Figs. 6(a) and 6(b).

However, Figs. 6(c) and 6(d) show that both the plants' extracts have the same degree of corrosion inhibiting actions to the steel rebars in concrete slabs at 1000, 2000, and 4000 ppm concentrations, which is a slightly lower inhibiting level than by 500-1000 ppm *Vitex negundo*, and slightly higher inhibiting actions than of 500 ppm *Catharanthus roseus* extract. Such result is in agreement with the previously reported works that anti-corrosion properties of the methanolic extract of *Vitex negundo* leaves to mild steel in biofuels reported higher than of the *Catharanthus roseus* extract due to the formation of stable and diffusion-barrier passive film from the potentiodynamic measurements [36].

https://doi.org/10.1051/e3sconf/202130202018

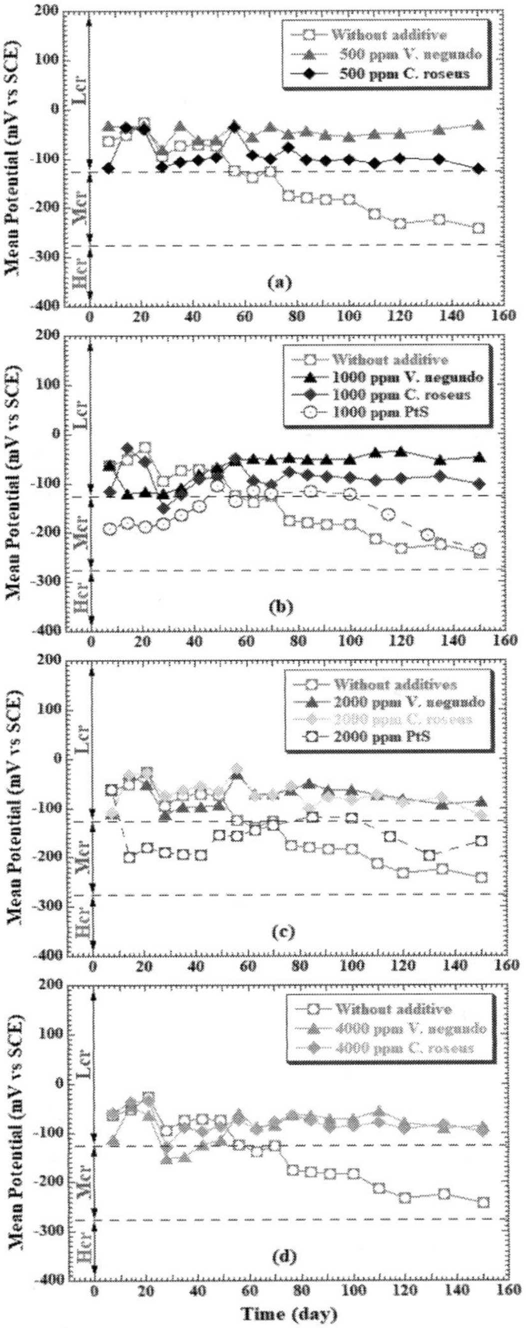

Fig. 6. The effects of (a) 500 ppm, (b) 1000 ppm, 2000 ppm, and 4000 ppm concentrations of the *V. negundo* and *C. roseus* plant extracts, and PtS waterproofing chemical in the concrete slab.

Such high anti-corrosion effect to the steel rebars in concrete slabs by low concentrations (i.e., 500 & 1000 ppm) of the *Vitex negundo* extract compared with the *Catharanthus roseus* extract could be accredited to have high amounts of phytoconstituents in the *Vitex negundo* extract than *Catharanthus roseus* extract. The high amounts of N-, S-, O-containing aromatic and π electron-rich compounds, which have higher basicity and electron density, are responsible for the formation of a protected diffusion-barrier passive film on the rebar surface by adsorption mechanism that isolates the steel rebars from the corrosive concrete environment [60, 61]. Apart from such causes, aromatic ring containing compounds with oxygen reported to be zone of corrosion inhibition [62]. Despite the fact, both the plant extracts have good anti-corrosion properties in all concentrations of 500-4000 ppm and thence could apply as green concrete additives to increase the corrosion resistance of the steel rebar. However, the anti-corrosion performance of the locally practiced waterproofing synthetic chemical, named PtS showed remarkably lower than the additions of plant extracts at the same concentrations, as displayed in Figs. 6(b) and b(c), based on the shifting of corrosion potential ($\phi_{corr.}$) values to a more positive direction than −126 mV (SCE).

4 Conclusion

The leaf extracts of *Vitex negundo* and *Catharanthus roseus* plants used in this study acted as a green-based corrosion inhibitor to increase the corrosion resistance properties of the reinforced steel rebars in concrete slabs. The plant extracts have sufficient anti-corrosion properties for the reinforced rebars, which is presumably due to the presence of abundant amounts of hetero-atoms and aromatic ring compounds as the major phytoconstituents in these two plant extracts.

The efficacious corrosion inhibiting the action of the methanolic extracts of *V. negundo* leaves to the reinforced steel rebars in concrete is especially found to be more effective green-based corrosion inhibitors than the *Catharanthus roseus* extracts at the concentrations of 500-1000 ppm. The present research work explored first time the beneficial uses of different green-based concrete additives obtained from Nepal origin *Vitex negundo* and *Catharanthus roseus* plants extract for controlling the steel corrosion mechanism in different types of concrete infrastructures.

Acknowledgment

The University Grants Commission-Nepal, Sanothimi, Bhaktapur of Nepal is highly acknowledged for providing the UGC Master Research Support (MRS-76/77-S&T-22) to AG for conducting this research work.

References

1. J.P. Broomfield, *Corrosion of steel in concrete: understanding, investigation, and repair* (2nd edition, CRC Press, Taylor & Francis Group, London, p. 296, 2019)
2. W. Green, F. Collins, M. Forsyth, Up to date overview of aspects of steel reinforcement corrosion in concrete. In *Brian Cherry paper series* (The Australasian Corrosion Association Inc., 2020)

3. ISO/TC 071, *Strategic business plan* (ISO Technical Committee, p. 11, 2016)
4. C.R. Gagg, Cement and concrete as an engineering material: An historic appraisal and case study analysis, Engineering Failure Analysis, **40**, (2014): 114-140
5. U.M. Angst, Challenges and opportunities in corrosion of steel in concrete, Materials and Structures, **51**, (2018): 4
6. R.W. Revie, H.H. Uhlig, *Corrosion and corrosion control: An introduction to corrosion science and engineering* (4th edition, John Wiley and Sons, Inc., Hoboken New Jersey, USA, p. 490, 2008)
7. J. Bhattarai, *Frontiers of corrosion science* (1st edition, Kshitiz Publication, Kirtipur, Kathmandu, Nepal, p. 304, 2010)
8. B. Cherry, W. Green, *Corrosion and protection of reinforced concrete* (1st edition, CRC Press, Taylor & Francis Group, p. 402, 2021)
9. J.O'M. Bockris, A.K.N. Reddy, *Modern electrochemistry: Electrodics in chemistry, engineering, biology, and environmental science* (2nd edition, **volume 2B**, Kluwer Academic/Plenum Publisher, New York, 2001)
10. NPL, *The corrosion of steel in concrete- Basic understanding, monitoring and corrosion control methods* (National Physical Laboratory, NPL Management Ltd., Middlesex, UK, p. 11, 2020)
11. S.A. Asipita, M. Ismail, M.Z.A. Majid, Z.A. Majid, C. Abdullah, J. Mirza, Green Bambusa arundinacea leaves extract as a sustainable corrosion inhibitor in steel-reinforced concrete, Journal of Cleaner Production, **67**, (2014): 139-146
12. C. Apostolopoulos, K. Koulouris, *Corrosion effect on bond loss between steel and concrete* (IntechOpen, p. 18, 2020)
13. E.C. Seyhan, F. Goodwin, I.-W. Huang, Corrosion protection of steel reinforcement by using surface applied corrosion inhibitors, MATEC Web Conference, **289**, (2019): 05001
14. K. Doi, S. Hiromoto, T. Shinohara, K. Tsuchiya, H. Katayama, E. Akiyama, Role of mill scale on corrosion behavior of steel rebars in mortar, Corrosion Science, **177**, (2020): 108995
15. R. Cantone, M. Fernández Ruiz, A. Muttoni, A detailed view on the rebar-to-concrete interaction based on refined measurement techniques, Engineering Structures, **226**, (2021): 111332
16. J.O. Okeniyi, J.O. Ikotun, E.T. Akinlabi, E.T. Okeniyi, Anticorrosion behaviour of *Rhizophora mangles* L. bark-extract on concrete steel-rebar in saline/marine simulating-environment, The Scientific World Journal, **2019**, (2019): 6894714
17. J.O. Okeniyi, E.T. Akinlabi, J.O. Ikotun, S.A. Akinlabi, E.T. Okeniyi, O.M. Elizabeth, *Rhizophora mangle* L. leaf biochemical characterization: Natural green total corrosion inhibition prospect on concrete steel-reinforcement in 3.5 % NaCl, Jurnal Teknologi (Sciences & Engineering), **81**, 1 (2019): 11-21
18. J.O. Okeniyi, C. Loto, A. Popoola, Effects of *Phyllanthus muellerianus* leaf-extract on steel-reinforcement corrosion in 3.5 % NaCl-immersed concrete, Metals, **6**, 11 (2016): 255
19. Y.P. Asmara, T. Kurniawan, A.G.E. Sutjipto, J. Jamiluddin Jafar, Application of plants extracts as green corrosion inhibitors for steel in concrete- a review, Indonesian Journal of Science & Technology, **3**, 2 (2018): 158-170
20. D.B. Left, M. Zertoubi, S. Khoudali, M. Azzi, New application of Chamaerops humilis L. extract as a green corrosion inhibitor for reinforcement steel in a simulated carbonated concrete pore solution, Portugaliae Electrochimica Acta, **36**, 4 (2018): 249-257
21. M. Ben Harb, S. Abubshait, N. Etteyeb, M. Kamoun, A. Dhouib, Olive leaf extract as a green corrosion inhibitor of reinforced concrete contaminated with seawater, Arabian Journal of Chemistry, **13**, 3 (2020): 4846-4856
22. M. Zhang, *Davidian involucrata* leaves extract as green corrosion inhibitor for low carbon steel in concrete pore solution containing chloride ions, International Journal of Electrochemical Science, **16**, (2021): 21042
23. G. Xie, L. Wei, Inhibitor effect of *Anthocleista djalonensis* extract on the corrosion of concrete steel reinforcement, International Journal of Electrochemical Science, **13**, (2018): 5311-5322.
24. V.S. Helbert, L. Gaillet, T. Chaussadent, V. Gaudefroy, J. Creus, Rhamnolipids as an eco-friendly corrosion inhibitor of rebars in simulated concrete pore solution: Evaluation of conditioning and addition methods, Corrosion Engineering, Science and Technology, **55**, 2 (2020): 91-102
25. R. Anitha, S. Chitra, V. Hemapriya, I.M. Chung, S.H. Kim, M. Prabakaran, Implications of eco-addition inhibitor to mitigate corrosion in reinforced steel embedded in concrete, Construction and Building Materials, **213**, (2019): 246-256
26. C.A. Loto, O.O. Joseph, R.T. LotoA.P.I. Popoola, Inhibition effect of *Vernonia amygdalina* extract on the corrosion of mild steel reinforcement in concrete in 3.5 M NaCl environment, International Journal of Electrochemical Science, **8**, 9 (2013): 11087-11100
27. R.A. Limco, H.P. Bacosa, A.A. Lubguban, J.S. Buluran, *Morinda citrifolia* (Noni) leaf extract as corrosion inhibitor for steel-reinforced concrete in saline environment, International Journal of Environmental Science and Technology, **17**, (2020): 4531-4540
28. J.A. Selvi, M. Arthanareeswari, T. Pushpamalini, S. Rajendran, T. Vignesh, Effectiveness of *Vinca rosea* leaf extract as corrosion inhibitor for mild steel in 1 N HCl medium investigated by adsorption and electrochemical studies, International Journal of

Corrosion and Scale Inhibition, **9**, 4 (2020): 1429-1443

29. R.M.A. Shahba, A.E.E. Fouda, A.E. El-Shenawy, A.S.M. Osman, Effect of *Catharanthus roseus* (Vince rosea) and turmeric (*Curcuma longa*) extracts as green corrosion inhibitors for mild steel in 1 M HCl, Materials Sciences and Applications, **7**, (2016): 654-671
30. K. Hema, B.R. Venkatraman, A. Subramania, Aqueous extract of *Vitex negundo* leaves as green corrosion inhibitor for the protection of carbon steel in 1 N HCl solution, Journal of Environmental Nanotechnology, **4**, 3 (2013): 42-50
31. N. Palaniappan, I. Cole, F. Caballero-Briones, S. Manickam, K.R. Justin Thomas, D. Santos, Experimental and DFT studies on the ultrasonic energy-assisted extraction of the phytochemicals of *Catharanthus roseus* as green corrosion inhibitors for mild steel in NaCl medium, RSC Advances, **10**, (2020): 5399-5411
32. M. Rana, S. Joshi, J. Bhattarai, Extract of different plants of Nepalese origin as a green corrosion inhibitor for mild steel in 0.5 M NaCl solution, Asian Journal of Chemistry, **29**, 5 (2017): 1130-1134
33. Savita, P. Mourya, N. Chaubedi, V.K. Singh, M.M. Singh, Eco-friendly inhibitors for copper corrosion in nitric acid: Experimental and theoretical evaluation, Metallurgical and Materials Transactions B, **47**, 1 (2016): 47-57
34. S. Sirajunnisa, M.I. Fazal Mohamed, A. Subramania, *Vitex negundo* leaves extract as green inhibitor for the corrosion of aluminium 1 N NaOH solution, Journal of Chemical and Pharmaceutical Research, **6**, (2014): 580-588
35. B.N. Subedi, K. Amgain, S. Joshi, J. Bhattarai, Green approach to corrosion inhibition effect of Vitex negundo leaf extract on aluminium and copper metals in biodiesel and its blend, International Journal of Corrosion and Scale Inhibition, **8**, 3 (2019): 744-759 (2019)
36. P. Katuwal, R. Regmi, S. Joshi, J. Bhattarai, Assessment on the effective green-based Nepal origin plants extract as corrosion inhibitor for mild steel in bioethanol and its blend, European Journal of Advanced Chemistry Research, **1**, 5 (2020): 1-13
37. N.R. Phulara, J. Bhattarai, Assessment on corrosion damage of steel reinforced concrete structures of Kathmandu Valley using corrosion potential mapping method, Journal of the Institute of Engineering, **15**, 2 (2019): 47-56
38. I. Laudari, N.R. Phulara, M. Gautam, J. Bhattarai, Evaluation on the corrosion condition of some steel-reinforced concrete infrastructures available in Pokhara valley of Nepal, Tribhuvan University Journal, (2021): in press
39. J. Bhattarai, D. Paudyal, K. P. Dahal, Study on the soil corrosivity towards the buried-metallic pipes in Kathmandu and Chitwan valley of Nepal. In *Proceedings of the 17th Asian-Pacific Corrosion Control Conference* (Paper No. 17039, 27-30 January, 2016, IIT Bombay, India, p. 12)
40. P.P. Bhandari, K.P. Dahal, J. Bhattarai, The Corrosivity of soil collected from Araniko Highway and Sanothimi areas of Bhaktapur, Journal of Institute of Science and Technology, **18**, 1 (2013): 71-77
41. K.P. Dahal, J.N. Timilsena, M. Gautam, J. Bhattarai, Investigation on probabilistic model for corrosion failure level of buried-pipelines in Kirtipur urban areas (Nepal), Journal of Failure Analysis and Prevention, **21**, 3 (2021): 914-926
42. K.P. Dahal, R.K. Karki, J. Bhattarai, Evaluation of corrosivity of soil collected from the central part of Kathmandu Metropolis (Nepal) to water supply metallic pipes, Asian Journal of Chemistry, **30**, 7 (2018): 1525-1530
43. Y.P. Dhakal, K.P. Dahal, J. Bhattarai, Investigation on the soil corrosivity towards the buried water supply pipelines in Kamerotar Town Planning areas of Bhaktapur, Nepal, Bibechana, **10**, (2014): 82-91
44. A. Poudel, K.P. Dahal, J. Bhattarai, A classification approach for corrosion rating of soil to buried water pipelines: A case study in Budhanilkantha-Maharajganj roadway areas of Nepal, World Journal of Applied Chemistry, **5**, 3 (2020): 47-56 (2020)
45. S.K. Regmi, K.P. Dahal, J. Bhattarai, Soil corrosivity to the buried-pipes used in Lalitpur, Kathmandu Valley, Nepal, Nepal Journal of Environmental Science, **3**, 1 (2015): 15-20
46. J. Bhattarai, An overview on the non-destructive in-depth surface analysis of corrosion-resistant films: A case study of W-xCr deposits in 12 M HCl solution, Bibechana, **18**, 1 (2021): 201-213
47. J. Bhattarai, Review on in-depth analysis of the passive films of W-xTi alloys by angle-resolved X-ray photoelectron spectroscopy, Science Journal of Chemistry, **8**, 2 (2020): 28-35
48. ASTM C876-15, Standard test method for corrosion potentials of uncoated reinforcing steel in concrete (ASTM International, West Conshohocken, PA, USA, p. 8, 2015)
49. ASTM C1582 / C1582M-11, Standard specification for admixtures to inhibit chloride-induced corrosion of reinforcing steel in concrete (ASTM International, West Conshohocken, USA, pp. 10, 2017)
50. T.M.B. Bandiola, Extraction and qualitative phytochemical screening of medicinal plants: a brief summary, International Journal of Pharmacy, **8**, 1 (2018): 137-143
51. A. Flores-Nicolas, M. Flores-Nicolas, J. Uruchurtu-Chavarin, Corrosion effect on reinforced concrete with the addition of graphite powder and its evaluation on physical-electrochemical properties, Revista ALCONPAT, **11**, 1 (2021): 18-33
52. N. Koirala, C. Dhakal, N.N. Munankarmi, S.W. Ali, A. Hameed, N. Martins, J. Sharifi-Rad, M. Imran, A.M. Arif, M.S. Hanif, B. Salehi, (2020). *Vitex*

negundo Linn: phytochemical composition, nutritional analysis, and antioxidant and antimicrobial activity. Cellular and Molecular Biology (Noisy-le-Grand, France), **66**, 4 (2020): 1-7

53. S. Saklani, A.P. Mishra, H. Chandra, M.S. Atanassova, M. Stankovic, B. Sati, M.A. Shariati, M. Nigam, M.U. Khan, S. Plygun, H. Elmsellem, H.A.R. Suleria, Comparative evaluation of polyphenol contents and antioxidant activities between ethanol extracts of Vitex negundo and Vitex trifolia L. leaves by different methods, Plants, **6**, 4 (2017): 45
54. S.C. Ohadoma, H.U. Michael, Effects of co-administration of methanol leaf extract of Catharanthus roseus on the hypoglycemic activity of metformin and glibenclamide in rats, Asian Pacific Journal of Tropical Medicine, **4**, 6 (2011): 475-477
55. M.K. Dwivedi, R. Shukla, N.K. Sharma, A. Manhas, K. Srivastava, N. Kumar, P.K Singh, Evaluation of ethnopharmacologically selected Vitex negundo L. for In vitro antimalarial activity and secondary metabolite profiling, Journal of Ethnopharmacology, **275**, (2021): 114076
56. A. Kalaiselvi, S.M. Roopan, G. Madhumitha, C. Ramalingam, G. Elango, Synthesis and characterization of palladium nanoparticles using Catharanthus roseus leaf extract and its application in the photo-catalytic degradation, Spectrochimica Acta- Part A: Molecular and Biomolecular Spectroscopy, **135**, (2015): 116-119
57. R. Ylmen, U. Jaglid, I. Panas, Monitoring early hydration of cement by ex situ and in situ ATR-FTIR– a comparative study, Journal of the American Ceramic society, **97**, 11 (2014): 3669-3675
58. E.V. Tararushkin, T.N. Shchelokova, V.D. Kudryavtseva, A study of strength fluctuations of Portland cement by FTIR spectroscopy, IOP Conference Series: Materials Science and Engineering, **919**, (2020): 022017
59. J. Higl, D. Hinder, C. Rathgeber, B. Ramming, M. Lindén, Detailed in situ ATR-FTIR spectroscopy study of the early stages of C-S-H formation during hydration of monoclinic C_3S, Cement and Concrete Research, **142**, (2021): 106367
60. A.A. Al-Amiery, M.H.O. Ahmed, T.A. Abdullah, T.S. Gaaz, A.A.H. Kadhum, Electrochemical studies of novel corrosion inhibitor for mild steel in 1 M hydrochloric acid, Results in Physics, **9**, (2018): 978-981
61. A. Kadhim, A.K. Al-Okbi, D.M. Jamil, A. Qussay, A.A. Al-Amiery, T.S. Gaaz, T. S., … M.H. Nassir, Experimental and theoretical studies of benzoxazines corrosion inhibitors, Results in Physics, **7**, (2017): 4013-4019
62. N.U. Shehu, U.I. Gaya, A.A. Muhammad, Influence of side chain on the inhibition of aluminium corrosion in HCl by α-amino acids, Applied Science and Engineering Progress, **12**, 3 (2019): 187-197

Viscosity improvement of recycled poly (ethylene terephthalate) from waste bottles by adding antioxidants and chain-extender

Isarankura Na Ayutthaya Siriorn[1,2,3] *, *Bunyarak* Munchumart [1], *Boondeacha* Natanicha [1], *Sangrawee* Nut[1], *Markmee* Nopparat[1], and *Haaian* Suchart[1].

[1]Department of Power Engineering Technology, College of Industrial Technology, King Mongkut's University of Technology North Bangkok, Bangkok, Thailand.10800
[2]Research Centre for Combustion Technology and Alternative Energy – CTAE, College of Industrial Technology, King Mongkut's University of Technology North Bangkok, Bangkok, Thailand. 10800
[3]Center for Design and Engineering for Railway and Modern Transport System, King Mongkut's University of Technology North Bangkok, Bangkok, Thailand. 10800

Abstract. This research work is focused on the viscosity improvement of recycled poly(ethylene terephthalate) (rPET) from waste bottles by adding antioxidants and multi-functional reactive polymer (Joncry ADR 4468, chain extender). The achieve the objective of this work the investigated was broken into two parts. The first part studied the effect of rPET viscosity after adding various types of antioxidants and stabilizer such as Irgafos®168, Tinuvin® 770, Irganox®1010. The second part observed the effect of viscosity after it was blended with chain extender at 0, 0.2, 0.4, 0.6 and 0.8 pph. rPET was then dried in the oven at 120 °C for 12 hrs, to deplete the moisture. Then, the dried rPET (mixed with the chemicals above) was extruded into a compound using a twin screw extruder. The shear viscosity of the extruded compound was then measured using a rotational rheometer at 270 °C. The results revealed that the addition of chain extender increased the shear viscosity and the tensile strength at break of rPET. Therefore, the chain extender interacted with chains, which could change the structure to be the longer chains, branching or network structures. These structures are entangled and interrupt the movement of the molecular chains. It can be concluded that the viscosity of rPET can be improved by adding a chain extender at 0.6 pph, and the antioxidants of Irgafos®168, Tinuvin®770 and Irganox®1010 at 0.2, 0.1 and 0.5 pph, respectively.

Keyword. rPET, Antioxidant, Chain-extender, Viscosity.

1 Introduction

Nowadays, the world produces around 300 million tons of plastic waste. One popular type of recycle plastic is PET waste bottles [1, 2]. PET or poly (ethylene terephthalate) is a kind of thermoplastic polymer that has a high molecular weight, which is used to produce drink and beverage bottles. The so-called "PET bottles" [3]. PET bottles are light weight bottles, which have high durability, toughness, good impact and pressure resistance, high gas permeability resistance, heat resistance and flame retardant [4, 5, 6]. PET is not a type of plastic that is easy to decompose. Therefore, recycling PET is a solution to make PET more environmentally friendly [7]. Presently, the recycled PET can be used to produce various type of products. Such as fibers, fabric, packaging, clothing, carpet, suitcase, and core materials [6].

For the recycle process, many types of PET bottles (such as clear bottles, colored bottles, thick bottles, and thin bottles etc.) is to wash and ground them into small pieces or flake. Then, remelt the PET flakes in an extrusion machine into rPET extrudate, then cut into pellets. However, the process of washing and grinding PET bottles into rPET flakes involves water and moisture and these are the main problem of the recycle process. Moisture and water will cause hydrolysis degradation during the extrusion process. Therefore, the molecular chain of rPET becomes shorter, decreasing the viscosity and melting strength of the polymer. This affects the formability of melted rPET, which could not be formed into a strong product [2, 4, 8, 9, 10, 11, 12]. To fix this problem, the rPET properties can be improved to better support the extrusion process. For example, by increasing the melted strength by increasing the molecular weight and the modifying the molecular structure into a branching structure by adding chain extenders [2, 4, 6, 12, 13].

Another way to increase the melt strength of rPET is by adding the antioxidants and stabilizers. Because, during the degradation process. The free-radicals, polymer radicals and peroxy radicals always occur. These radicals interact with the surrounding oxygen and hydrogen (from air and moisture) rapidly, which induced polymer chain cleavage quickly resulting in decreasing the melt strength, which can serve process-ability

* Corresponding author: siriorn.i@cit.kmutnb.ac.th

https://doi.org/10.1051/e3sconf/202130202019

limitation [14, 15, 16]. From this reason, the antioxidants and stabilizers were added in some research. However, rarely research investigated on the effect of antioxidant and stabilizer on the viscosity and tensile properties of rPET. So, the proper amount of these additives has not yet been investigated.

In this work focused on the improvement of process-ability of rPET by improve heat resistance and increased molecular chain of rPET during extrusion process are of interest. The aims of this work were to investigate the types and optimum concentration of antioxidant and stabilizers used during polymer processing. Afterwards, the concentration of chain extender was investigated. Lately, the mechanical properties such as tensile strength at break, elongation at break and Young's modulus were also interest.

2 Materials and methods

2.1 Materials

PET flakes from PET waste bottles were purchased from Thai Plastic Recycle Group Company Limited. Flakes and chemical structure and of PET as shown in Fig.1.

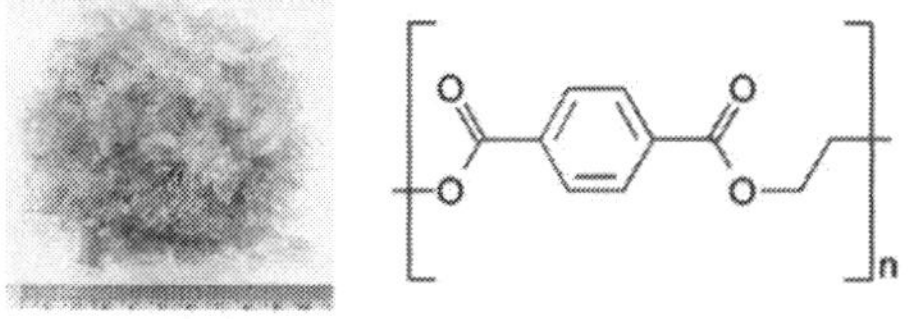

Fig.1 Flakes and chemical structure of PET.

2.2 Chemicals

Phenolic antioxidant (heat stabilizer): Pentaerythritol tetrakis(3-(3,5-di-tert-butyl-4-hydroxyphenyl) propionate (Irganox®1010), molecular weight 1178 g/mol. Hindered amine light stabilizer: Bis(2,2,6,6-tetramethyl-4-piperidyl) sebaceate (Tinuvin®770), molecular weight 481 g/mol; and antioxidant (peroxide decomposer type) Tris (2,4-di-tert-butylphenyl) phosphite (Irgafos®168), molecular weight 646.9 g/mol. The chemicals above were supplied from Merit Solution Company Limited. Chemical structure of Tinuvin®770, Irganox®1010, and Irgafos®168 as shown in Fig.2-4.

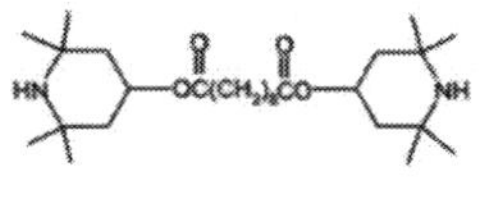

Fig.2 Chemical structure of Tinuvin®770.

Chain extender (multi-functional reactive polymer): Acrylic epoxy resin (Joncryl®ADR4468) was received from Chemical Innovation Company Limited. The chemical structure of Joncryl®ADR4468 was shown in Fig.5.

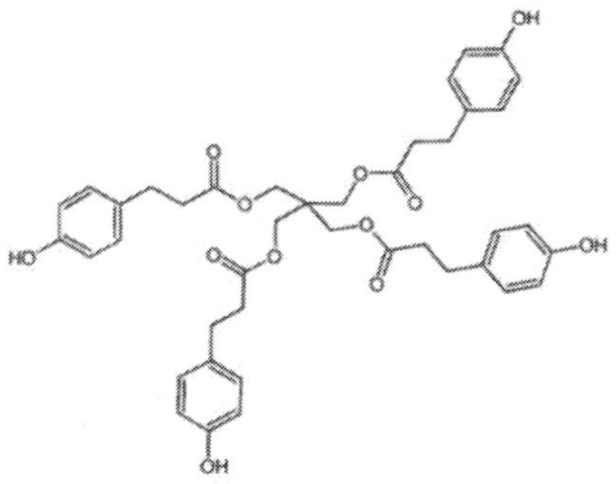

Fig.3 Chemical structure of Irganox®1010.

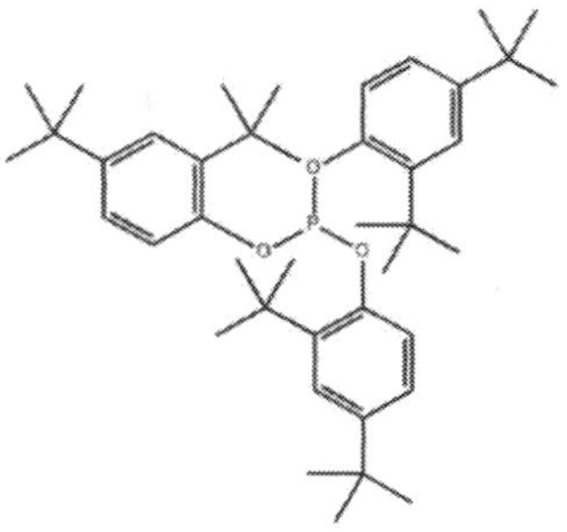

Fig.4 Chemical structure of Irgafos®168.

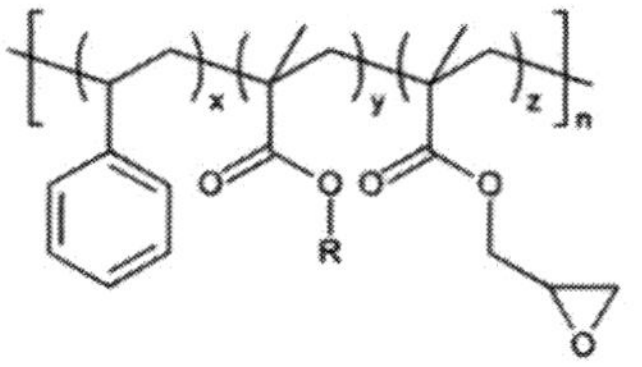

Fig.5 Chemical structure of Joncryl®ADR4468.

2.3 Equipment

A co-rotating twin screw extrusion machine (XINDA, SHJ-200) was used to prepare the rPET compounds. An oven with ventilation (Memmert, UN750plus) was used to dry the flakes and pellets of plastic. A hot-press compression molding machine (ENMACH, XH-406CEWP) was used to prepare the specimens. A rotational rheometer (Kinexus, Lab+) was used to measure the viscosity of rPET flakes and rPET compounds. A universal testing (Testometric, M500-25AT was used to measure the mechanical strength of specimens.

2.4 Preparation of rPET flakes

rPET flakes, which were cut and ground from PET waste bottles were dried in an oven (Memmert UN750plus) at 120 °C for 12 h. PET flakes should be dried to prevent hydrolysis degradation of PET during the recycle processing (extrusion process).

https://doi.org/10.1051/e3sconf/202130202019

2.5 Preparation of rPET Compounds

Dried rPET flakes, were mixed with antioxidants and a chain extender as shown in Table 1 and 2. Next, these formulas were compounded through the mixing of rPET flakes with various additives with the formula in accordance with Table 2. The compounds were mixed in a co-rotating twin screw extruder (XINDA, SHJ-200) with a twin rod die (2 mm. diameter each). The temperature profiles from the feed zone to the die zone were 180-180-200-220-220-200 °C, respectively. The screw rotating speed was 60 rpm. The extrudate rod was cooled under water and collected by a pulling unit. Then the compounds were cut into pellet form (Fig. 6).

Table 1. Antioxidants content in rPET compounds.

Formulas	Tinuvin®770 (pph)	Irgafos®168 (pph)	Irganox®1010 (pph)
rPET flake	-	-	-
Anti-1	-	-	0.5
Anti-2	0.1	0.2	0.5
Anti-3 (or Jon-0)	0.1	0.2	0.5

Table 2. Chain extender content in rPET compounds.

Compound formulas	rPET (pph)	Joncryl®ADR 4468
Pure-PET	100	-
Jon-0.0 *	100	0
Jon-0.2 *	100	0.2
Jon-0.4 *	100	0.4
Jon-0.6 *	100	0.6
Jon-0.8 *	100	0.8

Remark: * this formula has Irganox®1010 (0.5 pph), Tinuvin®770 (0.1 pph) and Irgafos®168 (0.2 pph).

2.6 Specimen preparation from rPET compounds

The pellet of rPET compound was dried in an oven at 120 °C for 12 h. Then the compound was fabricated into a specimen of (120 mm x 120 mm x 2 mm) with a hot-press hydraulic machine (ENMACH XH-406CEWP).

The details of the fabrication process (temperature and pressure) of the specimen are described below. The rPET compound was added into the mould. Then, the mould was put into the hot-press hydraulic machine using a temperature of 280 °C, and it was pressed at 0.5 MPa for 5 mins. Then, adjusted the pressure to 10 MPa for 2 mins. After that, released the pressure to 0 MPa for 10 second (the air or bubble inside the polymer melt was released at this stage). Next, increased pressure again to 10 MPa for 10 mins. Afterwards, transfer the mould to the cooling section at 10 MPa for 1 min. The specimens were removed from the mould and cut into dumbbell shapes by following the standard ASTM D638 type IV. The samples were tested by a universal testing machine to obtain the result of tensile strength at breakage, elongation at breakage and Young's modulus.

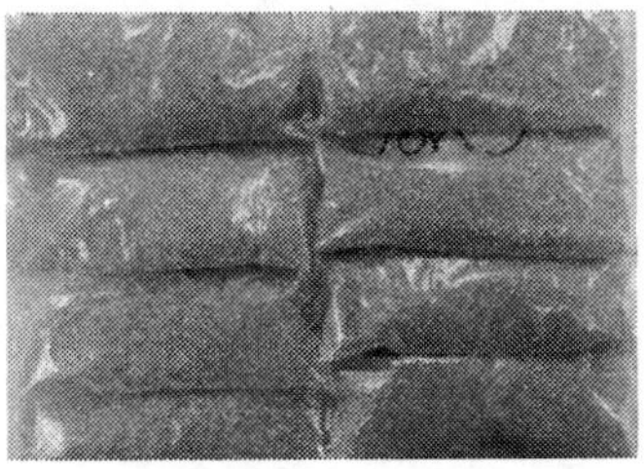

Fig.6 Pellets of rPET compounds

2.7 Testing and measurement methods

Viscosity measurement: rPET compound pellets (from Table 2) were dried at 120 °C for 12 h. Then, they were measured for viscosity by a rotational rheometer (Kinexus, Lab+) using cone 40 mm. diameter, at 270 °C. Mechanical testing: the samples with dumbbell shapes (ASTM D638 type IV) were tested by using a universal testing machine. Load cell and speed were 10 kPa and 5 mm/min, respectively.

3 Results and Discussions

3.1. Appearance of the rPET compounds

Fig.7 show the rPET extrudate from 1.2.2, which are Anti-0, Anti-1, Anti-2, Anti-3 (or Jon-0), Jon-1, Jon-2, Jon-3 and Jon-4.

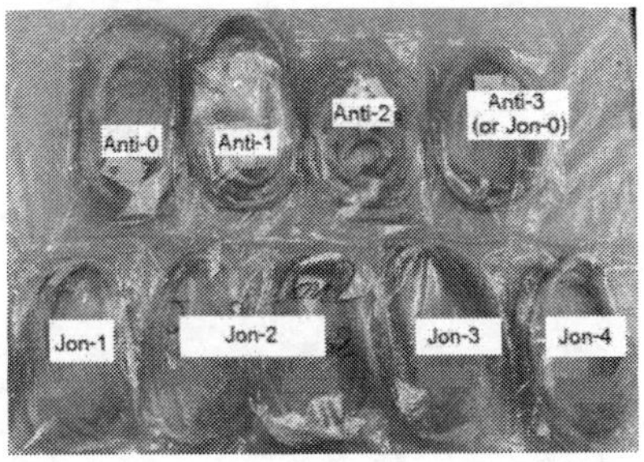

Fig.7 Extrudate of rPET compounds

It was found that the diameter of Jon-4 was bigger than other formulas. It was because, Jon-4 has the highest content of chain extender (Joncry®ADR4468), which could have more branches than others.

3.2 Viscosity of the rPET compounds

3.2.1 Viscosity of rPET compounds with various types of antioxidants

Fig.8 shows the comparative of shear viscosity of rPET compounds, including rPET flake, Anti-1, Anti-2, Anti-3 (or Jon-0), which were different in types and the content of antioxidant, including Irgafos®168, Tinuvin®770 and Irganox®1010.

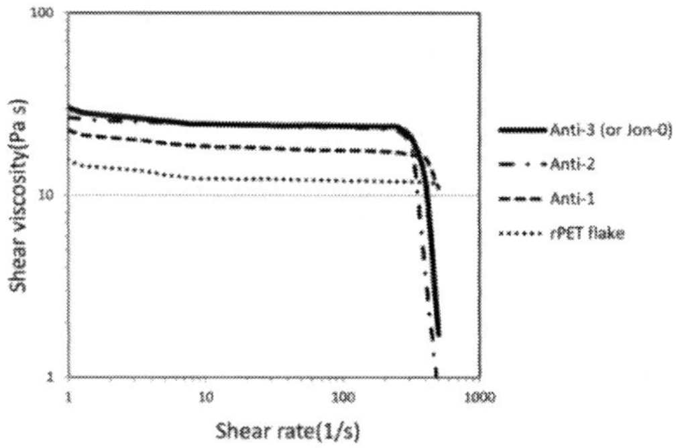

Fig.8 Viscosity value of rPET compounds (rPET flake, Anti-1, Anti-2, Anti-3 (or Jon-0).

Fig.7 shows viscosity value of rPET flake and various rPET compounds (Anti-1, Anti-2, Anti-3 (or Jon-0)). The non-Newtonian behaviour can be observed from all materials. The lowest viscosity was found in rPET flake. This flake came from the ground PET waste bottle directly without adding more additives. For Anti-1, the viscosity value increased when added 0.5 pph of Irganox®1010 (heat stabilizer).

Notably, when adding other types of additives, Anti-2 (which added Tinuvin®770 and Irgafos®168, at 0.1 and 0.2 pph, respectively) and Anti-3 (or Jon-0), which added Tinuvin®770, Irgafos®168 and Irganox®1010 equal to 0.1, 0.2 and 0.5 pph, respectively. It was found that they were of similar viscosity value. Therefore, the main additives that help to improve the viscosity value are Tinuvin®770 and Irgafos®168, at 0.1 and 0.2, respectively. It was because, Tinuvin®770 and Irgafos®168 are primary antioxidants (or radical scavenger) and peroxide decomposer, respectively. They can reduce the free radicals inside polymer effectively [14]. Meanwhile, Irganox®1010 was not significantly effective. It was because the molecular weight of Irganox®1010 is higher than Tinuvin®770 and Irgafos®168. Furthermore, the structure of Irganox®1010 is more complex and hinder than the others, which did not move rapidly (actively) to catch the radical in the polymer melt during the extrusion process. rPET is a kind of polyester that can degrade easily via hydrolysis reaction, which means it will break the chain into shorter chains and small radicals quickly. Hence, the addition of proper antioxidants will reduce the degradation rate of the polymer significantly.

3.2.2 Viscosity of rPET compounds with various content of chain extender

Fig.9 shows the comparative of shear viscosity of rPET compounds, including Jon-0, Jon-1, Jon-2, Jon-3 and Jon-4, which were different in the content of chain extender (Joncryl®ADR4468). All formulas were added Tinuvin®770, Irgafor®168 and Irganox®1010 at 0.1, 0.2 and 0.5 pph, respectively.

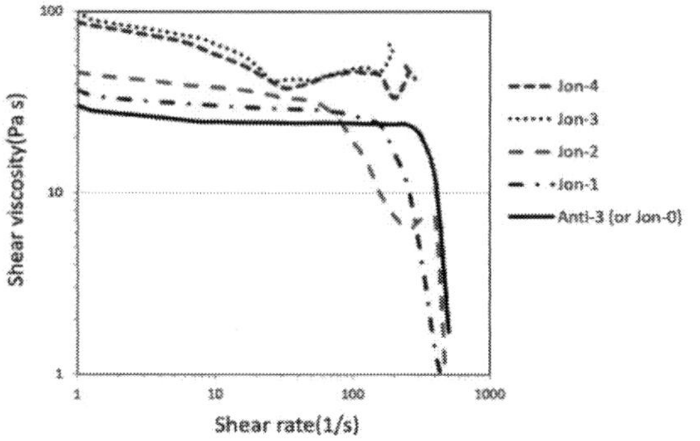

Fig.9 Viscosity value of rPET compounds (Jon-0, Jon-1, Jon-2, Jon-3 and Jon-4).

Fig.9 shows the high viscosity value of Jon-3 and Jon-4. As such these formulas have high concentration of chain extender when compared with other formulas. The chain extender interacted with the molecular chain, including short and long molecular chains. The interaction between rPET chains and the chain extender occurred at the hydroxy terminal end groups of PET and an epoxy-based chain extender, via esterification and etherification, respectively [13]. Consequently, rPET has a higher molecular weight and more branching, which increases the viscosity value.

3.3 Mechanical Properties

Fig.10, 11 and 12 shown the tensile strength at breakage, elongation at breakage and Young's modulus of various rPET compounds, which are Jon-0, Jon-1, Jon-2, Jon-3 and Jon-4. Fig.10, Jon-0, Jon-1 and Jon-2 show the same value of tensile strength at breakage. Therefore, the quantity of chain extender, 0.2 and 0.4 pph, were not significant enough to improve the chain branching and chain entanglement of polymer chains. As a result, it cannot improve the value of tensile strength at break of these specimens. On the other hand, it was found that the high value of the tensile strength at breakage were Jon-3 and Jon-4, which were 30.26±1.15 and 30.64±14.24 MPa, respectively. This is because these formulas have a high concentration of chain extender, which makes rPET higher in molecular weight and allows more branching. [4, 8, 9, 17]. Furthermore, the rPET chains have a higher content of chain entanglement resulting in the higher value of tensile strength at break. The higher in chain entanglement is related to the higher viscosity value of Jon-3 and Jon-4. As shown in Fig.9.

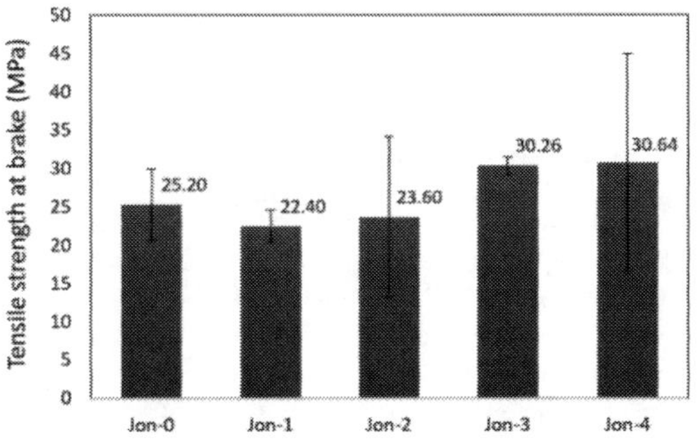

Fig.10 Tensile strength at break of various rPET compounds (Jon-0, Jon-1, Jon-2, Jon-3 and Jon-4).

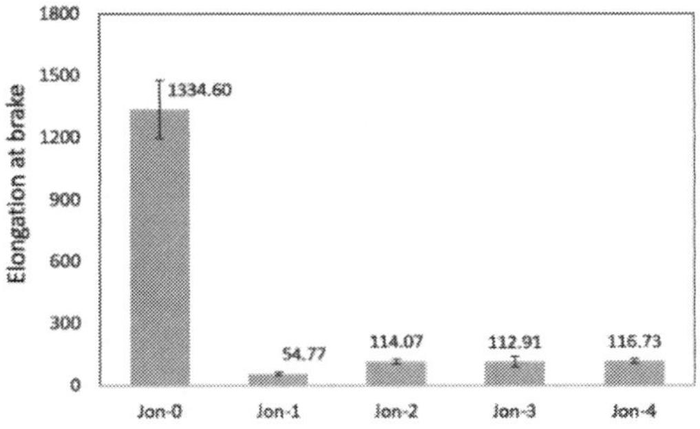

Fig.11 Elongation at break of various rPET compounds (Jon-0, Jon-1, Jon-2, Jon-3 and Jon-4).

From Fig.11 shows the elongation at breakage of rPET compounds with and without a chain extender. It was found the decrease of elongation at breakage values decreases rapidly after adding a chain extender, compared with rPET compound without a chain extender. Therefore, the chain extender interacted with chains, which could change the structure to be the longer chains, branching or network structures [4]. These structures are entangled and interrupt the movement of the molecular chain. However, with the higher concentration of a chain extender it was found a slightly increased of the elongation value, due to the branching effect of the compounds.

Fig.12 shows the Young's modulus of rPET compounds with and without a chain extender. It was found there was an increasing trend of the Young's modulus values of Jon-1, Jon-2 and Jon-3. In contrast, Jon-4 shows the rapid decrease of the modulus value after adding a chain extender of more than 0.6 pph. This was because the concentration of the chain extender is too high. It will change the chain structure of the compound to be a branching structure with the network structure [4]. If the compound has the network structure (gel content), the area of network will lock the chain, which means the chain cannot move or slide past each other.

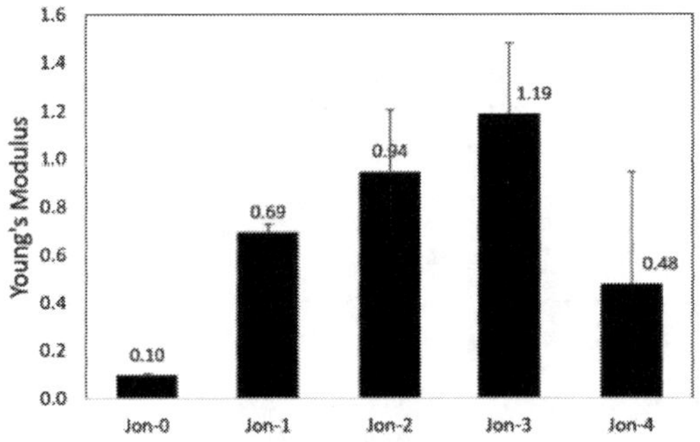

Fig.12 Young's modulus of various rPET compounds (Jon-0, Jon-1, Jon-2, Jon-3 and Jon-4).

4 Conclusion

The shear viscosity of rPET can be improved by adding the additives of two systems. The first system represents the rPET that adding of Tinuvin®770 and Irgafos®168 at 0.1 and 0.2 pph, respectively. The second system represents the rPET that adding of Tinuvin®770, Irgafos®168 and Irganox®1010 at 0.1, 0.2 and 0.5 pph, respectively. Moreover, the viscosity and mechanical properties (tensile strength at breakage, elongation at breakage and the Young's modulus) of rPET, can be improved by adding the chain extender (Joncryl®ADR4468) at 0.6 pph.

Acknowledgement

This research was funded by Thailand Science Research and Innovation Fund, and King Mongkut's University of Technology North Bangkok with Contract no. KMUTNB-MHESI-64-14.5). The authors would like to thank Research Center of Innovative Rubber (RIR), King Mongkut's University of Technology North Bangkok for supporting the fabrication processing and the characterization equipment. We would also like to thank Merit Solution Company Limited and Chemical Innovation Company Limited to support the antioxidants and chain extender, respectively.

References

1. T.M. Letcher, Plastic waste and recycling : Environmental Impact, Societal, Issues, Prevention, and Solutions (Academic Press, London, 2020)
2. V.A. Szabó, G. Dogossy, Investigation of flame retardant rPET foam, Periodic Polytechnica Mechanical Engineering, 2019
3. P. Kongsupapkul, K. Cheenkachorn, S. Tontisirin, Effects of MgO-ZSM-23 zeolite catalyst on the pyrolysis of PET bottle waste, KMUTNB international journal of applied science, 10, 3 (2017) : 205-211.
4. S. Makkam, W. Piewjan, W. Harnnarongchai, The effect of chain extender on rheological properties of

recycled poly (ethylene terephthalate), RMUTP Research Journal Humanities and Social Sciences, 8, 2 (2014) : 70-80.

5. C. Fan, C. Wan, F. Gao, C. Huang, Z. Xi, Z. Xu, L. Zhao, T. Liu, Extrusion foaming of poly(ethylene terephthalate) with carbon dioxide based on rheology analysis, Journal of Cellular Plastics, 0, 0 (2015) : 1-22
6. S. Yao, T. Guo, T. Liu, Z. Xi, Z. Xu, L. Zhao, Good extrusion foaming performance of long-chain branched PET induced by its enhanced crystallization property, Journal of Applied Polymer Science, (2020)
7. F. Ronkay, B. Molnár, F. Szalay, D. Nagy, B. Bodzay, I. E. Sajó, K. Bocz, Development of flame-retarded nanocomposites from recycled PET bottles for the electronics industry, Polymer, 11, 233 (2019)
8. L. Incarnato, L. Scarfalo, L. Maio, D. Acierno, Structure and rheology of recycled PET modified by reactive extrusion, Polymer, 41, 18 (2000): 6825-6831
9. R. Assadi, X. Colin, J. Verdu, Irreversible structure changes during PET recycling by extrusion, Polymers, 45, 13 (2004) : 4403-4412
10. L.K. Nait-Ali, X. Colin, A. Bergeret, Kinetic analysis and modelling of PET macromolecular changes during its mechanical recycling by extrusion, Polymer Degradation and Stability, 96, 2 (2011) : 236-246
11. F. Awaja, D. Pavel, Recycling of PET, European Polymer Journal, 41, 7 (2005) : 1453-1477
12. V.A. SzabÓ, G. Dogossy, Structure and properties of closed-cell foam prepared from rPET, IOP Conf. Series: Materials Science and Engineering, 426 (2018) :012043
13. M. Aksit, S. Gröschel, U. Kuhn, A. Aksit, K. Kreger, H-W. Schmidt, V. Altstädt, Low-density polybutylene terephthalate foams with enhanced compressive strength via a reactive-extrusion process, Polymers, 12, 2021 (2020)
14. S. Isarankura Na Ayutthaya, J. Wootthikanokkhan, Investigation of the photodegradation behaviors of an ethylene/vinyl acetate copolymer solar cell encapsulant and effects of antioxidants on the photostability of the material, Journal of Applied Polymer Science, 107 (2008) : 3853-3863
15. A. Gooneie, P. Simonetti, K. Salmeia, S. Gaan, R. Hufenus, M. Heuberger, Enhanced PET processing with organophosphorus additive : Flame retardant products with added-value for recycling, Polymer Degradation and Stability, 160 (2019) : 218-228
16. C. Huang, X. Dang, R. Bei, Y. Zhao, C. Li, Q. Chen, S. Wang, Effect of SiOx layer on preventing the migration of plasticizer and antioxidant from polyethylene terephthalate films, Iranian Polymer Journal, 28 (2019) : 173-182
17. M. Villalobos, A. Awojulu, G. Greeley, G. Turco, G. Deeter, Oligomer chain extenders for economic reprocessing and recycling of condensation plastics, Energy, 31, 15 (2006) : 3272-3234

Effect of Chelating Agents on Removal of Heavy Metal Cations of Cellulose-based Ion Exchange Resins from Water Hyacinth

Matcharee Suppapruek[1], *Poonsub* Threepopnatkul[1,*], *Amnard* Sittattrakul[1], and *Wanchai* Lerdwijitjarud[1]

[1]Department of Materials Science and Engineering, Faculty of Engineering and Industrial Technology, Silpakorn University, Nakhon Pathom, 73000, Thailand

Abstract. Presently, heavy metal pollution in natural water resources is considered very hazardous. The purpose of this research is to develop the novel cellulose-based ion exchange resins from water hyacinth and the effect of chelating agents on the removal of heavy metal cations from single-metal aqueous solutions has been investigated. Cellulose was prepared from the stem of water hyacinth by alkaline treatment process then cellulose-based ion exchange resins were prepared via esterification reaction between cellulose with two different chelating agents such as pyromellitic dianhydride and 3,3',4,4'- benzophenone tetracarboxylic dianhydride. The chemical structure, crystallinity index, surface morphology, and thermal stability of resins were characterized and analyzed with FTIR, XRD, SEM, and TGA, respectively. The results confirmed the esterification reaction between anhydride groups of chelating agents with hydroxyl groups of celluloses due to the apparent peak of the ester group. Heavy metal cations adsorptions were studied in optimum conditions and the residual concentration of heavy metal cations was measured by ICP. These resins showed high adsorption capacities of more than 98 mg/g for Pb^{2+}, Cd^{2+}, Cu^{2+}, and Ni^{2+}. Moreover, the adsorption process was controlled by the ion exchange mechanism. Therefore, the novel cellulose-based ion exchange resins could be suitable for the removal of pollutants from wastewater.

Keyword. Water hyacinth, Ion exchange resins, Heavy metal cations adsorption, Chelating agent, Dianhydride

1 Introduction

Heavy metal has been continuously polluted the natural water resources from various sources: agriculture and the rapid development of industrialization and urbanization. To make the pictures clear; for example, here are those industries that may have caused the pollution: electroplating, steel production, metal foundry, battery and leather manufacturing, and nuclear power operation. However, these issues have become so critical due to the impact on the ecosystem, environment, living well-beings, living organisms, and health hazards. These hazardous and adverse effects led to the necessity for the development of wastewater treatment methods for the removal of heavy metal cations from contaminated water to prevent these harmful consequences [1, 2].

Presently, there are numerous ways of water treatment for metal ion removal from wastewater such as adsorption, chelation ion exchange, chemical precipitation, coagulation, flocculation, activated carbon, electrolytic methods, reverse osmosis, and membrane separation [3]. Chelation ion exchange is one of the superior physicochemical methods for heavy metal ion disposal in comparison with other methods because it has several advantages of high removal efficiency, low costs, no chemical sludge and easy accessibility. In addition, this method filtrates those harmless ions and releases them into the environment, while removes the toxic metal ions out. More importantly, chelation ion exchange means that chelating polymers and heavy metal ions are bonding with coordination bonds.

Ion exchange has been widely practiced in wastewater treatment industry. Despite some commercial ion exchange resin cost relatively lower than other wastewater treatment methods. It is still considered a hefty price tag when it comes to the actual bill. Most of these resins are not eco-friendly because of the non-biodegradable matrix of the resins. Dowex50W and Amberlite IR-120 synthetic resins are commercial ion exchange resins. The structural polymeric backbone of the resins is styrene cross-linked with divinylbenzene functionalized with sulfonic acid, a functional group acted as ion exchange site [4, 5]. Furthermore, there are other commercial ion exchange resins such as Dow XUS 43578, Diaion CR11, Dowex HCR W2, and Purolite C160 [6]. Each resin has different matrix and its own active group acting as the ion exchange site. Therefore, the task of developing an eco-friendly bio-based ion exchange resin yet efficient enough to eliminate the

* Corresponding author: threepopnatkul_p@su.ac.th

heavy metals from wastewater has always been in the public interests and necessities.

One of the most interesting bio-based ion exchange resins is cellulose-based ion exchange resin due to its low cost, non-toxicity, high sorption capacity, biodegradation, and regenerating capability. Cellulose is a natural macromolecular compound and one of the most abundant renewable resources in nature. More importantly, it is a biopolymer, an eco-friendly material for environment. Moreover, cellulose possesses the hydroxyl groups (−OH) at the carbon atom positioned at 2nd, 3rd and 6th which are possibly active central sites for chemical modification and incorporation of effective chelating functional groups such as -SO_3H, -NH_2OH, and -COOH. These increase the adsorption capacity several times in the removal of heavy metal pollutants from the wastewater.

A variety of cellulosic biomass such as banana stem, sugarcane bagasse, moso bamboo, wood, cotton, reed, straw and other plants fibers has been modified. However, the interesting cellulose that is suitable for synthetic cellulose-based ion exchange resin is water hyacinth (Eichhornia crassipes). The water hyacinth is an aquatic weed that can be found in natural water resources. If they are so abundant in water resources, it will cause troubles and lead to hazard and wastewater.

During this time, chemical modification of cellulose mainly includes cellulose-based composites with organic polymers or hybrid materials, grafting copolymerization [1, 2] and the direct chemical functionalization of the hydroxyl groups on the cellulose such as esterification, etherification, and sulfonation [7-9]. The esterification is one of the most popular chemical modifications due to its easy accessibility and not complicated processes. It may choose chemicals that are cheaper and less toxic for production process.

Moreover, the cellulose derivatives were modified via esterification reaction. For example, hydroxyethyl cellulose was functionalized with phthaloyl chloride, isophthaloyl chloride and terephthaloyl choride [10]. However, the only disadvantage of cellulose derivatives is higher cost of the substrates because some cellulose derivatives are relatively expensive.

In addition, the cellulose is functionalized with commercial chelating agents such as D-penicillamine, meso-2,3-dimercaptosuccinic acid (DMSA), EDTA, and deferasirox [11]. There is also cellulose modification via esterification with carboxylic acid anhydride as a modifying agent or chelating agent. They are used in many fields of applications to deal with numerous issues of pollution.

Haier Chen et al. investigated the adsorption of Pb(II) from aqueous solution by mercerization of moso bamboo with chemically modified pyromellitic dianhydride (CMMB). The results clearly showed that the equilibrium adsorption capacity of CMMB adsorbent was about 2.7 times higher than activated carbon. This indicates that the CMMB adsorbent has good adsorption effect on Pb(II). After three adsorption–desorption cycles, the adsorption efficiency was decreased by maintaining the efficiency of above 80% [9]. However, the disadvantage is the fact that CMMB was synthesized from moso bamboo, a kind of winter bamboo and also a rare wood yet commercial one. Undoubtedly, moso bamboo and its derivatives like CMMB are extremely pricey. Simply because moso bamboo is not suitable for planting in tropical climate like Thailand. Moreover, 3,3′,4,4′ – benzophenone tetracarboxylic dianhydride (BPTCD) has been used to modify cellulose structure via esterification for multifunctional finishing or photoactive antibacterial cotton fabrics [12, 13]. Therefore, PMDA and BPTCD are our main focus of the chelating agents, the modified agents for esterification reaction with cellulose because the structure contains two carboxylic acid anhydride groups that could be efficient chelating functional groups for removing those heavy metal cations. These two compounds are relatively cheap and low toxicity. Thus, these synthetic ion exchange resins are eco-friendly and low cost.

The purpose of this work is to develop the novel cellulose-based ion exchange resins from water hyacinth by modifying with pyromellitic dianhydride (PMDA) and 3,3′,4,4′ – benzophenone tetracarboxylic dianhydride (BPTCD) as chelating agents. These resins were characterized by FTIR, XRD, SEM, and TGA analysis. Furthermore, the effect of chelating agents i.e., type and ratio on the removal of heavy metal cations such as Pb^{2+}, Cd^{2+}, Cu^{2+}, and Ni^{2+} from single-metal aqueous solutions had been investigated in optimum adsorption conditions.

2 Experimental

2.1 Materials

Cellulose biomass was prepared from the stem of water hyacinth (Eichhornia crassipes, Thailand) by alkaline treatment process. Pyromellitic dianhydride (PMDA), 3,3′,4,4′ - benzophenone tetracarboxylic dianhydride (BPTCD), lead nitrate ($Pb(NO_3)_2$), and nickel acetate tetrahydrate ($Ni(CH_3COO)_2{\cdot}4H_2O$) were purchased from Sigma Aldrich, Singapore. Cadmium nitrate tetrahydrate ($Cd(NO_3)_2{\cdot}4H_2O$) was obtained from KEMAUS, Australia. Copper sulfate pentahydrate ($CuSO_4{\cdot}5H_2O$) was purchased from HIMEDIA, India. Sodium hydroxide (NaOH) was purchased from QREC CHEMICAL, Thailand. Hydrochloric acid (HCl), acetone, ethanol and N,N-dimethylformamide (DMF) were obtained from RCI Labscan, Thailand. All other chemical reagents were analytical grade and they were used directly without further purification.

2.2 Preparation of cellulose from water hyacinth

Cellulose biomass was prepared from the stem of water hyacinth (Eichhornia crassipes) by the alkaline treatment process. Firstly, stems of water hyacinth were cut to small pieces and washed with ultrapure water several times then sun-dried outdoor for 48 hr. Afterward, the dried water hyacinth was ground until it became powder and was sieved through 140-mesh sieve. The dried powder of 20 g was soaked in 500 ml of 5%w/v NaOH solution at room temperature for 24 hr in order to remove hemicellulose, lignin, wax, and other impurities from biomass. After alkaline treatment, these powders were filtered and washed with ultrapure water until the pH was neutral at about 7-8. Finally, cellulose powder from water hyacinth was dried in a hot air oven at 70 °C for 24 hr [9].

2.3 Preparation of cellulose-based ion exchange resins

Cellulose-based ion exchange resins were prepared via esterification reaction between cellulose from water hyacinth with two different chelating agents (PMDA and BPTCD), as shown in Fig.1. Firstly, 5 g of cellulose and each chelating agent (1:1 and 1:3 mole fraction of repeating unit of cellulose per chelating agents) were dispersed in 250 ml DMF. Then, the reaction mixture was stirred at 70 °C for 20 hr. After esterification reaction, the materials were washed with DMF, ultrapure water, 0.1M NaOH, ethanol and acetone, respectively. These cellulose-based ion exchange resins (Cell-PMDA and Cell-BPTCD) were dried in a hot air oven at 60 °C for 24 hr. Finally, the dried resins were kept in desiccator and the percentage graft yield (P_g) was calculated by the following equation (1).

$$P_g = [(w_1 - w_2) / w_1] \times 100 \quad (1)$$

Where w_1 is the weight of cellulose-based ion exchange resins (g) and w_2 is the weight of cellulose from water hyacinth (g).

2.4 Characterizations

2.4.1 FTIR analysis

The chemical structure and functional groups of raw biomasses, cellulose from water hyacinth, and cellulose-based ion exchange resins were characterized with fourier transform infrared spectrometer; FTIR (VERTEX70, Bruker, USA). The FTIR spectra were recorded in the range of 4000-400 cm^{-1} and used a resolution of 4 cm^{-1}.

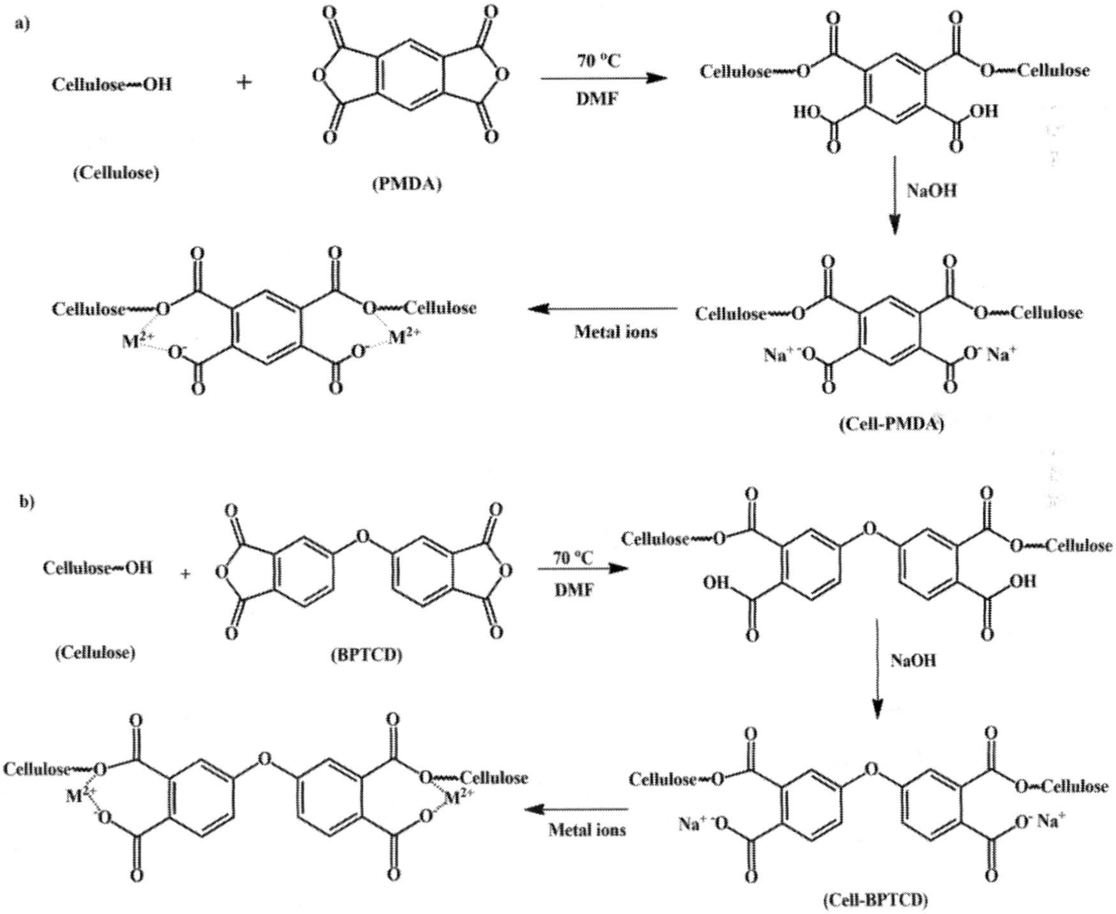

Fig. 1. Synthesis scheme and the adsorption mechanism of heavy metal cations on cellulose-based ion exchange resins for a) Cell-PMDA, and b) Cell-BPTCD resin.

2.4.2 XRD

The crystallinity index of samples was measured with X-ray diffractometer; XRD (MAXima_X XRD7000, Shimadzu, Japan). Cu Kα radiation was generated at 30 kV and 20 mA. XRD patterns were obtained with continuous scan mode at a scan speed of 5 degrees/min from 10° to 80°. The crystallinity index (*CrI*) was obtained by crystalline and amorphous intensity values obtained from the diffraction pattern. The *CrI* was calculated by the following equation (2) [1, 9, 14, 15].

$$\%CrI = [(I_{002} - I_{18^\circ}) / I_{002}] \times 100 \quad (2)$$

Where I_{002} is the maximum intensity at the (002) lattice diffraction at 2θ between 22° and 23° in the crystalline region. I_{18° is the intensity of the background scatter at 2θ = 18°.

2.4.3 SEM

The surface morphology and structure of samples were observed by using scanning electron microscope; SEM (TM3030, Hitachi, Japan) with an accelerating voltage of 15 kV. The cellulose powder before and after modification with esterification reaction were put on an aluminium stub and coated with a thin gold layer through sputtering process.

2.4.4 TGA

The thermal stability of cellulose and cellulose-based ion exchange resins from water hyacinth were investigated by thermogravimetric analyzer; TGA (Perkin Elmer TGA7, USA). The powdered samples about 3-7 mg were heated from 50 °C to 700 °C at a heating rate of 10 °C/min under N_2 atmosphere.

2.5 Adsorption experiments

The stock solution of 100 mg/L of Cu(II), Pb(II), Cd(II), and Ni(II) metal cations were prepared in distilled water by taking a definite amount of $CuSO_4 \cdot 5H_2O$, $Pb(NO_3)_2$, $Cd(NO_3)_2 \cdot 4H_2O$, and $Ni(CH_3COO)_2 \cdot 4H_2O$. Cellulose-based ion exchange resins were used as an adsorbent for the removal of each heavy metal cation from a single-metal aqueous solution.

Each heavy metal cation adsorption was studied under the optimal conditions by using initial metal concentration of 100 mg/L, solution pH ≈ 5, room temperature, stirring at 400 rpm and contact time of 180 min. Firstly, 50 mg of dried cellulose-based ion exchange resin was added to a 100 ml erlenmeyer flask containing 50 ml of a heavy metal aqueous solution. Then, the erlenmeyer flask was sealed and stirred at the room temperature. After adsorption, the residual concentration of heavy metal cation in the solution was measured by inductively coupled plasma spectrometer (ICP).

The adsorption experiment was repeated three times and the results have been given in terms of average values. The adsorption capacity of cellulose-based ion exchange resins at equilibrium (q_e, mg/g) was determined by the following equation (3).

$$q_e = [(C_0 - C_e) \times V] / W \quad (3)$$

Where C_0 and C_e is the initial and equilibrium concentration of adsorbate in heavy metal aqueous solutions (mg/L), V is the volume of solution (L) and W is the weight of the dried cellulose-based ion exchange resins (g).

3 Results and Discussion

3.1 Synthesis of cellulose-based ion exchange resins from water hyacinth

Before modification, the water hyacinth powder was immersed in 5 %w/v of NaOH to remove hemicellulose and lignin from biomass. Moreover, the alkaline treatment process also increased the specific surface area of fiber. As a consequently, hydroxyl groups of cellulose would be easily accessible [9, 16]. The hydroxyl groups (–OH) positioned at 2nd, 3rd, and 6th carbon atoms of cellulose ring structure were susceptible to be the active sites for chemical reaction [1]. PMDA and BPTCD were used as chelating agent to react with the hydroxyl groups of cellulose from water hyacinth at 70 °C for 20 hr in DMF. This reaction was the esterification between anhydride groups of chelating agents with hydroxyl groups of cellulose by obtaining the ester groups, cross-linking, and grafting on the cellulose materials. Then, cellulose-based ion exchange resins would be washed with saturated solution of NaOH to change the carboxylic group into carboxylate and enable them to increase their chelating ability as shown in Fig. 1.

Percentage graft yield (P_g) of cellulose-based ion exchange resins were compared in Table 1. These results showed that the percentage graft yield was increased dramatically with the raise of chelating ratio from 1 to 3 moles with the sacrifice of the cost because of the need for more chelating agents amount used.

Table 1. Percentage graft yield (P_g) of cellulose-based ion exchange resins.

Resins	Cellulose (mole)	Chelating agent (mole)		P_g (%)
		PMDA	BPTCD	
Cell-PMDA11	1	1	-	19.18
Cell-PMDA13	1	3	-	30.66
Cell-BPTCD11	1	-	1	20.96
Cell-BPTCD13	1	-	3	32.25

The presence of ester group and cross-linking including grafted chains on the cellulose from water hyacinth has been confirmed by the FTIR, XRD, SEM, and TGA techniques.

3.2 FTIR analysis

Different functional groups of untreated water hyacinth (WHC) were identified with FTIR as shown in Fig. 2(a). The FTIR spectra of WHC showed the typical characteristic peaks of lignocellulose that included cellulose, hemicellulose, and lignin. The broad band at 3401.42 cm^{-1} would be O-H stretching in hydroxyl group of biomasses. The peak at 2922.76 cm^{-1} corresponded to C-H stretching of the alkane. The peak at 1733.14 cm^{-1} was attributed to C=O stretching due to either acetyl or uronic ether linkages of carboxylic group in the ferulic and p-coumeric acids, two major phenolic acids in lignin biopolymer. The peak at 1637 cm^{-1} was associated with O-H bending due to water or humidity adsorption in the biomass sample. Additionally, the peak at 1428.35 cm^{-1} represented the $-CH_3$ asymmetric bending. The characteristic band at 1373.59 cm^{-1} corresponded to the C=C linkages which was presented in the guaiacyl ring of the lignin. The peaks at 1324.82 cm^{-1} normally indicated $-CH_2$ wagging which was typically presented in the hemicellulose and cellulose. The peak at 1249.93 cm^{-1} represented the C-O-C asymmetric stretching of aryl-alkyl ether in the lignin structure. The peak at 1050.93 cm^{-1} exhibited the C–O stretching of secondary alcohols and ethers functional group in hemicellulose and cellulose. Furthermore, the peak at 897 cm^{-1} explained the C-O-C symmetric stretching of the β-(1, 4)-glycosidic linkage in cellulose and xylose unit in hemicellulose [1, 14, 15, 17-20].

Fig. 2(b) showed the FTIR spectrum of cellulose from water hyacinth after alkaline treatment process. This FTIR spectrum was similar to untreated WHC. Only some major peaks were changed and shifted in the terms of shape, location, and transmittance of the FTIR spectral bands. The disappearance of C=O stretching at 1733.14 cm^{-1} peak from the alkali pretreated WHC sample indicated that lignin was removed from water hyacinth and obtained the effective delignification. Moreover, the reduction of peak intensity for O-H bending, C=C stretching, C-O-C asymmetric stretching, and C–O stretching exhibited at 1625.28, 1370.64, 1245.46, and 1060.50 cm^{-1}, respectively. It related to decreasing C=C bond of guaiacyl ring, C-O-C bond of aryl-alkyl ether in lignin and C–O bond of xylan in hemicellulose due to the dissolution of hemicellulose and lignin in NaOH solution. Therefore, the remaining was mostly cellulose [14, 18]. Furthermore, the FTIR spectrum obtained was somewhat similar to commercial cellulose [2, 17, 20].

Fig. 2(c-f) illustrated the FTIR spectrum of cellulose-based ion exchange resins that were synthesized via esterification by using different types and ratio of chelating agents (PMDA and BPTCD). After modification of cellulose from water hyacinth, the FTIR spectra of Cell-PMDA11, Cell-PMDA13, Cell-BPTCD11, and Cell-BPTCD13 resins showed additional peaks of C=C stretching (aromatic) about 1500 cm^{-1} and C=O stretching of ester group at 1727.92, 1727.79, 1726.91, and 1727.52 cm^{-1}, respectively [9, 21-23]. The ratio of chelating agent was increased from 1 to 3 moles that affected on the increment of C=O stretching. Moreover, the bands of O-H stretching and O-H bending at 3404 and 1625 cm^{-1} showed relatively broader peak with lesser intensity. These results confirmed the esterification reaction between anhydride groups of chelating agents with hydroxyl groups of cellulose due to the apparent peak of the ester group [22]. Furthermore, the ester and carboxylate bands in FTIR spectra indicated that the PMDA and BPTCD was introduced into cellulose via the formation of the ester bond with the consequent release of a carboxyl group [9].

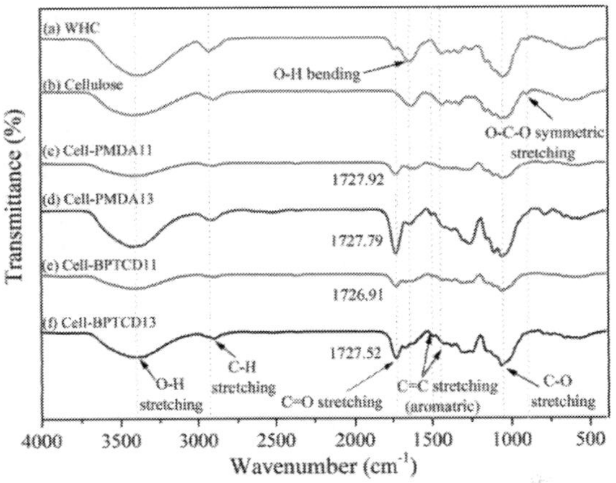

Fig. 2. FTIR spectra of (a) Water hyacinth (WHC), (b) Cellulose from water hyacinth, (c) Cell-PMDA11, (d) Cell-PMDA13, (e) Cell-BPTCD11, and (f) Cell-BPTCD13.

3.3 X-ray diffraction analysis

X-ray diffraction patterns of untreated and treated water hyacinth and cellulose-based ion exchange resins were shown in Fig. 3. The crystallinity indexes (*CrI*) were represented in Table 2. The XRD pattern of WHC in Fig. 3(a) exhibited the diffraction peaks at 15.20°, 21.84°, 28.54°, and 40.76°, while the diffraction peaks of alkali pre-treated WHC in Fig. 3(b) appeared at 15.04° and 22.34° which corresponded with (101) and (002) crystal planes of cellulose, respectively [2]. However, the disappearance of two diffraction peaks at 28.54° and 40.76° had proved that the alkali-treatment process could improve the rearrangement of cellulose chain. Moreover, the crystallinity indexes (*CrI*) of untreated and alkali-treated WHC were 55.17% and 64.47%, respectively. The results indicated that the alkaline treatment could increase the crystallinity index of cellulose by removing the amorphous hemicellulose and lignin from biomass. Furthermore, alkali treatment would more likely expose hydroxyl group of the water hyacinth-cellulose on surface. This led to more attachments of functional group and subsequently the higher selectivity for metal ion adsorption. Previous studies had shown results similarly and reported [9, 14, 15].

Fig. 3(c-f) showed the XRD patterns of cellulose-based ion exchange resins. These resins had shown similar diffraction patterns and characteristic peak at 2θ about 16° and 22°. These two diffraction peaks shifted and the intensity decreased when compared with cellulose from water hyacinth. This result indicated that the substrate had been changed after esterification and chemical modification [9]. Moreover, the crystallinity index (*CrI*) of Cell-PMDA11, Cell-PMDA13, Cell-BPTCD11, and Cell-BPTCD13 were 56.41%, 51.47%, 49.38%, and 48.65%, respectively. These results showed that the crystallinity index decreased and the amorphous region increased with esterification and higher chelating agent ratio. Because it raised the amount of grafting and crosslinking that broke the crystallinity formation of cellulose [1, 2, 9]. Furthermore, Cell-BPTCD possesses the crystallinity index less than Cell-PMDA because BPTCD structure is relatively larger and more steric effect than PMDA structure, which inhibited the crystalline packing of cellulose. These results confirmed that cellulose from water hyacinth was grafted and crosslinked with PMDA or BPTCD, which were associated with the percentage graft yield and FTIR result described earlier.

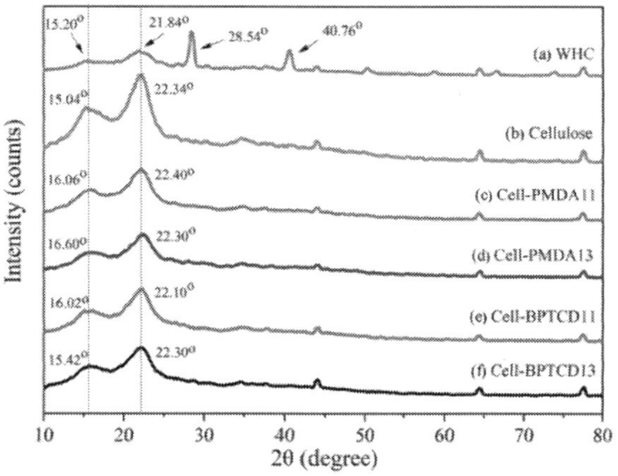

Fig. 3. XRD pattern of (a) Water hyacinth (WHC), (b) Cellulose from water hyacinth, (c) Cell-PMDA11, (d) Cell-PMDA13, (e) Cell-BPTCD11, and (f) Cell-BPTCD13.

Table 2. The Crystallinity index (*CrI*) of untreated and treated water hyacinth and cellulose-based ion exchange resins.

Resins	*CrI* (%)
Water hyacinth (WHC)	55.17
Cellulose from water hyacinth	64.47
Cell -PMDA11	56.41
Cell -PMDA13	51.47
Cell -BPTCD11	49.38
Cell -BPTCD13	48.65

3.4 Surface morphology analysis

The surface morphology of untreated water hyacinth (WHC) was shown in Fig. 4(a). The untreated WHC showed a firm, smooth surface and there were plenty of minute impurities on the surface. On the other hand, alkali pretreated WHC sample exhibited highly ordered structure, clearly surface, and rougher than untreated WHC because some hemicellulose and lignin dissolved in 5%w/v of NaOH solution until most of the structure remains were cellulose from water hyacinths, as shown in Fig. 4(b). Moreover, there were spaces between fibers that might facilitate the chances of a chemical reaction between hydroxyl functional groups on cellulose with the chelating agent. As a result, there were multiple active sites of functional groups, which could chelate with metal ions immensely [1, 9, 14, 18].

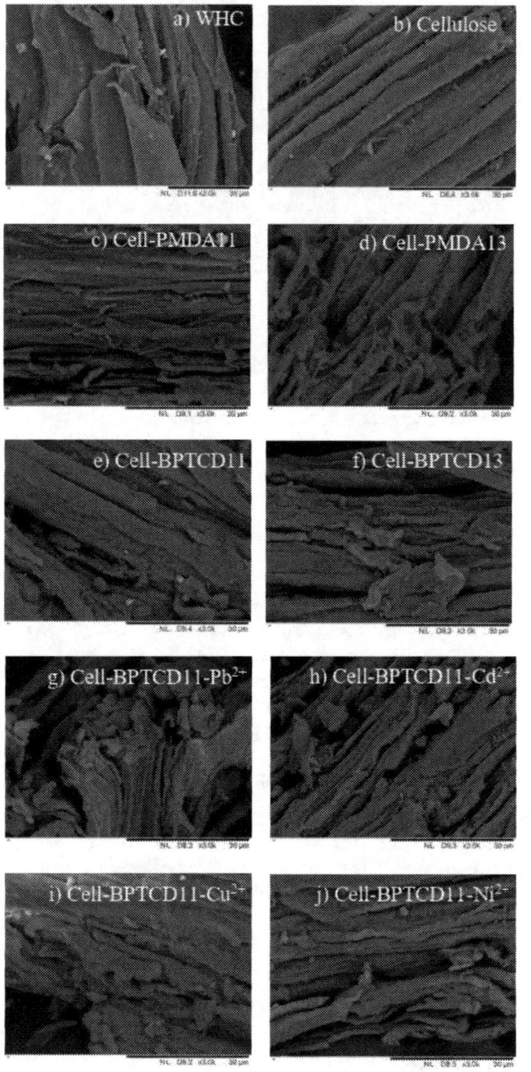

Fig. 4. Surface morphology of a) Water hyacinth (WHC), b) Cellulose from water hyacinth, c) Cell-PMDA11, d) Cell-PMDA13, e) Cell-BPTCD11, f) Cell-BPTCD13, and (g-j) After adsorption of Pb^{2+}, Cd^{2+}, Cu^{2+}, Ni^{2+}, respectively.

Fig. 4(c-f) showed surfaces of cellulose-based ion exchange resin such as Cell-PMDA11, Cell-PMDA13, Cell-BPTCD11, and Cell-BPTCD13, respectively. The results shown were cluster structure, irregular and rougher surface than cellulose from water hyacinth. The rough surface was increased with the increase of chelating agent ratio. These results confirmed grafting or crosslinking on cellulose surface, which correlated with the FTIR analysis. The surface morphologies of cellulose-based ion exchange resin with heavy metal ions adsorption were represented in Fig. 4(g-j). It was clear from the SEM images that surfaces of cellulose-based ion exchange resin became rougher and more agglomerated due to adsorption or ion exchange mechanism of heavy metal cations [1, 2]. Therefore, it could be confirmed that cellulose-based ion exchange resins adsorbing heavy metal cations in the contaminated wastewater effectively.

3.5 Thermogravimetric analysis

The TGA curves of untreated water hyacinth (WHC) were shown in Fig. 5(a), which illustrated the thermal degradation of untreated WHC in four stages. The first stage of weight loss occurred below 100 °C because of water elimination and volatile evaporation from the sample. The second stage of weight loss between the range of 130 - 220 °C was attributed to hemicellulose decomposition, followed by thermal degradation of cellulose at temperature range of 220 - 385 °C in the third stage. Finally, it was the lignin decomposition at the range of 415 - 525 °C since lignin was more stable and only decomposed at higher temperatures [1, 24].

On the other hand, The TGA curve of alkali-treated WHC in Fig. 5(b) represented being after alkaline treatment with only the two stage of thermal degradation. The first stage at lower than 100 °C which caused by water evaporation and the latter stage of decomposition was cellulose decomposition at temperature range of 230 - 390 °C [2]. These results indicated that the alkali-treated WHC had higher thermal stability when compared with untreated WHC due to the disposal of amorphous hemicellulose and lignin via dissolution in NaOH solution. Moreover, this result confirmed that the major remaining after alkaline treatment was cellulose from water hyacinth and it was consistent with the FTIR, XRD and SEM results.

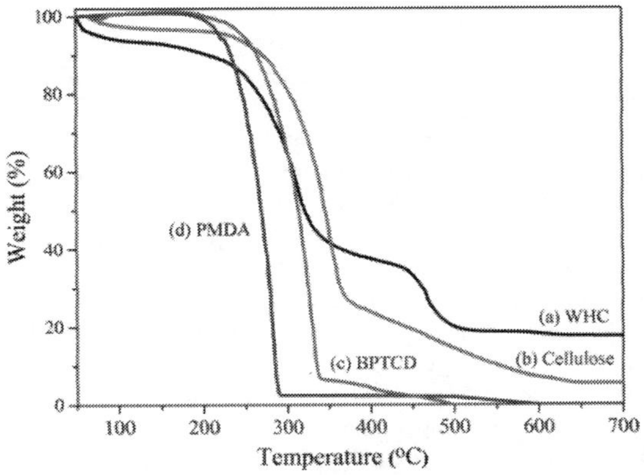

Fig. 5. TGA curve of raw materials (a) untreated WHC, (b) Cellulose, (c) BPTCD, and (d) PMDA.

Fig. 5(c-d) displayed the TGA curves of BPTCD and PMDA as chelating agent. The result showed that BPTCD and PMDA had lower thermal stability than WHC-cellulose and they showed only one stage of thermal decomposition. The thermal degradation of BPTCD occurred at temperature range of 187 - 344 °C [25], while PMDA had the thermal degradation at 180 - 290 °C [23, 26].

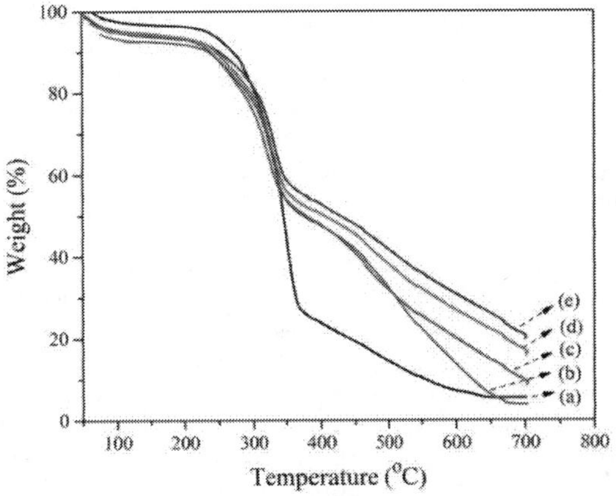

Fig. 6. TGA curve of (a) WHC-cellulose, (b) Cell-BPTCD11, (c) Cell-PMDA11, (d) Cell-BPTCD13, and (e) Cell-PMDA13

TGA analysis was employed to compare thermal stability of WHC-cellulose and cellulose-based ion exchange resins as shown in Fig. 6. The WHC-cellulose illustrated the highest decomposition onset temperature and small amount of degraded residues as shown in Fig. 6(a), while TGA curves of the cellulose-based ion exchange resin showed that all resins had similar thermal degradation behavior with three stages of weight loss, which it contradicted to WHC-cellulose. The first stage was due to the loss of water and volatile contents in the resins, followed by the degradation of ester crosslinks and the third stage was mainly contributed to the degradation of the cellulose chains as shown in Fig. 6(b-d). The main decomposition stage of resins was evidently extended and the rate was reduced. In addition, residual increased as compared with WHC-cellulose. These TGA results exhibited that the cellulose-based ion exchange resins had lower thermal stability than WHC-cellulose since the formation of unstable ester linkage formed during crosslinking or esterification reaction, as previously noted in the literature for the related ester products [12, 13, 23]. Moreover, the thermal degradation residue of the cellulose-based ion exchange resins was increased with the increment of PMDA or BPTCD ratio. Whereas the decomposition temperature decreased slightly due to the raising of carbon-carbon network, thermo-oxidative stability, and heat resistant properties of the crosslinker. In addition, the PMDA and BPTCD acted as chelating agents, grafting and crosslinking groups on the cellulose that increased carbonization rate of the cellulose during thermal degradation and led to high degradation residues [2, 12]. Despite the thermal stability of cellulose-based ion exchange resins decreased when compared to WHC-cellulose. However, it was clear that these resins were heat stable, having good thermal stability, not being affected at the applicable temperatures.

3.6 Adsorption capacity of heavy metal cations on cellulose based ion exchange resins

The adsorption capacity of heavy metal cations on cellulose-based ion exchange resins as shown in Fig.7 indicated the influences of type and ratio of chelating agents affected on metal ion adsorption in single-metal aqueous solution. Nevertheless, these resins illustrated the high adsorption capacities of more than 98 mg/g for heavy metal cation. While the Cell-BPTCD11 resin showed the highest adsorption capacity for Pb^{2+}, Ni^{2+} and Cd^{2+}, Cu^{2+} of 99.91, 99.88, 99.83, and 99.79 mg/g, respectively.

However, the Cell-BPTCD possessed slightly higher adsorption capacity than the Cell-PMDA because the BPTCD chelating agent on the cellulose had equivalent carboxylic acid group with the PMDA, which could form the chelate compounds, ion exchange, or electrostatic interaction with a heavy metal cation. Another reason is the BPTCD structure has relatively larger and more steric effect than the PMDA structure. This led to larger porous and more free volume inside the resin. Thus, heavy metal cations had higher diffusion rate and ion mobility [10]. It meant that heavy metal cations could easily diffuse into the structure of resin. As a result, this adsorption took place faster and more efficient. These results indicated that the adsorption process was controlled by the ion exchange mechanism as shown in Fig. 1.

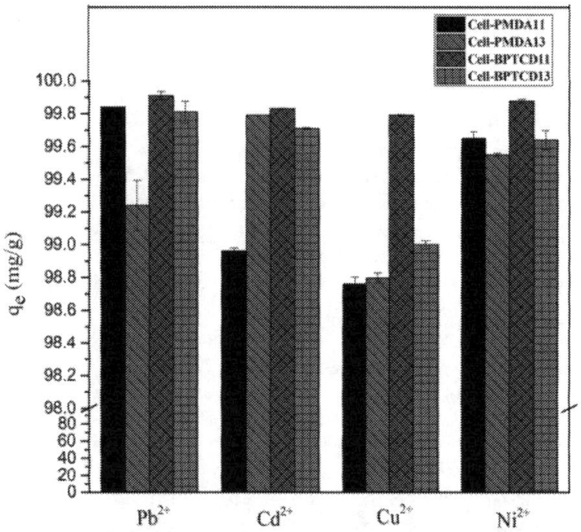

Fig. 7. Adsorption capacity of heavy metal cation on Cell-PMDA11, Cell-PMDA13, Cell-BPTCD11, and Cell-BPTCD13 resins. (Initial concentration 100 mg/L; contact time 180 min; stirring rate 400 rpm; room temperature; pH 5; resin 50 mg)

Moreover, the adsorption capacity of heavy metal cation on Cell-BPTCD resin decreased with the amplification of BPTCD ratio from 1 to 3 moles. The reason was the number of grafting or ester crosslinking was so excessive with the increase of BPTCD ratio. This led to the decrease of swelling ability that including matrixes of resins becoming narrow and tight. Therefore, it could intercept the movement and then heavy metal cation in aqueous solution could not diffuse into the resin. As a result, the adsorption capacity was declined [10, 27].

The adsorption capacity of Pb^{2+} and Ni^{2+} on Cell-PMDA resin decreased with the increment of PMDA ratio due to excessive grafting and crosslinking. While the adsorption capacity of Cd^{2+} and Cu^{2+} on Cell-PMDA resin increased with the rise of PMDA ratio due to the augmentation of chelating functional group. Also, it was found that each metal ion adsorption capacity on these resins was unequal which depended on the size and charge density of each metal ion [4].

The Cell-BPTCD11 resin was the focus of this study because it illustrated the highest adsorption capacity of heavy metal cation. The adsorption capacity for each cation increased in the following sequences: $Pb^{2+} > Ni^{2+} > Cd^{2+} > Cu^{2+}$. Among those ions with the same charge, the tendency of adsorption capacity increased with the radius reduction of heavy metal cations in water (hydrated radius) in the order: $Pb^{2+} < Ni^{2+} < Cu^{2+} < Cd^{2+}$ as shown in Table 3. Due to electrostatic effect and the metal binding strength increases with the decrement of the hydrated metal cation radius [4, 27, 28]. However, the actual ion exchange of $-COO^-Na^+$ group with metal cations also depend on the other factor such as pH, charge density and ionic strength that had influenced on adoption capacity. Moreover, Pb^{2+} in the aqueous solution possessed the smallest hydrated radius when compared with other cations as seen in Table 3. It meant that Pb^{2+} ions could easily diffuse into the resin and could chelate with chelating functional group in the resin with superiority [27].

Table 3. Ionic radius [29, 30], charge density [29], hydratedradius [30] and the adsorption capacity of heavy metal cations on Cell-BPTCD11 resin.

Metal cations	Ionic radius (A°)	Charge density	Hydrated radius (A°)	q_e (mg/g)
Pb^{2+}	1.32	0.21	4.01	99.91
Ni^{2+}	0.70	1.45	4.04	99.88
Cd^{2+}	0.97	0.52	4.26	99.83
Cu^{2+}	0.72	1.28	4.19	99.79

Furthermore, the costs of chelating agents for esterification reaction with cellulose from water hyacinth were compared with commercial cation exchange resin. PMDA and BPTCD are undoubtedly cheaper than many commercial cation exchange resins i.e., Dowex50WX8, Amberchrom 50WX8 and Chelex100 resins. PMDA and BPTCD are cheaper than Dowex50WX8 about 3.3 times and 5 times, respectively [31]. On the other hand, the PMDA and BPTCD are still considered pricey when compared to some commercial resins such as AmberliteIR120, DiaionCR11 and Purolite C160 resins. For example, the cost of PMDA and BPTCD are more expensive than AmberliteIR120 about 3.8 times and 2.8 times, respectively [31]. However, the relatively high cost is feasible because these cellulose-based ion exchange resins are biodegradable and high adsorption capacity while most of these commercial resins are not eco-friendly because of the non-biodegradable matrix of the resins. The structural polymeric backbone of these commercial resins is styrene cross-linked with divinylbenzene functionalized with sulfonic acid [4, 5, 31]. Moreover, the removal of heavy metals from aqueous solution by using Cell-PMDA and Cell-BPTCD resins was found to be cheaper than using the membrane separation method such as Empore™ SPE Disks about 5.7 times and 7.7 times, respectively [31]. Therefore, the results clearly showed that these cellulose-based ion exchange resins are also eco-friendly and relatively low cost.

The reuse of the adsorbents is very important for keeping the adsorption cost-effective process. There is also the possibility to reuse these cellulose-based ion exchange resins. It is consistent with other researches that investigated other cellulose-based ion exchange resins [2, 9, 10]. After three adsorption-desorption cycles, the adsorption efficiency of chemically modified moso bamboo (CMMB) was decreased to approximately 80% and above [9]. While after five cycles, the acrylamide/acrylic acid cellulose hydrogel still retained the relatively high adsorption capacity, which indicated the good reusability of the modified cellulose hydrogels [2]. The desorption and reusability of cellulose-based ion exchange resins from water hyacinth would be further studied.

4 Conclusions

These novel cellulose-based ion exchange resins from water hyacinth were successfully prepared via esterification with either PMDA or BPTCD. The ester crosslinking or grafting increased with the rise of chelating agent ratio from 1 to 3 moles which correlated to the percentage graft yield results. The FTIR, XRD, SEM, and TGA analysis confirmed that cellulose was functionalized with PMDA or BPTCD. The peak of the ester group had appeared on FTIR spectra of cellulose-based ion exchange resins. SEM images indicated the newly synthetic resins were cluster structure, irregular and rougher surface. Moreover, XRD results showed the decrease of crystallinity index after cellulose modification because the ester crosslinking and grafting broke the crystallinity of the cellulose. The thermal stability of cellulose-based ion exchange resins decreased, but they were still good thermal stability. Furthermore, the Cell-BPTCD possessed higher adsorption capacity than the Cell-PMDA. If the chelating ratio was to be increased, it could possibly decrease the adsorption capacity. However, these resins displayed high adsorption capacities of more than 98 mg/g for heavy metal cation, at which the Cell-BPTCD11 resin showed the highest adsorption capacity for Pb^{2+}, Cd^{2+}, Cu^{2+}, and Ni^{2+}. The adsorption process was governed by the ion-exchange mechanism and electrostatic interactions. Therefore, these novel cellulose-based ion exchange resins could potentially be employed for the removal of heavy metal cations from contaminated wastewaters. Furthermore, these resins are also eco-friendly and relatively low cost.

Acknowledgement

This research was financially supported by the Department of Materials Science and Engineering, Faculty of Engineering and Industrial Technology, Silpakorn University, Thailand.

References

1. A. Guleria, G. Kumari, E. Limac, Cellulose-g-poly-(acrylamide-co-acrylic acid) polymeric bioadsorbent for the removal of toxic inorganic pollutants from wastewaters, Carbohydrate Polymers, **228**, (2020): 115396
2. B. Zhao, H. Jiang, Z. Lin, S. Xu, J. Xie, A. Zhang, Preparation of acrylamide/acrylic acid cellulose hydrogels for the adsorption of heavy metal ions, Carbohydrate Polymers, **224**, (2019): 115022
3. Y. Li, Y. Wen, L. Wang, J. He, S. Al-Dey, M. El-Newehy, J. Yue, B. Ding, Simultaneous visual detection and removal of lead(II) ions with pyromellitic dianhydride-grafted cellulose nanofibrous membranes, Journal of Materials Chemistry A, **3**, 35(2015): 18180-18189
4. E. Pehlivan, T. Altun, The study of various parameters affecting the ion exchange of Cu^{2+}, Zn^{2+}, Ni^{2+}, Cd^{2+}, and Pb^{2+} from aqueous solution on Dowex 50W synthetic resin, Journal of Hazardous Materials, **134**, 1(2006): 149-156
5. A. Demirbas, E. Pehlivan, F. Gode, T. Altun, G. Arslanc, Adsorption of Cu(II), Zn(II), Ni(II), Pb(II), and Cd(II) from aqueous solution on Amberlite IR-120 synthetic resin, Journal of Colloid and Interface Science, **282**, 1(2005): 20-25
6. S. Edebali, E. Pehlivan, Evaluation of chelate and cation exchange resins to remove copper ions, Powder Technology, **301**, (2016): 520-525
7. C. Dong, H. Zhang, Z. Pang, Y. Liu, F. Zhang, Sulfonated modification of cotton linter and its application as adsorbent for high-efficiency removal of lead(II) in effluent, Bioresource Technology, **146**, (2013): 512-518
8. C. Dong, F. Zhang, Z. Pang, G. Yang, Efficient and selective adsorption of multi-metal ions using sulfonated cellulose as adsorbent, Carbohydrate Polymers, **151**, (2016): 230-236
9. H. Chen, Y. Cheng, Z. Zhu, H. He, L. Zhang, N. Li, Y. Zhu, Adsorption of Pb(II) from Aqueous Solution by Mercerized Moso Bamboo Chemically Modified with Pyromellitic Dianhydride, Journal of Environmental Engineering, **146**, (2020): 04019127
10. P. Prasannam, Removal of heavy metals from solution using ion exchangers based on hydroxyl cellulose, Materials Science and Engineering, **1**, (2004): 30-43
11. V. Kumar, A. Kumar, S.K. Singh, M. Kumar, S. Kumar, D. Kumar, R. Singh, Chelating Drug Therapy, Austin Journal of Genetics and Genomic Research, **2**, 1(2015): 1-3
12. K.H. Hong and G. Sun, Photoactive antibacterial cotton fabrics treated by 3,3′,4,4′-benzophenonetetracarboxylic dianhydride, Carbohydrate Polymers, **84**, 3(2011): 1027-1032

13. A. Hou, G. Sun, Multifunctional finishing of cotton fabrics with 3,3′,4,4′-benzophenone tetracarboxylic dianhydride: Reaction mechanism, Carbohydrate Polymers, **95**, 2(2013): 768-772
14. S. Das, A. Bhattachary, S. Haldar, A. Ganguly, S. Gud, Y.P. Tinga, P.K. Chatterjee, Optimization of enzymatic saccharification of water hyacinth biomass for bio-ethanol: Comparison between artificial neural network and response surface methodology, Sustainable Materials and Technologies, **3**, (2015): 17-28
15. A. Ganguly, S. Das, A. Bhattacharya, A. Dey, P. K. Chatterjee, Enzymatic hydrolysis of water hyacinth biomass for the production of ethanol: optimization of driving parameters, Indian journal of experimental biology, **51**, (2013): 556-566
16. T. Pusic, A.M. Grancaric, I. Soljacic, V. Ribitsch, The effect of mercerisation on the electrokinetic potential of cotton, Coloration Technology, **115**, 4(1999): 121-124
17. B. Abderrahim, E. Abderrahman, A. Mohamed, T. Fatima, T. Abdesselam, O. Krim, Kinetic thermal degradation of cellulose, polybutylene succinate and a green composite: Comparative Study, World Journal of Environmental Engineering, **3**, 4(2015): 95-110
18. K. Das, P. Ghosh, A. Dey, A. Ganguly, S. Das, P. K. Chatterjee, Studies on the Optimization of Phenolics during Production of Xylitol from Water Hyacinth, European Journal of Biotechnology and Bioscience, **3**, 4(2015): 25-33
19. N. Atikah, M. Nor, W. Aida, W. Mustapha, O. Hassan, Deep Eutectic Solvent (DES) as a Pretreatment for Oil Palm Empty Fruit Bunch (OPEFB) in Sugar Production, Procedia Chemistry, **18**, (2016): 147-154
20. D. Gaspar, S.N. Fernandes, A.G. de Oliveira, J. G. Fernandes, P. Grey, R.V. Pontes, L. Pereira, R. Martins, M.H. Godinho, E. Fortunato, Nanocrystalline cellulose applied simultaneously as the gate dielectric and the substrate in flexible field effect transistors, Nanotechnology, **25**, (2014): 094008
21. I. Kavianinia, P.G. Plieger, N.G. Kandile, D.R.K. Hardinga, New hydrogels based on symmetrical aromatic anhydrides: Synthesis, characterization and metal ion adsorption evaluation, Carbohydrate Polymers, **87**, 1(2012): 881-893
22. A. Tripathi, G.N. Parsons, O.J. Rojas, S.A. Khan, Featherlight, Mechanically Robust Cellulose Ester Aerogels for Environmental Remediation, ACS Omega, **2**, 8(2017): 4297-4305
23. F. Afinjuomo, T.G.Barclay, Y. Songa, A. Parikh, N. Petrovsky, S. Garg, Synthesis and characterization of a novel inulin hydrogel crosslinked with pyromellitic dianhydride, Reactive and Functional Polymers, **134**, (2019): 104-111
24. G. Luo, P.J. Strong, H. Wang, W. Ni, W. Shi, Kinetics of the pyrolytic and hydrothermal decomposition of water hyacinth, Bioresource Technology, **102**, 13(2011): 6990-6994
25. M.G. Junior, K.M. Novack, V.R. Botaro, Bamboo pulp treated with NaOH and benzophenone tetracarboxylic dianhydride as reinforcement of new polymerics materials, Revista Latinoamericana de Metalurgia y Materiales, **34**, 2(2013): 196-208
26. Y. Zhang, M. Zhang, Y. Zhang, X.W. Chen and J. H. Wang, Green and catalyst-free preparation of triazinyl polyimide for the efficient adsorption of glycoproteins, RSC Advances, **6**, 51(2016): 46002-46007
27. T.T. Hong, H. Okabe, Y. Hidaka, B.A. Omondi, K. Hara, Radiation induced modified CMC-based hydrogel with enhanced reusability for heavy metal ions adsorption, Polymer, **181**, (2019): 121772
28. M. Maslova, V. Ivanenko, N. Yanicheva, L. Gerasimova, The effect of heavy metal ions hydration on their sorption by a mesoporous titanium phosphate ion-exchanger, Journal of Water Process Engineering, **35**, (2020): 101233
29. S. Huang, C. Min, Y. Liao, H. Sun, Y. Jiang, Poly(1-amino-5-chloroanthraquinone): Highly selective and ultrasensitive fluorescent chemosensor for ferric ion, Journal of Fluorescence, **23**, (2013) :621-627
30. E. R. Nightingale, Phenomenological Theory of Ion Solvation : Effective Radii of Hydrated Ions, The Journal of Physical Chemistry, **63**, 9(1959): 1381-1387
31. Sigma Aldrich Co.,Ltd, Cation exchange resin, June 25, (2021), Available from https://www.sigmaaldrich.com/TH/en/search/cation-exchange-resin

https://doi.org/10.1051/e3sconf/202130202021

Application of Biochar for Cadmium Stabilization in Contaminated Paddy Soil

Marisa Oranratmanee and *Warapong* Tungittiplakorn*

Faculty of Applied Science, King Mongkut's University of Technology North Bangkok, Thailand

Abstract. Cadmium contamination in rice fields near zinc mines in Mae Sot District, Tak Province has been a persistent problem for decades. The contamination covers a vast area, making several soil remediation methods, such as soil washing or excavation impractical. Phytoremediation would also take several years and interfere with farming. However, there are recent reports on the potential of biochar in cadmium stabilization which resulted in reduced cadmium uptake by rice. Therefore, in this study, several biomaterials were investigated to produce the most suitable biochar for cadmium stabilization in paddy soil in Mae Sot. Rice husk, rice straw and bagasse were selected because of their availability in Mae Sot. Cadmium adsorption efficacy of biochar made from these biomaterials were analyzed using the adsorption isotherm. The results showed that cadmium adsorption by biochar from bagasse, rice husk, and rice straw were in accordance with Freundlich adsorption isotherm equation. By comparing the distribution coefficient, the adsorption efficacy of the biochar was ranked as rice straw> bagasse> rice husk. Therefore, rice straw biochar, which had the highest cadmium adsorption efficacy, was selected for stabilization experiments. After 30 days of incubating contaminated soil with rice straw biochar, the amount of extractable Cd using $CaCl_2$ and EDTA were reduced significantly. The optimal application rate of rice straw biochar was 5% with the incubation period of 20 days. This warrants the next phase of this study which will be conducted in the field experiments in Mae Sot.

Keyword. Adsorption Isotherm, Cadmium, Stabilization, Biochar, Paddy, Mae Sot District

1 Introduction

Cadmium contamination in rice fields near zinc mines in Mae Sot District, Tak Province was reported by the International Water Management Institute (IWMI) and the Pollution Control Department (PCD). [1, 2]. According to the reports, the contamination covered a large area of paddy field downstream from the mines (approximately 6 square kilometres) as shown in **Fig. 1** and **2.** Cadmium concentration in paddy soil and rice grains in the contaminated area were between 0.5 to 284 mg/Kg of soil and 0.05 to 7.7 mg/Kg of rice, respectively. These contaminations pose a health risk to human because harvested rice contained cadmium that exceeded CODEX standard of 0.2 mg Cd / Kg of rice. Besides, there is also contamination in other agricultural products such as soybeans, mung beans, corn and sugarcane. [3]. Cadmium contamination could cause respiratory and circulatory disorders, including an orthopedic disorder known as Itai-Itai. [1 ,4]. The contamination in soils also affected soil fertility and agricultural productivity. Hence, it is imperative to remediate the contaminated area with appropriate and effective methods.

There are several methods for remediation of heavy metals in contaminated soil. Excavation can quickly restore soil that is contaminated with a wide variety of contaminants. However, this method is limited by the cost of excavation, transportation, and disposal, therefore not suitable for a large contaminated area such as Mae Sot. Soil washing, which leaches contaminants from the soil with chemicals, can be done quickly but is also limited by the amount chemicals required and their effects on soil's fertility and pH. This method can subsequently affect soil organisms and its suitability for agriculture. Biological method such as phytoremediation is relatively safe, but limited by slow plant growth and uptake [5]. Thus, this

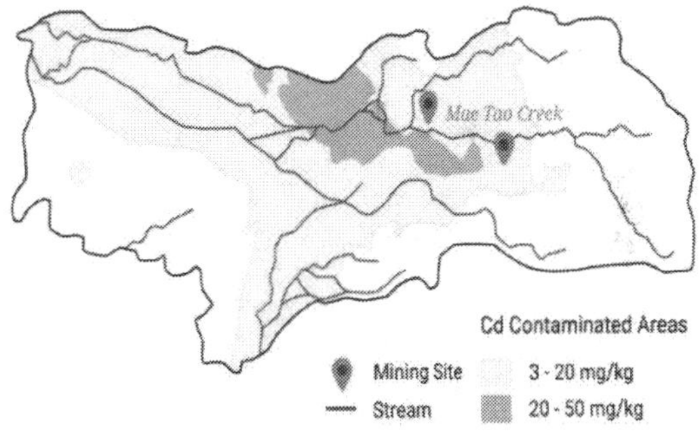

Fig. 1. Contaminated area in Mae Sot District, Tak Province [2].

* Corresponding author: warapong.t@sci.kmutnb.ac.th

 https://doi.org/10.1051/e3sconf/202130202021

method would take several years to reduce the contaminant concentration to acceptable level.

Fig. 2. Cadmium contamination in rice fields in Phra That Pha Deang Sub-district, Mae Sot District, Tak Province.

Cadmium stabilization using clay minerals or biochar have been shown in several research for their potential for treating heavy metal-contaminated soil [4, 6, 7]. Treated clay minerals were shown to adsorb heavy metals effectively from aqueous solution and immobilize heavy metals in contaminated soil [8, 9]. Application of wheat straw derived biochar was shown to reduce the bioavailability of Cd and Pb in paddy soil and limit Cd uptake by rice [7, 10]. Bamboo and rice straw biochar decreased the amount of extractable heavy metals (Cd, Cu, Pb and Zn). [11]. Acacia biochar decreased the accumulation of As and Mn in Napier grass grown in contaminated mine tailing. [12]. Moreover, rice straw biochar was shown to reduce the accumulation of Cd, Zn, Mn and Ni in roots, stems and seeds of wheat. [13].

However, unlike clay minerals, biochar can be produced from various agricultural wastes by pyrolysis process. It can also be used as a soil amendment material because of its high organic carbon, the ability to retain minerals, and its water holding capacities [14]. Rice straw and bamboo biochar were found to increase the growth and yield of wheat and rice, respectively [13, 15]. Moreover, pine pellet biochar was shown to increase the growth of cucumber plug seedling in both stem and root. [16]. Therefore, biochar can be applied to contaminated soil to decrease heavy metal uptake by crops while increasing the growth and yield of the crops. This allows farmers to grow crops while remediating the soil, making this method highly suitable for solving the cadmium contamination problem in Mae Sot.

Biochar is black carbon which is very stable and can be retained in the soil for a long time. It has a large surface area with numerous functional groups which are able to adsorb heavy metals. Biochars produced from different biomass and pyrolysis temperature have different physical and chemical properties, as shown in the Table 1. The mechanisms of biochar adsorption include electrostatic attraction, ion exchange, complexation and co-precipitation at the surface of biochar as shown in **Fig. 3.** Plant uptake generally requires Cd to be in the exchangeable form such as Cd^{2+}, thus Cd could be stabilized by reducing the amount of exchangeable Cd by means of biochar adsorption. Additional factors include physical and chemical properties of biochar which are greatly influenced by the production temperature, the application rate of biochar in the soil, and the pH of the soil [17].

Table 1. Properties of different biochars.

Type of Biochar	Pyrolysis Temp (°C)	pH	Surface area (m^2/g)	CEC (cmol/kg)	Ref
Bagasse	450	6.5	3.04	17.58	
Napier grass	450	9.86	11.68	18.61	[12]
Acacia wood	450	9.13	25.55	23.83	
Bamboo	750	9.5	907.4	15	[11]
Rice straw	500	10	36.7	45	
Rice Husk	700	8.5	25.161	-	[18]
Empty fruit bunch	700	9.4	1.89	-	
Populus wood	500-550	8.2	15	-	
Carozo corn	400-500	10.3	22.3	-	[19]
Sewage sludge	500-550	8.7	31.4	-	
Scots pine	450	8.56	9.16	3.41	
Scots pine	700	8.52	10.4	2.40	[20]
Silver birch	450	8.69	5.92	5.09	
Silver birch	700	9.27	7.17	5.71	

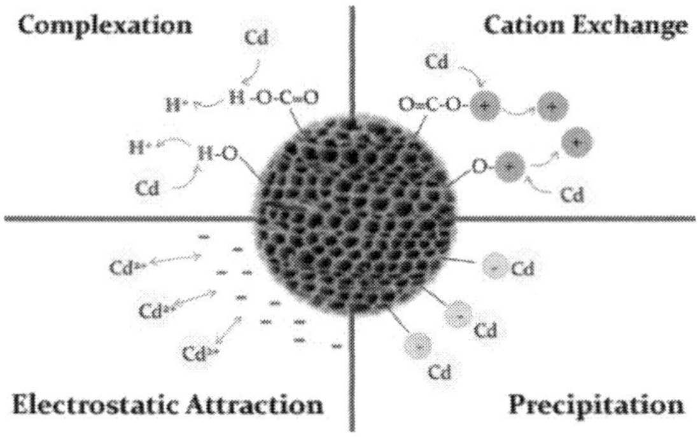

Fig. 3. Biochar - Heavy metals adsorption mechanism.

The purpose of this research is to study the efficacy of biochar made from different types of agricultural wastes in stabilizing cadmium in paddy soil. Bagasse, rice husk, and rice straw were used to produce biochar because of their availability in Mae Sot District, Tak Province. The ability of biochar to adsorb and reduce exchangeable cadmium in paddy soil was tested in lab-scale experiments. Biochar with the highest efficacy for cadmium adsorption and stabilization from this study will be applied in a field experiment in Mae Sot to provide the most effective and practical remediation method for rice farmers in the area.

2 Experimental Methods

2.1. Soil and biochar preparation

Cadmium contaminated soil samples and rice straw were obtained from a paddy field in Mae Tao Sub-district, Mae Sot District, Tak Province as shown in **Fig. 4**. Bagasse biochar (BB), rice husk biochar (RHB), and rice straw biochar (RSB) were produced by pyrolysis process at 450 °C for 4 hours. The process was conducted in a 200-liter steel furnace at Huai Sai Royal Development Study Center as seen in **Fig. 5.** Biochar was then ground and sifted to 250 μm. The pH of paddy soil and biochar were determined according to the USEPA Method 9045D by mixing soil or biochar with water at 1:1 ratio. The pH of the supernatant was then measured with a Mettler Toledo F20 pH meter. The pH of Cd-contaminated paddy soil, uncontaminated paddy soil, BB, RHB, and RSB were determined to be 7.50, 5.23, 9.31, 7.80, and 10.15, respectively.

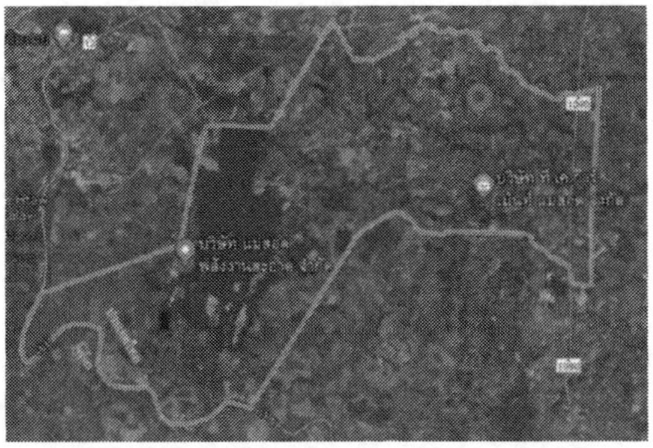

Fig. 4. The location of the sampling site in Mae Tao Sub-district, Mae Sot District, Tak Province. The red line outlines the boundary of Mae Tao Sub-district.

Fig. 5. Biochar steel furnace.

2.2 Cadmium adsorption

To study the efficacy of cadmium adsorption on biochar, 1 g of uncontaminated paddy soil sample was added to 20 mL of cadmium solution (prepared from cadmium chloride dissolved in 0.01M $CaCl_2$) at 5 concentrations: 10, 15, 20, 25 and 30 mg/L. To compare the effect of biochar concentration, the concentration of biochar added to each vial was varied at 0, 10, 20 and 100% of soil sample. The mixtures were shaken at 120 rpm for a period of 24 hours. The supernatant was then filtered with No.42 Whatman filter. Cadmium concentration and pH of the supernatant were determined by Atomic Absorption Spectroscopy (Agilent 240FS) and pH meter, respectively. The amount of adsorbed cadmium was determined using Equation (1).

$$q_e = \left(\frac{(C_0 - C_e) \times V}{W}\right) \tag{1}$$

where q_e is the amount of Cd adsorbed by biochar, C_0 and C_e are the initial and equilibrium concentration of Cd in the supernatant, respectively [21]. Adsorption isotherm was plotted between q_e and C_e.

2.3 Cadmium extraction

To study the stabilization effect of biochar on cadmium-contaminated soil, biochar with the highest efficacy for cadmium adsorption was incubated with 200 g of contaminated paddy soils for 30 days with water saturation. The amount of biochar was varied at 5 and 7 % (w/w). Soil samples were collected at 0, 10, 20 and 30 days. Cd was extracted by adding 2 g of soil in 15 mL of 0.01M $CaCl_2$ and 15 ml of 0.05M EDTA and shaken at 120 rpm for a period of 24 hours [11, 14]. The concentration of extracted Cd was determined with Atomic Absorption Spectroscopy.

3 Results and Discussion

3.1 Cadmium adsorption isotherm

The cadmium adsorption isotherms of paddy soil, BB, RHB, and RSB at concentration of 10, 20 and 100% are shown in **Fig. 6-9**. Least-square linear and non-linear regression were used to fit the adsorption isotherms with linear and Freundlich adsorption models, respectively. The goodness of fit of the experimental data to the adsorption models were compared using the standard error of the estimate (SE). The parameters of the adsorption models are shown in Table 2.

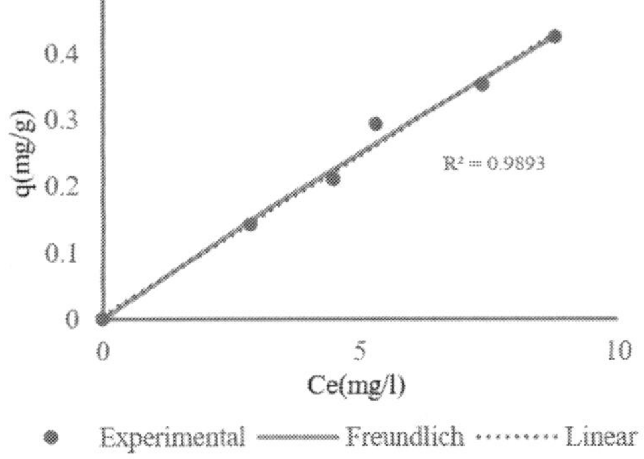

Fig. 6. Adsorption Isotherm of paddy soil.

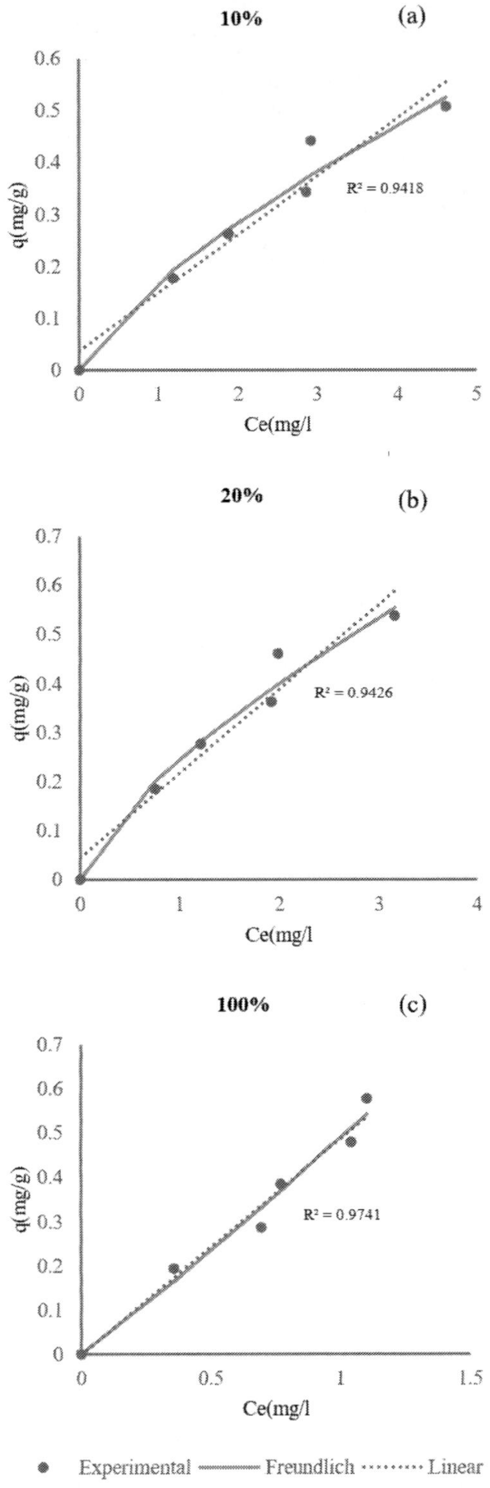

Fig. 7. Adsorption Isotherm of BB with concentration of biochar at a) 10, b) 20 and c)100%

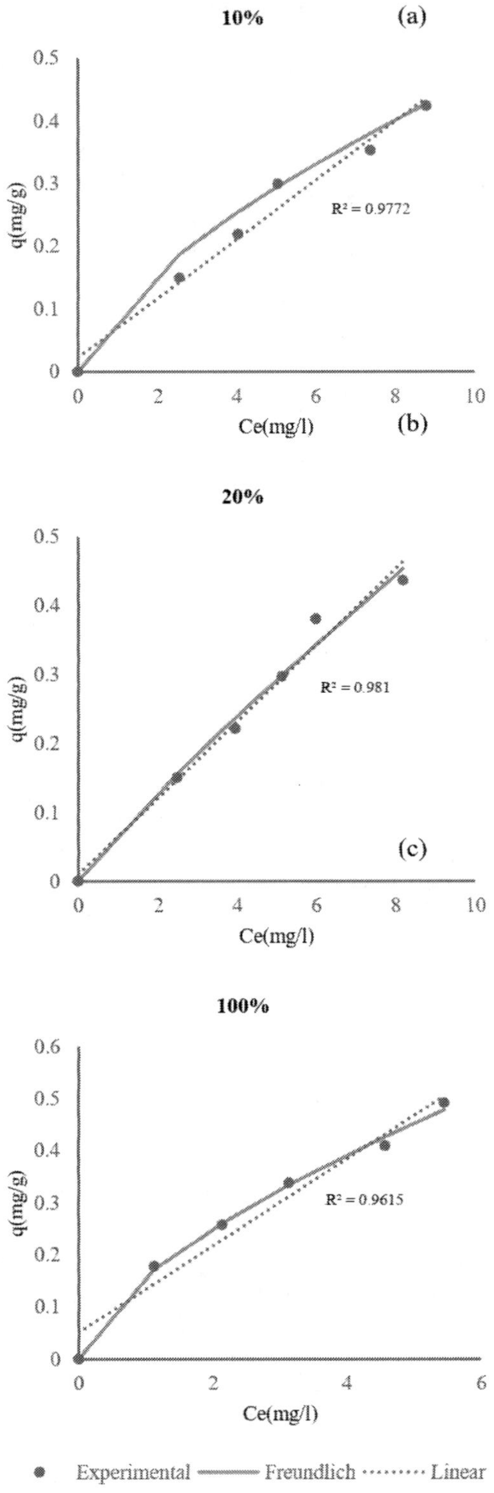

Fig. 8 Adsorption Isotherm of RHB with concentration of biochar at a) 10, b) 20 and c)100%

https://doi.org/10.1051/e3sconf/202130202021

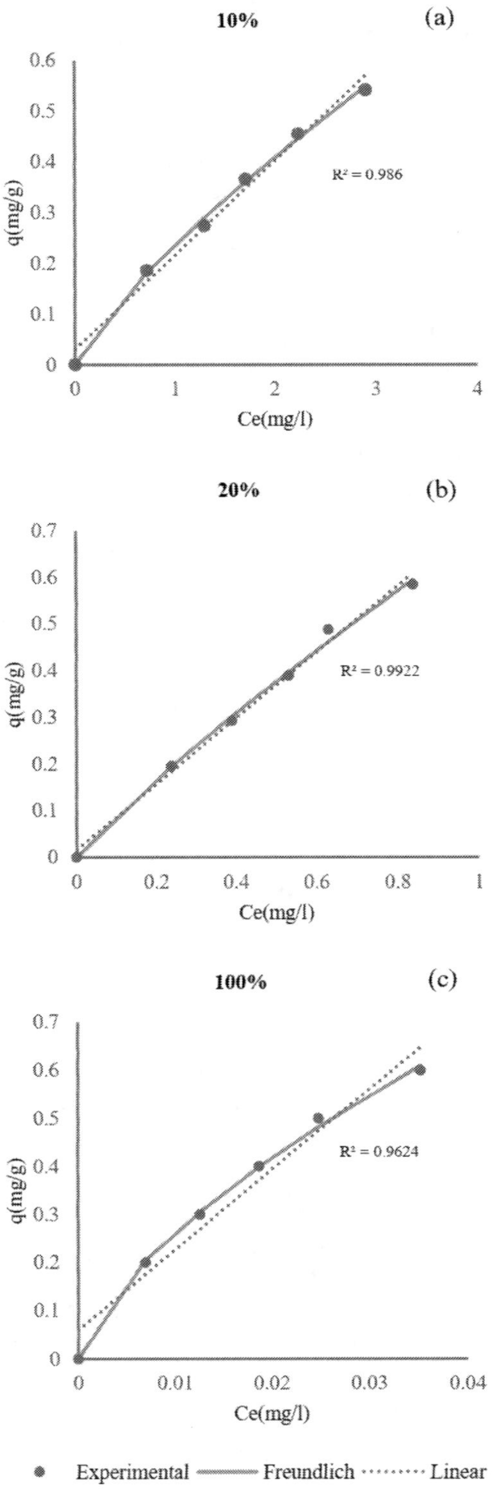

Fig. 9. Adsorption Isotherm of RSB with concentration of biochar at a) 10, b) 20 and c)100%

Table 2. Adsorption isotherm parameters and standard error of estimate.

Biochar	conc. (%)	Linear model		Freundlich model		
		K_d	SE	K_F	1/n	SE
Paddy Soil	100	0.0484	0.018	0.055	0.944	0.017
BB	10	0.1125	0.050	0.171	0.735	0.038
	20	0.1719	0.052	0.245	0.710	0.036
	100	0.4889	0.037	0.489	1.066	0.037
RHB	10	0.0471	0.026	0.101	0.662	0.021
	20	0.0554	0.025	0.070	0.889	0.022
	100	0.083	0.038	0.160	0.645	0.012
RSB	10	0.186	0.026	0.236	0.790	0.010
	20	0.7052	0.021	0.694	0.882	0.015
	100	16.677	0.047	5.790	0.673	0.011

Table 3. The pH of paddy soil, BB, RHB and RSB.

Biochar	conc. (%)	pH
Paddy Soil	100	5.47
BB	10	6.01
	20	6.49
	100	8.59
RHB	10	5.78
	20	5.91
	100	7.53
RSB	10	6.76
	20	7.05
	100	9.90

3.2 Comparison of cadmium adsorption efficacy

Cadmium adsorption efficacy of paddy soils and biochar can be determined by comparing the constants K_d in linear adsorption isotherm or K_F in Freundlich isotherm depending on which model has the best fit. By comparing the SE, it was found that the adsorption isotherm of paddy soil and biochar fit better with Freundlich adsorption isotherm model. The K_F value showed that all types of biochar had more ability to adsorb cadmium than paddy soil. Moreover, higher concentration of biochar increased the adsorption efficacy. These results corresponded well to the results reported by Maichu et al., [14]. By comparing K_F, it was found that RSB has the highest affinity for cadmium and RHB has the lowest affinity for cadmium. Furthermore, the K_F of RSB is almost 11 times higher than BB and 36 times higher than RHB. This means that RSB has the highest efficacy for cadmium adsorption. The reason that RSB has very high cadmium adsorption efficacy could be due to its high surface area and CEC as shown in Table 1. High pH could also contribute to the precipitation of cadmium. The pH of

 https://doi.org/10.1051/e3sconf/202130202021

RSB was 9.90 as can be seen in Table 3. When RSB was mixed with soil, it raised the pH of the soil from 5.47 to 7.05. With higher pH, the CEC of the soil could have been increased as well [22]. These mechanisms have also been reported by other research. [23, 24].

3.3 Cadmium extraction

From the results of previous experiment, RSB, which had the highest efficacy for cadmium adsorption, was selected to be tested for cadmium stabilization efficacy. From the results shown in **Fig. 10** and **11**, it can be seen that the value of $CaCl_2$ extracted Cd from cadmium contaminated soil before the treatment was 1.05 ± 0.20 mg/kg. When the soil was mixed with RSB at 5 and 7 %, the amount of $CaCl_2$-extracted Cd was immediately reduced to 0.67 ± 0.41 and 0.49 ± 0.22 mg/kg, respectively, and further reduced to 0.18 ± 0.00 and 0.10 ± 0.01 mg/kg in 20 days. After that the amount of $CaCl_2$-extracted Cd remained almost unchanged. Similarly, the amount of EDTA-extracted Cd before the treatment was 9.23 ± 1.38 mg/kg. After the application if RSB at 5 and 7 %, the amount of EDTA-extracted Cd was immediately reduced to 7.47 ± 1.86 and 4.72 ± 0.19 mg/kg, respectively, and further reduced to 2.35 ± 0.20 and 2.21 ± 0.22 mg/kg in 10 days. After that the amount of EDTA-extracted Cd remained almost unchanged. These results are consistent with the results reported by Lu et al. [11] that RSB could decrease the amount of extractable Cd in soil.

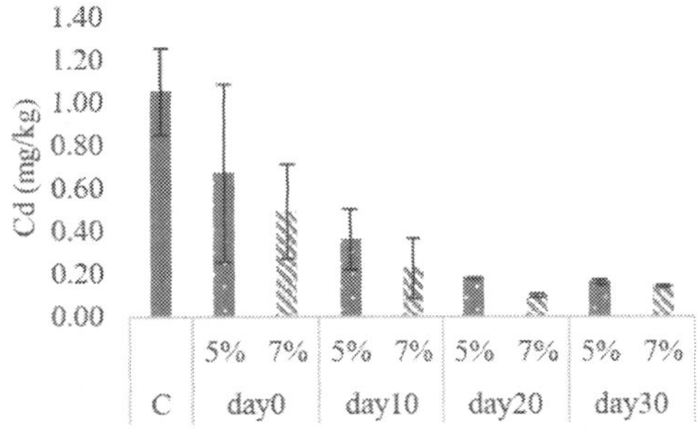

Fig. 10. Concentration of $CaCl_2$-extracted Cd after incubation of 30 days.

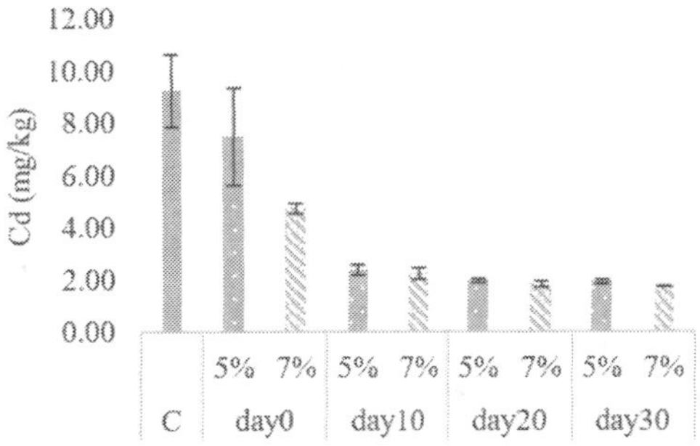

Fig. 11. Concentration of EDTA-extracted Cd after incubation of 30 days.

The concentration of $CaCl_2$ and EDTA-extracted Cd in the soil decreased with longer curing time. This is a result of cadmium adsorption on the biochar surface. The increase in the amount of biochar in the contaminated soil resulted in greater adsorption of cadmium because of the increase of absorption sites on biochar surface. The results at day 20 showed less adsorption rate, indicating that adsorption sites were mostly occupied or saturated. As a result, the adsorption rate decreased. However, the reduction of Cd from the soluble form in saturated soil condition could also contribute to the decrease of extractable Cd as shown in Eh-pH diagram of Cd suggested by Langmuir et al. [25].

According to the results, after incubation of 30 days, RSB at 5 and 7 % was able to reduce the amount of $CaCl_2$ extracted Cd by 84.84 and 90.85 % and the amount of EDTA-extracted Cd by 79.44 and 81.26 %. However, when comparing the variance, there was no difference in the results of the experiments on day 20 and 30. This means that the incubation of 10 - 20 days is adequate to achieve the maximum cadmium stabilization effect by RSB at the concentration used in this study.

4 Conclusion

Cadmium adsorption of paddy soil, BB, RHB, and RSB complied with Freundlich adsorption isotherm. The value of K_F constant in the adsorption isotherm equation showed that the efficacy of each type of biochar for Cd adsorption could be ranked as RSB> BB> RHB> paddy soil. Cadmium stabilization efficacy was indicated by the quantity of $CaCl_2$ and EDTA-extracted Cd. It was found that the higher application rate and longer curing time resulted in lower amount of $CaCl_2$ and EDTA-extracted Cd. RSB at 5% was the optimal application rate which could reduce $CaCl_2$ and EDTA-extracted Cd by 83.15 and 79.00 %, respectively, with an optimal incubation period of 20 days.

Acknowledgement

This research was funded by the Agricultural Research Development Agency.

References

1. R.W. Simmons, P. Pongsakul, D. Saiyasitpanich and S. Klinphoklap, Elevated levels of cadmium and zinc in paddy soils and elevated levels of cadmium in rice grain downstream of a zinc mineralized area in Thailand: Implications for public health, Environmental Geochemistry and Health, **27** (2005): 501–511
2. Pollution Control Department, Survey and Assessment of Cadmium Distribution and Sources of Contamination in Mae Tao River Basin, Mae Sod District, Tak Province. (2007)
3. B. Prapakdee and J. Sangthong, An environmentally friendly approach to cadmium contaminated areas... From the test bottle to the real space, Environmental Journal, **20**, 1 (2016): 1-13

4. J. Jiang, et al., Immobilization of Cu(II), Pb(II) and Cd(II) by the addition of rice straw derived biochar to a simulated polluted Ultisol, Journal of Hazardous Materials, **229-230** (2012): 145-150
5. S. Khaokaew, and G. Landrot, A field-scale study of cadmium phytoremediation in a contaminated agricultural soil at Mae Sot District, Tak Province, Thailand: (1) Determination of Cd-hyperaccumulating plants, Chemosphere, **138** (2015): 883-887
6. F. Jing, Biochar effects on soil chemical properties and mobilization of cadmium (Cd) and lead (Pb) in paddy soil, Soil Use and Management, **36**, 2 (2020): 320-327
7. L. Cui, Continuous immobilization of cadmium and lead in biochar amended contaminated paddy soil: A five-year field experiment, Ecological Engineering, **93** (2016): 1-8
8. P. E. Dim and M. Termtanun, Treated Clay Mineral as Adsorbent for the Removal of Heavy Metals from Aqueous Solution, Applied Science and Engineering Progress, (2021)
9. A. Usman, Y. Kuzyakov and K. Stahr, Effect of Clay Minerals on Immobilization of Heavy Metals and Microbial Activity in a Sewage Sludge-Contaminated Soil (8 pp), The Journal of Soils and Sediments, **245–252** (2005)
10. F. Jing, Effects of wheat straw derived biochar on cadmium availability in a paddy soil and its accumulation in rice, Environmental Pollution, **257** (2020): 113592
11. K. Lu, X. Yang, G. Gielen, N. Bolan, Y.S. Ok, N.K. Niazi, S. Xu, G. Yuan, X. Chen, X. Zhang, D. Liu, Z. Song, X. Liu and H. Wang, Effect of bamboo and rice straw biochars on the mobility and redistribution of heavy metals (Cd, Cu, Pb and Zn) in contaminated soil, Journal of Environmental Management, **186** (2017): 285-292
12. A. Kowitwiwat and P. Sampanpanish, Comparison of physical and chemical properties of Bagasse, Napier grass and Acacia wood derived biochar on site contaminated, 18th National Environment Symposium. Environmental Engineering Society of Thailand, (1-10), (2019).
13. T. Abbas, M. Rizwan, S. Ali, M. Zia-ur-Rehman, M.F. Qayyum, F. Abbas, F. Hannan, J. Rinklebe, Y.S. Ok, Effect of biochar on cadmium bioavailability and uptake in wheat (Triticum aestivum L.) grown in a soil with aged contamination, Ecotoxicology and Environmental Safety, **140** (2017): 37–47
14. P. Maichu, P. Pinnoi and S. Jaroonwarasate, Comparison The Effects of Napier grass biochar and Bagasse biochar on absorption of cadmium contaminated soil, A Special Project Submitted in Partial Fulfillment of the Requirement for the Degree of Bachelor of Science (Environmental Chemistry), Faculty of Science, King Mongkut's Institute of Technology Ladkrabang (2018)
15. D. Dong, Q. Feng, K. McGrouther, M. Yang, H. Wang and W. Wu, Effects of biochar amendment on rice growth and nitrogen retention in a waterlogged paddy field, The Journal of Soils and Sediments, **15** (2015): 153–162
16. G. Ma, H. Mao, Q. Bu, L. Han, A. Shabbir and F. Gao. Effect of Compound Biochar Substrate on the Root Growth of Cucumber Plug Seedlings, Agronomy, **10** (2020): 1-14
17. T.G. Ambaye, M. Vaccari, E.D. van Hullebusch, et al, Mechanisms and adsorption capacities of biochar for the removal of organic and inorganic pollutants from industrial wastewater, International Journal of Environmental Science and Technology, (2020)
18. A.W. Samsuri, F. Sadegh-Zadeh and B.J. Seh-Bardan, Characterization of biochars produced from oil palm and rice husks and their adsorption capacities for heavy metals, International Journal of Environmental Science and Technology, (2013). doi: 10.1007/s13762-013-0291-3
19. J. Zhao, XJ. Shen, X. Domene, et al, Comparison of biochars derived from different types of feedstock and their potential for heavy metal removal in multiple-metal solutions, Scientific Reports, **9**, 9869 (2019)
20. J. Komkiene and E. Baltrenaite, Biochar as adsorbent for removal of heavy metal ions [Cadmium(II), Copper(II), Lead(II), Zinc(II)] from aqueous phase, International Journal of Environmental Science and Technology, **13**, (2016): 471–482
21. N. Puangpun, I. Suwimon and K. Piyamongkala, Adsorption Soluble Cutting Fluid Emulsion by Modified Chitosan with SLES, Applied Science and Engineering Progress, **12**, 4 (2019): 243–252
22. D.L. Sparks, *Environmental Soil Chemistry (Second Edition),* (2003): 133-186
23. X. Cui, H. Hao, C. Zhang, Z. He and X. Yang, Capacity and mechanisms of ammonium and cadmium sorption on different wetland-plant derived biochars, Science of The Total Environment, **539** (2016): 566-575
24. J. Ma, W. Huang, X. Zhang, Y. Li, and N. Wang, The utilization of lobster shell to prepare low-cost biochar for high-efficient removal of copper and cadmium from aqueous: Sorption properties and mechanisms, Journal of Environmental Chemical Engineering, **9**, 1 (2021): 104703
25. D. Langmuir, C. Paul, V. Bernard and C. Rufus, Issue Paper on the Environmental Chemistry of Metals, (2004). [Online]. Available from: https://www.researchgate.net/figure/Eh-pH-diagram-for-the-system-Cd-O-2-CO-2-S-H-2-O-for-ECd-10-7-mol-kg-ES-10-5_fig18_237330649 [Accessed 24 May 2021]

Statistical design and optimization of nutritional value production by an oleaginous yeast *Yarrowia lipolytica* cultured in industrial – waste molasses

Benjamaporn Wonganu [1,2*], *Sasithorn* Kongruang[1], *and Ratiya* Charoensakdi[1]
[1]Department of Biotechnology, Faculty of Applied Science King Mongkut's University of Technology North Bangkok Bangkok, Thailand
[2]Biorefinery and Process Automation Engineering Center King Mongkut's University of Technology North Bangkok Bangkok, Thailand

Abstract. Abundant by-products from sugar mills as industrial-waste molasses can be used as a carbon source in yeast culture media. *Yarrowia lipolytica* is an interesting yeast used as a candidate for cultivation in molasses medium. Here, we used response surface methodology to derive a statistical model for the individual and interactive effects of pH, temperature, and shaking speeds on the production of yeast cells. Cultivation conditions of yeast were optimized using Design Expert based on a 2^3 factorial central composite design (CCD) for maximum yeast cell production. Optimal conditions for maximum *Y. lipolytica* 5151 cell masses were as follows: pH, 6.45; temperature, 30°C; Shaking speed, 165 rpm. The Design Expert represented the maximal numerical solution with a predicted cell mass production level at 8.96 g/L. The experimental production of *Y. lipolytica* 5151 cell mass yielded 8.27 g/L that is 7.67% deviated from the model. Whereas, the model of TISTR 5621 was not adequate for prediction. Yeasts cultured under statistic prediction provide 55.94% and 51.25% of total protein. Amino acid content and vitamin B1 (1.06 mg and 1.47 mg per 100 g of dried *Y. lipolytica* 5151 and 5621, respectively) provided the relevant information for an alternative supplement in aquatic feed.

Keyword. Molasses; An oleaginous yeast; statistic design

1 Introduction

Yarrowia lipolytica is classified as an oleaginous yeast that has a high ability to produce lipids or triacylglycerides (TAGs) depended on a culture condition. It becomes to be an attractive alternative source to decrease the dependency on vegetable and mineral oil. Moreover, it was used as a non-pathogenic microorganism for lipid production applied as components in coatings, paints, personal care products, production of fine chemicals, and biodiesel [1]. *Y. lipolytica* can produce and accumulate 30-43% lipid in the cell, and it also can produce and secret a high level of protein [2, 3].

A study in a yeast species known as *saccharomyces cerevisiae* as a nutritional yeast containing protein, vitamins, minerals, and antioxidants. However, the composition and nutritive value of *Y. lipolytica* biomass are still limited. Due to the ability to secrete protein at a high level depending on the carbon source in contrast to *S. cerevisiae*, *Y. lipolytica* has become the most interesting organism for many applications in producing and secretion of several organic acids and fragrances. [4, 5]

The development of an economic culture medium for *Y. lipolytica* needs to obtain a high quantity of its biomass. The satisfied medium must compose of the basic requirements for cell biomass. The carbon substrate is a main role in biosynthesis and energy generation [6]. There are various carbon sources in microbial culture. Since *Y. lipolytica* can use unusual carbon sources, such as hydrocarbons, several studies focused on increasing yeast biomass by using different carbon sources. Juszczyk et al. [7] used waste derived from the production of ethyl esters of polyunsaturated fatty acids of flaxseed oil to produce biomass of *Y. lipolytica*. Wrobel-Kwiatkowska et al. [8] developed *Y. lipolytica* media containing different kinds of honey for product improvement of Kynurenic acid. Media development for *Y. lipolytica* growth is relevant because media cost is a significant problem of the large-scale yeast cultivation for industrial applications. Therefore, Molasses is a by-product of sugar processing, and it also can be an efficient, low-cost carbon source in a *Y. lipolytica* culture.

* Corresponding author: Benjamaporn.w@sci.kmutnb.ac.th

 https://doi.org/10.1051/e3sconf/202130202022

The response surface methodology (RSM) is a mathematical and statistical technique to estimate the experimental variability, confirm the adequacy between the proposed model and the experimental data, predict the observed response and construct the results in a 3D empirical model [9]. Due to the tedious task of the classical "one-factor-at-a-time" method in investigating interactions between variables, RSM method has been applied in biological research, food, and feed technology applications. This study aims to optimize the best conditions for *Y. lipolytica* TISTR 5151 and 5621 to grow in industrial-waste molasses using the Design expert program. *Y. lipolytica* could be an alternative source or ingredient for fish and shrimp feeds.

2 Materials and Methods

2.1 Materials

Y. lipolytica TISTR 5151 and 5621 were purchased from Thailand Institute of Scientific and Technological Research (TISTR), Thailand. Peptone, Yeast extract, Malt extract, Agar, Hydrochloric acid, Sodium hydroxide used for cell culture and analysis were of analytical grade. They were purchased from Sigma, USA, and HiMedia, India. Molasses was supported by a local sugar factory in Suphanburi province, Thailand. Design Expert ver. 10.0.2 was used in the optimization analysis of the yeast cell growth.

2.2 Yeast cells preparations

Y. lipolytica TISTR 5151 and 5621 were grown on modified media composed of 5.6% molasses as a carbon source, 0.5% peptone, 0.3% yeast extract, and 0.3% malt extract. The yeasts were cultured in three various optimized factors (pH, temperature, and shaking speed). Moisture contents from a moisture analyzer were relative to yeast cell dry weight values, as the following equation shows below [10].

$$\%\ \text{Moisture content} = \frac{\text{Initial weight (g)} - \text{Final weight (g)}}{\text{Initial weight of sample (g)}}$$

2.3 Experimental Design setup for yeast cell growth optimization

In the research, three selected factors were pH, temperature, and shaking speed. These factors were used to determine their effects on two strains of yeast cell growth in the modified culture medium containing molasses as a carbon source. The experiments used a central composite design. Table 1 shows the factors and their levels where the high level indicates the highest range of the factors, and the low level reveals the lowest of the factors. Experiment design Tables 2 and 3 of *Y. lipolytica* TISTR 5151 and 5621 were constructed using the Design Expert Software v.10.0.2, and the experiment data was analyzed using the same software.

2.4 Nutritional value analysis of dry yeasts.

Free amino acid contents and vitamin B1 were determined from each yeast cells. HPLC technique was applied to the quantification of each amino acid and vitamin B1 of 100g of each dry yeast [11, 12].

Table 1. The levels of selected factors

Factors	Low level	High level
pH	5	9
Temperature (Celsius)	28	32
Shaking speed (rpm)	150	250

Table 2. Experimental setup of *Y. lipolytica* TISTR 5151 that has been constructed by using 2^3 factorial analyses by Design Expert software v.10.0.2

Run	Factor1 A: pH	Factor2 B: temperature (Celsius)	Factor3 C: Shaking speed (rpm)
1	7	30	200
2	7	30	200
3	7	28	200
4	7	30	200
5	9	28	250
6	9	32	250
7	7	30	150
8	5	30	200
9	5	28	150
10	9	28	150
11	7	30	200
12	9	30	200
13	5	32	250
14	7	30	200
15	7	30	200
16	7	30	250
17	7	32	200
18	5	32	150
19	5	28	250
20	9	32	150

Table 3. Experimental setup of *Y. lipolytica* TISTR 5621 that has been constructed by using 2^3 factorial analyses by Design Expert software v.10.0.2

Run	Factor1 A: pH	Factor2 B: temperature (Celsius)	Factor3 C: Shaking speed (rpm)
1	7	30	200
2	7	30	200
3	7	32	200
4	5	32	150
5	7	30	200
6	9	28	250
7	7	30	150
8	9	30	200
9	7	28	200
10	7	30	200
11	9	32	150
12	7	30	200
13	5	32	250
14	7	30	250
15	5	28	250
16	5	30	200
17	5	28	150
18	9	32	250
19	7	30	200
20	9	28	150

3 Results and discussions

3.1 Screening of Factors Affecting on *Y. lipolytica* TISTR 5151 and 5621 Growth

pH, temperature, and shaking speed were applied to study the practical effect on yeast growth. 2^3 level factorial method in Design Expert was used to optimize the condition to obtain the highest biomass yields. Tables 4 and 5 show 20 runs of experiments that reveal obtained yeast cell dry weight results. The response was analyzed using the ANOVA component of central composite design (CCD) found in response surface methodology of the Design-Expert software v10.0.2 at 95% confidence level. For *Y. lipolytica* TISTR 5151, the highest yield of biomass was obtained at pH7, 30°C, and 200 rpm of shaking speed, whereas the condition for the highest yield of biomass of *Y. lipolytica* TISTR 5621 at pH5, 32°C, and 150 rpm of shaking speed. The design matrix for CCD experiments and the experiment results and the predicted responses for Cell Dry Weight (CDW) of Y. lipolytica TISTR 5151 and 5621 were regressed using a quadratic polynomial equation and linear equation for TISTR 5151 and 5621, respectively. The two regression equations were as follows:

CDW of *Y.lipolytica* TISTR 5151 =
$-23.12384 - 0.22690A + 1.64501B - 0.00191783C + 0.010257AB - 0.000289708AC + (0.000239625BC - 0.00205265A^2 - 0.028965B^2 - 0.00000947090C^2$

CDW of *Y. lipolytica* TISTR 5621 $= 0.75398 - 0.00282833A + 0.00377B - 0.0003456C$

where A is pH, B is the temperature (Celsius), and C is shaking speed (rpm)

Table 4. Experimental results of *Y. lipolytica* TISTR 5151 cell dry weight on different factors using Design Expert software v.10.0.2 based on 2^3 level factorial CCD

Run	Factor1 A: pH	Factor2 B: Temperature (Celsius)	Factor3 C= Shaking speed (rpm)	Response Cell Dry Weight (CDW) (10 g/L)
1	7	30	200	0.891
2	7	30	200	0.891
3	7	28	200	0.721
4	7	30	200	0.891
5	9	28	250	0.504
6	9	32	250	0.828
7	7	30	150	0.876
8	5	30	200	0.877
9	5	28	150	0.733
10	9	28	150	0.764
11	7	30	200	0.896
12	9	30	200	0.895
13	5	32	250	0.748
14	7	30	200	0.891
15	7	30	200	0.891
16	7	30	250	0.865
17	7	32	200	0.837
18	5	32	150	0.797
19	5	28	250	0.767
20	9	32	150	0.814

Table 5. Experimental results of *Y. lipolytica* TISTR 5621 cell dry weight on different factors using Design Expert software v.10.0.2 based on 2^3 level factorial CCD.

Run	Factor1 A:pH	Factor2 B:Temperature (Celsius)	Factor3 C= Shaking speed (rpm)	Response Cell Dry Weight (CDW) (10 g/L)
1	7	30	200	0.765
2	7	30	200	0.765
3	7	32	200	0.746
4	5	32	150	0.845
5	7	30	200	0.765
6	9	28	250	0.695
7	7	30	150	0.771
8	9	30	200	0.784
9	7	28	200	0.824
10	7	30	200	0.765
11	9	32	150	0.842
12	7	30	200	0.765
13	5	32	250	0.784
14	7	30	250	0.813
15	5	28	250	0.799
16	5	30	200	0.732
17	5	28	150	0.787
18	9	32	250	0.766
19	7	30	200	0.764
20	9	28	150	0.794

3.2 Analysis of Variance (ANOVA) for yeast cell dry weight

The analysis of variance (ANOVA) for the CCD experiment was performed to estimate the coefficient of the model and determine the significance of each parameter (Table 6). The F-values and P-values were used to identify the effect of each factor on obtained cell dry weight in each strain of *Y. lipolytica*. The model of *Y. lipolytica* TISTR 5151 exhibited correlation with the experimental data with high F-value and P-value < 0.05, which implied the significance of the model. In contrast, low F-value and $P > 0.05$ of *Y. lipolytica* TISTR 5621 model indicated that this model was not significant. By inspecting each F-values and P-values of the Y. lipolytica TISTR 5151 model, it can be seen that the effect on biomass of Y. lipolytica cultured in media containing molasses carbon source depended on the temperature. Different strains in *Y. lipotica* had various physiological and metabolism properties and required other conditions to grow.

The previous studies were found that high temperature affects sugar alcohol production [13]. In TISTR 5151, R^2 value (0.90) indicated that the model was reliable for predicting biomass. The lack of fit F-value which was 1.87 (P-value $0.1577 < 0.05$), also replied that the model was adequate for prediction.

Three-dimensional response surface plots were constructed to show the effect of pH, temperature and shaking speed on cell dry weight of *Y. lipolytica* TISTR 5151 (Figure 1). The impact of individual variables is paired within the experiment range. This graphical visualization allows investigating the relationships between the experimental levels of each factor and the response. There was a large interactive effect of temperature and another factor on cell dry weight. At moderate temperature, the strongest yeast biomass increases *Y. lipolytica* TISTR 5151 cultured in the media containing molasses as a carbon source. When dry cell weight was considered, the interactions between pH and temperature and the interaction between shaking speed and temperature are strong (Figure 1A and 1C). In contrast, it is not been shown in the interaction between shaking speed and pH (Figure 1C).

The optimal levels of those factors can be deduced from a 3D response and the equation obtained from regression analysis. The model predicted that the maximum cell dry weight (8.96 g/mL) was located at A= 6.45, B = 30 Celsius, and C= 165 rpm. The experiments were performed in the predicted cell culture condition to examine the validity of this model. The average obtained biomass was 8.27 g/L. It was close to the predicted values and supported the suitability of the model in this study.

Table 6. Analysis of variance (ANOVA) for the experimental results of the CCD

Source	*Y. lipolytica* TISTR 5151					*Y. lipolytica* TISTR 5621				
	SS	df	SM	F-value	P-value	SS	df	SM	F-value	P-value
Model	0.16	9	0.017	9.43	0.0008	0.0039	3	0.0013	1.09	0.3815
A	0.0042	1	0.0043	2.33	0.1577	0.00032	1	0.00032	0.27	0.6103
B	0.033	1	0.033	17.80	0.0018	0.00057	1	0.00057	0.48	0.4983
C	0.00017	1	0.00017	0.0094	0.7650	0.00298	1	0.00298	2.52	0.1319
AB	0.013	1	0.013	7.33	0.0221	-	-	-	-	
AC	0.0067	1	0.0067	3.65	0.0850	-	-	-	-	-
BC	0.0046	1	0.0046	2.50	0.1450	-	-	-	-	-
A^2	0.00019	1	0.00019	0.10	0.7573	-	-	-	-	-
B^2	0.037	1	0.037	20.08	0.0012	-	-	-	-	-
C^2	0.0015	1	0.0015	0.84	0.3813	-	-	-	-	-
Lack of fit	0.018	0.0037	0.0037	1.87	0.1577	0.019	11	0.0017	2.86	0.003
R^2=0.90						R^2= 0.16				

A, pH; B, temperature; C, shaking speed
*P < 0.05 are significant
SS: Sum of squares; df: degree of freedom; SM: mean squares

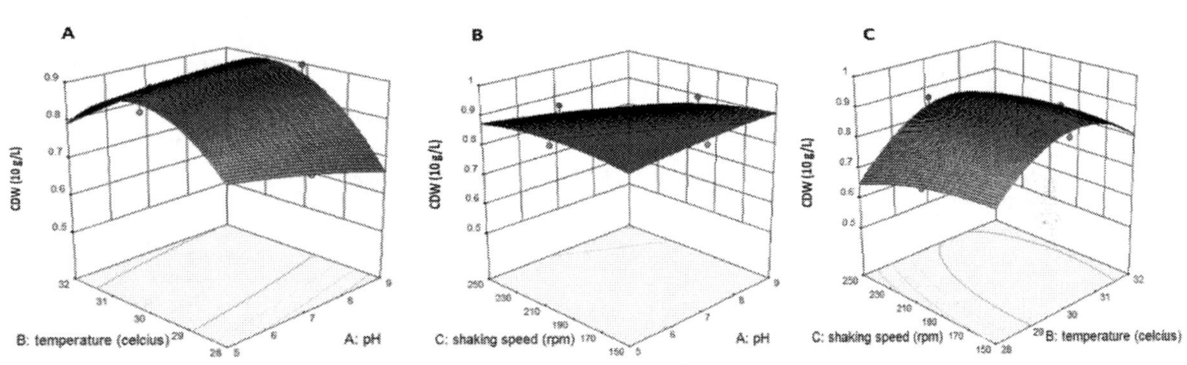

Figure 1. Response surfaces (A) A 3D Graphical combination between pH and temperature to the cell dry weight of *Y.lipolytica* TISTR 5151 (B) A 3D Graphical combination between pH and shaking speed to the cell dry weight of *Y.lipolytica* TISTR 5151 (C) A 3D Graphical combination between temperature and shaking speed to the cell dry weight of *Y.lipolytica* TISTR 5151.

Table 7. Validation of the optimization values predicted by RSM and the experiment

Factor A pH	Factor B Temperature (Celsius)	Factor C Shaking speed (rpm)	Predicted Cell dry weight (g/mL)	Cell dry weight from the experiment (g/mL)		
				Run 1	Run 2	Run 3
6.45	30	165	8.96	8.84	8.09	7.89
				Average: 8.27, Error: 7.67%		

Table 8. Percent composition of yeast biomass and yeast protein concentrates

Components*	*Y. lipolytica* TISTR 5151	*Y. lipolytica* TISTR 5621	Method
Total protein (%)	55.94	51.25	AOAC 2010.AOAC official method 984.13
Total lipid (%)	0.73	1.54	AOAC 2010.AOAC official method 2003.05
Ash (%)	7.43	6.47	AOAC 2010.AOAC official method 942.05
Total fiber (%)	0.61	0.58	AOAC 2010.AOAC official method 962.09

* Percentages on a dry basis

3.3 Yeast composition, amino acid, and vitamin B1 content

The biomass compositions are evaluated for two *Y. lipolytica* different strains, TISTR 5151 and 5621. Protein concentrations of whole yeast biomass TISTR 5151 and TISTR 5621 cultured in the predicted condition were 55.94 % and 51.25%, respectively (Table 8). The amino acid profiles of yeast biomass in both strains are shown in Table 9. The results of amino acid content in dry yeast cells in both yeast strains were not significantly different. The high content of Lysine is found in yeast biomass (Table 9). This amino acid is essential dietary protein for the growth of catla (*Catla catla*), the major South Asian carp [14], and Nile tilapia [15, 16]. Aquatic feeds for farmed fish and shrimp need each amino acid to depend on marine species [17]. Using *Y. lipolytica* dry cell in feed formulation can be an alternative approach to increase essential amino acids for aquatic feeds. Moreover, Yeast cells also contained vitamin B1 (Thiamin) as another supplement for fish feed (Table 10) [18].

Table 9. Amino acid composition analysis for the yeast biomass

Amino acids (mg/100 g protein)	*Y. lipolytica* TISTR 5151	*Y. lipolytica* TISTR 5621
Alanine	6,140	4,512
Arginine	2,481	2,532
Aspartic acid	3,782	4,514
Cystine	1,515	1,546
Glutamic acid	6,785	5,992
Glycine	6,765	4,482
Histidine	889	1,054
Isoleucine	1,456	1,861
Leucine	2,396	2,573
Lysine	5,347	4,921
Methionine	717	782
Phenylalanine	1,367	1,646
Proline	2,501	2,431
Serine	2,631	2,189
Threonine	1,813	2,201
Tryptophan	971	714
Tyrosine	1,132	1,221
Valine	1,689	2,264

Table 10. Vitamin B1 analysis for the yeast biomass

Vitamin B1 (mg/100 g protein)	*Y. lipolytica* TISTR 5151	*Y. lipolytica* TISTR 5621
	1.06	1.47

4 Conclusion

Different *Y. lipolytica* isolates derived from Thailand Institute of Scientific and Technological Research might require various factors for maximizing yeast biomass. RSM is a powerful method for program optimization and identification of the relative significance of different factors, interactions between factors, and optimal level of test variable. The method has been successfully applied to optimize critical medium components for mycelium biomass of *Lentinus squarrosulus* [19], obtain the optimum temperature and drying time in making cocktail yeast molds [20], and maximize biomass of *Purpureocillium lilacinum* KU8 [21]. *Y. lipolytica,* a non-pathogenic organism, is considered as a model for a study of protein secretion, hydrophobic substrate utilization, and peroxisome biogenesis, dimorphism and mitochondrail complexI [22]. Protein, amino acid content, and vitamin B1 were found in *Y. lipolytica* dry cell, indicating that the dry cell yeast might be applied as a significant protein-rich ingredient in aquafeeds. Nutritional quality of *Saccharomyces cerevisiae, Cyberlindnera jadinii, Kluyveromyces marxianus, Blastobotrys adeninivorans,* and *Wicker-hamomyces anomalus* had been studied to use them as a replacement for fishmeal and soy protein in the diets of Atlantic salmon [23].

This study has been successfully optimized critical growth conditions in using molasses as a carbon source for *Y. lipolytica* varieties cultivated in Thailand. This alternative medium could reduce media cost that a significant issue for large-scale yeast cultivation.

Acknowledgment

This research was supported by the basic research funding in 2016 of Faculty of Applied Science, King Mongkut's University of Technology North Bangkok. We acknowledge Miss Mananya Thongchay and Miss Chutarat Posila for any research help

References

1. A. Beopoulos, J. Verbeke, F. Bordes, M. Guicherd, M. Bressy, A. Marty, J.M. Nicaud, Metabolic engineering for ricinoleic acid production in the oleaginous yeast Yarrowia lipolytica, Applied Microbiology and Biotechnology, **98**, 1 (2014): 251-62
2. S. Papanikolaou, G. Aggelis, Lipid production by Yarrowia lipolytica growing on industrial glycerol in a single-stage continuous culture, Bioresource Technology, **82**, 1 (2002): 43-9
3. J.-M. Nicaud, Yarrowia lipolytica, Yeast, **29**, 10 (2012): 409-418
4. P. Saçakl1, B.H. Koksal, A. Ergun, B. Özsoy, Usage of brewer's yeast (Saccharomyces cerevisiae) as a replacement of vitamin and trace mineral premix in broiler diets, Revue de Médecine Vétérinaire, **164**, 1 (2013): 39-44
5. M.T. Pacheco, G.M. Caballero-Córdoba, V.C. Sgarbieri, Composition and nutritive value of yeast biomass and yeast protein concentrates, Journal of Nutritional Science and Vitaminology (Tokyo), **43**, 6 (1997): 601-12
6. G.M. Walker, Fermentation (Industrial) : media for industrial fermentations, (E.o.f.m. editor (Ed.) Carl A. Batt. 2nd. ed. London : Academic Press,2014): 769-777
7. P. Juszczyk, W. Rymowicz, A. Kita, A. Rywińska, Biomass production by Yarrowia lipolytica yeast using waste derived from the production of ethyl esters of polyunsaturated fatty acids of flaxseed oil, Industrial Crops and Products, **138** (2019): 111590
8. M. Wróbel-Kwiatkowska, W. Turski, P. Juszczyk, A. Kita, W. Rymowicz, Improved Production of Kynurenic Acid by Yarrowia lipolytica in Media Containing Different Honeys, Sustainability, **12**, 22 (2020): 9424
9. L.A. Sarabia, M.C. Ortiz, 1.12 - *Response Surface Methodology* (S.D. Brown, R. Tauler, B. Walczak (Eds.), Comprehensive Chemometrics, Elsevier, Oxford, (2009): 345-390
10. E. Li, R. Mira de Orduña Heidinger, A rapid method for the determination of microbial biomass by dry weight using a moisture analyser with an infrared heating source and an analytical balance, Letters in applied microbiology, **50** (2010): 283-8
11. P. Herbert, P. Barros, N. Ratola, A. Alves, HPLC Determination of Amino Acids in Musts and Port Wine Using OPA/FMOC Derivatives, Journal of Food Science, **65**, 7 (2000): 1130-1133
12. F. Batifoulier, M.A. Verny, E. Chanliaud, C. Rémésy, C. Demigné, Effect of different breadmaking methods on thiamine, riboflavin and pyridoxine contents of wheat bread, Journal of Cereal Science, **42**, 1 (2005): 101-108
13. S. Hackenschmidt, F. Bracharz, R. Daniel, A. Thürmer, S. Bruder, J. Kabisch, Effects of a high-cultivation temperature on the physiology of three different Yarrowia lipolytica strains, FEMS Yeast Research, **19**, 7 (2019): 31605534
14. J. Ravi, K.V. Devaraj, Quantitative essential amino acid requirements for growth of catla, Catla catla (Hamilton), Aquaculture, **96**, 3 (1991): 281-291
15. C.B. Santiago, R.T. Lovell, Amino Acid Requirements for Growth of Nile Tilapia, The Journal of Nutrition, **118**, 12 (1988): 1540-1546
16. R.d.S. Romaneli, do Nascimento, T. Matias Torres, Gous, R. Mervyn, Reis, M. de Paula, Mansano, C. Fernando Menegasso, Khan, K. Ullah, Sakomura, N. Kazue, & Fernandes, J. Batista Kochenborger., Response of Nile tilapia (Oreochromis niloticus) to lysine: Performance, body composition, maintenance and efficiency of utilization, Aquaculture, **538** (2021): 736522
17. A.J.P. Nunes, M.V.C. Sá, C.L. Browdy, M. Vazquez-Anon, Practical supplementation of shrimp and fish feeds with crystalline amino acids, Aquaculture, **431** (2014): 20-27
18. F.G. Laura Gasco, G. Maricchiolo, L. Genovese, S. Ragonese, T. Bottari, G. Caruso, *Supplementation of Vitamins, Minerals, Enzymes and Antioxidants* (Feeds for the Aquaculture Sector. SpringerBriefs in Molecular Science. Springer, Cham, 2018): 63-103
19. R. Ahmad, N.K.N. Al-Shorgani, A.A. Hamid, W.M.W. Yusoff, F. Daud, Optimization of medium components using response surface methodology (RSM) for mycelium biomass and exopolysaccharide production by *Lentinus squarrosulus*, Advances in Bioscience and Biotechnology, **4**, 12 (2013): 1079-1085
20. R. Rahmawati, Hunaefi, D., Basriman, I., Saputra, D., Apriliani, A.A. and Jenie, B.S.L., Optimization of temperature and drying time of indigenous cocktail yeast mold culture using response surface methodology (RSM), Food Research, **4**, 2 (2020): 389 - 395
21. M. Mousumi Das, M. Haridas, A. Sabu, Process development for the enhanced production of bio-nematicide *Purpureocillium lilacinum* KU8 under solid-state fermentation, Bioresource Technology, **308** (2020): 123328
22. J.M. Nicaud, Yarrowia lipolytica, Yeast, **29**, 10 (2012): 409-418
23. J.O. Agboola, M. Øverland, A. Skrede, J.Ø. Hansen, Yeast as major protein-rich ingredient in aquafeeds: a review of the implications for aquaculture production, Reviews in Aquaculture, **13**, 2 (2021): 949-970

AUTHOR INDEX

AUTHOR INDEX

EDP Sciences – Web of Conferences
17, Avenue du Hoggar
Parc d'Activité de Courtabouf
BP 112
F-91944 Les Ulis Cedex A
France

ISBN 978-1-7138-3648-3

CURRAN ASSOCIATES INC.
proceedings.com

9 781713 836483